Twentieth Century Harmonic Analysis – A Celebration

NATO Science Series

A Series presenting the results of scientific meetings supported under the NATO Science Programme.

The Series is published by IOS Press, Amsterdam, and Kluwer Academic Publishers in conjunction with the NATO Scientific Affairs Division

Sub-Series

I. **Life and Behavioural Sciences**	IOS Press
II. **Mathematics, Physics and Chemistry**	Kluwer Academic Publishers
III. **Computer and Systems Science**	IOS Press
IV. **Earth and Environmental Sciences**	Kluwer Academic Publishers

The NATO Science Series continues the series of books published formerly as the NATO ASI Series.

The NATO Science Programme offers support for collaboration in civil science between scientists of countries of the Euro-Atlantic Partnership Council. The types of scientific meeting generally supported are "Advanced Study Institutes" and "Advanced Research Workshops", and the NATO Science Series collects together the results of these meetings. The meetings are co-organized bij scientists from NATO countries and scientists from NATO's Partner countries – countries of the CIS and Central and Eastern Europe.

Advanced Study Institutes are high-level tutorial courses offering in-depth study of latest advances in a field.
Advanced Research Workshops are expert meetings aimed at critical assessment of a field, and identification of directions for future action.

As a consequence of the restructuring of the NATO Science Programme in 1999, the NATO Science Series was re-organized to the four sub-series noted above. Please consult the following web sites for information on previous volumes published in the Series.

http://www.nato.int/science
http://www.wkap.nl
http://www.iospress.nl
http://www.wtv-books.de/nato-pco.htm

Twentieth Century Harmonic Analysis – A Celebration

edited by

James S. Byrnes

Prometheus Inc.,
Newport, RI, U.S.A. and
University of Massachusetts at Boston,
Boston, MA, U.S.A.

Kluwer Academic Publishers

Dordrecht / Boston / London

Published in cooperation with NATO Scientific Affairs Division

Proceedings of the NATO Advanced Study Institute on
Twentieth Century Harmonic Analysis – A Celebration
Il Ciocco, Italy
2–15 July 2000

A C.I.P. Catalogue record for this book is available from the Library of Congress.

ISBN 0-7923-7168-2 (HB)
ISBN 0-7923-7169-0 (PB)

Published by Kluwer Academic Publishers,
P.O. Box 17, 3300 AA Dordrecht, The Netherlands.

Sold and distributed in North, Central and South America
by Kluwer Academic Publishers,
101 Philip Drive, Norwell, MA 02061, U.S.A.

In all other countries, sold and distributed
by Kluwer Academic Publishers,
P.O. Box 322, 3300 AH Dordrecht, The Netherlands.

Printed on acid-free paper

Printed in the Netherlands.

Dedication

This book is dedicated to the memory of Alan David Fisher, a kind and gentle soul who brought joy to all who knew him.

Preface

The chapters in this volume were presented at the July 2000 NATO Advanced Study Institute on *Twentieth Century Harmonic Analysis – a Celebration.* The conference was held at the beautiful Il Ciocco resort near Lucca, in the glorious Tuscany region of northern Italy. Once again we gathered at this idyllic spot to explore and extend the reciprocity between mathematics and engineering. The dynamic interaction between world-renowned scientists from the usually disparate communities of pure mathematicians and applied scientists, which occurred at our 1989, 1991, 1992, and 1998 ASI's, continued at this meeting.

Almost exactly one century ago harmonic analysis entered a (still continuing) golden age, with the emergence of many great masters throughout Europe. *Some* of these illustrious names were: Hardy, Littlewood, Landau, van der Corput, Hadamard, de la Vallée-Poussin, Tchebychev, Bernstein, Markov, Fejér, Riesz, Pólya, Szegö, etc. They created a wealth of profound analytic methods which were so successfully exploited and further developed by succeeding generations. This flourishing of concrete harmonic analysis is, today, as lively as ever, with such contemporary great analysts as ASI lecturers and authors Havin, Kahane, Shapiro, Weiss, Salem Prize winner Körner, and Nobel Laureate Hauptman.

In addition to its own ongoing internal development and its basic role in other areas of mathematics (number theory, differential equations, probability, statistics), physics and chemistry (visible light and infrared optics, crystallography, wave phenomena), financial analysis (time series), medicine (tomography, brain and heart wave analysis), and biological signal processing, harmonic analysis has made fundamental contributions to essentially all twentieth century technology-based human endeavors. This includes telephone, radio, television, radar, sonar, satellite communications, medical imaging, the internet, and multimedia. This ubiquitous nature of the subject is touched upon by many of the authors.

Thus, the ASI and this volume are intended not only to promote the infusion of new mathematical tools into applied harmonic analysis, but also to fuel the development of applied mathematics by providing opportunities for young engineers, mathematicians, and other scientists to learn more about problem areas in today's technology that might benefit from new mathematical insights.

Many of the world's harmonic analysis experts were principal speakers at the ASI, and their chapters appear in this volume. These renowned scientists address their talks and chapters to an audience which consists of a broad spectrum of pure and applied mathematicians, as well as a diverse group of engineers and scientists. Thus, the reader has the opportunity to learn or reinforce fundamental concepts from the individuals who have played a major role in the ongoing flourishing of harmonic analysis, and to see them discuss in accessible terms their profound contributions and ideas for future research.

Victor Havin begins this book by clearly describing, both heuristically and precisely, how one should view "smallness" in the context of the uncertainty principle. His chapter gives the reader a fundamental understanding of one of the major themes of classical Fourier analysis, and also discusses important applications of the uncertainty principle to physics and engineering.

Harold Shapiro beautifully shows how operator theory, involving genuinely infinite-dimensional constructions, has profound applications to concrete classical harmonic analysis problems. A wonderful case in point is D. Sarason's recognition of the fundamental role played by commutativity in understanding (and getting radically new proofs of) classical interpolation and moment problems, like those usually associated with the names of Pick-Nevanlinna, Caratheodory, and Schur.

Jean-Pierre Kahane offers clear and concise insight into the interactions between Baire's category theorem, Lebesgue's measure theory, and trigonometric series.

A beautiful exposition of the current state of the art in Gabor theory is presented by Guido Janssen. He details the key role played by the Gabor frame operator associated with the set of elementary signals being used in the expansion of a given signal. Working in the time, frequency, time-frequency, and Zak transform domains, he addresses the basic problems of whether there is a Gabor frame and how to compute the dual frame.

While describing the many fascinating twists and turns that can occur when rearranging orthogonal series, Tom Körner clarifies the surprising phenomenon that wavelet expansions continue to work well under decreasing rearrangements whereas Fourier series do not. His focus is on the work of Olevskiĭ, Tao, and others in this area.

Stephane Jaffard clearly describes the interplay between function spaces, wavelet expansions, and multifractal analysis. For example, he shows how refinements of the numerical techniques introduced to compute turbulence spectra have led to the introduction of new function spaces, which turn out to be the right setting to determine the fractal dimensions of graphs, and offer natural extensions of the Besov spaces to negative p's.

Richard Tolimieri and Myoung An offer significant insight into the emergence at IBM of Fourier transform algorithms, a critical step in advancing the widespread use of digital computers in scientific and technological applications. They show how early efforts focused on reducing "expensive" multiplications at the cost of increasing additions, and how the recent importance of FPGA's and reconfigurable hardware has renewed the need to reduce multiplication counts. Throughout they weave in the interactions between these fascinating computational techniques and harmonic analysis.

Herb Hauptman describes an application of harmonic analysis which is yielding profound medical benefits daily, and for which he won the 1985 Nobel Prize in chemistry. Namely, he shows how the known atomicity of crystal structures and the redundancy of the observed magnitudes of the normalized structure factors of the X-ray diffraction pattern render the classical phase problem of X-ray crystallography solvable. He goes on to describe his new "Shake-and-Bake" algorithm, a completely automatic solution of the phase problem for structures containing as many as 1000 atoms when data are available to atomic resolution.

By combining the concept of frames with the Zak transform, Josh Zeevi elucidates the use of localized bases or frames in the representation, processing, compression and transmission of speech, images and other natural nonstationary signals.

Babar Saffari offers a fascinating chapter on a fascinating subject: extremal problems involving polynomials, trigonometric polynomials, and exponential sums. The depth and beauty of this "elementary" subject come alive in his exposition.

By considering function algebras, formal power series, and operator algebras, including a quantitative treatment of the Weiner-Pitt-Sreider phenomenon for convolution measure algebras on locally compact abelian groups, Nikolai Nikolski gives an in-depth survey of recent results on the phenomenon of the "invisible spectrum" for Banach algebras.

As Hugh Montgomery makes clear, a wide variety of questions of harmonic analysis arise naturally in various contexts of analytic number theory. In his chapter, Hugh gives a clear exposition of a number of examples of this type.

By focusing on two major and central mathematical areas, the radar ambiguity function and radar waveform design, Bill Moran beautifully brings forth the intimate relationship between harmonic analysis and signal processing. It is difficult to imagine a more perfect exposition for the mathematician interested in real-world signal design and analysis.

In the past 20 years or so, wavelets have emerged as one of the central topics in both pure and applied harmonic analysis. Guido Weiss and Edward Wilson present a broad overview of many aspects of the underlying mathematical theory. A highlight is their recently obtained characterization of "all" wavelets – an important new result.

The concluding chapter consists of several lesser known (as compared, for example, to the Riemann hypothesis) but certainly worthwhile unsolved problems and conjectures of current appeal in harmonic analysis, to challenge the interested twenty-first century reader.

Numerous giants in the field, including about half of the lecturers and authors, have recently retired or will do so in the near future. A major purpose of the ASI was to afford them the opportunity to join together to share their profound wisdom with the many future stars of pure and applied harmonic analysis. A second purpose was to produce this book for current and future generations, highlighting their thoughts and insights.

The cooperation of many individuals and organizations was required in order to make the conference the success that it was. First and foremost I wish to thank NATO, and especially Dr. F. Pedrazzini and his staff, for the initial grant and subsequent help. Financial support was also received from the Air Force Office of Scientific Research (Dr. Jon Sjogren), the European office of the Office of Naval Research (Dr. Igor Vodyanoy), the Raytheon Company (Dr. Philip W. Cheney, Chief Scientist), the National Science Foundation, the Australian Research Council, Philips Research Laboratories, the European Mathematical Society, the University of Massachusetts at Boston, and Prometheus Inc. This additional support is gratefully acknowledged.

I wish to express my sincere appreciation to my assistants, Marcia Byrnes and Gerald Ostheimer, for their invaluable aid. I am also grateful to Kathryn Hargreaves, our TEXnician, for her superlative work in preparing this volume. Finally, my heartfelt thanks to the Il Ciocco staff, especially Bruno Giannasi and Alberto Suffredini, for offering an ideal setting, not to mention the magnificent meals, that promoted the productive interaction between the participants of the conference. All of the above, the other speakers, and the remaining conferees, made it possible for our Advanced Study Institute, and this volume, to fulfill the stated NATO objectives of disseminating advanced knowledge and fostering international scientific contacts.

January 14, 2001   *Jim Byrnes*, Newport, Rhode Island

Contents

Dedication   v

Preface   vii

Part 1. The Papers   1

On the Uncertainty Principle in Harmonic Analysis, *V.P. Havin*   3

Operator Theory and Harmonic Analysis, *H.S. Shapiro*   31

Probabilities and Baire's theory in harmonic analysis, *J.-P. Kahane*   57

Representations of Gabor frame operators, *A.J.E.M. Janssen*   73

Does Order Matter, *T.W. Körner*   103

Wavelet expansions, function spaces and multifractal analysis, *S. Jaffard*   127

Some Plots of Bessel Functions of Two Variables, *F.A. Grünbaum*   145

Lesser Known FFT Algorithms, *R. Tolimieri and M. An*   151

The Phase Problem of X-ray Crystallography, *H.A. Hauptman*   163

Multiwindow Gabor-type Representations
and Signal Representation by Partial Information, *Y.Y. Zeevi*   173

Some polynomial extremal problems which emerged in the twentieth century,
Bahman Saffari   201

The Problem of Efficient Inversions and Bezout Equations, *N. Nikolski*   235

Harmonic Analysis as found in Analytic Number Theory, *H.L. Montgomery*   271

Mathematics of Radar, *B. Moran*   295

The Mathematical Theory of Wavelets, *G. Weiss and E.N. Wilson*   329

Part 2. Problems   367

Assorted Problems, *Various Authors*   369

How to Use the Fourier Transform in Asymptotic Analysis, *V. Gurarii, et al.*   387

Index   403

Part 1

The Papers

On the Uncertainty Principle in Harmonic Analysis

V.P. Havin

Department of Mathematics and Mechanics
St. Petersburg State University
Stary Peterhof
St. Petersburg 198904 Russia
havin@havin.nsr.pu.ru

ABSTRACT. The Uncertainty Principle (UP) as understood in this lecture is the following informal assertion: *a non-zero "object" (a function, distribution, hyperfunction) and its Fourier image cannot be too small simultaneously.* "The smallness" is understood in a very broad sense meaning fast decay (at infinity or at a point, bilateral or unilateral), perforated (or bounded, or semibounded) support etc. The UP becomes a theorem for many "smallnesses" and has a multitude of quite concrete quantitative forms. It plays a fundamental role as one of the major themes of classical Fourier analysis (and neighboring parts of analysis), but also in applications to physics and engineering. The lecture is a review of facts and techniques related to the UP; connections with local and non-local shift invariant operators are discussed at the end of the lecture (including some topical problems of potential theory). The lecture is intended for the general audience acquainted with basic facts of Fourier analysis on the line and circle, and rudiments of complex analysis.

Introduction

This lecture is devoted to the following phenomenon known as the Uncertainty Principle (the UP):

it is impossible for a non-zero function f and its Fourier image $\hat{f}$ to be too small simultaneously. In other words, the approximate equalities $f \approx g, \hat{f} \approx \hat{g}$ cannot hold at the same time and with a high degree of accuracy unless f and g are identical. Gaining some "certainty" about f (in the form of a good approximation g) we have to pay by the uncertainty about $\hat{f}$, since the error $\hat{f} - \hat{g}$ is bound to be considerable. The term is borrowed from quantum mechanics where it is usually understood as the Heisenberg inequality for

3

J.S. Byrnes (ed.), Twentieth Century Harmonic Analysis - A Celebration, 3–29.
© 2001 *Kluwer Academic Publishers. Printed in the Netherlands.*

the wave function, but in the present text it is interpreted in a much less definite sense; this very vagueness makes it flexible and susceptible to a multitude of rigorous interpretations (or refutations) depending on a concrete kind of respective "smallness" of f and $\hat{f}$ mentioned in its statement. Our UP can be patently wrong (e.g. if the sizes of f and $\hat{f}$ are measured in the L^2-norm and in many other cases); this means the UP can be sometimes overcome, and "small" non-zero pairs $(f, \hat{f})$ *may* exist, this fact being also one of our themes. Nevertheless the UP plays an outstanding role in harmonic analysis and its applications to physics and engineering. But these applications won't be discussed here. We treat the UP as a phenomenon of pure mathematics, or, to be more precise, classical Fourier analysis (mainly on $\mathbb{R}$ and the unit circle $\mathbb{T}$). Our theme is very vast and can be looked at from many points of view; ours will be that of pure analysis. The concrete forms of the UP to be considered here pertain mostly to quasianalyticity, approximation theory, and, first of all, to complex analysis, an abundant source of concrete manifestations (and disprovements) of the UP. Among the omissions of this lecture are the operator theoretic approach to the UP (commutation relations) and the modern time-frequency approach. But even after we have confined our discussion to the purely analytic aspects of the phenomenon we are still left with a huge mass of facts, techniques, and approaches. Thus the choice of what is to be discussed was inescapable and difficult. It was a compromise dictated by what I know (or don't), time and size limitations, my personal predilections, but also by my desire to publicize impressive results obtained in the eighties and nineties by my (partly former) colleagues from St. Petersburg, although a good deal of the subsequent text is quite old and classical.

To describe the organization of the lecture let us first introduce some notation. Let X denote $\mathbb{R}$ or $\mathbb{T} = \{\xi \in \mathbb{C} : |\xi| = 1\}$; m will stand for Lebesgue measure on X; we always normalize m on $\mathbb{T} : m(\mathbb{T}) = 1$; sometimes we write $|A|$ in place of $m(A)$. The Fourier transform $\hat{f}$ of a function $f \in L^1 = L^1(X, m)$ is understood as $\hat{f}(\xi) = (2\pi)^{-1} \int_X f(t) \exp(-it\xi) dt$ where $\xi \in \mathbb{R}$ or $\mathbb{Z}$ (i.e. $\xi \in \hat{X}$), but different normalizations of $\hat{f}$ can also occur here and there. We assume the reader is acquainted with Fourier analysis of (tempered) distributions. In particular our most frequent symbols related to a distribution T on X will be supp T (= *the closed support of* T) and spec $T = $ supp $\hat{T}$, the spectrum of T. Now we try to bring some order into the heap of "smallnesses" to be used below (see the statement of the UP above). We compose a list of properties of a function (or distribution) on X:

$S_1(f)$ ("fast bilateral decay of f at infinity"): f is defined on $\mathbb{R}$ or $\mathbb{Z}$ and satisfies

$$f(t) = O(M(t)), |t| \to +\infty$$

where M is a given majorant, $\lim_{|t| \to +\infty} M(t) = 0$:

$S_2(f)$: replacing $|t| \to +\infty$ in $S_1(f)$ by $t \to +\infty$ (or $t \to -\infty$) we get fast decay at $-\infty$ (or $+\infty$);

$S_3(f)$ ("a deep zero of f at a point $x_0 \in X$"): $f(t) = O(M(t)), t \to x_0$, where $\lim_{t \to x_0} M(t) = 0$;

$S_4(T)$ ("a sparse support") : $X \setminus \operatorname{supp} T$ is non-empty and more or less "rich" (say, consists of many long intervals or arcs),
but very interesting forms of the UP arise even when $X \setminus \operatorname{supp} T$ is connected, this is why the following three properties are stated separately:

$S_5(T)$ ("a gap in the support"): $\operatorname{supp} T$ omits a non-degenerate interval (or an arc) of X;

$S_6(T)$ ("a bounded support"): $\operatorname{supp} T$ is bounded (this case refers to $X = \mathbb{R}$);

$S_6(\hat{T})$ means that T is "band limited".

$S_7(T)$ ("a sembounded support"): $\operatorname{supp} T \subset [0, +\infty)$ (or $\subset (-\infty, 0]$); the spectral property $S_7(\hat{T})$ is absolutely fundamental for our subject and deserves special attention.

To conclude this list we have to define the so-called logarithmic integral $\mathcal{L}(f)$ of a function f defined on X:

$$\mathcal{L}(f) = \int_{\mathbb{T}} \log|f|\, dm, \text{ if } X = \mathbb{T}; \mathcal{L}(f) = \int_{\mathbb{R}} \log|f|\, d\Pi, \text{ if } X = \mathbb{R}$$

where Π is the Poisson measure $\pi^{-1}(1 + x^2)^{-1}m$. Our last "smallness condition" means the geometric mean of f (w.r. to m on $\mathbb{T}$ or Π on $\mathbb{R}$) is zero; it looks less natural than its predecessors, but in fact it is *responsible* for many variants of the UP, and is omnipresent in many books on Fourier analysis, complex analysis and probability:

$$S_{\mathcal{L}}(f) : \mathcal{L}(f) = -\infty.$$

A discrete logarithmic integral $\mathcal{L}(f)$ of a function defined on $\mathbb{Z}$ will also play a role: this time $\mathcal{L}(f) = \sum_{n \in \mathbb{Z}} \frac{\log|f(n)|}{1+n^2}$.

Now we are in a position to specify (slightly) our main question:

given j and $k = 1, 2, \ldots, 7$ or $\mathcal{L}$, is it true that $S_j(T)$ and $S_k(\hat{T})$ imply the complete vanishing of T?

The conditions S_j being still vague, the answer to such an "$(S_j, \hat{S}_k)$-question" depends heavily on the concrete relations between the time condition $S_j(f)$ and the frequency condition $S_k(\hat{f})$; some combinations of j and k may even not admit a satisfactory answer at all. But a remarkable fact is *the existence* of good answers to many such questions, the answers being sharp and verifiable. The list of the S_k-conditions looks dull (something like bookkeeping) and formal, but the diversity, variety and beauty of the ideas and tools required to answer at least some of our questions is quite amazing. Note that "qualitative" questions $(S_k, \hat{S}_j)$ usually entail some "quantitative" problems resulting in useful and explicit *estimates*.

The lecture consists of three parts. Part 1 is a collection of results not requiring any use of complex analyticity; we try to sketch (or at least allude to) some proofs (when they are simple). This policy becomes almost impossible in Part 2 based on complex analyticity. The "complex" proofs usually involve a good portion of hard analysis, so Part 2 is mainly a collection of results accompanied by some comments (kind of a guided tour). Part 3 is devoted to some remote repercussions of the UP: description of symbols of local and non-local shift invariant operators; some closely related topical problems stemming from potential theory are also discussed.

Our theme is present more or less explicitly in any course on Fourier analysis ([Z, Ba, Katz, Kah]), quasianalyticity ([M]), and complex analysis ([Bo, Du, Pr, Ho, Ga, Koo1]). The books [PW], [B], [L], [Lev], [KahS], [DeBr], [Koo2], [Koo3], [Car], [Carl], and [Nik] are especially close to our theme and have influenced our exposition in many ways.

The present lecture is mainly based on [HJ] just reproducing some of its parts in a very compressed form; we often refer to the bibliography therein. I also want to mention the long article [Na] with its impressive amount of excellent results.

1. The UP without Complex Variables

1.1. *On functions with semibounded spectra*

Let L be a linear shift invariant operator on the time line $\mathbb{R}$ defined on a vector space of functions (or distributions). The generic form of L is the convolution with a function (distribution) a:

$$Lf = a * f; \quad a = L(\delta).$$

We may interpret L as a device transforming inputs f into outputs Lf. We say L obeys the causality principle (or is *causal*) if $Lf|(-\infty, t_0)$ for any given moment t_0 depends only on $f|(-\infty, t_0)$ ("no output without an input"), or what is the same $a|(-\infty, 0) = 0$. In the Fourier coordinates the action of L becomes

$$\hat{L}f = \hat{a} \cdot \hat{f}$$

which means Lf is a frequency filter. The causality imposes severe restrictions on the spectral characteristic $\hat{a}$ of L : $\hat{a}$ cannot suppress too many frequencies unless $L = 0$. This phenomenon stems from the analytic continuability of $\hat{a}$ into a half-plane (due to the semiboundedness of supp a). This complex variable explanation will be one of the themes of Part 2. Actually one can explain many properties of $\hat{a}$ in the causal case staying on the line and ignoring the existence of the complex plane. One of these properties is *the Jensen inequality* for plus-functions (i.e. for functions with positive spectra).

It is convenient to interchange time and frequency lines and concentrate on the objects with semibounded *spectra* (rather than *supports*). We start with the periodic case: suppose $f \in L^p(\Pi, m) = L^p(\Pi)$, $1 \leqslant p \leqslant +\infty$, so that $\hat{f}$ lives on $Z : \hat{f}(n) = \int \bar{z}^n \, dm$, $n \in Z$. We say that f is in the Hardy class $H^p(\mathbb{T})$ if its spectrum is non-negative: spec $f \subset Z_+$ (sometimes we call such an f a plus-function).

For a probability measure μ in a measure space X we put

$$\mathcal{A}_\mu(f)(=\mathcal{A}(f)) = \int_X |f|d\mu, \quad \mathcal{G}_\mu(f)(=\mathcal{G}(f)) = \exp\int_X \log|f|d\mu,$$

thus defining *the arithmetic* and *geometric* means of f. Note that the meaning of the two "smallnesses"

$$\mathcal{A}(f) = 0 \quad \text{and} \quad \mathcal{G}(f) = 0 (\Leftrightarrow \mathcal{L}(f) = \int_X \log|f|d\mu = -\infty)$$

is very different: the first just means $f = 0$ a.e. whereas the second is implied by $\mu(\{f = 0\}) > 0$ or (depending on μ) by a fast decay of f at a point of X. By the Jensen inequality for the means we always have

(1) $$\mathcal{G}(f) \leqslant \mathcal{A}(f).$$

Let us now go back to $X = \Pi$, $\mu = m$. Clearly, $|\hat{f}(0)| = |\int f dm| \leqslant \mathcal{A}(f)$ for any $f \in L^1(\Pi)$. But another Jensen inequality asserts that

(2) $$|\hat{f}(0)| \leqslant \mathcal{G}(f), \quad \text{if} \quad f \in H^1(\mathbb{T})$$

This crucial fact has many far-reaching implications pertaining to the UP. So, for example,

$$f \in H^1(\mathbb{T}) \ \& \ \mathcal{G}(f) = 0 \Rightarrow f = 0$$

($\mathcal{G}(f) = 0$ kills $\hat{f}(0)$, but then it kills any $\hat{f}(n)$). But then

$$f \in H^1(\mathbb{T}) \ \& \ |\{f = 0\}| > 0 \Rightarrow f = 0$$

(total absence of negative frequencies is not compatible with vanishing on a set of positive length). For a proof of (2) see, e.g., [HJ], p. 34; it is quite short and elementary.

By a Möbius change of variables we get the following version of (2): *If $f \in L^1(\mathbb{R})$ and* spec $f \subset [0, +\infty)$, *then*

(3) $$\left|\int_{\mathbb{R}} f d\Pi\right| \leqslant \mathcal{G}_\Pi(f)$$

(f is regarded here as a tempered distribution, so $\hat{f}$ and spec $\hat{f}$ make sense; Π is the Poisson measure, see the Introduction). In particular (3) is valid for any plus-function $f \in L^p(\mathbb{R})$ (w.r. to m), $1 \leqslant p \leqslant +\infty$ (i.e. if $f \in H^p(\mathbb{R}) = \{f \in L^p(\mathbb{R}) : \text{spec } f \subset [0, +\infty)\}$). It is easy to deduce from (3) that $f \in H^p(\mathbb{R}) \ \& \ \mathcal{G}_\Pi(f) = 0 \Rightarrow f = 0$ so that if $f \in H^p(\mathbb{R})$, $f \neq 0$, then it cannot decay too fast at $+\infty$ or $-\infty$ or at any finite point, and it cannot vanish on a set of positive length. This last property is stable in the following sense:

THE THEOREM ON TWO CONSTANTS. *For any* $f \in H^{\infty}(\mathbb{R})$ *and any Lebesgue measurable* $S \subset \mathbb{R}$

(4) $$|(f * \Pi)(x)| \leqslant (\|f\|_{\infty,S})^{\Pi_x(S)} (\|f\|_{\infty,S'})^{\Pi_x(S')}, x \in \mathbb{R}.$$

$S' = \mathbb{R}\backslash S, \Pi_x(E) = \Pi(E - x)).$

The proof is a straightforward combination of two Jensen inequalities (1) and (3). Note that $\Pi_x(S)$ is the angle (divided by π) under which S is seen from $x+i$, and $f*\Pi_x = P(f)(x)$ is the Poisson integral of f (i.e. the bounded solution of the Dirichlet problem for the upper half-plane $\mathbb{C}_+$ with the boundary function f) computed at $x + i$. Clearly, $P(f) = 0$ implies $f = 0$; (4) shows that if a plus-function f is globally bounded (say, $|f| \leqslant 1$) and very small on S' (say, $|f| \leqslant \varepsilon$), then $P(f)$ is small globally: $|P(f)(x)| \leqslant \varepsilon^{\Pi_x(S)}$ for any real x.

We will also need an integral version of this result: *suppose* $\gamma > 0$, *and* $\Pi_x(S) \geqslant \gamma$ *for any* $x \in \mathbb{R}$; *if* $f \in H^2(\mathbb{R})$, *then*

(4') $$\int_{\mathbb{R}} |P(f)|^2 dm \leqslant 2(\int_S |f|^2 dm)^{\gamma} \|f\|_2^{2(1-\gamma)}$$

where $\| \ \|_2$ *denotes the* $L^2(m)$-*norm* ([HJ], p. 40).

The logarithmic integral $\mathcal{L}_{\mu}(f)$ figuring in $\mathcal{G}_{\mu}(f)$, $\mathcal{L}_{\mu}(f) = \int_X \log|f| d\mu = \log \mathcal{G}_{\mu}(f)$ for $X = \Pi$, $\mu = m$ or $X = \mathbb{R}$, $\mu = \Pi$ plays an outstanding role in many problems concerning the UP (not only for semibounded spectra!). The two conditions

$$\mathcal{L}(f) = -\infty \quad \text{and} \quad \mathcal{L}(f) > -\infty$$

define two separate realms: in the first one the rule of the UP is indisputable whereas in the second it can be sometimes resisted (see [HJ], but especially [Koo2]).

1.2. Hilbert Space methods

1.2.1. *Annihilating pairs of sets.* For a function $f \in L^2(\mathbb{R}^d) = L^2$ the set $\{x \in \mathbb{R}^d : f(x) \neq 0\}$ is called *the essential support of* f and denoted by ess supp f; it is defined up to a set of zero Lebesgue measure. *The essential spectrum of* f is defined as ess supp $\hat{f}$ and denoted by ess spec f ($\hat{f}$ is understood in accordance with the Plancherel theorem).

A pair (S, Σ) of Lebesgue measurable sets in $\mathbb{R}^d$ is said to be *annihilating* (or a-*pair*) if

(5) $$f \in L^2, \text{ess supp } f \subset S, \text{ess spec } f \subset \Sigma \Rightarrow f = 0.$$

The following property of (S, Σ) is more interesting: we say that (S, Σ) is *a strong* a-pair if

$$(6) \qquad \int_{\mathbb{R}^d} |f|^2 \leqslant c(S, \Sigma)\left(\int_{S'} |f|^2 + \int_{\Sigma'} |\hat{f}|^2\right)$$

for any $f \in L^2$ (A' denotes $\mathbb{R}^d \setminus A$). The annihilation property (5) of a strong a-pair is "stable": (5) only means that *vanishing* of $f|S'$ and $\hat{f}|\Sigma'$ implies *global vanishing* of f whereas (6) says that *the smallness* of $f|S'$, $\hat{f}|\Sigma'$ implies *the global smallness of f*.

The d-dimensional Lebesgue measure of a set $A \subset \mathbb{R}^d$ will be denoted by $|A|$.

The following version of the UP can be proved using only the basic properties of the Fourier transform and very general properties of projectors in a Hilbert space:

THE AMREIN-BERTHIER THEOREM.. *If*

$$|S| + |\Sigma| < +\infty,$$

then (S, Σ) is a strong a-pair.

Note that the sets $S, \sum$ are not supposed to be bounded. We are going to sketch a proof based on two orthogonal projectors P_S, $\hat{P}_\Sigma$ of L^2:

$$P_S f = \chi_S f, \mathcal{F}(\hat{P}_\Sigma f) = \chi_\Sigma \hat{f}$$

where χ_A denotes the characteristic function of the set $A \subset \mathbb{R}^d$ and $\mathcal{F}$ is the Fourier transform in $\mathbb{R}^d$ (duly normalized to define a unitary operator in L^2). The proof is sketched in 2.3 after some preparation in 2.2.

1.2.2. *Positive angle between two subspaces.* Let us now forget the concrete nature of these projectors and move to an abstract Hilbert space H; let (M, N) be a pair of its closed subspaces. We denote by P and Q the projectors of H onto M, N (resp.). We are interested in the following property of the pair (M, N) (or (P, Q)):

$$(7) \qquad \|h\|^2 \leqslant c(M, N)(\|P^\perp h\|^2 + \|Q^\perp h\|^2) \text{ for any } h \in H,$$

where $P^\perp = I - P, Q^\perp = I - Q$ project onto the orthogonal complements of M and N (resp.). Clearly, (7) is an abstract form of (6). It can be given several equivalent forms:

$\qquad$ (a) $\|PQ\|(= \|QP\|) < 1$;

$\qquad$ (b) $\sup\{| < m, n > | : m \in M, n \in N, \|m\| \leqslant 1, \|n\| \leqslant 1\} < 1$;

$(8) \qquad$ (c) $M \cap N = \{0\}$, and $M + N$ is closed;

$\qquad$ (d) $\|P^\perp n\| \geqslant c\|n\|$ for any $n \in N$ (or, equivalently, $\|Q^\perp n\| \geqslant c\|m\|$
$\qquad \qquad$ for any $m \in M$), $c > 0$.

If M, N are of finite dimension, then all these properties just mean $M \cap N = \{0\}$, but in general this last property alone does not imply (8) which is often expressed as "the positivity of the angle between M and N " (look at (b)). A proof of the equivalence of (7) and all properties in (8) is, e.g., in [HJ], p. 80.

The following general observation is crucial for the Amrein-Berthier theorem: *If $M \cap N = \{0\}$, and PQ is compact, then* (7) *holds.* Indeed, PQ being compact there is a unit vector $v \in H$ such that $\|PQv\| = \|PQ\|$; if $\|PQ\| = 1$, then $1 = \|PQv\| \leqslant \|Qv\| \leqslant \|v\| = 1$ whence $\|Qv\| = \|v\|$ and $Qv = v$, so that $v \in N$ and $\|Pv\| = \|v\|, v \in M$ whereas the only element of $M \cap N$ is zero.

1.2.3. Let us now return to $L^2 = L^2(\mathbb{R}^d)$ and put $P = P_S, Q = \hat{P}_\Sigma, M = \{f \in L^2 :$ ess supp $f \subset S\}, N = \{f \in L^2 :$ ess spec $f \subset \Sigma\}$. If $|S|, |\Sigma|$ are finite, then PQ becomes an integral operator in L^2 with the kernel $H(x, y) = c\chi_S(x)\hat{\chi}_\Sigma(y - x)$ which is Hilbert-Schmidt: by Plancherel

$$\iint |H(x, y)|^2 dx dy = c^2 \int \chi_S^2 \cdot \int \chi_\Sigma^2 = c^2|S||\Sigma| < +\infty.$$

If $f \in M \cap N$ then f is an eigenvector of PQ corresponding to the eigenvalue 1, so $M \cap N$ is finite dimensional. Moreover, its dimension can be estimated by $|S||\Sigma|$:

$$(9) \qquad \dim(M \cap N) \leqslant \iint |H(x, y)|^2 dx dy = c^2|S||\Sigma|.$$

Using this estimate it is not hard to prove that $M \cap N = \{0\}$ (i.e. that (S, Σ) is an a-pair), and thus complete the proof of the theorem. Suppose $\varphi \in M \cap N, \varphi \neq 0$, so that $S_0 \subset S, 0 < |S_0|$ where $S_0 =$ ess supp φ. For a vector $v \in \mathbb{R}^d$ put $\varphi_v(x) = \varphi(x - v)$. For a $v_1 \in \mathbb{R}^d$ the essential support S_1 of φ_{v_1} (i.e. $S_0 + v_1$) sticks out of S_0 (slightly):

$$0 < |S_1 \setminus S_0| < \varepsilon_1$$

where $\varepsilon_1 > 0$ is arbitrary (we are using the finiteness of $|S_0|$); functions φ, φ_{v_1} are linearly independent. Then we find $v_2 \in \mathbb{R}^d$ so as to make $S_2 = S_1 + v_1 =$ ess supp $\varphi_{v_1 v_2}$ stick out of $S_1 \cup S_0$ (slightly):

$$0 < |S_2 \setminus (S_1 \cup S_0)| < \varepsilon_2;$$

$\varphi, \varphi_{v_1}, \varphi_{v_1 v_2}$ are linearly independent. Continuing this process we arrive at an infinite linearly independent sequence of *shifts* of φ:

$$(10) \qquad \varphi, \varphi_{v_1}, \varphi_{v_1 v_2}, \dots$$

and the sequence of sets $S_0, S_0 + v_1, S_0 + v_1 + v_2, \dots$ whose union S^* is of finite measure if only $\sum \varepsilon_j < +\infty$. The essential spectrum being shift invariant the sequence (10) is in $M^* \cap N, M^* = P_{S^*}(L^2), N = \hat{P}_\Sigma(L^2)$ which is impossible $(\dim(M^* \cap N) \leqslant c^2|S^*||\Sigma|)$.

1.2.4. *Some remarks on strong annihilation.* Using the equivalence of (7) and (8) and staying in an abstract Hilbert space H we may solve a series of problems pertaining to the UP. For example, if (8) holds, then the operator $v \to (Pv, Qv)$ maps H onto $M \times N$. This means that whenever (S, Σ) is a strong a-pair the following system of equations (with "the unknown" $r \in L^2$) is solvable:

$$r|S = p|S, \hat{r}|\Sigma = q|\Sigma$$

for any $p, q \in L^2$. Another example is the following problem: describe the image of the unit ball of H under the mapping $h \to (\|Ph\|, \|Qh\|) \in \mathbb{R}^2$. This problem can be solved quite explicitly for many pairs (P, Q) such that PQ is compact; the result is a quantitative refinement of the Amrein-Berthier theorem (the Slepian-Pollack inequality): roughly speaking the point $(\|Ph\|, \|Qh\|)$ of the square $[0, 1] \times [0, 1]$ cannot get too close to the vertex $(1, 1)$. For a pair of sets S, Σ of finite measure the Slepian-Pollack inequality answers the following question: suppose $h \in L^2, \|h\| = 1$, and $\int_S |h|^2 = \alpha$ with a given $\alpha \in (0, 1)$; how large can $\int_\Sigma |\hat{f}|^2$ be ? It turns out that the least upper bound of this "spectral energy carried by Σ" is one if $\alpha \leqslant c(S, \Sigma) < 1$, but it *does depend* on $\alpha \in (c(S, \Sigma), 1)$ remaining *strictly less* than one.

1.2.5. *The Paneah Theorem.* The definition of a strong a-pair suggests the following general question: given a class s of measurable sets $S \subset \mathbb{R}^d$ find the class

$$\hat{s} = \{\Sigma \subset \mathbb{R}^d : (S, \Sigma) \text{is a strong a-pair for any } S \in s\}.$$

Denoting by s_{fin} the class of all sets in $\mathbb{R}^d$ of finite Lebesgue measure we may restate the Amrein-Berthier theorem:

(11) $$\hat{s}_{fin} \supset s_{fin}.$$

It is known that this inclusion is strict; I do not know any satisfactory and complete description of $\hat{s}_{fin}$. Let us turn instead to an important example of s when $\hat{s}$ admits a complete and explicit description: $s = s_b = $ the class of all *bounded* measurable sets in $\mathbb{R}^d$. We say that a Lebesgue measurable set $E \subset \mathbb{R}^d$ is *relatively dense* (at infinity) (or belongs to s_{rd}) if there exists a cube $K \subset \mathbb{R}^d$ and a number $\gamma > 0$ such that $|(K + x) \cap E| \geqslant \gamma$ for any $x \in \mathbb{R}^d$. A typical example (for $d = 1$) is the union of all intervals of a given positive length centered at equidistant points nh where $h > 0$ is fixed and $n \in \mathbb{Z}$. The rd-sets $E \subset \mathbb{R}$ can be *characterized* by the following property: the observer moving along the line $y = 1$ sees E all the time under an angle exceeding a positive number $\sum$; in other words

(12) $$\Pi_x(E) = \frac{1}{\pi} \int_E \frac{dt}{1 + (x - t)^2} \geqslant \sigma \text{ for any } x \in \mathbb{R}.$$

Denote by S'_{rd} the set of all complements of the rd-sets. The following theorem refers to $d = 1$ (i.e. to $\mathbb{R}$).

THE PANEAH THEOREM. $\hat{s}_b = s'_{rd}$.

(Paneah proved $\hat{s}_b \subset s'_{rd}$ for any dimension; the inverse inclusion in any dimension was proved by Logvinenko and Sereda later.) Here we sketch a short proof of a part of the Paneah theorem ([JH, Gor, HJ]), namely $s'_{rd} \subset \hat{s}_b$.

First note that the Poisson integral $P(\varphi) = \frac{1}{\pi}\varphi * \frac{1}{1+x^2}$ of $\varphi \in L^2$ is again in L^2, since

$$(13) \qquad\qquad (\hat{P}\varphi)(\xi) \equiv e^{-|\xi|}\hat{\varphi}(\xi)$$

(where $\hat{\varphi}(\xi) = (2\pi)^{-1} \int\limits_{\mathbb{R}} \varphi(x) \exp(-is\xi)dx$) whence

$$\|P(\varphi)\|_2 \leqslant \|\varphi\|_2.$$

If spec $\varphi \subset [0, l]$, then an inverse estimate can be obtained:$|\xi|$ in (13) becomes ξ so that

$$|\hat{\varphi}(\xi)| = e^{\xi}|P\hat{(\varphi)}(\xi)| \leqslant e^{l}|P\hat{(\varphi)}(\xi)|,$$

and

$$(14) \qquad\qquad \|\varphi\|_2 \leqslant e^{l}\|P(\varphi)\|_2$$

by Plancherel. Suppose now $f \in L^2(\mathbb{R})$, spec $f \subset [a, b]$; but then (14) is applicable to $\varphi = e^{-iax}f$ (with $l = b - a$). Applying the integral form of the two constants theorem to an rd-set E (see (4') and (12)) we get

$$\|f\|_2^2 = \|\varphi\|_2^2 \leqslant e^{2l} \cdot 2 \left(\int\limits_E |\varphi|^2\right)^{\Sigma} \|\varphi\|_2^{2(1-5)} = 2e^{2l}\left(\int\limits_E |f|^2\right)^{\Sigma} \|f\|_2^{2(1-5)}$$

whence $\|f\|_2^2 \leqslant 2e^{2l/\Sigma}\int_E |f|^2$. We have thus proved (8d) for $P = P_S, S = E'$, and $Q = \hat{P}_{[a,b]}$ which means that $(S, [a, b])$ is a strong a-pair.

1.2.6. *Periodic case: strong annihilation of supports omitting a set of positive length and sparse spectra.* The problem setting of 1.2.1 has obvious $L^2(\mathbb{T})$-parallels, the corresponding definitions of a-pairs and strong a-pairs $(S, \Sigma), S \subset \mathbb{T}, \Sigma \subset \mathbb{Z}$ being essentially the same. E.g., (S, Σ) is a strong a-pair if

$$\int\limits_{\mathbb{T}} |f|^2 dm \leqslant c \left(\int\limits_{S'} |f|^2 dm + \sum_{n\in\Sigma'} |\hat{f}(n)|^2\right) \text{ for any } f \in L^2(\mathbb{T}).$$

This case can be also included into the general scheme of 1.2.2.

Let $s_{\mathbb{T}}$ be the class of all sets $S \subset \mathbb{T}$ satisfying $m(S) < 1$; denote $\hat{s}_{\mathbb{T}}$ by SPARSE. A deep and difficult result on SPARSE is due to Mikheev who proved that

$$\Lambda^*(2) \subset SPARSE \subset \Lambda(2)$$

where $\Lambda^*(2), \Lambda(2)$ are certain classes of rarefied sets of integers ($\Lambda(2)$ is very familiar to the specialists; see [HJ], p. 102-110 for definitions). Here we only mention that it is unknown whether $\Lambda^*(2) = \Lambda(2)$. It is known however that the finite unions of lacunary sets are in

$\Lambda^*(2)$ (a set A of positive integers is called *lacunary* if $\sup\{m/n : m, n \in A, m < n\} < 1$; a set A of integers is called lacunary if $A \cap (0, +\infty)$ and $\{|n| : n \in A, n < 0\}$ are lacunary). The strong annihilation of pairs (S, Σ) with $m(S) < 1$ and lacunary Σ (but not a finite union of lacunary sets) was proved by Zygmund.

1.3. Review of some "non-complex" results

The main source of concrete forms of the UP is, of course, Complex Analysis. This was avoided (or carefully masked) in the preceding parts of the lecture. Before turning to the powerful complex machinery I just want to mention some more forms of the UP susceptible to other methods.

1.3.1. *The Annrein-Berthier Theorem revisited.* To describe a new approach to this theorem we start with a proof (due to Benedicks) of an L^1-analog of annihilation of pairs $(S, \Sigma), S, \Sigma \subset \mathbb{R}, |S| + |\Sigma| < +\infty$: *if* $f \in L^1(\mathbb{R})(= L^1), f|S' = 0, \hat{f}|\Sigma' = 0$, *then* $f = 0$. The proof is based on two facts:

   (i) If $\Sigma \subset \mathbb{R}, |\Sigma| < +\infty$, then for m-almost all $h \in (0, +\infty)$ almost all points of the lattice $(kh)_{k\in\mathbb{Z}}$ (i.e. all but a finite number) avoid Σ ([HJ], p. 456)
   (ii) For $f \in L^1(\mathbb{R})$ put

$$p(t) = \sum_{k\in\mathbb{Z}} f(t + k)$$

   the series converges in $L^1((-A, A))$ for any $A > 0$ thus defining a 1-periodic function p summable on $(0, 1)$, *the periodization of f*. Put $S =$ ess supp $f, \tilde{S} =$ ess supp $\cap (0, 1)$

it is easy to see that

(15) $$m(\tilde{S}) \leqslant m(S).$$

The ε-compression f_ε of f is defined by $f_\varepsilon(x) = f(x/\varepsilon)$; by $p^{(\varepsilon)}$ we denote the periodization of f_ε. The k-th Fourier coefficient $p_k^{(\varepsilon)}$ of $p^{(\varepsilon)}$ (w.r. to the system $(\exp(2\pi ikx))$) is $\varepsilon\hat{f}(\varepsilon k)$. Hence, by (i), $p^{(\varepsilon)}$ is a trigonometric polynomial of period $1/\varepsilon$ for almost all ε provided $f|\Sigma' = 0, |\Sigma| < \infty$. So for any $f \in L^1(\mathbb{R})$ and $\varepsilon \to 0$

$$(p^{(\varepsilon)})_{1/\varepsilon} \to f \text{ in } L^1((-A, A))$$

([HJ], p. 458). Now we are ready to complete the proof: a sequence $(p^{(\varepsilon_k)})_{\frac{1}{\varepsilon_k}}$ of trigonometric polynomials (with $\varepsilon_k \to 0$) tends to f in L^1_{loc}; but if ε_k is small, then $p^{(\varepsilon_k)}$ vanishes on $(0, 1) \setminus \tilde{S}_{\varepsilon_k}$, a set of *positive length*, since $|\tilde{S}_{\varepsilon_k}| \leqslant \varepsilon_k|S|$ (by (15)) whence $p^{(\varepsilon_k)} \equiv 0$.

   In the same spirit, but in a much more quantitative way Nazarov found an explicit estimate of the constant c in the Amrein-Berthier inequality (6). The abstract proof discussed in 1.2.2-1.2.3 did not yield any information on c; it was not even clear whether c depends on $|S|, |\Sigma|$ rather than on S, Σ.

THE NAZAROV THEOREM. *The Amrein-Berthier inequality (6) (for $d = 1$) holds with* $c = A \exp A|S||\Sigma|$ *where A is an absolute constant.*

Writing (6) for the Gauss function $f(x) = \exp(-x^2/2)$, $S = \Sigma = [-N, N]$ we find $c \geqslant \exp A'|S||\Sigma|$ for an absolute A'. Note that in fact c tends to one as $|S||\Sigma| \to 0$ which fact can be easily deduced from the abstract geometric considerations of 1.2.2 and the estimate $\|P_S \hat{P}_\Sigma\|^2 \leqslant const|S||\Sigma|$; for $|S||\Sigma|$ bounded off zero the first factor A in Nazaror's estimate can be dropped.

An elementary probabilistic analysis of "random lattices" in the spirit of Benedicks argument led Nazarov to a "finite" version of the UP which is interesting in itself. In a particular case it was discovered by Turan in the fifties.

THE TURAN LEMMA. *Let P be a trigonometric polynomial*

$$(P(\xi) = \sum \hat{P}(n)\xi^n, \xi \in \mathbb{T}),$$

spec P *being a finite set of integers. Put* ord P = card spec P. *There exists an absolute constant C such that*

(16)
$$\max_{\mathbb{T}} |P| \leqslant (C/m(\gamma))^{ordP} \max_{\gamma} |P|$$

for an arbitrary arc $\gamma \subset \mathbb{T}$.

Note that $ordP \leqslant degP = \max\{|n| : n \in \text{spec } P\}$, and (16) is an essentially non-linear result, since the set of all P's with a given $ordP$ is not a linear space. Turan's original proof was based on some explicit interpolation formulas. Nazarov needed (16) not for *arcs* γ, but for arbitrary compact subsets of $\mathbb{T}$, and he succeeded in proving (16) for this more general situation which required a new approach involving the Kolmogorov weak type estimate of the Hilbert transform. He proved along the way that if spec $P \subset [-M, M]$, then for any $t > 0$

$$|\{\zeta \in \mathbb{T} : |P'(\zeta)| \geqslant tM|P(\zeta)|\}| \leqslant C_{abs}/t,$$

so that the Bernstein norm estimate of the derivative of a trigonomtric polynomial P of degree M ($\max_{\mathbb{T}} |P'| \leqslant M \max_{\mathbb{T}} |P|$) holds *pointwise* off a set of small measure ($\leqslant C_{abs}/t$) with tM in place of M.

1.3.2. *The F. and M. Riesz Theorem.* originally appeared and was perceived as a fact of Complex Analysis, but later it was given several proofs not using analytic functions. The theorem states that a charge (= a complex valued Borel measure) on $\mathbb{R}$ or $\mathbb{T}$ with positive spectrum (a plus-charge) is m-absolutely continuous. Of those non-complex proofs I mention here only one due to $A.B.$ Aleksandrov and J. Shapiro and based on peculiarities of the L^p-metric with $p \in (0, 1)$ restricted to trigonometrical polynomials $\sum_{n \geqslant 0} c_n z^n$ with non-negative spectrum. This approach is applicable to the charges on a multidimensioinal torus ([HJ], p. 41-50). An interesting quantitative version of the F. and M. Riesz theorem is due to Pigno, Smith ([HJ], p. 23-28).

1.3.3. *The De Leeuw-Katznelson Theorem.* The Fourier coefficients $\hat{\mu}(n)$ of a plus-charge on $\mathbb{T}$ tend to zero as $|n| \to +\infty$ (an immediate corollary of the F. and M. Riesz theorem). This property is stable: the De Leeuw-Katznelson theorem states that for any $\varepsilon > 0$ there is a $\delta > 0$ such that for any plus-charge μ on $\mathbb{T}$ with $var\,\mu \leqslant 1$

$$\limsup_{n \to -\infty} |\hat{\mu}(n)| < \delta \Rightarrow \limsup_{n \to +\infty} |\hat{\mu}(n)| < \varepsilon.$$

The proof is quite "real" ([HJ], p. 29-31).

1.3.4. *Spectral decay of singular charges.* An m-singular charge μ (on $\mathbb{T}$) is highly concentrated; the UP suggests its Fourier image $\hat{\mu}$ should be "spread", but to what extent? E.g., is it possible for $\hat{\mu}$ to tend to zero (in which case we call it an r-charge in honour of Rajchman)? The answer is "Yes." In this connection I mention a beautiful result due to Salem (preceded by a more elementary partial result due to Bari) characterizing the Cantor subsets of $\mathbb{T}$ whose Cantor measure is an r-measure ([HJ], p. 63, 86). Another device to produce singular r-measures are infinite Riesz products ([Z, HJ]; a very nice treatment of the Riesz products is in [Pey]).

An obvious spectral obstacle for a charge μ on $\mathbb{T}$ to be singular is the inclusion $\hat{\mu} \in l^2(\mathbb{Z})$. *The Ivashev-Musatov Theorem* asserts that this fact is sharp: for any "nice" non-negative function Φ defined on $[0, +\infty)$ with $\sum_{1}^{\infty} \Phi^2(n) = +\infty$ there exists a non-zero m-singular positive measure μ on $\mathbb{T}$ with compact support such that $|\hat{\mu}(n)| \leqslant \Phi(|n|)$ for any $n \in \mathbb{Z}$ (we are not in a position to discuss here "the nicety" of Φ; dropping the compactness of supp μ from the statement above we may just assume Φ to be decreasing ([Kor]). The proof is based on ingenious asymptotic estimates of oscillating integrals in the spirit of the Van-der-Corput lemmas.

1.3.5. *Deep zero & sparse spectrum.* Suppose $f \in C(\mathbb{T}), \varepsilon > 0, f(t) = O(\exp(-|t - 1|^{-(1+\varepsilon)}))$ as $t \to 1$ ("a deep zero at 1") and $\sum_{n \in \mathbb{Z}} |n|^{\varepsilon-1/2} < +\infty$ ("a sparse spectrum"); then $f = 0$. This is a very particular case of the Mandelbrojt theorem. It was given a purely "real" proof by Belov ([HJ], p. 80-85). This proof is interesting in itself providing some useful quantitative relations. But the complex approach to the Mandelbrojt theorem results in its much stronger forms and seems to be the only way to prove its sharpness.

2. Complex Methods

2.1. *Introductory Remarks*

2.1.1. Partial sums of the Fourier series on $\mathbb{T}$ are rational functions, and partial Fourier integrals on $\mathbb{R}$ are entire functions; both live not only on $\mathbb{T}$ or $\mathbb{R}$, but in the whole ambient plane $\mathbb{C}$. Leaving $\mathbb{R}$ and $\mathbb{T}$ for $\mathbb{C}$ we get a vast new perspective making many manifestations of the UP just some uniqueness theorems of Complex Analysis. A (very primitive) example is this: if the support of a non-zero charge μ on $\mathbb{R}$ is bounded, then spec μ is not, since $\hat{\mu}$ is an entire function. We can of course, strengthen this trivial remark replacing the boundedness of supp μ by fast decay of μ at infinity (say, by the convergence of $\int_{\mathbb{R}} (\exp c|t|) d|\mu|(t)$ for a $c > 0$ entailing the analyticity of $\hat{\mu}$ in the strip $\{|Imz| < c\}$. The wealth of subtle uniqueness theorems of Complex Analysis yields far more precise and profound forms of the UP.

The complex approach gives a new explanation of the UP phenomena for plus-charges on $\mathbb{R}$ and $\mathbb{T}$. Suppose a function f summable on the time axis $\mathbb{R}$ has no past, that is $f|(-\infty, 0) = 0$. Then its Fourier integral

$$\hat{f}(\xi) = (2\pi)^{-1} \int_{\mathbb{R}} f(t) \exp(-it\xi) dt$$

makes sense not only for *real* ξ, but also for any $\xi \in \mathbb{C}_-$, the open lower-half-plane, and f is *analytic there* (the condition $f \in L^1(\mathbb{R})$ is not essential, we might speak of a charge, an L^p-function, or a distribution living in the future, i.e. supported by $(0, +\infty)$). Reversing the order of time and frequency and changing sign in the exponent we conclude that any plus-function (distribution) f (i.e. when spec $f \subset [0, +\infty)$) is in a way extendable to the upper half-plane $\mathbb{C}_+$ and this is (heuristically) a *complete* characterization of the plus-functions: if f is extendable from $\mathbb{R}$ to a function analytic in $\mathbb{C}_+$ satisfying some growth conditions, then f is a plus-function (this is not a theorem, but rather a useful heuristic principle).

An analogous description of the plus-functions on $\mathbb{T}$ is even more obvious. A Fourier series $\sum_{n \geqslant 0} \hat{f}(n) z^n$ lacking negative harmonics becomes a power series converging in the open unit disc $\mathbb{D}$, and the interpretation of a plus-function f on $\mathbb{T}$ as a boundary trace of its sum looks very plausible (and, similarly $\sum_{n<0} \hat{f}(n) z^n$ becomes a Laurent series in $\mathbb{C} \setminus (\mathbb{D} \cup \mathbb{T})$).

These remarks are meaningful and rich in consequences for *arbitrary* functions (not just for plus- or minus-functions): a more or less arbitrary function f on $\mathbb{R}$ can be written as $\int_{-\infty}^{+\infty} \hat{f}(\xi) e^{it\xi} d\xi$ (in a sense) whence

(17) $f = f_+ - f_- \text{where} f_+(t) = \int_0^{+\infty} \hat{f}(\xi) e^{it\xi} d\xi, f_-(t) = - \int_{-\infty}^0 \hat{f}(\xi) e^{it\xi} d\xi,$

so that $f_\pm$ are plus- and minus-functions extendable to the respective half-planes. An analogous decomposition is valid for functions (and distributions) on $\mathbb{T}$: if, say, $f \in L^1(\mathbb{T})$, then putting $f_+ = \sum_{n \geqslant 0} \hat{f}(n) z^n, f_- = - \sum_{n<0} \hat{f}(n) z^n$ we get $f = f_+ - f_-$, a formal equality to be duly interpreted which is quite possible in many cases.

Let us remember the following form of the UP: a continuous non-zero plus-function on $\mathbb{T}$ cannot vanish on a set of positive length (see section 1.1 of Part 1). This fact becomes now a boundary uniqueness theorem for functions analytic in $\mathbb{D}$ and continuous up to $\mathbb{T}$, and a conformal mapping of $\mathbb{C}_+$ onto $\mathbb{D}$ immediately yields an analogous result for the plus-functions on the line.

2.1.2. The complex point of view gives a new insight into the notions of *support* and *spectrum*. The support of a (say, summable) function f on $\mathbb{R}$ can be characterized as the set of singularities of its Cauchy potential Φ,

$$\Phi(\xi) = \frac{1}{2\pi i} \int_{\mathbb{R}} \frac{f(t)dt}{t - \xi}, \quad (\xi \in \mathbb{C} \setminus \mathbb{R})$$

since $f(t) = \lim_{\varepsilon \downarrow 0}(\Phi(t + i\varepsilon) - \Phi(t - i\varepsilon))$ a.e. on $\mathbb{R}$. But spec $f = $ supp $\hat{f}$, and (if $\hat{f}$ is summable and in many other cases) spec f is the set of singularities of the function Φ, analytic in $\mathbb{C} \setminus \mathbb{R}$ defined as $\Phi(\zeta) = (2\pi i)^{-1} \int_{\mathbb{R}} (\hat{f}(\xi)/(\xi - \zeta)) d\xi$ which is readily seen to be

$-(2\pi)^{-1} \int\limits_{0}^{+\infty} f(t) \exp(-it\zeta)dt$ for $Im\zeta < 0$ and $(2\pi)^{-1} \int\limits_{0}^{+\infty} f(t) \exp(-it\zeta)dt$ for $Im\zeta > 0$.

A similar description of supp f for $f \in L^1(\mathbb{T})$ is obvious: it coincides with the set of singularities of Φ: $\zeta \to (2\pi i)^{-1} \int_{\mathbb{T}} f(t)(t - \zeta)^{-1}dt(|\zeta| \neq 1)$, or the complement of the largest open set $O \subset \mathbb{T}$ such that Φ is analytic in $\mathbb{D} \cup O \cup \{|\zeta| > 1\}$.

2.1.3. Versions of the UP obtained by the complex tools are very often based on the following fact: *suppose $F \neq 0$ is analytic in a domain $O \subset \mathbb{C}$; then $\log|F|$ is subharmonic in O:*

(18)
$$\log|F(a)| \leqslant \int\limits_{\mathbb{T}} \log|F(a + rz)|dm(z)$$

provided the disc $\{|z - a| \leqslant r\}$ is in O (this is Jensen inequality (2)). The subharmonicity is akin to convexity, and (18) implies a certain rigidity of $|F|$: the smallness of $|F|$ on a small (but solid) part P of O makes $|F|$ small on $O \setminus P$ as well. A rigorous statement of this kind is the two constants inequality (4). The subharmonicity of $\log|F|$ for an analytic F entails the following extremely useful heuristic principle: *if a non-zero analytic function is not too big (globally), then it cannot be too small (even locally).* If for example supp $f \subset [-\sigma, \sigma], f \in L^1([-\sigma, \sigma])$, then $|\hat{f}(\zeta)| \leqslant const \exp(\sigma|\zeta|), \zeta \in \mathbb{C}$, which is a global growth restriction imposed onto entire function $\hat{f}$; an appropriate form of the Jensen inequality forbids $\hat{f}$ to decay too fast along $\mathbb{R}$ or to have too many zeros.

To conclude these introductory remarks let me mention the very special plasticity of *the formulas of Complex Analysis.* So, for example, a band limited function has at least four faces: it is a trigonometric integral $\int\limits_{-\sigma}^{\sigma} \hat{f}(\xi) \exp(i\xi t)d\xi$, but also a power series, or a contour integral, the Borel transform of the entire function $\hat{f}$ (with a freedom to deform the contour), or an infinite canonical product.

Now we start our guided tour around (some) applications of the complex tools to the UP.

2.2. *Fast decay of f and f̂ at infinity*

A pair (M, N) of positive functions on $(0, \infty)$ is called *sufficient* if

(19) $|f(t)| \leqslant M(|t|) \& |\hat{f}(\xi)| \leqslant N(|\xi|) (t, \xi \in \mathbb{R}) \Rightarrow f = 0.$

A complete description of sufficient pairs $(e^{-At^p}, e^{-B\xi^r})$ is due to Morgan who proved in 1934 that such a pair is sufficient if $1/p + 1/r < 1$ (not depending on $A, B > 0$) and found the conditions to be imposed on (A, B) to make the pair sufficient for given p, r satisfying $1/p + 1/r = 1$. His main tool was the Phragmen-Lindelöf theorem (a far reaching generalization of the maximum modulus principle). Using another complex tool (the Carleman formula for a contour integral involving the logarithm of a function analytic in $\mathbb{C}_+$) Dzhrbashyan obtained some sufficiency criteria (replacing the pointwise majorization (19) by integral estimates); he also got a description of some "unilaterally sufficient" pairs (that is those (M, N) for which $f = 0$ follows from the inequalities in (19) if both (or one) of them are fulfilled only on the ray $(0, +\infty)$). The following elegant result is due to Beurling: if $f \in L^1(\mathbb{R})$, and $\iint_{\mathbb{R} \times \mathbb{R}} |f(t)||\hat{f}(\xi)|e^{|t||\xi|}dtd\xi < +\infty$, then $f = 0$ ([B]). The results of this section (with their proofs and references) can be found in [HJ], p.128-137; we also recommend Nazarov's article [Na] containing a new approach to sufficient pairs.

2.3. *Deep zero & fast decay at infinity*

Let H be a non-negative function on $[0, +\infty)$. In this section we call it *sufficient* if

$$f(t) \cdot t^n = O(1) \ (|t| \to \infty, n \in \mathbb{Z}_+) \ \& \ |\hat{f}(t)| \leqslant H(|t|) \ (t \in \mathbb{R}) \Rightarrow f = 0.$$

If $H(t)t^n = O(1) \ (t \to +\infty)$ for any $n > 0$, then "the depth of zero" of f at the origin just means $f^{(n)}(0) = 0, n \in \mathbb{Z}_+$. This is actually a classical quasianalyticity problem related to the moment problem and weighted polynomial approximation.

A *necessary* condition for H to be sufficient is $\mathcal{L}(H) = -\infty \ (\mathcal{L}(H) = \int_0^{+\infty} \log H d\Pi$, see Part 1 section 1.1); this "quantitative" condition is *sufficient* if H satisfies a "qualitative" *regularity condition* (which cannot be dropped), e.g., if H is logarithmically convex $(H(x) = \exp(-x/\log(x + 1))$ or $\exp(-x/\log(x + 1)\log\log(x + 1)\ldots)$ are sufficient (see [Koo2], [HJ]). This combination of a quantitative condition (19) and a qualitative one (the regularity of the majorant) is very typical for many forms of the UP.

2.4. *Deep zero & sparse spectrum*

Here we return to the Mandelbrojt theorem (see 1.3.5 of Part 1) and briefly discuss its proof due to Levin (this proof results actually in a much stronger theorem which we won't state here, see [L], [HJ]). The strategy of the proof is this: suppose $f \in L^1(\mathbb{T})$ has a deep zero at 1 (say, $\int_0^\varepsilon |f(e^{it})|dt = O(e^{-\varepsilon^{-\rho}})$ as $\varepsilon \downarrow 0$; ρ is positive); look at the Laplace transform

Φ of the periodic function $\varphi, \varphi(t) = f(e^{it})(t \in \mathbb{R}); \Phi(p) = \int_0^{+\infty} \varphi(t)e^{-pt}dt$ coincides in

$\{Rep > 0\}$ with the meromorphic function $\sum_{n\in\mathbb{Z}} \hat{f}(n)(p - in)^{-1}$ of a very tempered growth off the union of the discs $\{|z - in| < 1/4\}$; the deep zero of f at 1 (that is the deep zero of φ at the origin) can be translated as the fast decay of $|\Phi(\xi + i\eta)|$ as $\xi \uparrow +\infty$ (uniformly in η). The Poisson-Jensen formula

$$\int_0^R (N(t)/t)dt = \int_{\mathbb{T}} \log|\Phi(R\zeta)|dm(\zeta) - \log|\Phi(0)|$$

where $N(t) = n(t) - p(t), n(t), p(t)$ being, resp., the numbers of zeros and poles in $t\mathbb{D}$, implies the estimate

$$\int_0^R (p(t)/t)dt \geqslant -\left(\int_{\mathbb{T}\cap\{Re\zeta>0\}} + \int_{\mathbb{T}\cap\{Re\zeta<0\}}\right)\log|\Phi(R\zeta)|dm(\zeta) + \log|\Phi(0)|;$$

the first integral tends to $-\infty$ and so does the whole bracket, since the modulus of the second integral grows too slowly; if Φ is regular at the origin and $\Phi(0) \neq 0$ (which we may assume if $\varphi \neq 0$), then $p(t)/t$ has to be big for arbitrarily large values of t, and an excessive sparseness of spec f (= the set of poles of Φ) is impossible. We can now go back: given a sufficiently sparse set Λ of integers we construct a meromorphic function $\Phi = 1/B$ where B is a suitable infinite product (an entire function) vanishing exactly at the points in, $n \in \Lambda$, and growing fast enough off the discs $\{|z - in| < a\}$ for a positive a; applying the Riemann-Mellin inversion formula for the Laplace transform to Φ we obtain a periodic φ and then $f \in C(\mathbb{T})$ with spec $f = \Lambda$ and a deep zero at 1; this is how the sharpness of the Mandelbrojt theorem is proved.

2.5. Semibounded spectra

The complex point of view sheds a new light upon the functions with semibounded spectra. Here we can only mention the highly developed theory of the Hardy classes H^p whose main objects are L^p-functions with non-negative frequencies.

Let f be an L^p-function $(1 \leqslant p \leqslant +\infty)$ on $X = \mathbb{T}$ or $\mathbb{R}$ with a non-negative spectrum: spec $f \subset [0, +\infty)$ (if $p > 2, X = \mathbb{R}$, then spec f is the support of the distribution $\hat{f}$); then we say $f \in H^p(X)$, the Hardy space on X. Another object related to $H^p(X)$ is a Banach space $H^p(O)$ of functions analytic in $O(-\mathbb{D}$ for $X = \mathbb{T}, \mathbb{C}_+$ for $X = \mathbb{R})$ satisfying a certain L^p-growth restriction. It turns out that any $F \in H^P(O)$ has finite boundary values along the normals m -a.e. on X thus defining a function $F^* \in H^p(X)$,

$$F^*(t) = \lim_{r\uparrow 1} F(rt) \quad (t \in \mathbb{T}), F^*(t) = \lim_{\varepsilon\downarrow 0} F(t + i\varepsilon) \quad (t \in \mathbb{R}).$$

The mapping $F \to F^*$ takes $H^P(O)$ isometrically onto $H^p(X)$, so that the $L^p(X)$-functions with no negative frequencies can be *identified* with the *analytic* functions in $H^p(O)$. This close connection makes it possible to understand *completely* many forms of the UP for the plus-functions (including continuous and smooth plus-functions, see [Du], [Pr], [Ho], [Ga], [Koo1], [HJ]; these forms are usually sharp in contrast with other spectral "smallnesses" (e.g., for the band limited functions, see section 2.6 below). As an example we consider here a complete and quite satisfactory description of the moduli of the $H^p(X)$-functions.

THEOREM. *Let $h \geqslant 0$ be a non-zero function on X. The following are equivalent: (i) $h = |f|$ where $f \in H^p(X)$; (ii) $h \in L^p(X,m)$ and $\mathcal{L}(h) > \infty$ (see section 1.1 in Part 1).*

Thus the convergence of the logarithmic integral $\mathcal{L}(h)$ is the *only* smallness restriction for the equation $|f| = h, f \in H^p(X)$ to be solvable. Its necessity follows immediately from the Jensen inequality; its sufficiency can be proved by an explicit construction: if $h \in L^p(\mathbb{T}, m)$ and $\mathcal{L}(h) > -\infty$, then $\mathrm{Ext}\, h : z \mapsto \exp \int_{\mathbb{T}} \log h(\zeta) \frac{\zeta+z}{\zeta-z} dm(\zeta)$ is in $H^p(\mathbb{D})$ and $|(\mathrm{Ext}\, h)^*| = h$ a.e. on $\mathbb{T}$; $\mathrm{Ext}\, h$ is the so-called *outer* (or *exterior*) function corresponding to h. An analogous formula can be written for $X = \mathbb{R}$ as well.

The conditions of the solvability of the equation $|f| = h$ with an unknown band limited function f (i.e. with a *bounded* and not just semibounded spectrum) can hardly be expressed in palpable terms (see however [Dy] for some useful results in this direction). In the next section in place of the equation $|f| = h$ we turn to non-trivial band limited solutions of *the inequality $|f| \leqslant h$.*

2.6. *Fast decay at infinity and bounded spectrum*

Let h be a non-negative function defined on $\mathbb{R}$. We call it *a Beurling-Malliavin majorant* (BM-*majorant*) if there exists a non-zero function f with a bounded spectrum such that

$$(20) \qquad\qquad |f| \leqslant h.$$

A bounded set being semibounded we immediately conclude that any BM-majorant h satisfies $\mathcal{L}(h) > -\infty$. But in contrast with section 2.5 this condition is far from being sufficient: to guarantee the solvability of (20) with a band limited $f \neq 0$ we have to impose some *regularity* conditions on h to moderate its oscillations at infinity. The reason is simple: a band limited function is entire and of finite degree (i.e. $f(t) = O(\exp \sigma |t|), t \in \mathbb{C}, |t| \to +\infty$): the Poisson-Jensen formula shows that the zeros of f tend to run away with a certain speed from any (big) disc (so that the number of zeros of f in $r\mathbb{D}$ is $O(r)$ as $r \uparrow +\infty$). And if, say, $h(\sqrt{n}) = 0, n = 1, 2, \ldots$, or even if $h(\sqrt{n})$ tends to zero fast enough, then (20) implies $f = 0$; but this behavior of h is very well compatible with $\mathcal{L}(h) > -\infty$ if only the slopes of the pits on the graph of h over $\sqrt{n}$ are steep, i.e. if h oscillates intensely. (There are, of course, even much more obvious obstacles for an h with $\mathcal{L}(h) > -\infty$ to be a BM-majorant: for example $h(t) = \exp(-1/\sqrt{|t|})$ is not a BM-majorant. But we concentrate now on BM-majorants bounded off zero on any bounded interval.)

Assume $\mathcal{L}(h) > -\infty$; the characterization of the oscillations of h at infinity compatible with h being a BM-majorant is a very hard problem. A remarkable breakthrough is due to Beurling and Malliavin. Their work [BM1] describing a large class of BM-majorants is deep and difficult involving a good deal of potential theory and complex analysis. Here I state only one corollary: *Suppose h is bounded and strictly positive; if $\mathcal{L}(h) > -\infty$ and $\log h$ satisfies a Lipschitz condition, then h is a BM-majorant.*

Another (and even more famous) corollary is the so-called Beurling-Malliavin multiplier theorem which I won't state here. Subsequent proofs, simplifications, and approaches to these corollaries are due to Koosis, Kargaev, and Nazarov see [Koo2, Koo3]; one more proof is in [DeBr]. But the original result still seems to remain the most general (it is also exposed in [HJ], p. 306-369).

Note that if h is even and decreasing on $[0, +\infty)$ (no oscillations at all), then $\mathcal{L}(h) > -\infty$ is sufficient for h to be a BM-majorant. This fact is relatively simple and known actually for a long time before the Beurling-Malliavin theorem had been proved (see the references and a proof in [HJ], p. 276).

Another "whale", the second Beurling-Malliavin theorem, can be only named here. It is devoted to the following form of the UP: bounded support & missing frequencies (characterization of the discrete sets $\Lambda \subset \mathbb{R}$ such that $\hat{f}|\Lambda = 0$ for a function $f \neq 0$ concentrated on $[-\sigma, \sigma], \sigma > 0$; see [BM2, Koo2, HJ]). Denoting by χ_A the characteristic function of a set A we may rephrase the problem: describe the BM-majorants of the form $\chi_{\mathbb{R} \setminus \Lambda}$.

2.7. *Four theorems on the unilateral decay*

This series of theorems starts historically with the following result due to Levinson and Cartwright:

(I) Suppose $f \in L^1(\mathbb{T})$ satisfies

$$(21) \qquad |\hat{f}(n)| \leqslant h(|n|) \text{ for all negative integers } n,$$

$h : [1, +\infty) \to (0, +\infty)$ being a decreasing function ("unilateral decay of $\hat{f}$"). If

$$(22) \qquad \sum_{n=1}^{\infty} \frac{\log h(n)}{n^2} = -\infty,$$

then f cannot vanish identically on a non-degenerate arc unless $f = 0$.

In 1960 Beurling proved the following result which is in fact much stronger than (I): for a finite charge μ on $\mathbb{R}$ and $A > 0$ put $\rho_\mu(A) = (var\mu)([A, +\infty))$;

(II) if

$$(23) \qquad \int_1^{+\infty} \frac{\log \rho_\mu(A)}{A^2} dA = -\infty,$$

then $\hat{\mu}$ cannot vanish identically on a set of positive length unless $\mu = 0$.

This time "the unilateral decay" refers to "the object" μ and the spectral smallness means vanishing on a large set. The remarkable feature of this result is the total absence of the regularity conditions (cf. section 2.3 and 2.6). The strategy of the proof is killing the Cauchy potential $C(\mu)(\zeta) = \int_{\mathbb{R}} (t - \zeta)^{-1} d\mu(t) (\zeta \notin \mathbb{R})$ for any μ satisfying (23) with $\hat{\mu}$ vanishing on a set of positive length. The Levinson-Carwright theorem (I) follows from (II) very easily.

(III) (The Volberg Theorem) Suppose the conditions of (I) are fulfilled and h satisfies some supplementary regularity conditions (not to be stated here); then $\mathcal{L}(f) > -\infty$ unless $f = 0$.

The conclusion of this theorem is much stronger than in (I), but it does not imply (I) because of those unnamed regularity conditions (their sharp form is due to J. Brennan, see [Koo2, HJ]); the regularity of h in (I) is its mere decrease. Theorem (III) was conjectured by Dyn'kin in 1975; its proof is based on Dyn'kin's theory of pseudoanalytic continuation and delicate estimates of pseudoanalytic functions in the so-called boundary layers.

(IV) (The Borichev Theorem) The last result of this series is due to Borichev and looks (at first glance) even stronger than (III). It is applicable not only to *functions* on $\mathbb{T}$, but to *distributions* and even to *hyperfunctions*. Any two-sided sequence $(a_n)_{n \in \mathbb{Z}}$ of complex numbers such that $\limsup_{|n| \to \infty} |a_n|^{1/|n|} \leqslant 1$ generates two analytic functions g_+, g_-:

$$g_+(\zeta) = \sum_{n \geqslant 0} a_n \zeta^n (|\zeta| < 1), g_-(\zeta) = -\sum_{n < 0} a_n \zeta^n (|\zeta| > 1).$$

If $|a_n| = O(|n|^m)$ for a positive m, then $\sum_{n \in \mathbb{Z}} a_n \zeta^n$ is the Fourier series of a distribution T on $\mathbb{T}$ and $\operatorname{supp} T$ is the complement of the largest open part of $\mathbb{T}$ across which g_+ is analytically extendable to $-g_-$. The Borichev theorem asserts in particular that if

$$\lim_{n \to -\infty} \log |a_n|/h(|n|) = -\infty, \limsup_{n \to +\infty} \log |a_n|/h(n) < +\infty,$$

and h satisfies (22) and some regularity conditions (again !) then it is impossible for the non-tangential limits of g_+ and g_- to coincide on a subset of $\mathbb{T}$ of positive length unless $a_n \equiv 0$. In fact, [Bor, BorV] contain much stronger results involving the divergence of a logarithmic integral (to be defined properly, since $\sum a_n \zeta^n$ is not a function on $\mathbb{T}$, and $g_\pm$ are not bound to possess boundary values on $\mathbb{T}$).

For the most popular classical majorants

$$h(n) = \exp(-cn/\log n \cdot \log \log n \cdot \ldots)$$

each of these four theorems is a step forward compared with the preceding one. But in general none of them implies the rest (because of the discrepancies in the regularity conditions). Their proofs are different, and it is unclear whether they can be encompassed by a single statement and proof.

2.8. *Spectral gap & sparse support*

We say a tempered distribution T on $\mathbb{R}$ has a *spectral gap* if $\mathbb{R} \setminus \operatorname{spec} T$ contains a non-degenerate interval. The Beurling theorem ((II) in section 2.7 above) forbids the fast unilateral decay of a function (or a charge) with a spectral gap. But now we are going to discuss *the sparseness of the support* of a charge with a spectral gap. One more Beurling theorem gives an answer: *Suppose $S \subset \mathbb{R}$ is a closed set such that*

$$\int_{\mathbb{R}} \frac{dist(x, S)}{1 + x^2} dx = +\infty.$$

Then any non-zero charge μ supported by S has no spectral gap ([B, Koo2, HJ]).

A proof due to Koosis is based on the Pollard approach to weighted approximation and the Bernstein band limited function $\cos \sigma \sqrt{(z - x_0)^2 - R^2}$ peaking at $x_0 \in \mathbb{R}$ and bounded by one off $(x_0 - R, x_0 + R)$. This result is only an illustration. The problem to characterize the support carrying a charge with a spectral gap has impressive connections with potential theory, weighted approximation by polynomials and entire functions of finite degree (see the results by Levin, Akhiezer & Levin, Kargayev, Benedicks, Koosis, De Branges, Levin & Logvinenko & Sodin quoted in [HJ], p.375). Here we only mention that the Beurling theorem of this section is sharp in the following sense: the spectral gap cannot be replaced in its statement by a set of positive length; this was predicted by Koosis and proved by Kargaev's counterexample (his original construction was simplified by Kislyakov and Nazarov, see [HJ], p.520).

We conclude this section by the following problem posed by Sapogov: is there a set $A \subset \mathbb{R}$ of finite length whose characteristic function χ_A has a spectral gap? The answer is yes, it is due to Kargayev who constructed such an A as the union of disjoint intervals I_n gravitating to n as $|n| \to +\infty$; their endpoints can be computed by the Newton-Kantorovich method in $l^2(\mathbb{Z})$ which yields alot of information on I_n (see [HJ], p.376-392 and the paper by Kargaev & Volberg quoted there).

2.9. *Sparse support & unilateral decay*

A compact set $K \subset \mathbb{T}$ is called *spacious* if it carries a non-zero charge μ such that

$$(24) \qquad |\hat{\mu}(n)| = O(|n|^{-m}) \ (n \to -\infty)$$

for any $m > 0$. The characterization of "bilaterally spacious" sets K (i.e. carrying a charge $\mu \neq 0$ satisfying (24) with $|n| \to \infty$) is easy: they are just the sets with interior points. The unilateral character of (24) makes the description of spacious sets a much more delicate task. It is obvious that the length of a spacious K is positive, since any μ satisfying (24) is m-absolutely continuous by the F. and M. Riesz theorem. But this condition is not sufficient.

Denote by $\mathcal{L}(K)$ the set of all components of $\mathbb{T} \setminus K$ and call

$$\sum_{l \in \mathcal{L}(K)} |l| \log |l|$$

the entropy of K.

THE HRUŠČEV THEOREM. *K is spacious iff it contains a compact subset of positive length and finite entropy (see* [Hru, HJ]*).*

This is a difficult result. It uses among other things some variants of the Khinchin-Ostrowski theorem on normal families of functions analytic in a disc, delicate estimates of outer functions and a clever construction of a special measure on $\mathbb{T}$ (its simplified version due to N. Makarov is in [HJ]). The Hruščev theorem stated above is just a representative of a long series of his results on "sparse supports & unilateral decay" including many concrete sorts of "decay" and various "objects" (not necessarily charges); only a part of the results of [Hru] is in [HJ].

3. Local and Antilocal Convolutions

This part is devoted to a form of the UP for shift invariant linear operators; we call it *antilocality*. The most interesting examples and problems come from potential theory and are discussed in section 3.3. A class of antilocal operators is the theme of section 3.2, in section 3.1 we discuss *local* ("almost differential") operators as opposed to the antilocality of sections 3.2 and 3.3. In this part everything is closely related to the UP of Parts 1 and 2.

3.1. *Local and completely local convolutions*

We denote by $\mathcal{D}'(\mathbb{R}^d)$ the set of all distributions in $\mathbb{R}^d$. Let $\mathcal{K}$ be a linear operator mapping a linear set $dom\mathcal{K} \subset \mathcal{D}'(\mathbb{R}^d)$ into $\mathcal{D}'(\mathbb{R}^d)$. We call it *local* if it does not increase the support:

$$T \in dom\mathcal{K} \Rightarrow \operatorname{supp} \mathcal{K}T \subset \operatorname{supp} T,$$

or what is the same $\mathcal{K}(T)|O$ depends only on $T|O$ for any open $O \subset \mathbb{R}^d$ and $T \in dom\mathcal{K}$. A typical example is any linear differential operator with C^∞-coefficients; this is in a sense the only possible example: it can be proved under some mild conditions to be imposed on $dom\mathcal{K}$ (but not in general !) that local operators are differential (the Peetre theorem).

A local operator $\mathcal{K}$ reproduces any *open* zero set E of a distribution:

(25) $$T \in dom\mathcal{K}, T|E = 0 \Rightarrow \mathcal{K}T|E = 0.$$

Suppose $dom\mathcal{K}$ and $im\mathcal{K} = \mathcal{K}(dom\mathcal{K})$ consist of *locally summable functions* so that $T|E$ and $\mathcal{K}T|E$ make sense for $T \in dom\mathcal{K}$ and any Lebesque measurable set $E \subset \mathbb{R}^d$. Then we call $\mathcal{K}$ *completely local* if (25) holds for any such E (not only open). Any linear differential operator whose domain consists of sufficiently smooth functions is completely local.

Now we turn to the shift invariant operators $\mathcal{K}$ on $\mathbb{R}$: Let K be a Lebesque measurable function on $\mathbb{R}$. Consider the convolution operator $\mathcal{K}$

$$(26) \qquad dom\mathcal{K} = \{f \in L^2 : K\hat{f} \in L^2\}, \hat{\mathcal{K}f} = \widehat{Kf} \ (f \in dom\mathcal{K}).$$

De Branges found a complete characterization of the symbols of local operators.

THE DE BRANGES THEOREM. *(a) Suppose K is a restriction to $\mathbb{R}$ of an entire function k of the Cartwright class and degree zero (that is $k(\zeta) = O(\exp \varepsilon|\zeta|), |\zeta| \to +\infty$, for any $\varepsilon > 0$, and $\mathcal{L}(k) < +\infty$); then $\mathcal{K}$ is local;*
(b) suppose $dom\mathcal{K}$ contains a non-zero function vanishing on a non-degenerate interval; if $\mathcal{K}$ is local, then $K = k|\mathbb{R}$ for an entire function k of the Cartwright class and degree zero.

In other words local shift invariant operators with a sufficiently rich domain are precisely "the almost differential linear operators with constant coefficients"; they can be written formally as $t \to \sum_{k=0}^{\infty} a_k t^k$ where $\sum_{k=0}^{\infty} a_k \zeta^k$ represents a slowly growing entire function, not too far from a polynomial. This result is only a particular case of a much more precise theorem due to De Branges and describing the symbols of the so-called σ-local convolutions ([DeBr, HJ]).

In the extreme case of a polynomial symbol K our operator $\mathcal{K}$ becomes a usual linear differential operator with constant coefficients defined on the Sobolev space $W_2^n = \{f \in L^2 : f \in C^{(n-1)}, f^{(n-1)}$ absolutely continuous, $f^{(n)} \in L^2, n = degK\}$. In this case $\mathcal{K}$ is not just *local*, but *completely local*. De Branges posed the following question: suppose $\mathcal{K}$ is *almost* differential (as in his theorem above); is it completely local ? A counterexample was constructed by Kargayev. He actually showed that $f \to \sum_{k=0}^{\infty} a_k f^{(k)}$ can be not completely local even for an entire symbol $k(\zeta) = \sum_{k=0}^{\infty} a_k \zeta^k$ of *order* zero (i.e. $k(\zeta) = O(\exp |\zeta|^{\varepsilon})$ for any $\varepsilon > 0$), see [HJ],p.482-484.

3.2. Complete antilocality

The results of section 3.1 are in fact closely related to the themes of Part 2, but only with respect to the tools. We turn now to an opposite property of certain convolutions (the antilocality) which is *itself* a form of the UP for a shift invariant operator forbidding us to know too much on the operator. Our theme here is the compulsory increase of the support of a non-zero function under certain convolution operators.

A linear operator $\mathcal{K}$ defined on $dom\mathcal{K} \subset \mathcal{D}'(\mathbb{R}^d)$ with values in $\mathcal{D}'(\mathbb{R}^d)$ is called *antilocal* if

$$T \in dom\mathcal{K}, T \neq 0 \Rightarrow \operatorname{supp} \mathcal{K}(T) \supset \mathbb{R}^d \setminus \operatorname{supp} T$$

which means that the following UP holds: for any non-empty open $E \subset \mathbb{R}^d$ and $T \in dom\mathcal{K}$

$$(27) \qquad T|E = \mathcal{K}(T)|E = 0 \Rightarrow T = 0,$$

forbidding the simultaneous vanishing of T and $\mathcal{K}(T)$ on a solid (=open) set. This UP is valid for some interesting convolution operators and is akin to its "harmonic" prototype. The simplest example is the Hilbert transform (in $\mathbb{R}$) taking $f \in L^2$ to $\tilde{f}$, $\tilde{f}(x) = p.v. \int\limits_{\mathbb{R}} f(t)/(t -$

$x)dt$; in spectral terms this means $\hat{\tilde{f}}(\xi) = c \cdot sgn\xi \cdot \hat{f}(\xi)$. If $f, \tilde{f}$ both vanish on an open $E \subset \mathbb{R}$, then the function $\varphi : \zeta \mapsto \int\limits_{\mathbb{R}} f(t)/(t - \zeta)dt$ analytic in the domain $\mathbb{C}_+ \cup E \cup \mathbb{C}_-$ vanishes on E and is thus identically zero; being the jump of φ as its argument crosses $\mathbb{R}$, $f = 0$ a.e. on $\mathbb{R}$. It is not hard to see that the logarithmic potential $f \mapsto f * \log|x|$ and the M.Riesz potentials $f \mapsto f * |x|^{-\alpha}$ $(0 < \alpha < 1)$ are antilocal. But the Hilbert transform and the logarithmic potential enjoy in fact a much stronger uniqueness property: they are *completely antilocal*. We say that $\mathcal{K}$ is *completely antilocal* if (27) holds not only for any open E, but for any set E of positive Lebesgue measure in $\mathbb{R}^d$ (we assume that $dom\mathcal{K}$ and $im\mathcal{K}$ consist of locally summable functions). It is sometimes quite hard to prove (or disprove) that an antilocal operator is completely antilocal; for the Hilbert transform this property coincides with the UP for $H^2(\mathbb{R})$, see section 1.1 of Part 1.

The symbols of local operators described by the De Branges theorem of section 3.1 consist of a single analytic block being polynomials or entire functions. It turns out that many symbols consisting of *two different* "analytic blocks" define an antilocal (or even a completely antilocal) operator.

Let K be a Lebesgue measurable function on $\mathbb{R}$; $b, c \in \mathbb{R}, b < c$. Suppose K coincides on $(c, +\infty)$ with a rational function r, and $|\{\xi : \xi \leqslant b, r(\xi) = K(\xi)\}| = 0$. Then we call K a *semirational symbol* and r its rational part. A typical example is $K(\xi) = sgn\xi \cdot r(\xi)$ where r is rational (if $r \equiv 1$, then we get the Hilbert transform).

Define $\mathcal{K}$ by (26). This operator is completely antilocal for many semirational symbols, e.g., for any $K(\xi)$ of the form $sgn\xi/q(\xi)$ where q is a polynomial. But for general semirational symbols (27) is only proved under a supplementary smoothness conditions to be imposed on T ; (27) can be restored for *all* $T \in dom\mathcal{K}$ provided E satisfies an extra "entropy condition" as in the Hruščev theorem (section 2.9 of Part 2), and it is unknown whether these supplementary conditions can be dropped ([HJ], p.484-488 and the references therein including papers by Havin, Joericke, Makarov, and Ch.Bishop). For example it is unknown whether (27) is true for $K(\xi) = sgn\xi \cdot (\xi - i)/(\xi - 2i)$ (it is true if we assume $T \in W_2^1$).

Another interesting antilocal convolution is the M. Riesz potential

$$(\mathcal{K}(f) = f * |x|^{-\alpha}, \alpha \in \mathbb{R}, \alpha \neq -2, -4, \ldots)$$

whose symbol $c|\xi|^{\alpha-1}$ also consists of two *different* analytic pieces. As to the *complete* antilocality, it is only known that for $\alpha \in (0, 1)$ the following property holds: *if*

$$(28) \qquad \int\limits_{|t|>1} |f(t)||t|^{-\alpha}dt < +\infty,$$

(29) $$E \subset \mathbb{R}, |E| > 0, f|E = (f * |x|^{-\alpha})|E = 0,$$

and f satisfies a Hölder condition (depending on α) near E,then f = 0 (see [HJ], p.499-508 where all real values of α are also considered). Thus an extra smoothness condition emerges here once again although the proof is quite different from the proof of the UP for semirational symbols. But in this case these conditions cannot be dispensed with.Using a method due to Bourgain & Wolff a non-zero continuous function f on $\mathbb{R}$ has been constructed in [BH] which satisfies (28) and (29).

3.3. *A uniqueness problem for the Newton potentials*

An extremely interesting example of the antilocal behaviour can be observed on the Newton convolution in $\mathbb{R}^2$, i.e. on the operator

$$f \to U^f, \ U^f(x) = \int\limits_{\mathbb{R}^2} \frac{f(y)dy}{|x - y|}, x \in \mathbb{R}^2$$

(or, more generally, on the M.Riesz potentials)

$$U^f_\alpha(x) = \int\limits_{\mathbb{R}^d} \frac{f(y)dy}{|x - y|^{d-\alpha}}, \alpha \neq 2, 4, \ldots, x \in \mathbb{R}^d).$$

The antilocality of the Newton potential U can be interpreted as a uniqueness property of the solutions of the Cauchy problem for the Laplace equation in the upper half-space $\mathbb{R}^3_+$, and it is very close to a boundary uniqueness property of harmonic (=divergence- and curl-free) vector fields in $\mathbb{R}^3_+$, a three dimensional analog of the boundary uniqueness theorem for functions analytic in the upper half-plane $\mathbb{C}_+$ (= a UP for plus-functions, see section 1.1 of Part 1 and section 2.5 of Part 2).

A remarkable construction due to Bourgain and Wolff [BW] (preceded by a breakthrough in [W] and some simplifications due to Aleksandrov and Kargayev [AK]) has disproved the complete antilocality of the Newton potential in $\mathbb{R}^2$: there exists a continuous non-zero function f in $\mathbb{R}^2$ such that $U^f = f = 0$ on a set of positive area in $\mathbb{R}^2$. It is, however, unknown whether such an example is possible with a smooth (say C^1, not to mention C^∞) function f. The one-dimensional construction of [BH] related to the M.Riesz potentials suggests that the answer may be negative. The antilocality properties of the Newton potentials are one of the themes of [HJ] (see p.488-508, and the references to the papers of Mergelyan, Landis, M.M.Lavrentjev, N.Rao, and Maz'ya & Havin).

Acknowledgement

This text was written during my stay at the Department of Mathematical Sciences of NTNU (Trondheim) in May 2000.I want to express my sincere gratitude to the Department for its hospitality and excellent working atmosphere. I am also indebted to Yu.Lyubarskii for useful discussions which have influenced some parts of the text and to E.Malinnikova for detecting and removing some misprints.

References

[AK] Aleksandrov A.B., Kargaev P.P., *Hardy classes of functions harmonic in the upper halfspace.* Algebra i Analiz, 5, N2, 1993, p.1–73 (Russian; English translation: St.Petersburg Math.J., 5, N2, 1994, p.663–718).

[B] Beurling A., *Collected works*, vols I-II. Birkhäuser, Boston, 1989.

[Ba] Bari N.K. *A treatise on trigonometric series*, vols I-II. Pergamon-Press, Oxford,1964.

[BH] Beliaev D.B., Havin V.P., *On the uncertainty principle for the M.Riesz potentials.* Preprint, Department of Math, Royal Institute of Technology, Stockholm, TRITA-MAT-1999-24 (Dec 1999), p.1–18.

[Bo] Boas R.P., *Entire Functions*, Academic Press, NY, 1954.

[Bor] Borichev A.A., *Boundary uniqueness theorems for almost analytic functions and asymmetric algebras.* Mat.Sbornik, 136, N3, 1988 (Russian).

[BorV] Borichev A.A., Volberg A.L., *Uniqueness theorems for almost analytic functions.* Algebra i Analiz, 1, N1, 1989, p. 146–177. (Russian; English translation: Leningrad Math.J.,1, N1,1990, p. 157–192.)

[BM1] Beurling A., Malliavin P., *On Fourier transforms of measures with compact support.* Acta Math., 107, 1962, p. 291–302.

[BM2] Beurling A., Malliavin P., *On the closure of characters and the zeros of entire functions.* Acta Math., 118, N1-2, 1967, p. 79–93.

[BW] Bourgain J., Wolff T. *A remark on gradients of harmonic functions in dimension 3.* Colloq.Math., 60/61, N1, 1990, p. 253–260.

[Car] Carleson L., *Selected problems on exceptional sets.* Van Nostrand, Princeton, New Jersey, 1967.

[Carl] Carleman T., *L'intégrale de Fourier et quelques questions qui s'y rattachent.* Almkvist und Wiksels boktr., Uppsala, 1944.

[DeBr] De Branges L. *Hilbert spaces of entire functions.* Prentice Hall, Englewood Cliffs (N.J.), 1968.

[Du] Duren P.L. *Theory of H^p spaces.* Academic Press, NY, 1970.

[Dy] Dyakonov K.M., *Moduli and arguments of analytic functions from the subspaces invariant for the backward shift operator.* Sibirsk. mat. zhurn., 31, N6, 1990, p. 64–79 (Russian).

[Ga] Garnett J., *Bounded analytic functions.* Academic Press, NY, 1981.

[Gor] Gorin E.A., *Some remarks in connection with a problem of B.P.Paneah on equivalent norms in spaces of analytic functions.* Teoriya funktsii, funkt. analiz i ih prilozhenia, 44, 1985, p. 23–32 (Russian).

[HJ] Havin V., Jöricke B., *The uncertainty principle in harmonic analysis.* Springer-Verlag, NY, 1994.

[Ho] Hoffman K., *Banach spaces of analytic functions.* Prentice Hall, Englewood Cliffs. N.J., 1962.

[Hru] Hruščev S.V., *Simultaneous approximation and removable singularities of integrals of the Cauchy type.* Trudy MIAN, 130, 1978, p. 124–195 (Russian).

[JH] Jöricke B., Havin V., *Traces of harmonic functions and comparison of norms of analytic functions*, Math. Nachr., 98, 1980, p. 269–302 (Russian).

[Kah] Kahane J.-P., *Séries de Fourier absolument convergentes.* Springer-Verlag, NY, 1970.

[KahS] Kahane J.-P., Salem R., *Ensembles parfaits et séries trigonométriques.* Paris, Hermann, 1963.

[Katz] Katznelson Y., *An introduction to harmonic analysis (Second edition).* Dover, NY, 1976.

[Koo1] Koosis P., *Introduction to H^p spaces.* Cambridge University Press, Cambridge, 1980.

[Koo2] Koosis P., *The logaritmic integral,vols.I-II*. Cambridge University Press, Cambridge, 1988, 1992.

[Koo3] Koosis P., *Leçons sur le théorème de Beurling et Malliavin.* Les publications CRM, Montréal, 1996.

[Kor] Körner T.W., *Uniqueness for trigonometric series.* Ann. of Math., 126, N1, 1987, p. 1-34.

[L] Levin B.Ya., *Distribution of zeros of entire functions.* AMS, Providence, 1964.

[Lev] Levinson N., *Gap and density theorems.*Amer.Math.Soc.Coll. Publ., NY, 1940.

[M] Mandelbrojt Sz., *Séries adhérentes, régularisation des suites, applications.* Gauthier-Villars, Paris, 1952.

[Na] Nazarov F., *Local estimates of exponential polynomials and their applications to the uncertainty principle.* Algebra i analiz, 5, N4, 1993, p. 3–66 (Russian; English translation: St.Petersburg Math.J., 5, N4, 1994, p. 229–286).

[Nik] Nikolskii N.K. *Treatise on the shift operator.*Springer-Verlag, NY, 1986.

[PW] Paley R.E.A.C., Wiener N.,*Fourier transform in complex domain.* AMS Colloq.Publ.XIX, Providence, 1934.

[Pey] Peyriére J., *Etude de quelques propriétés de produits de Riesz.* Ann. Inst. Fourier, 25, N2, 1975, p. 127–169.

[Pr] Privalov I.I., *Boundary properties of analytic functions.* GITTL, Moscow, 1950 (Russian).

[W] Wolff T., *Counterexamples with harmonic gradients in* $\mathbb{R}^3$. In Essays on Fourier analysis in honor of Elias M.Stein, Princeton, NJ, 1995, p. 321–384.

[Z] Zygmund A., *Trigonometric series.* Vol. I-II. Cambridge Univ. Press, Cambridge, London, NY, 1959.

Operator Theory and Harmonic Analysis

H.S. Shapiro

Department of Mathematics
Royal Institute of Technology
S-100 44 Stockholm
Sweden
shapiro@math.kth.se

ABSTRACT. There are very close ties between the two subjects of the title. In the early days of operator theory, this was mainly manifested in applications of harmonic (and complex) analysis to operator theory. For example, von Neumann proved that the norm of $p(T)$, where T is any contraction on Hilbert space and p a polynomial, cannot exceed the maximum of $|p(z)|$ for z on the unit circle, as an application of complex analysis. In like manner, Stone's theorems on groups of unitary operators were proved with the aid of Bochner's theorem characterizing those functions on the circle (or on the reals) which are the Fourier transforms of positive measures. The spectral theorem itself was obtained by various routes based on Bochner's theorem, the trigonometric and/or algebraic moment theorem, etc.

What is of more recent date (and the main focus of this exposition) are deep and interesting applications of operator theory to classical analysis. A breakthrough here was Bela Sz.-Nagy's discovery that every linear operator on a Hilbert space has a unitary dilation; without here entering into details this implies that many properties of general contractions can be deduced from corresponding properties of operators in the much nicer class of *unitary* ones, to be sure on a vastly bigger space but nonetheless allowing many nontrivial deductions. For example, von Neumann's theorem mentioned earlier is thus reduced to the corresponding, and much easier, problem where T is unitary. We can here truly say that operator theory, involving genuinely infinite-dimensional constructions, starts to have applications to concrete classical problems. A beautiful case in point is D. Sarason's recognition of the fundamental role played by commutativity in understanding (and getting radically new proofs of) classical interpolation and moment theorems, like those usually associated with the names of Pick-Nevanlinna, Caratheodory, and Schur. The abstract kernel of this approach was further clarified and generalized as the so-called "commutant lifting theorem" of Sz.-Nagy and Foias. This in turn led to new and far-reaching extensions of the classical results *e.g.* to matrix-valued functions, which also have important applications *e.g.* in control theory.

J.S. Byrnes (ed.), Twentieth Century Harmonic Analysis - A Celebration, 31–56.

1. Introduction

Since the dawning of functional analysis in the early years of the twentieth century, it has had much interaction with, and inspiration from, harmonic analysis. *Formal trigonometric series* were the original impetus to various notions of generalized functions (distributions, hyperfunctions), absolutely convergent trigonometric series were the original model for what was to become the theory of normed rings (Banach algebras), and so on. Indeed, it is well known that the Lebesgue integral itself was first conceived in connection with the study of trigonometric series.

In this talk I want to focus on a part of this interaction, that between harmonic analysis (understood here in a broad sense, so as to encompass e.g. Hardy spaces of analytic functions on the unit disk) and *linear operators in Hilbert space*. That these have close links is evident if one considers (we shall further develop this point shortly) that the solution to the trigonometric moment theorem is *grosso modo* equivalent to the spectral theorem for unitary operators. The "common denominator" in this case is *positive definiteness*: characteristic for a Hilbert space H is its inner product, a positive definite Hermitian form on $H \times H$; whereas positive definite functions on *groups* (and semigroups) are a central notion in harmonic analysis.

Generally speaking, the interplay between classical analysis and functional analysis begins by someone taking a "second look" at some classical theorem, and finding that it contains the seeds of something more general. One postulates some abstract object having only part of the features of the original classical one, and tries to deduce analogous results. This procedure does not always lead to fruitful generalizations, but in rare cases "a miracle happens" and the general theory returns more than was put into it (thus, for example, the theory of commutative Banach algebras, originally modelled on absolutely convergent trigonometric series, turns out to be the tool *par excellence* for studying spectral decomposition of normal operators on Hilbert space).

In most cases there occurs no such miracle, but none the less the perspective opened up by abstract thinking may focus researchers upon new kinds of questions which in turn stimulate the development of classical analysis, and channel thought into new pathways. Thus, for example, study of the invariant subspace problem for general linear operators on Hilbert space has not thus far "paid off" by e.g. clarifying the structure of non-normal operators in a way that could be compared with what the spectral theorem accomplishes, and may never do so. But, there has been much valuable "fallout" from the massive efforts that have been expended on this problem. For example, finding the invariant subspaces for just one concrete integral operator (the so-called *Volterra operator*) has led to a new and beautiful proof of a deep "classical" theorem of Titchmarsh on the support of a convolution of two functions. Quest of invariant subspaces for *subnormal operators* has stimulated profound researches into approximation by rational functions in the complex plane, which already has produced results of independent importance.

The present talk is intended to present, for non-specialists, a small but hopefully interesting body of results illustrating the aforementioned interplay. I assume familiarity with the notion of a Hilbert space, and shall adopt the following notations and conventions.

All Hilbert spaces are *complex* and *separable*. In a Hilbert space H, $\langle f, g \rangle$ denotes the inner product of elements f and g. When several Hilbert spaces are involved, a subscript as in $\langle f, g \rangle_H$ may be used to specify in which Hilbert space the inner product is intended. By $\|f\|$ (or $\|f\|_H$) the norm of f is denoted. An *operator* T between Hilbert spaces H and K refers to a *continuous linear* map from H into K; when $K = H$, we say T is an operator on H. The set of all operators from H to K is denoted $L(H, K)$, and $L(H, H)$ is usually abbreviated $L(H)$.

The *span* of a subset E of H denotes the set of finite linear combinations of elements of E with complex coefficients. The *closed span of E* is the closure of this set.

Other notions like subspace, adjoint, etc, which will be used are completely standardized, and the reader may refer to any of the textbooks such as [AkGl], [DuSc], [Ma] or [RiSz]. Subspaces will be tacitly assumed to be *closed*. More specialized notions and notations will be defined as needed.

The following *basic geometric result* will be needed several times.

LEMMA 1.1. *Let H, H' be Hilbert spaces, and E any subset of H. Suppose φ is any injective map from E into H', such that*

$$(1.1) \qquad\qquad \langle \varphi f, \varphi g \rangle_{H'} = \langle f, g \rangle_H$$

for all f, g in E. Then, there is a continuous linear map V from the closed span of E onto the closed span of $\varphi(E)$, which moreover is an isometry, such that $V|_E = \varphi$.

For the proof, see e.g. [AkGl, p. 77].

Outline of the talk. In § 2 we shall deduce the spectral theorem for unitary operators, at least in a special form, from the Herglotz-F. Riesz-Toeplitz characterization of the Fourier coefficients of positive measures on the circle, and discuss "the basic à priori inequality" that results therefrom and that underlies the *functional calculus*.

In § 3 we discuss the theorem of Wold-von Neumann-Kolmogorov on the structure of isometries, and the application of this to Beurling's invariant subspace theorem, and to problems of extrapolation and prediction for stationary random processes.

In § 4 we discuss the theorems of B. Sz.-Nagy on *isometric lifting* and *unitary dilation* of contractive operators, with applications (von Neumann inequality, mmean ergodic theorem).

In § 5 we present some further ramifications of dilation theory, notably the "commutant lifting theorem" and discuss its application to problems of interpolation (of the type of the Pick-Nevanlinna problem) and moment theorems.

For the most part, we shall not give detailed proofs, but try to convey some of the main ideas and techniques needed in the proofs. We cannot give a complete bibliography (the Pick-Nevanlinna theorem alone would involve us with hundreds, if not several thousands, of references) but believe that the references we give, together with references in those works, suffice to give the interested reader a good orientation in the literature.

2. The spectral theorem and harmonic analysis

The starting point of our story will be the following theorem, discovered independently by G. Herglotz, F. Riesz and O. Toeplitz ([He, Ri, To]).

THEOREM 2.1. *Given a sequence $\{c_n\}_{n=\infty}^{\infty}$ of complex numbers, a necessary and sufficient condition that there exists a bounded non-negative measure μ on the (Borel sets of the) unit circle $\mathbb{T}$ satisfying*

$$(2.1) \qquad \int e^{-in\theta} d\mu(\theta) = c_n \quad , \qquad n \in \mathbb{Z}$$

is: for every $N \geqslant 0$, and all choices of complex numbers $t_0, t_1, \ldots, t_N$ we have

$$(2.2) \qquad \sum_{j,k=0}^{N} c_{j-k} t_j \bar{t}_k \geqslant 0.$$

REMARKS.

(i) Condition (2.2) can also be formulated so that the principal finite sections of the infinite matrix

$$(2.3) \qquad \begin{bmatrix} c_0 & c_1 & c_2 & c_3 & \cdots \\ c_{-1} & c_0 & c_1 & c_2 & \cdots \\ c_{-2} & c_1 & c_0 & c_1 & \cdots \\ \cdots & \cdots & \cdots & \cdots & \cdots \end{bmatrix}$$

are non-negative definite matrices. The matrix (2.3) is called a *Toeplitz matrix*; this designates a matrix whose entries are constant along any diagonal parallel to the main diagonal.

(ii) From (2.2) it follows readily that $c_{-n} = \bar{c}_n$ for all n (in particular, c_0 is real (and non-negative)). These relations also follow at once from (2.1).

(iii) This theorem is nowadays seen as a special case of a more general theorem of Bochner about positive functions on locally compact Abelian groups, see e.g. [Ru].

(iv) In some variants one replaces (2.1) by the equivalent statement that the harmonic function

$$(2.4) \qquad u(r,\theta) := \sum_{n=-\infty}^{\infty} c_n r^{|n|} e^{in\theta}$$

in the open unit disk $\mathbb{D}$ is non-negative.

We refer for the proof of Theorem 2.1 to [Zy, p. 138].

One of the well-known deductions of the *spectral theorem for unitary operators* is based on Theorem 2.1. Let us trace the main ideas. Recall first that an operator U in $L(H)$ is *unitary* if it is a bijection of H that preserves inner products: $\langle Uf, Ug \rangle = \langle f, g \rangle$ for all f, g in H (or, what comes to the same, $U^*U = UU^* = I$, the identity operator).

DEFINITION 2.2. *The unitary operator U has simple spectrum if there exists a vector $f \in H$ such that the linear manifold spanned by $\{U^n f : n \in \mathbb{Z}\}$ is dense in H.*

THEOREM 2.3. (spectral theorem for unitary operators with simple spectrum). *If $U \in \mathcal{L}(H)$ is unitary with simple spectrum, there is a bounded positive measure μ on $\mathbb{T}$ such that the linear operator M on $L^2(\mathbb{T}, d\mu)$ defined by*

$$(2.5) \qquad M : \varphi \mapsto z\varphi \quad , \qquad \varphi \in L^2(\mathbb{T}, d\mu)$$

where $z = e^{i\theta}$ denotes a generic point of $\mathbb{T}$, is unitarily equivalent to U.

If H and K are Hilbert spaces, operators $A \in L(H)$ and $B \in L(K)$ are *unitarily equivalent* if and only if there is a unitary map $U \in L(H, K)$ such that the diagram

$$(2.6) \qquad \begin{array}{ccc} H & \xrightarrow{A} & H \\ {\scriptstyle U}\downarrow & & \downarrow{\scriptstyle U} \\ K & \xrightarrow{B} & K \end{array}$$

commutes i.e. $BU = UA$. (This is also expressible as "U intertwines A and B"). The analogous weaker relation when U is merely assumed to have an inverse in $L(K, H)$ is called *similarity*.

Remark. Observe that M has a simple spectrum, since trigonometric polynomials are dense in $L^2(\mathbb{T}, \mu)$.

Proof of Theorem. Define

$$(2.7) \qquad c_n = \langle U^n f, f \rangle \quad , \qquad n \in \mathbb{Z}$$

where $f \in H$ is such that $\{U^n f\}_0^\infty$ span H. Then, $\{c_n\}$ satisfy (2.2), since

$$\sum_{j,k=0}^N c_{j-k} t_j \bar{t}_k = \sum_{j,k=0}^N \langle U^{j-k} f, f \rangle t_j \bar{t}_k$$

$$= \left\| \sum_{j=0}^N t_j U^j f \right\|^2 \geqslant 0.$$

Hence, by Theorem 2.1, there is μ in $M^+(\mathbb{T})$ (the class of bounded positive measures on $\mathbb{T}$) such that (2.1) holds, and hence

$$\langle U^n f, f \rangle = \int e^{-in\theta} d\mu(\theta) \quad , \qquad n \in \mathbb{Z}.$$

Writing $n = j - k$, this says that $\langle U^j f, U^k f \rangle$ equals the inner product of $e^{-ij\theta}$ with $e^{-ik\theta}$ in $L^2(\mathbb{T}, d\mu)$. By virtue of Lemma 1.1, there exists a unitary map V from H onto $L^2(\mathbb{T}, d\mu)$ such that

$$VU^j f = e^{-ij\theta} \quad , \quad j \in \mathbb{Z}.$$

Then, if M_- denotes multiplication by $e^{-i\theta}$ on $L^2(\mathbb{T}, d\mu)$, the diagram

$$
\begin{array}{ccc}
H & \xrightarrow{\;U\;} & H \\
\downarrow{\scriptstyle V} & & \downarrow{\scriptstyle V} \\
L^2(\mathbb{T}, d\mu) & \xrightarrow[M_-]{} & L^2(\mathbb{T}, d\mu)
\end{array}
$$

commutes (it is enough to check that $M_- Vg = VUg$ holds for a set of g whose linear combinations are dense in H, e.g. for $g = U^n f$ with $n \in \mathbb{Z}$, and that is immediate). This proves the theorem (with M_- instead of M, but clearly M_- and M are unitarily equivalent).

COROLLARY 2.4. *For U as in Theorem 2.3, if* $p(z) = \sum_{-N}^{N} a_j z^j$, *then*

$$(2.8) \qquad \|p(U)\| \leqslant \max_{z \in \mathbb{T}} |p(z)|.$$

Proof. Because of the unitary equivalence, it is enough to show that $\|p(M)\| \leqslant \max |p(e^{i\theta})|$, i.e. that $\|p(M)g\| \leqslant \max |p(e^{i\theta})| \cdot \|g\|$ for all $g \in L^2(\mathbb{T}, \mu)$. But $p(M)g = p(e^{i\theta})g$ so

$$(2.9) \qquad \|p(M)g\|^2 = \int |p(e^{i\theta})|^2 |g(e^{i\theta})|^2 d\mu \leqslant \left(\max |p(e^{i\theta})|^2 \right) \|g\|^2$$

which completes the proof.

Now, what about a unitary operator that does not have simple spectrum? Suppose f_1 is any unit vector in H, and let H_1 denote the closed span of $\{U^n f_1 : n \in \mathbb{Z}\}$. If $H_1 = H$ we are in the case already treated, so suppose the contrary. Then, pick a unit vector f_2 in $H \ominus H_1$ (the orthogonal complement of H_1) and let $H_2 :=$ closed span of $\{U^n f_2 : n \in \mathbb{Z}\}$. After at most a countable number of steps in this way, we obtain a decomposition of H as the sum of mutually orthogonal subspaces H_j such that:

(i) H_j is invariant for U and $U^* = U^{-1}$ (in other words, the H_j *reduce* U).
(ii) The restriction of U to H_j (also called the *part of U in H_j*) is unitarily equivalent to the operator "multiplication by z" on $L^2(\mathbb{T}, d\mu_j)$ for some measure $\mu_j \in M^+(\mathbb{T})$.

Therefore, introducing a new Hilbert space K, the *direct sum* of the $L^2(\mathbb{T}, d\mu_j)$ (whose elements are thus vectors $\varphi = (\varphi_1, \varphi_2, \dots)$ where $\varphi_j \in L^2(\mathbb{T}, d\mu_j)$), and with $\|\varphi\|^2 := \sum \int |\varphi_j|^2 d\mu_j$ it is easy to check that V is unitarily equivalent to "multiplication by z" on K. Then (once some boring questions about measurability etc. are disposed of) we can deduce (now, with no hypothesis of "simple spectrum"):

THEOREM 2.5. *If U is unitary, then*

(2.10)
$$\|p(U)\| \leqslant \max \left| p(e^{i\theta}) \right| =: \|p\|_\infty$$

for every Laurent polynomial $p(z) = \sum_{-N}^{N} c_k z^k$.

Remarks. We could call (2.10) the *fundamental à priori inequality* for unitary operators. It is the "high ground" from which one can (by many laborious steps, to be sure) build up the full spectral theorem for unitary operators (which we have no need to formulate here; for this whole development the reader is referred to [RiSz, § 109] or [AkGl, Chapter 6]).

The "model" we have constructed for U, as "multiplication by z on K" is not a fully satisfactory one (even though it suffices for the deduction of the important Theorem 2.5) because K is not a uniquely determined or "canonical" space, in general many different choices for the H_j are possible. This is at bottom the problem called "spectral multiplicity" and can be dealt with by methods in [DuSc, Chapter 10] or [Ha1]. One can also bypass it, and use Theorem 2.5 as the point of departure for the full spectral theorem (see remarks and references below).

A parallel development can be done for self-adjoint operators: if $T = T^*$, F.J. Murray showed by an elementary argument that *for every polynomial $f(\lambda) = \sum_{k=0}^{N} a_k \lambda^k$ we have* (λ being here restricted to real values)

(2.11)
$$\|f(T)\| \leqslant \max |f(\lambda)|, |\lambda| \leqslant \|T\|.$$

See [Ma, p. 100], where (2.11) is proven, and from it the spectral theorem deduced, following an argument of Eberlein. Alternatively, the spectral theorem for self-adjoint operators can be deduced from that for unitary operators (and vice versa) by using "Cayley transforms".

Let us also observe that the spectral theorem for several *commuting* unitary operators can be deduced from the *multivariable generalization* of the Herglotz-F. Riesz-Toeplitz theorem (itself a special case of Bochner's theorem) by adapting the argument we have given above in the case of one operator. Again, using Cayley transforms this can be carried over to several commuting self-adjoint operators. (The case of *two* commuting self-adjoint operators is equivalent to the case of one "normal" operator, i.e. an operator commuting with its adjoint).

We shall not dwell on the matter. There are a great many known proofs and variants of the spectral theorem. One of the most powerful methods known for proving them uses the theory of commutative Banach algebras to obtain the *Gelfand-Naimark theorem*, giving spectral resolutions for *commuting families of normal operators* ([DuSc, Chapter 9]. The "complete spectral theorem" of von Neumann gives a canonical representation of a normal operator as a "direct integral" independently of any multiplicity assumptions regarding the spectrum.

It is characteristic that, regardless of the particular features of these various approaches, they are all deeply influenced by ideas originating in harmonic analysis. In the opposite direction, the spectral theorem has applications to harmonic analysis: the reader is invited to deduce the Herglotz-F. Riesz-Toeplitz theorem from Theorem 2.3.

Before ending this section, a few words are in order about "functional calculus". For any operator T, with spectrum $\sigma(T)$, there is an obvious way to define $f(T)$, where f is any *rational function with no poles on $\sigma(T)$* (for example, for $f(z) = (z - a)^{-1}$, $f(T)$ is defined to be $(T - aI)^{-1}$, where I denotes the identity operator, which is well defined for $a \in \mathbb{C} \setminus \sigma(T)$). We thus get a "functional calculus", that is a map $f \mapsto f(T)$ which is a *continuous algebra-homomorphism* from the set of rational functions with no poles on $\sigma(T)$, to $L(H)$. Moreover, a more general calculus (the "Riesz-Dunford calculus") can be built up, based on Cauchy integrals, the "domain" of which is *all functions f holomorphic on a neighborhood of $\sigma(T)$*. One of the great thematic problems of operator theory is to enlarge the domain of functional calculus beyond these holomorphic f. But, this cannot be done so as to apply to all operators. For special classes of operators T, however, one can widen the class of f such that $f(T)$ has meaning. This is closely related to estimates for $\|f(T)\|$, when f is rational; we shall return to this point in § 4 in connection with von Neumann's inequality. For example, the estimate (2.10) makes possible a useful definition of $f(U)$ for any unitary operator U, for any f continuous on the unit circle and even (ultimately) for any bounded Borel function on $\sigma(U)$; full details of this are in the already cited textbooks.

3. Isometries, and the Wold decomposition

Once we leave the category of normal operators, we cannot expect unitarily equivalent models so simple and transparent as "multiplication by some bounded function on an $L^2(\mu)$ space". Nevertheless there are important non-normal operators that impose themselves on us, and much of the operator theory of recent years is dedicated to finding structural properties, or models of some kind, for various classes of non-normal operators. In this section we discuss *isometries*. A fundamental theorem attributed to von Neumann, Wold, and Kolmogorov (independently) is

THEOREM 3.1. ("Wold decomposition"). *Let V be an isometry on a Hilbert space H. Then, there is a (uniquely determined) decomposition of H as the direct sum of two mutually orthogonal subspaces $H = H_1 \oplus H_2$ such that*

(i) *H_1 and H_2 reduce V.*
(ii) *The part of V in H_1 is unitary.*
(iii) *The part of V in H_2 is unitarily equivalent to a block shift.*

Before proceeding, let us explain the terminology. If K is any Hilbert space, we may construct from it a new one $\widetilde{K}$ whose elements are sequences from K, that is vectors

$$\tilde{k} := (k_0, k_1, k_2, \dots) \quad , \quad k_j \in K$$

such that $\|\tilde{k}\|_{\tilde{K}}^2 := \sum_0^\infty \|k_j\|^2 < \infty$. The map $S \in L(\tilde{K})$ taking $\tilde{k}$ to $(0, k_0, k_1, \dots)$ is called the *shift operator* on $\tilde{K}$. It is an isometry, but certainly (assuming K has dimension at least one) it is not surjective, and so not unitary. It is not hard to show that, for any two Hilbert spaces K_1 and K_2, the shifts on $\tilde{K}_1$ and $\tilde{K}_2$ are unitarily equivalent if and only if dim K_1 (the dimension of K_1) equals dim K_2. Thus, each shift has associated to it a unique unitary invariant, a positive integer or $+\infty$ which is the dimension of the space K from which $\tilde{K}$ is formed. It is called the *multiplicity* of the shift. Since some authors use the term to denote exclusively the shift of multiplicity one, we use the term *block shift* to denote a general shift.

A shift (or block shift) not only is not unitary, it is as far from unitarity as an isometry can be. Indeed, clearly

$$\lim_{n \to \infty} S^n \tilde{k} = 0 \quad , \qquad \text{for all } \tilde{k} \in \tilde{K}.$$

We may remark, in passing, that this relation is the basis for an abstract (axiomatic) definition of a block shift (cf. [GoGoKa]). The block shift as defined above is also called *unilateral*; this is something of a misnomer—the same shift is called *bilateral* if the ambient Hilbert space of sequences is *doubly infinite* (like $(\dots k_{-1}, k_0, k_1, \dots))$ in which case it is unitary.

If we examine the proof of Theorem 3.1 (which we shall not give here, see [GoGoKa, p. 654]) we note the following features:

a) H_1 is the intersection of the ranges of all V^n, $n \geqslant 1$.
b) H_2 is a block shift of dimension dim W, where W is the kernel of V^*. Moreover, H_2 is the direct sum of the subspaces $W, VW, V^2W, \dots$ which are mutually orthogonal. (For this reason, W has been christened "wandering subspace" belonging to V, by Halmos. It is also called the *defect space* of V for the obvious reason that, being the orthogonal complement of the range of V, it indicates how far V falls short of being unitary.)

Of course, H_1 can be $\{0\}$ (and then V is a ("pure") shift, or H_2 can be $\{0\}$ (and then V is unitary, and $W = \{0\}$).

Theorem 3.1 has some remarkable consequences, and we shall discuss two of these, plus a remarkable generalization obtained recently by S. Shimorin.

First of all, following P. Halmos [Ha2] we shall outline a deduction from Theorem 3.1 of Beurling's famous invariant subspace theorem. (A subspace M of H is *invariant* for the operator T, if $TM \subset M$.) Before formulating it, let us recall some definitions and notations. By $L^p(\mathbb{T})$ for $p > 0$ we denote the usual Lebesgue space of measurable functions on the unit circle $\mathbb{T}$, endowed with the norm $\| \ \|_p$ where

$$\|f\|_p = \left(\frac{1}{2\pi} \int_{\mathbb{T}} |f(e^{i\theta})|^p d\theta \right)^{1/p} \quad , \qquad \|f\|_\infty = \text{ess sup} \, |f(e^{i\theta})|.$$

For $p \geqslant 1$, $L^p(\mathbb{T})$ is a Banach space. By $H^p(\mathbb{T})$, for $p \geqslant 1$ we designate the subspace of $L^p(\mathbb{T})$ consisting of those functions whose negatively indexed Fourier coefficients are all zero. As is well known ([Du], [Ga]), each function in $H^p(\mathbb{T})$ is the nontangential limiting value almost everywhere of a holomorphic function g in the open unit disk $\mathbb{D}$ such that

$$\varlimsup_{r \to 1} \left(\frac{1}{2\pi} \int_0^{2\pi} |g(re^{i\theta})|^p d\theta \right)^{1/p} =: \|g\|_{H^p(\mathbb{D})}$$

is finite. The correspondence of g with its boundary values induces an isometric isomorphism between $H^p(\mathbb{D})$ and $H^p(\mathbb{T})$. For all of this see e.g. [Du], to which we also refer for certain special terminology like *inner* and *outer* functions, etc.

The shift on the space of square summable sequences

$$a = \{a_0, a_1, a_2, \dots\}$$

of complete numbers is, in an obvious way resulting from Parseval's identity, unitarily equivalent to the operator "multiplication by $e^{i\theta}$" on $H^2(\mathbb{T})$, and also to "multiplication by z" on $H^2(\mathbb{D})$. To avoid excessive pedantry we shall feel free to use the term "shift operator" to denote any of these, according to the context of the discussion. Now, we state

THEOREM 3.2. (A. Beurling, 1949). *The invariant subspaces of the shift operator on* $H^2(\mathbb{T})$ *are precisely the sets*

(3.1) $$\varphi H^2 := \left\{ \varphi f : f \in H^2(\mathbb{T}) \right\}$$

where φ is a unimodular function in $H^\infty(\mathbb{T})$ ("inner function").

Proof. It is clear that a set of the form (3.1) is a vector subspace of $H^2(\mathbb{T})$, also it is closed (being the range of the isometric operator "multiplication by φ") and invariant with respect to the shift (here, multiplication by $e^{i\theta}$). The deeper part of the theorem is the converse direction, whose proof à la Halmos [Ha2] we now proceed to sketch.

Thus, let M be a (proper) invariant subspace. Then, restricted to M, the shift is an isometry which we denote by V, and to this we apply Theorem 3.1 with M in the role of the Hilbert space H there. It is easy to check that the intersection of the ranges of V^n for $n = 1, 2, \dots$ is $\{0\}$, so V is unitarily equivalent to a block shift. It is intuitively plausible that the multiplicity of this block shift is one, but that is no tautology, and indeed verification of this is the heart of the proof. We must show that the wandering space W is one-dimensional. To this end, let φ denote any vector in W, which in our context is identified as $M \ominus e^{i\theta} M$. Thus, φ is in M, and orthogonal to $e^{i\theta} f$ to every f in M.

We can in particular choose $f = e^{ik\theta} \varphi$ for $k = 0, 1, 2, \dots$ and this gives

$$\int_{\mathbb{T}} |\varphi|^2 e^{in\theta} d\theta = 0 \quad , \quad n = 1, 2, \dots .$$

By complex conjugation this holds also for all negative integers n, hence $|\varphi(e^{i\theta})|^2$ is constant almost everywhere on $\mathbb{T}$.

Thus, W is a closed vector subspace of $H^2(\mathbb{T})$ with the remarkable property that every function in it has (almost everywhere on $\mathbb{T}$) constant absolute value. The reader should have no trouble supplying the deduction, from this, that dim W is at most one (hence, exactly one, since otherwise V would be a unitary map of M onto itself, which we have ruled out).

We conclude that every element f of $H^2(\mathbb{T})$ is uniquely representable in the form

$$\sum_{n=0}^{\infty} w_n e^{in\theta}$$

where $w_n \in W$ and $\sum \|w_m\|^2 < \infty$. But, $w_n = c_n \varphi$ for some fixed inner function φ, where $\sum |c_n|^2 < \infty$. Hence

$$f\left(e^{i\theta}\right) = \varphi\left(e^{i\theta}\right) \sum_{0}^{\infty} c_n e^{in\theta}$$

which completes the proof of the theorem.

One advantage of this proof is that it can be adapted to prove generalizations of Theorem 3.2, pertaining to Hardy spaces of *vector-valued functions* (due to Lax and Halmos, see [GoGoKa] for details and references). These generalized versions are important for applications e.g. in systems and control theory, as well as multivariate stochastic processes and seem very difficult to prove by classical arguments.

Another, very important, application of Theorem 3.1 is to *wide-sense stationary random sequences*. Indeed, this was the source of the interest in Theorem 3.1 on the part of Wold, and of Kolmogorov. For lack of space, we cannot develop this here in much detail, but refer for the background to [IbRo] and the appendix by Peller and Khruschev to [Ni].

We consider a "random sequence" $\{X_n\}_{n\in\mathbb{Z}}$. The X_n are elements of an $L^2(\Omega, d\rho)$ space for some measure space Ω and probability measure ρ on it (we omit the σ-algebra of subsets of Ω from the notation). We assume the X_n all have mean zero, and variance 1. Moreover, the sequence $\{X_n\}$ is "stationary" in the sense that the *covariance* $E(X_m \overline{X}_n)$, with E denoting mean value, depends only on $n - m$.

For many of the purposes of statistics, we can ignore the underlying "sample space" $(\Omega, d\rho)$ and study the more abstract model, whereby: $\{X_n\}_{n\in\mathbb{Z}}$ *are unit vectors in some Hilbert space H, and $\langle X_m, X_n \rangle$ depends only on $m - n$*. We may also suppose that H is the closed span of $\{X_n\}$ (otherwise, simply re-define H).

Denoting $\langle X_m, X_n \rangle := c_{m-n}$, one easily checks that $\{c_k\}_{k\in\mathbb{Z}}$ satisfy the hypotheses of Theorem 2.1, hence there is a positive measure μ on $\mathbb{T}$ satisfying $\int e^{ik\theta} d\mu(\theta) = c_k$ for all $k \in \mathbb{Z}$, and hence

$$\langle X_m, X_n \rangle = \int e^{i(m-n)\theta} d\mu(\theta) = \langle e^{im\theta}, e^{in\theta} \rangle_{L^2(\mathbb{T}, d\mu)}.$$

Thus, by Lemma 1.1, *there is a unitary map U from H onto $L^2(\mathbb{T}, d\mu)$ carrying X_n onto $e^{in\theta}$, for all $n \in \mathbb{Z}$.*

This makes possible the transformation of certain "statistical" questions concerning $\{X_n\}$ into purely "analytical" ones in $L^2(\mathbb{T}, d\mu)$. Consider, for example, the problem of *prediction*. Let us denote, for any integer m, the closed span of $\{X_n\}_{n \leqslant m}$ by $\mathbb{P}_m$ ("the past up to time m"). Suppose we "know" X_n for all $n < 0$, and wish to "predict" the value of X_0. A very widely used method is to use a "least-squares estimate". That is, use as the "estimate" for X_0 the random variable $\widehat{X}_0$ which is defined as the orthogonal projection of X_0 on $\mathbb{P}_{-1}$. If one assumes (as we shall) that the above unitary map U is "known" (also called "the spectral representation of the random sequence"), then this problem can be transformed to the $L^2(\mathbb{T}, d\mu)$ context, where it is a standard type of approximation problem (weighted L^2 approximation by trigonometric polynomials) and can be tackled by traditional methods (Gram-Schmidt orthogonalization, etc.). For example, the variance of the prediction error when the estimator $\widehat{X}_0$ is employed, $\|X_0 - \widehat{X}_0\|^2$ can in principle be computed as the squared distance from the constant function 1 to the closed span of $\{e^{in\theta}\}_{n<0}$ in $L^2(\mathbb{T}, \mu)$. Hopefully, this conveys a little of the flavor of "prediction theory".

Now, there is an important unitary map V of H: the map defined for all n by $VX_n = X_{n-1}$ is easily seen to extend by linearity and continuity to a unitary map of H (which we continue to denote by V). (In the "spectral model" this becomes multiplication by $e^{-i\theta}$).

Now, $\mathbb{P}_0$ is invariant for V, and $V \mid \mathbb{P}_0$ is an isometry of $\mathbb{P}_0$. Hence, by Theorem 3.1, we get a Wold decomposition of $\mathbb{P}_0$ reducing $V \mid \mathbb{P}_0$. What are the two components into which it resolves?

Let us look first at the extreme cases:

(i) $V \mid \mathbb{P}_0$ is unitary; in this case, by $V\mathbb{P}_0 = \mathbb{P}_0$, so $\mathbb{P}_0 = \mathbb{P}_{-1}$, and by stationarity, all $\mathbb{P}_n, n \in \mathbb{Z}$ are equal. This is the hallmark of a *purely deterministic process*. By means of the spectral model, we know this happens when, and only when, in the Lebesgue decomposition of the spectral measure μ,

$$d\mu = d\mu_s + wd\theta$$

(where μ_s is singular with respect to $d\theta$, and $w \in L^1(\mathbb{T}, d\theta)$) we have $\int_\mathbb{T} \log wd\theta = -\infty$. This is a famous theorem due to G. Szegö, M. Krein and A. Kolmogorov.

(ii) $V|\mathbb{P}_0$ is a block shift; in this case, $\bigcap_{n \leqslant 0} \mathbb{P}_n$ (called the "remote past") reduces to $\{0\}$. The best least squares estimate of X_0 on the basis of "old" information $\mathbb{P}_{-m}$, where m is a large positive integer, tends in norm to 0 as $m \to \infty$. The variance of the prediction error tends to 1. In other words, the best least-squares estimate of X_0 on the basis of very old observations is simply its mean value (here assumed to be zero). Such a process is called *purely indeterminate*.

The general case is an amalgam of (i) and (ii), that is, the Wold decomposition in the stochastic model says:

There is a unique splitting $X_n = X'_n + X''_n$ where $\{X'_n\}$ is purely determinate and $\{X''_n\}$ is purely indeterminate. Their closed spans H', H'' are orthogonal complements in H.

In the spectral model, the splitting can be described explicitly: it mirrors the Lebesgue decomposition of $d\mu$, such that when $\int \log w d\theta = -\infty$, $w d\theta$ is grouped together with $d\mu_s$ and we have a purely determinate process.

The prediction problem is only one of a great many problems about stationary sequences, involving questions of mixing, regularity and so forth. These translate into interesting and sometimes very deep, in some cases still unsolved problems of harmonic analysis. For further information see [IbRo], and the Peller-Khruschev appendix to [Ni].

In closing this section, let us remark that there has been much study in recent years of the invariant subspaces of the operator "multiplication by z" on the *Bergmann space* of the disk $\mathbb{D}$, that is, the space $AL^2(\mathbb{D})$ of analytic functions on $\mathbb{D}$ square integrable with respect to area measure. Here the situation is much more complicated than in the corresponding H^2 scenario of Beurling's theorem. For one thing, "multiplication by z" is now a contractive mapping, but not isometric. It turns out that there are invariant subspaces M for which $M \ominus zM$ is not one-dimensional; the dimension of this subspace (analogous to the "wandering subspace" in our discussion of Beurling's theorem) may be any positive integer, or even infinity.

Nevertheless, it has been proved by Aleman, Richter and Sundberg [AlRiSu], that the space $W := M \ominus zM$ still has the property that, together, $W, zW, z^2W, \ldots$ span M. This remarkable and deep result inspired S. Shimorin [Sh] to discover a new general theorem implying a Wold-type decomposition for certain classes of operators (including, of course, all isometries, but also many others) which in particular yields the main result of [AlRiSu].

4. Dilation theory

Let us start by reviewing some definitions. If $A \in L(H)$ and $A' \in L(H')$, where H, H' are Hilbert spaces with $H \subset H'$, A' is said to *lift* A (or, be a *lifting* of A) if the diagram

(4.1)

$$
\begin{array}{ccc}
H' & \xrightarrow{A'} & H' \\
P \downarrow & & \downarrow P \\
H & \xrightarrow{A} & H
\end{array}
$$

commutes, P being the orthogonal projector of H' on H; that is,

$$(4.2) \qquad AP = PA'.$$

Applying A to both sides of (4.2) from the left, gives $A^2P = APA' = PA'A' = P(A')^2$, and now a simple inductive argument shows

$$(4.3) \qquad A^nP = P(A')^n \quad , \quad n = 1, 2, \ldots .$$

If $T \in L(H, H')$, the operator PT (in $L(H)$) is called the *compression of* T *to* H. It is easy to check that this is equivalent to, denoting PT by A,

$$(4.4) \qquad \langle Th_1, h_2 \rangle = \langle Ah_1, h_2 \rangle \text{ for all } h_1 \text{ and } h_2 \text{ in } H.$$

If A' is a lifting of A to H', then from (4.3)

$$(4.5) \qquad \langle A^n h_1, h_2 \rangle = \langle A^n P h_1, h_2 \rangle = \langle P(A')^n h_1, h_2 \rangle = \langle (A')^n h_1, h_2 \rangle$$

for all pairs h_1, h_2 in H. Taking $n = 1$ and comparing with (4.4) shows: *if A' is a lifting of A, then A is the compression to H of the restriction $A' \mid H$.* But, lifting implies more, namely that A^n is, for every positive integer n, the compression of $(A'|H)^n$ to H.

The converse is not true, and the last relation turns out to be important enough to be given a name:

DEFINITION 4.1. *If $A \in L(H)$ and $A' \in L(H')$, where H, H' are Hilbert spaces with $H \subset H'$, A' is a dilation of A if and only if A^n is, for every positive integer n, the compression of $(A' \mid H)^n$ to H, or what is equivalent*

$$(4.6) \qquad \qquad \langle (A')^n h_1, h_2 \rangle = \langle A^n h_1, h_2 \rangle$$

for every pair h_1, h_2 in H and every positive integer n.

Remarks. Formerly, the term "power dilation" was sometimes used to denote this relation.

If H is an invariant subspace of A', and A is the part of A' in H, then clearly A' is a dilation of A.

Thus, "A' is a dilation of A" is implied by, but in general strictly weaker than each of the assertions "A' is an extension of A" and "A' is a lifting of A".

A landmark discovery in operator theory, due to B. Sz.-Nagy, is

THEOREM 4.2. *Every contraction (i.e. operator of norm at most one) on a Hilbert space has a dilation that is unitary.*

One can formulate this theorem more precisely: if $A \in L(H)$, then there is a unitary dilation U of A to some, in general larger Hilbert space H', with the further property that the closed span of $\{U^n H\}$ for n in $\mathbb{Z}$ is H'. This unitary dilation is, in a natural sense *minimal* in that, roughly speaking, H' is no larger than it has to be and moreover, modulo a natural concept of *isomorphisms of unitary dilations*, this minimal one is unique. (For details, we refer to [SzFo].)

To get a feeling for this remarkable result, the reader is urged to try to construct a unitary dilation (u.d.) of the operator A that is identically zero on H. Any u.d. U must satisfy $\langle U^n h_1, h_2 \rangle = 0$ for all h_1, h_2 in H, from which it follows that the subspaces $\{U^n H\}$ for $n \in \mathbb{Z}$ are mutually orthogonal, i.e. H is a "doubly wandering" subspace for U in the larger space H' where U operates, so H' must be quite large!

Thanks to Theorem 4.2 some results known for unitary operators can be carried over to contractions. Thus,

COROLLARY 4.3. *(von Neumann's inequality) Let A be a contraction on the Hilbert space H, and p any polynomial, $p(\lambda) = \sum_{n=0}^{N} a_n \lambda^n$ with complex coefficients. Then*

$$(4.7) \qquad \qquad \|p(A)\| \leqslant \max |p(\lambda)| \qquad \lambda \in \mathbb{C}, |\lambda| \leqslant 1.$$

Proof. Let $U \in L(H')$ be any unitary dilation of A. Then, for every h_1, h_2 in H we have

$$\left| \langle p(U)h_1, h_2 \rangle \right| = \left| \langle p(A)h_1, h_2 \rangle \right|.$$

Taking the supremum of the right side over all unit vectors h_2 in H gives

(4.8) $$\|p(A)h_1\| \leqslant \|p(U)h_1\| \leqslant \|p(U)\| \, \|h_1\|$$

(the second norm being in the space H').

But, $\|p(U)\|$ does not exceed the maximum of $|p(\lambda)|$ for λ on the unit circle, as shown in § 2. Using this in (4.8) and maximizing over unit vectors h_1 in H gives (4.7), completing the proof.

Remarks. On the basis of (4.7) one can enlarge the functional calculus for contractions A, to encompass all operators $f(A)$ with f continuous on $\overline{\mathbb{D}}$ and in $H^\infty(\mathbb{D})$. This is, in turn, close to von Neumann's notion of "spectral sets", see [RiSz], [SzFo].

One can give proofs of (4.7) using only complex analysis and elementary Hilbert space theory, see [RiSz]. Here we have shown its derivation from Theorem 4.2 to illustrate a typical application of that theorem. Another very nice application is the *mean ergodic theorem*, see [RiSz].

We shall not prove Theorem 4.2, referring to [SzFo] or [GoGoKa], but will give a few indications. First of all, Theorem 4.2 follows from

THEOREM 4.4. *Every contractive operator on a Hilbert space has an isometric lifting.*

To see why this implies Theorem 4.2 we require

LEMMA 4.5. *Every isometric operator on a Hilbert space has a unitary extension.*

Assuming this for the moment, suppose A is a contraction on H and V is some isometric lifting to H'. Let U be a unitary extension of V to H''. Then, in view of earlier remarks, U is a dilation of V, and V is a dilation of A. It follows readily that U is a dilation of A.

We'll give an informal proof of Lemma 4.5. Let V be an isometry on a Hilbert space H. By virtue of the Wold decomposition there is a splitting of $H = H_1 \oplus H_2$ reducing V, such that $V \mid H_1$ is unitary and $V \mid H_2$ is unitarily equivalent to a "unilateral block shift", that is to the operator

$$S : w := (w_0, w_1, w_2, \dots) \mapsto (0, w_0, w_1, \dots)$$

on the Hilbert space of *one-sided sequences* of elements of some Hilbert space W, with norm of w equal to $(\sum_0^\infty \|w_j\|^2)^{1/2}$. (In fact, W can be taken as the wandering space $H \ominus VH$.) Now, S has an obvious unitary extension $\widetilde{S}$, namely the map of the Hilbert space of *two-sided sequences* $\tilde{w} := (\dots w_{-1}, w_0, w_1, \dots)$ given by the formula

$$\left(\widetilde{S}\tilde{w} \right)_n = \tilde{w}_{n-1} \quad , \quad n \in \mathbb{Z},$$

the "bilateral shift". Indeed, $\widetilde{S}$ is unitary and, restricted to the subspace for which all w_i with $i < 0$ vanish (which in an obvious way is identified with the above space of one-sided sequences), $\tilde{S}$ coincides with S. Thus, finally, the operator equal to V on H_1 and $\tilde{S}$ on the appropriate space is a unitary extension of, strictly speaking, an operator unitarily equivalent to V. And, it is easily checked that if an operator has a unitary extension, then so does every operator unitarily equivalent to it (this is because a unitary operator always has a unitary extension to every larger Hilbert space).

Theorem 4.4 also allows a strengthened version, whereby the isometric lifting is to a "minimal" space, and this "minimal isometric lifting" is essentially unique, for details we refer again to [SzFo] and [GoGoKa].

As to the proof of Theorem 4.4, we won't give it in full but illustrate one of the main ideas by sketching the proof of an earlier (weaker) version due to Halmos, [1] whose pioneering writings have been instrumental for modern developments in operator theory:

THEOREM 4.6. *Every contractive operator on a Hilbert space is the compression of a unitary operator.*

Proof. Let $A \in L(H)$, $\|A\| \leqslant 1$. Then, $I - A^*A$ and $I - AA^*$ are self-adjoint operators with spectrum contained in the set $\mathbb{R}^+$ of non-negative real numbers. By the basic functional calculus for self-adjoint operators, they have "positive square roots", that is, there exist self-adjoint operators D_A and D_{A^*} with spectrum in $\mathbb{R}^+$ such that

$$(4.9) \qquad \begin{aligned} D_A^2 &= I - A^*A \\ (D_{A^*})^2 &= I - AA^*. \end{aligned}$$

One calls D_A the *defect operator* associated to A. Observe that it is O if and only if A is an isometry, and both D_A and D_{A^*} are O if and only if A is unitary. Now, we have

$$A(A^*A)^n = (AA^*)^n A$$

and so, for every polynomial $p(t) = c_0 + c_1 t + \cdots + t^m$ with real coefficients $Ap(A^*A) = p(AA^*)A$. Here we can let p run through a sequence $\{p_j\}$ of polynomials converging uniformly to $t^{1/2}$ on $[0, 1]$. Then, by functional calculus, $\|p_j(A^*A) - D_A\|$ and $\|p_j(AA^*) - D_{A^*}\|$ tend to zero, and we obtain the *fundamental intertwining identity*

$$(4.10) \qquad AD_A = D_{A^*}A.$$

It is now easy to prove Theorem 4.6. Let A be a contraction on H. Then, we can define an operator on $H \oplus H$ by means of the "block matrix"

$$(4.11) \qquad U := \begin{bmatrix} A & D_{A^*} \\ -D_A & A^* \end{bmatrix}$$

---

[1] A weaker version of Theorem 4.6 with "univary" replaced by "isometric" was discovered earlier by G. Julia, see [SzFo], p. 51, for the references. I am indebted to N.K. Nikolski for pointing this out to me.

with the convention that U maps an element $\left[\begin{smallmatrix} h_1 \\ h_2 \end{smallmatrix}\right]$ of $H \oplus H$ to

$$\left[\begin{array}{ccc} Ah_1 & + & D_{A^*} h_2 \\ -D_A h_1 & + & A^* h_2 \end{array} \right].$$

Clearly $U \in L(H \oplus H)$ and let us now verify that it fulfills the requirements of the theorem.

First of all, its compression to H (here identified as the subset of $H \oplus H$ consisting of elements $\left[\begin{smallmatrix} h_1 \\ 0 \end{smallmatrix}\right]$ with $h_1 \in H$) is A. Indeed, the first component of $U \left[\begin{smallmatrix} h_1 \\ 0 \end{smallmatrix}\right]$ is Ah_1. As for unitarity, U^* is represented by the block matrix

$$U^* = \left[\begin{array}{cc} A^* & -D_A \\ D_{A^*} & A \end{array} \right]$$

so, at least formally, the rules of matrix multiplication give

$$U^* U = \left[\begin{array}{cc} A^* A + D_A^2 & A^* D_{A^*} - D_A A^* \\ D_{A^*} A - A D_A & D_{A^*}^2 + A A^* \end{array} \right].$$

The diagonal elements are I by virtue of (4.9), (4.10) while the element in row 2, column 1 vanishes by (4.11). The remaining off-diagonal element is the adjoint of this one, hence O. This completes the proof, for people who believe in block matrices. One can of course paraphrase all these calculations by introducing the orthogonal projector from $H \oplus H$ to $H \oplus \{0\}$ and avoiding block matrices, and it is perhaps an instructive exercise for the reader to carry out the proof in this way too. But, block matrices are a very convenient notational device.

This completes the proof of Theorem 4.6. Actually, we have proved a bit more: the unitary operator lives on $H \oplus H$. Moreover, if A is a contractive operator from H to K where K is any Hilbert space (so that $A^* \in L(K, H)$) then $A^* A \subset L(H)$, $AA^* \in L(K)$ and we can define defect operators $D_A \subset L(H)$ and $D_{A^*} \in L(K)$ as before. It is then easy to verify that *mutatis mutandis* the construction of a unitary operator as above goes through. For details see [GoGoKa] or [FoFr]. This is a highly nontrivial result even for finite matrices. It implies given any $m \times n$ matrix which is contractive from $\mathbb{C}^m$ to $\mathbb{C}^n$, we can embed it as the upper left corner of a unitary matrix of size $(m + n) \times (m + n)$ (see (4.12)).

The reader who has followed thus far should now have no trouble following the standard proofs of Theorems 4.4 and 4.2.

5. The further development of dilation theory—especially applications to interpolation problems

Since the discovery of the results reported on in the preceding section, they have continued to play a major role in the development of operator theory, especially to so-called "functional models" for contractions. This is beyond the scope of the present talk, see [Pe] (Douglas' article) and [SzFo]. But we do wish, in closing, to say something about one remarkable development of dilation theory with spectacular applications.

We shall introduce this by going back to the landmark paper [Sa1] which triggered this development. Sarason cast a new light on classical moment and interpolation problems. We can illustrate his approach with the example of the classical *interpolation problem of Pick and Nevanlinna* (henceforth abbreviated "PN problem"). In its simplest and purest form, this is:

(PN) *Given are distinct points* $z_1, \ldots, z_n$ *in the open disk* $\mathbb{D}$, *and complex numbers* $w_1, \ldots, w_n$. *Find necessary and sufficient conditions that there exist a holomorphic function* f *bounded in modulus by 1 in* $\mathbb{D}$ *(we denote this class henceforth by B) satisfying*

$$(5.1) \qquad f(z_j) = w_j \qquad j = 1, 2, \ldots, n.$$

There are many further questions that naturally arise from this one: If such f exists, is it unique? What can be said about functions in $H^\infty(\mathbb{D})$ *of least norm* satisfying (5.1)? If there is more than one f in B satisfying (5.1), can one describe the *totality* of these functions? And, how about the analogous problem with infinitely many points $\{z_j\}$? What happens if some z_j are permitted to coincide, say $z_1 = z_2$ and one prescribes the functional $f'(z_1)$ in (5.1) to replace the redundant $f(z_2)$? And so on. All of these variants have been studied, as well as many others (especially, the case where $H^\infty(\mathbb{D})$ is replaced by an analogous class of *vector valued* functions, which is important in applications). We cannot here enter into all these, but will give references at the end of this section. The remarkable thing is that the path trodden by Sarason has turned out to be fruitful in the study of all these variants, and moreover is virtually the only approach known to yield results for some of the vector-valued generalizations of (PN).

The original question was answered by the following theorem of Pick:

THEOREM 5.1. *Under the hypotheses of (PN) a necessary and sufficient condition for the existence of f in B satisfying (5.1) is that the matrix*

$$(5.2) \qquad \left[\frac{1 - \bar{w}_j w_k}{1 - \bar{z}_j z_k} \right]_{j,k=1}^n$$

be non-negative definite.

Remarks. To get some feeling for the problem let us look first at simple cases. For $n = 1$ the problem is trivial: the desired f exists if and only if $|w_1| \leqslant 1$. Moreover, if $|w_1| = 1$, the solution is unique (the constant function equal to w_1), whereas if $|w_1| < 1$, there are infinitely many solutions, the totality of which is easily described (we shall return to this point in a moment). (The reader is invited to examine carefully the case $n = 2$, and compare with the Schwarz Lemma.)

It is remarkable that, by a recursive algorithm first proposed by I. Schur the general problem (PN) can be reduced to the trivial case $n = 1$. Indeed, whenever $f \in B$, $a \in \mathbb{D}$ and $f(a) = b$, either

(i) $|b| = 1$ and $f \equiv b$

(ii) $|b| < 1$, and in this case

(5.3)
$$g(z) := \frac{f(z) - b}{1 - \bar{b}f(z)}$$

is in B, and vanishes at $z = a$, so that

(5.4)
$$h(z) := g(z)\left(\frac{1 - \bar{a}z}{z - a}\right) = \frac{f(z) - b}{1 - \bar{b}f(z)} \cdot \frac{1 - \bar{a}z}{z - a}$$

is again a function in B. Thus, if (PN) is solvable and $|w_n| < 1$, then, taking $a = z_n$, $b = w_n$ in (5.4) we see that

(5.5)
$$\frac{f(z) - w_n}{1 - \bar{w}_n f(z)} \cdot \frac{1 - \bar{z}_n z}{z - z_n} =: F(z)$$

is again in B. Moreover

(5.6)
$$F(z_j) = \frac{w_j - w_n}{1 - \bar{w}_n w_j} \cdot \frac{1 - \bar{z}_n z_j}{z_j - z_n} \qquad (j = 1, 2, \ldots, n - 1).$$

We also see from (5.5) that

(5.7)
$$F(z_n) = f'(z_n) \cdot \frac{1 - |z_n|^2}{1 - |w_n|^2}.$$

Thus, if $w_1, \ldots, w_n$ are admissible data for $f(z_1), \ldots, f(z_n)$ with $f \in B$ so are the n numbers appearing on the right hand sides of (5.6) and (5.7). Now, suppose we are able to solve the $(n-1)$-point version of (PN) for $F \in B$ satisfying (5.6). Then, from (5.5) solved for f we obtain a function in B. Indeed, solving (5.3) for f gives

(5.8)
$$f(z) = \frac{g(z) + b}{1 + \bar{b}g(z)}$$

which is in B if g is, so the solution to (5.5) is

(5.9)
$$f(z) = \frac{G(z) + w_n}{1 + \bar{w}_n G(z)}$$

where

(5.10)
$$G(z) := \left(\frac{z - z_n}{1 - \bar{z}_n z}\right) F(z)$$

is in B.

Recapitulating: If (5.6) is solvable with $F \in B$, then f defined by (5.9) and (5.10) is in B, and it is easy to check from these formulae that f satisfies all the n conditions (5.1). *Therefore*: If $|w_n| < 1$, the (PN) problem reduces to one with $n - 1$ points; whereas if $|w_n| = 1$ we see at a glance that the problem is solvable if and only if all remaining w_j are equal to w_n; but of course if $|w_n| > 1$ there is no solution. Let us call the last two cases *trivial cases*. We have thus arrived at *Schur's algorithm*:

If $|w_n| \geqslant 1$ we are done, trivially. If $|w_n| < 1$, we construct the new numbers

$$W_j := \frac{w_j - w_n}{1 - \bar{w}_n w_j} \cdot \frac{1 - \bar{z}_n z_j}{z_j - z_n} \qquad (j = 1, 2, \ldots, n - 1).$$

If $|W_{n-1}| \geqslant 1$ we are done, trivially; if $|W_{n-1}| < 1$ we reduce to a problem for $n - 2$ points by the analogous formulae applied to $W_1, \ldots, W_{n-1}$; and so on.

We therefore compute recursively a sequence of complex numbers W_n, $W_{n-1}, \ldots$ and (PN) is solvable if and only if either all of these "Schur parameters" remain in $\mathbb{D}$, or one of them is on the unit circle and the remaining data at that point are equal to this one.

Volumes have been written about the Schur algorithm and various generalizations, an excellent source is [FoFr] and we'll give other references. Nevertheless, although this adequately answers (PN) for small n, it does *not*, at least in an obvious way, yield the elegant theorem 5.1, to whose proof we now return.

Following in Sarason's footsteps: if we are to find an operator-theoretic proof of Theorem 5.1, which involves an unknown function in H^∞, we must start by "encoding" such functions as linear operators on a suitable Hilbert space. Let us try $H^2 = H^2(\mathbb{T})$ as our Hilbert space. Then, any $\varphi \in H^\infty(\mathbb{T})$ induces a linear operator on H^2 by associating with it the *multiplication operator* $M_\varphi : g \mapsto \varphi g$ on H^2. It is easy to show the operator norm $\|M_\varphi\|$ equals $\|\varphi\|_\infty$. Moreover, these multiplication operators are distinguished within $L(H^2)$ by a simple property: denoting by M_z the "shift" on L^2 (i.e. multiplication by the independent variable) we have

LEMMA 5.2. (Sarason).) *If* $A \in L(H^2)$ *and* A *commutes with* M_z, *then* $A = M_\varphi$ *for some* $\varphi \in H^\infty$.

The converse is obvious. The reader should be able to prove Lemma 5.2, starting with the simple observations that A commutes with M_p for every polynomial p: $A(pf) = p(Af)$ for all $f \in H^2$, hence $Ap = \varphi p$ for every polynomial p, where $\varphi := A1$. The main point is to deduce that this φ, so far only known to be in H^2, is in H^∞.

Now that we have a nice way to encode H^∞ as operators on H^2, we need a suitable way to encode the data (5.1). Sarason had the inspired idea to use for this purpose the *Toeplitz operator with symbol* $\bar{f}$. For any $\varphi \in L^\infty(\mathbb{T})$, "multiplication by φ" is a bounded operator on $L^2(\mathbb{T})$. Its compression to the subspace $H^2(\mathbb{T})$ is called the *Toeplitz operator with symbol* φ, and usually denoted T_φ. Thus

$$(5.11) \qquad T_\varphi g = P M_\varphi g = P(\varphi g) \qquad , \qquad \varphi \in L^\infty(\mathbb{T}), g \in H^2(\mathbb{T})$$

where P denotes orthogonal projection of L^2 on H^2. It is easy to check the properties

$$(5.12) \qquad \|T_\varphi\| \leqslant \|\varphi\|_\infty \qquad \text{(with equality if } \varphi \in H^\infty\text{)}.$$

(5.13) $$T_\varphi^* = T_{\bar\varphi}.$$

See, e.g. [Do], [Ni] for extensive discussion of this important class of operators. Now, another important property of Toeplitz operators *with conjugate-analytic symbol* is this: let, for $\zeta \in \mathbb{D}$,

(5.14) $$k_\zeta(z) = \frac{1}{1 - \bar\zeta z}.$$

This is the "representing element" for the functional $f \mapsto f(\zeta)$ on $H^2(\mathbb{D})$, i.e.

(5.15) $$\langle f, k_\zeta \rangle = f(\zeta) \qquad , \qquad f \in H^2(\mathbb{D}).$$

Crucial for us is the identity

(5.16) $$T_{\bar\varphi} k_\zeta = \overline{\varphi(\zeta)} k_\zeta \qquad , \qquad \varphi \in H^\infty.$$

The proof is easy, to verify (5.16) it suffices to show, for each $f \in H^2$, that both sides have the same inner product with f. Now,

$$\langle T_{\bar\varphi} k_\zeta, f \rangle = \langle k_\zeta, T_\varphi f \rangle = \langle k_\zeta, \varphi f \rangle = \overline{\varphi(\zeta)} \, \overline{f(\zeta)} = \overline{\varphi(\zeta)}, \langle k_\zeta, f \rangle,$$

proving (5.16). Hence, Sarason's reformulation of (PN) is:

Given are distinct points $z_1, \ldots, z_n$ in $\mathbb{D}$ and complex numbers

$$w_1, \ldots, w_n$$

Find necessary and sufficient conditions that there exist a linear operator T on H^2 such that

(5.17) $$\|T\| \leqslant 1$$

(5.18) $$T \text{ commutes with the shift } M_z$$

and

(5.19) $$T^* k_{z_j} = \bar w_j k_{z_j} \qquad (j = 1, 2, \ldots, n).$$

In view of our preceding discussion, it is clear that this problem is completely equivalent to (PN). Actually, it is slightly more convenient to replace T by its adjoint, and then (denoting T^* by S) the above conditions become

$$\|S\| \leqslant 1$$

$$S \text{ commutes with the } \textit{backward shift } M_z^* = T_{\bar z} \text{ on } H^2$$

$$S k_{z_j} = \bar w_j k_{z_j} \qquad (j = 1, 2, \ldots, n).$$

Now, let N denote the span of the vectors k_{z_j} $(j = 1, 2, \ldots, n)$. If S exists satisfying 5, 5 and 5 then $SN \subset N$ so the norm of $S \mid N$ is at most one, that is

(5.20) $$\left\| \sum_{j=1}^n \bar w_j t_j k_{z_j} \right\|^2 \leqslant \left\| \sum_{j=1}^n t_j k_{z_j} \right\|^2$$

holds for all complex n-tuples $\{t_j\}$. Expanding the squares and using the identity $\langle k_{z_i}, k_{z_j} \rangle = k_{z_i}(z_j) = 1/(1 - \bar{z}_i z_j)$, and a little manipulation, one sees that (5.20) is equivalent to the non-negative definite character of the matrix (5.2). This already shows the *necessity* of that condition for (PN), and now we turn to the more difficult issue of *sufficiency*. So, suppose (5.2) is non-negative definite, which is equivalent to (5.20), i.e. to the assertion

(5.21) $S \mid N$ has norm at most one.

Thus, $S \mid N$ *is a contractive operator on* N, *and it commutes with the restriction to* N *of the backward shift* $T_{\bar{z}}$ (indeed, each operator admits all the k_{z_i} as eigenvectors).

To finish the proof it suffices to show:

(E) $S \mid N$ has an extension to H^2 that is a contraction, and commutes with $T_{\bar{z}}$.

Indeed, if (E) is proven, and the extension is denoted $\widetilde{S}$, then $\widetilde{S}$ commutes with $T_{\bar{z}}$ so $(\widetilde{S})^*$ commutes with $(T_{\bar{z}})^* = M_z$ and hence is of the form T_f for some f in H^∞ by Lemma 5.2.

Since T_f (which is simply multiplication by f) is a contraction, $\|f\|_\infty \leqslant 1$. Thus, $\widetilde{S} = T_{\bar{f}}$ and so (using 5):

$$\bar{w}_j k_j = \widetilde{S} k_j = T_{\bar{f}} k_j = \overline{f(z_j)} k_j$$

so $f(z_j) = w_j$ $(j = 1, 2, \ldots, n)$, and this completes the proof.

Sarason was aware that there was more at stake here than just another solution to some classical interpolation problems. To grasp what this is, write H (a general Hilbert space) in place of H^2, and consider a subspace N that is invariant for an operator A on H (in PN the role of A is played by $T_{\bar{z}}$). Let now $S \in L(N)$ commute with $A \mid N$. And we ask:

(EG) *Under these conditions, can S be extended to all of H as a linear operator with the same norm, which moreover commutes with A?*

That is, at bottom the fundament of Pick's theorem is something very general, and geometric: an extension property for linear operators from a subspace of a Hilbert space to the whole space with preservation of (i) the norm and (ii) a commutation relation. Now examples show that this is not always possible. But, after the appearance of Sarason's paper, Foias and Sz.-Nagy, using dilation theory, succeeded to show that (EG) *has an affirmative answer whenever A is a co-isometry on H*, that is, A^* is an isometry. This covers the case discussed by Sarason, since there A^* is the shift. (Sarason was able to prove (E) by *ad hoc* methods; the Foias-Sz.-Nagy result is a new milestone, and opened up a fast-developing branch of operator theory.) The assertion that the extension asked for by (EG) exists when A is a co-isometry is sometimes called the "commutant lifting theorem" (CLT) because, in terms of the adjoint operators, it becomes a question of *lifting* (rather than extending) an operator commuting with an *isometry*. This in turn has a further generalization to the *intertwining lifting theorem*. Also, the CLT turns out to be intimately connected to the possibility of simultaneous unitary dilation of two commuting operators (Ando's theorem) and von-Neumann inequalities for polynomial functions of two commuting contractions. For all this, see [FoFr], [SzFo] and [Sz].

For all historical background concerning (PN) and related problems, an excellent source is [FrKi], which includes reprints of the fundamental papers of Herglotz, Schur, Pick and Nevanlinna, as well as a masterful historical review by B. Fritzsche and B. Kirstein and a thorough bibliography. The monograph [DuFrKi] is devoted to matricial generalizations. Other valuable references will be given in our Appendix.

A1. **Acknowledgements.** It is a pleasure to thank Bernd Kirstein, Karim Kellay, and Mihai Putinar for valuable advice concerning the selection of material, and the literature. I regret that lack of space has made it unfeasible to take up some very important topics that relate to the current material, especially Hankel and Toeplitz operators. (See [Sa2], [Po1], [Ni] and [BoeKa] for this material, and references to other relevant literature). I also express my gratitude to Siv Sandvik, who produced this fine LaTeX manuscript from my hieroglyphics, under time pressure.

A.2. **Suggested reading.** The reader who wishes to learn more about the material we have discussed is referred to the following literature, which supplements the sources we already have named.

Stationary random sequences, prediction theory: [Lam, Ya, HeSz, HeSa, Sa3], [Ho, Chapter 4].

Operator-theoretic approaches to interpolation problems. Here we have barely scratched the surface. One alternative approach, building on the theory of *Krein spaces* (a generalization of Hilbert spaces, where the inner product is given by an *indefinite* Hermitian form) to (PN) and related problems is due to Ball and Helton. A very nice introduction to this theory is Sarason's lectures in the volume [Po2]. Also recommended from the same volume is N.K. Nikolskii's lectures on Hankel and Toeplitz operators, where the pace is more leisurely than in the comprehensive, standard monograph [Ni]. See also N.J. Young's article in [Po2]. The monograph [Po1] is also recommended, and [Sa2].

An important topic we have not mentioned is the class of Hilbert spaces of entire functions due to de Branges [Br], which are becoming increasingly popular as new applications and connections are discovered —for instance, at the time of this writing to "frames" and sampling theory by J. Ortega-Cerda and K. Seip.

The collection [Pe] is a gold mine. It contains an excellent introduction by Sarason to invariant subspaces, as well as A.L. Shields' survey of *weighted* shift operators, and R.G. Douglas' survey article on *canonical models* for operators, giving the "state of the art" of dilation theory and its ramifications as of 1970. The conference volume [Lan] contains an article of Sarason wherein he describes, with his usual clarity, the relation of *moment theorems* to operator theory in Hilbert space. On this score see also the monograph [RoRo] which gives a compact and unified framework for the application of Hilbert space methods (especially commutant lifting) to problems of interpolation both in $\mathbb{D}$ and $\overline{\mathbb{D}}$, moment theorems, Loewner's theorem on monotone matrix functions, and more. For interpolation of vector and matrix valued functions, see [DuFrKi], [FoFr] and [BaGoRo].

Fairly recently, theorems of Pick-Nevanlinna type in two or more variables have started to appear, a pioneering role having been played by Jim Agler. See [AgMc] for an account up to 1997. A spectacular application of operator theory to "hyperbolic geometry" was given in 1990 by Agler [Ag], who proved, using dilation techniques, the celebrated theorem of L. Lempert, that the Carathéodory and Kobayashi metrics agree on bounded convex domains. (To find out what those terms mean, the reader may consult the nice introductory book [Kr]).

A lively interest in generalizations of Beurling's theorem as well as of (PN), often in combination, and using tools from dilation theory, persists up to the present day. See, for example [Qu], [McTr] and [GrRiSu]. This work has been catalyzed in part also by new vistas and rich structure encountered in the recent work on Bergmann spaces, starting with the ground breaking paper of Hedenmalm [He]. We refer the reader to a forthcoming monograph by Hedenmalm *et al.* The current work [AlRiSu2], with a wealth of interesting results for classical analysts as well as a good bibliography, is also recommended.

Applications of Toeplitz matrices and operators in science and engineering are many. An interesting variant of the Carathéodory-Fejér problem important for engineering applications due to T. Georgiou is [Ge] and was followed up in later work, see [ByLi] for a recent account. Those are also beautiful applications to statistical physics, a good survey article is [Boe].

References

[Ag]   Agler, Jim, *Operator theory and the Carathéodory metric*, Invent. Math. **101** (1990) 483–500.

[AgMc]   Agler, Jim and John E. McCarthy, *Complete Nevanlinna-Pick kernels*, preprint, 1997.

[Ak]   Akhieser, N.I., *The Classical Moment Theorem*, Hafner, New York, 1965.

[AkGl]   Akhieser, N.I. and I.M. Glazman, *Theory of Linear Operators in Hilbert Space*, Ungar, New York, 1963.

[AlRiSu]   Aleman, A., S. Richter and C. Sundberg, *Beurlings theorem for the Bergman space*, Acta Math. **177** (1996) 275–310.

[AlRiSu2] Aleman, A., S. Richter and C. Sundberg, *The majorization function, and the index of invariant subspaces in the Bergmann spaces*, preprint, June 2000.

[BaGoRo] Ball, J., F. Gohberg and L. Rodman, *Interpolation of Rational Matrix Functions*, Birkhäuser, 1990.

[Be]   Beurling, A., *On two problems concerning linear transformations in Hilbert space*, Acta Math. **81** (1949) 239–255.

[BeChRe] Berg, Christian, J. Christensen and P. Ressel, *Harmonic Analysis on Semigroups*, Springer-Verlag, 1984.

[ByLi]   Byrnes, C. and A. Lindquist, *On the partial stochastic realization problem*, IEEE Trans. Automatic Control, Vol. 42, No. 8, Aug. 1997, pp. 1049–1070.

[Boe]   Böttcher, A., *The Onsager formula, the Fisher-Hartwig conjecture, and their influence on research into Toeplitz operators*, J. Statistical Phys. **78** (1995) 575–584.

[BoeKa]   Böttcher, A. and Yu. Karlovich, *Carleson Curves, Muckenhoupt Weights, and Toeplitz Operators*, Birkhäuser, 1997.

[Br]   de Branges, L., *Hilbert Spaces of Entire Functions*, Prentice-Hall, 1968.

[Do]   Douglas, R., *Banach Algebra Techniques in Operator Theory*, Academic Press, 1972.

[DuFrKi] Dubovoj, V., B. Fritzsche and B. Kirstein, *Matricial Version of the Classical Schur Problem*, B.G. Teubner Verlag, 1992.

[DuSc]   Dunford, N. and J. Schwartz, *Linear Operators, Part 2*, Wiley-Interscience 1958.
[Du]   Duren, P.L., *Theory of H^p Spaces*, Academic Press, 1970.
[FoFr]   Foias, C. and A. Frazho, *The Commutant lifting Approach to Interpolation Problems*, Birkhäuser, 1990.
[FrKi]   Fritzsche, B. and B. Kirstein, ed., *Ausgewählte Arbeiten zu den Ursprüngen der Schur-Analysis*, Teubner-Archiv zur Mathematik, Band 16, B.G. Teubner Verlag, 1991.
[Ga]   Garnett, J., *Bounded Analytic Functions*, Academic Press, 1981.
[Ge]   Georgiou, J.T., *Partial realization of covariance sequences*, Ph.D. dissertation, Univ. Florida, Gainesville, 1983.
[GoGoKa] Gohberg, I., S. Goldberg and M. Kaashoek, *Classes of Linear Operators*, 2 vol., Birkhäuser, 1990.
[GrRiSu] Greene, D., S. Richter and C. Sundberg, *The structure of inner multipliers on spaces with complete Nevanlinna-Pick kernels*, preprint, June 2000.
[GrSz]   Grenander, U. and G. Szegö, *Toeplitz Forms and Their Applications*, Calif. Monographs in Math. Sci. **2**, 1958.
[Ha1]   Halmos, P., *Introduction to Hilbert Space and the Theory of Spectral Multiplicity*, Chelsea, New York, 1951.
[Ha2]   Halmos, P., *A Hilbert Space Problem Book*, 2 ed., Springer-Verlag, 1982.
[Ha3]   Halmos, P., *Shifts on Hilbert Spaces*, J. Reine Angew. Math. **208** (1961) 102–112.
[He]   Hedenmalm, H., *A factorization theorem for square area integrable analytic functions*, J. Reine Angew. Math. **422** (1991) 45–68.
[HeSa]   Helson, H. and D. Sarason, *Past and future*, Math. Scand. **21** (1967) 5–16.
[HeSz]   Helson, H. and G. Szegö, *A problem of prediction theory*, Ann. Mat. Pura Appl. **51** (1960) 107–138.
[Ho]   Hoffman, K., *Banach Spaces of Analytic Functions*, Prentice-Hall, 1962.
[IbRo]   Ibragimov, I. and Yu. Rozanov, *Gaussian Stochastic Processes*, Springer-Verlag 1978.
[Kr]   Krantz, S., *Complex Analysis—The Geometric Viewpoint*, Carus Math. Monographs **23**, Math. Assoc. of America, 1990.
[Lam]   Lamperti, *Stochastic Processes*, Springer-Verlag, 1977.
[Lan]   Landau, H., ed., *Moments in Mathematics*, Proc. Symposia Appl. Math. **37**, AMS, 1987.
[Ma]   Maurin, K., *Methods of Hilbert Spaces*, Monografie Mat., Tom 45, PWN Publishers, Warsaw, 1967.
[McTr]   McCullough, S. and T. Trent, *invariant subspaces and Nevanlinna-Pick kernels*, preprint.
[Ne]   Nevanlinna, R., *Über beschränkte analytische Funktionen*, Ann. Acad. Sci. Fennicae **32**.7 (1929) 1–75.
[Ni]   Nikolskii, *Treatise on the Shift Operator*, Springer-Verlag, 1986.
[Pe]   Pearcy, C., ed., *Topics in Operator Theory*, Math. Surveys, No. 13, AMS, 1974.
[Pi]   Pick, G., Über die Beschränkungen analytischer Funktionen, welche durch vorgegebene Funktionswerte bewirkt werden, Math. Annalen **77** (1916) 7–23.
[Po1]   Power, S., *Hankel Operators on Hilbert Space*, Research Notes in Math. **64**, Pitman, 1982.
[Po2]   Power, S., *Operators and Function Theory*, proc. of NATO Adv. Study Inst., D. Reidel publishers, 1985.
[Qu]   Quiggin, P., *For which reproducing kernels is Pick's theorem true?*, Int. Equ. Op. Theory **16** (1993) 244–246.
[Ri]   Riesz, F., *Über ein Problem von Carathéodory*, J. Für Math. **146** (1916) 83–87.
[RiSz]   Riesz, F. and B. Sz-Nagy, *Leçons d'Analyse Fonctionelle*, 3 ed., Akad. Kiado, Szeged, 1955.
[Ro]   Rozanov, Yu., *Stationary Stochastic Processes*, Moscow, Fizmatgiz 1963 (Russian).
[RoRo]   Rosenblum, M. and J. Rovnyak, *Hardy Classes and Operator Theory*, Oxford, 1985.

[Ru]　　Rudin, W., *Fourier Analysis on Groups*, Second printing, Interscience, 1967.

[Sa1]　　Sarason, D., *Generalized interpolation in H^∞*, Trans. AMS **127** (1967) 179–203.

[Sa2]　　Sarason, D., *Function Theory on the Unit Circle*, Notes, Virginia Polytech. Inst., 1978.

[Sa3]　　Sarason, D., *Addendum to "Past and future"*, Math. Scand. **30** (1972) 62–64.

[Sc]　　Schur, I., *Über Potenzreihen, die im Innern des Einheitskreises beschränkt sind*, J. für Math. **147** (1917) 205–232 and ibid. **148** (1918) 122–145.

[Sh]　　Shimorin, S., *Wold-type decompositions and wandering subspaces for operators close to isometries*, preprint, Lund University, No. 1999:8.

[Sz]　　Sz.-Nagy, B., *Unitary Dilations of Hilbert Space Operators and Related Topics*, CBMS Regional Conf. Series in Math. **19**, AMS, 1974.

[SzFo]　　Sz.-Nagy, B. and C. Foias, *Analyse Harmonique des Opérateurs de l'Espace de Hilbert*, Akad. Kiado, Budapest, 1967.

[To]　　Toeplitz, O., *Über die Fouriersche Entwickelungen positiver Funktionen*, Rend. Palermo **32** (1911) 191–192.

[Wo]　　Wold, H., *A Study in the Analysis of Stationary Time Series*, Stockholm, 1938.

[Ya]　　Yaglom, A., *An Introduction to the Theory of Stationary Random Functions*, Prentice-Hall, 1962.

[Zy]　　Zygmund, A., *Trigonometric Series*, Cambridge, 1968.

Probabilities and Baire's theory in harmonic analysis

J.-P. Kahane

Batiment 425 (mathématique)
Université Paris-Sud
91405 Orsay Cedex
France
`jean-pierre.kahane@math.u-psud.fr`

ABSTRACT. This is an expository paper, with an emphasis on the history of the subject. It consists of three main parts:
　　history, terminology, and examples (Sections 1 and 2),
　　functions and series (Sections 3, 4, and 5),
　　thin sets (Sections 6, 7, 8, 9, and 10).
I chose the topics and references according to my own interest, but I decided not to develop what I wrote recently elsewhere [28], so that I was very brief at the end, in particular, on topics I like best.

1. Lebesgue, Baire, and trigonometric series

Lebesgue's measure theory and Baire's category theorem are both a hundred years old [1,2,43,48]. From the very beginning, they were linked with trigonometric series. Baire was inspired by series of continuous functions: His book "Leçons sur les fonctions discontinues" begins with the example of functions defined as sums of everywhere convergent trigonometric series, the main object of Riemann's thesis on trigonometric series [3,53]. The theory of the Lebesgue integral, as we now know it, was first presented in Lebesgue's book "Leçons sur les séries trigonométriques" [44]. Both books resulted from courses given at the Collège de France. Baire and Lebesgue had been elected "Peccot lecturers," which was an opportunity given to young mathematicians to address younger students. In fact, Denjoy attended Baire's lectures and wrote part of Baire's book (1905), and Fatou attended Lebesgue's lectures and contributed to Lebesgue's book. Fatou wrote his thesis, "Séries trigonométriques et séries de Taylor," at the same time (1906) [14].

J.S. Byrnes (ed.), Twentieth Century Harmonic Analysis - A Celebration, 57–72.
© 2001 *Kluwer Academic Publishers. Printed in the Netherlands.*

The relation between harmonic analysis and integration theory is completely natural, since the first instance of harmonic analysis is the computation of Fourier coefficients using the Fourier formulas, and these formulas involve an integral. The integral in question was not the same for Dirichlet, Riemann, Lebesgue, and Denjoy, but it was either applied (in the case of Dirichlet and Lebesgue) or designed (in the case of Riemann and Denjoy) for the specific purpose of giving a sense to the Fourier formulas. In particular, the second Denjoy totalization was invented specifically to compute the Fourier coefficients of functions considered by Riemann, namely, the sums of everywhere convergent trigonometric series. Riemann stated, without giving a proof, that the coefficients must to tend to zero. Cantor proved this, and moreover, he proved that the coefficients are well defined when the function is given. (This is Cantor's uniqueness theorem.) Denjoy, using the Lebesgue integral, the Baire category theorem, and the transfinite induction of Cantor, provided a way to compute these coefficients. This work was sketched in a Comptes-rendus note in 1921 and was later developed in a series of books between 1941 and 1949 [9, 10].

This Denjoy integral, called "the second totalization" (the first was intended to compute the primitive of the most general derivative [8]), is worth mentioning here because it involves both the Lebesgue integral, which is the prototype of a probability measure, and the Baire category theorem. Furthermore, it performs the most elementary step of harmonic analysis perfectly. However, the Denjoy integral plays no role in the rest of the paper. What is needed from now on is Lebesgue measure on $[0, 1]$, or any equivalent notion of a probability measure, and the Baire theorem expressed as follows:

Baire's category theorem. Let X be a complete metric space, and let G_n, $n = 1, 2, \ldots$, be a sequence of dense, open subsets of X. Then the intersection of the G_n, $\cap_{n=1}^{\infty} G_n$, is dense in X. ($\cap_{n=1}^{\infty} G_n$ is a G_δ set.[1])

Note that Baire could not express his theorem in these terms; the notions of metric space and complete metric space appeared later. They appear at the very beginning of Banach's book, "Théorie des opérations linéaires" (1932), and they are followed immediately by Baire's theorem [4]. The Polish mathematicians of that time were experts in using Baire's theorem in a variety of situations. It is used in Banach's book to prove the Banach–Steinhaus theorem, which was a new way to look at the old "principle of condensation of singularities." In turn, the Banach–Steinhaus theorem is used to show that there exists a continuous function whose Fourier series diverges at a point, the du Bois–Reymond phenomenon [11]. I discovered the power of Baire's theorem in harmonic analysis when Katznelson used it to prove that only analytic functions operate on the Wiener algebra, $A(\mathbb{T}) = \mathcal{F}l^1(\mathbb{T})$, the space of continuous functions on $\mathbb{T}$ whose Fourier series converge absolutely. This result is the converse of the Wiener–Lévy theorem, which asserts that analytic functions do operate on the algebra $A(\mathbb{T})$ [34].

---

[1] Any countable intersection of open sets is called a G_δ set, and any countable union of closed sets is called an F_δ set

I shall not develop the complete role of measure and topology in harmonic analysis, but rather consider how they are involved in several specific questions.

2. Almost sure and quasi sure; examples

In the terminology of Baire, the countable unions of nowhere dense sets are called sets of the first category. In Bourbaki's terminology, they are called meager sets. They are the analogues of null sets in measure theory.

The complements of sets of the first category (meager sets) are called sets of the second category. They are the analogues of sets of full measure, and they can be defined as sets that contain a dense G_δ set. When a property holds on a dense G_δ set it is called *generic* or *quasi sure*. This is the analogue of almost sure in probability theory. We also say that the property holds *quasi everywhere*, the analogue of almost everywhere, or that it is enjoyed by *quasi all* points, which is the analogue of almost all. The abbreviations q.s. and q.e. are parallel to a.s. and a.e.

The analogy goes further. There is a theorem by Kuratowski and Ulam that is analogous to Fubini's theorem, and it is convenient for us to state it in the following form:

The Kuratowski-Ulam theorem [28]. Let X and Y be Baire spaces (for example, complete metric spaces). Suppose, moreover, that Y is separable, that is, there is a countable base of open sets. Let A be a dense G_δ set in $X \times Y$. For each $x \in X$, define $A(x, \cdot) = \{y \in Y \mid (x, y) \in A\}$. Then $A(x, \cdot)$ is a dense G_δ set in Y for quasi all $x \in X$.

It is important to realize that almost sure and quasi sure properties can be very different, that null sets can be sets of the second category, and that sets of full measure can be meager sets. Here are a few examples.

1. Let $X = \{-1, 1\}^{\mathbb{N}}$. X is both a complete metric space and a probability space, when equipped with the usual probability. Let (c_n), $n \in \mathbb{N}$, be a positive sequence, and consider the series $\sum \pm c_n$ indexed by $(\pm) \in X$. Then

$$\sum \pm c_n \text{ converges a.s.} \iff \sum c_n^2 < \infty,$$
$$\sum \pm c_n \text{ converges q.s.} \iff \sum c_n < \infty.$$

The first result is a theorem of Rademacher, Khintchin, and Kolmogorov. The second is an easy exercise: When $\sum c_n = \infty$, the subset Y of X on which some partial sum of each series $\sum \pm c_n$, $(\pm) \in Y$, exceeds a given number N, is open and dense. Therefore the series is divergent on a dense G_δ set of X. In other words, the series $\sum \pm c_n$ diverges q.s. as soon as any one of them diverges.

2. Let $X = \{-1, 1\}^{\mathbb{N}}$ as before, but now consider the Dirichlet series $\sum_{n=1}^{\infty} \pm n^{-s}$, $s = \sigma + it$. These series converge when $\sigma > 1$. We write $f(s)$ either for their sum when

$\sigma > 1$ or for their analytic continuation. Choosing $(\pm) \in X$, the natural boundary of $f(s)$ is the line $\sigma = 1$ q.s. and the line $\sigma = 1/2$ a.s.

It is more interesting to consider $\sum_{n=1}^{\infty} \pm ((2n-1)^{-s} - (2n)^{-s})$, with $(\pm) \in X$. Then the natural boundary of the corresponding function, which we denote by $f_1(s)$, is $\sigma = 0$ q.s. and $\sigma = -1/2$ a.s. Furthermore, the order of $f_1(s)$,

$$\mu(\sigma) = \{\inf a \mid f(\sigma + it) = O(|t|^a), \ |t| \to \infty\},$$

is $(1-\sigma)^+ = \max\{0, 1-\sigma\}$ on $(0, \infty)$ q.s. and $(1/2 - \sigma)^+$ a.s. If the difference $(2n-1)^{-s} - (2n)^{-s}$ is replaced by the differences of higher orders,

$$\left((4n-3)^{-s} - (4n-2)^{-s})\right) - \left((4n-1)^{-s} - (4n)^{-s}\right),$$

$$\left((8n-7)^{-s} - (8n-6)^{-s})\right) - \cdots + \left((8n-1)^{-s} - (8n)^{-s}\right),$$

and so on, we obtain functions $f_2(s), f_3(s), \ldots$, that are defined on larger and larger domains. The functions $f_1(s), f_2(s), f_3(s), \ldots$ are all of the form $\sum_{n=1}^{\infty} \pm n^{-s}$ for $\sigma > 1$, although the domain of the coefficients $(\pm)$ becomes restricted as the process continues. Furthermore, $\mu(\sigma) = (1-\sigma)^+$ q.s. and $\mu(\sigma) = (1/2 - \sigma)^+$ a.s. on $\mathbb{R}$ where these functions are defined. By taking successive blocks of terms from the series representing the functions $f_1(s), f_2(s), f_3(s), \ldots$, it is possible to define an entire function $f_\infty(s)$ that it is represented by a series of the form $\sum_{n=1}^{\infty} \pm n^{-s}$ when $\sigma > 1$, and such that $\mu(\sigma) = (1-\sigma)^+$ q.s. and $\mu(\sigma) = (1/2 - \sigma)^+$ a.s. on $(-\infty, \infty)$. Many variations are possible, and the quasi-sure constructions play an important role in the study of convergence and summability properties of products of Dirichlet series [23, 32, 51].

3. Let $X = \mathbb{T}$ and consider the Hardy-Weierstrass function

$$f(t) = \sum_{n=1}^{\infty} 2^{-n} \cos(2\pi 2^n t).$$

Geza Freud made a careful study of this function. First, as already noticed by Hardy, it is nowhere differentiable. However, $\Delta f(t) = f(t+h) - f(t)$ is $O(|h|)$ as $h \to 0$ for some t, called *slow points*. Actually, the set of slow points is both a null set and a meager set, but it has Hausdorff dimension 1. The modulus of continuity of f is $O(h \log \frac{1}{h})$, meaning that $|\Delta f(t)| < C|h| \log \frac{1}{|h|}$ for all t. However, there are *rapid points* t such that

$$\overline{\lim_{h \to 0}} \ \frac{|\Delta f(t)|}{|h| \log \frac{1}{|h|}} > 0,$$

and indeed, quasi all t are rapid points. On the other hand,

$$0 < \overline{\lim_{h \to 0}} \ \frac{|\Delta f(t)|}{|h| \left(\log \frac{1}{|h|} \log \log \log \frac{1}{|h|} \right)^{1/2}} < \infty$$

for almost all t.

We see from these examples that the Baire approach emphasizes divergence, singularities, and large values, while the probabilistic approach favors convergence, smoothing, and regularizing effects. We shall encounter these ideas throughout the paper. However, the first effect of both approaches is to convert rather strange phenomena into familiar ones, to tame monsters, or to generate monsters in a familiar way. I shall illustrate this aspect by considering nondifferentiable continuous functions and noncontinuable Taylor series, before considering other questions.

3. Nowhere-differentiable functions

Nowhere-differentiable continuous functions and noncontinuable Taylor series (i.e., analytic functions whose domain of existence is a disc) were both discovered by Weierstrass. I remind you of what Hermite said about nowhere-differentiable functions in a letter to Stieltjes: "Je me détourne avec horreur et effroi de cette plaie lamentable des fonctions continue qui n'ont pas de dérivée."

Nevertheless, such functions attracted the attention of the physicist Jean Perrin when he observed the trajectories of Brownian particles, and they became part of Wiener's program to build a mathematical theory of Brownian motion, where the realizations are a.s. continuous and nowhere differentiable. Wiener quoted Jean Perrin several times, and his program was only achieved in 1933 with the help of Paley and Zygmund. Nowadays the local behavior of Brownian motion—replete with the concepts of a strong form of nowhere differentiability, modulus of continuity, average behavior (law of the iterated logarithm), rapid points, and slow points—is well understood (see [24] or [29] for references).

The same is true for the Baire point of view. Let us consider $C(I)$, the space of real continuous functions on the interval $I = [0, 1]$. Given any sequence $A_n \to 0$ and an integer ν, the set of $f \in C(I)$ with the property that

$$\exists n > \nu : \forall k (= 0, 1, \ldots, n - 1), \quad \frac{1}{n}\left| f\left(\frac{k+1}{n}\right) - f\left(\frac{k}{n}\right)\right| > A_n$$

is a dense, open subset of $C(I)$, which we denote by G_ν. If $f \in \cap_{\nu=1}^{\infty} G_\nu$, then

$$\varlimsup_{h \to 0} \frac{|f(t + h) - f(t)|}{\varphi(|h|)} = \infty \quad \text{for all } t \in [0, 1]$$

whenever φ is an increasing function such that $\varphi(1/h) = o(A_n)$ as $n \to \infty$. This is a very strong form of nowhere differentiability. In particular, the Hölder exponent of quasi all f is 0 at every point. Quasi surely in $C(I)$, the multifractal analysis is trivial. To have an interesting multifractal analysis as a generic phenomenon, more restricted classes of functions should be considered, as Stéphane Jaffard has done [17].

4. Random Taylor series: continuation, convergence, and divergence

Weierstrass used a lacunary trigonometric series to construct a continuous nowhere-differentiable function. Later, he constructed a noncontinuable Taylor series using the same idea. Poincaré and Hadamard also used lacunary Taylor series to construct noncontinuable functions. Then, in 1896, Borel issued a strange statement, namely, that in general, a Taylor series is not continuable across its circle of convergence.

Borel had in mind a probabilistic interpretation of this statement. He spoke of arbitrary coefficients (coefficients quelconques) and clearly thought of random, independent phases. In fact, he stated and used a first version of the Borel–Cantelli lemma for that purpose. But Borel's theory of countable probabilities was not yet born, and "in general" could not have a precise meaning in 1896.

Steinhaus gave a rigorous interpretation of probabilistic concepts by means of Lebesgue measure on $I = [0, 1]$. Using binary expansions, he first transferred Lebesgue measure to the space $\{0, 1\}^{N}$, then to the same space written as $\{0, 1\}^{N^2}$, and finally to I^{N}. In this way he obtained the so-called Steinhaus sequences $(\omega_n) \in I^{N}$, for which he proved a zero-one law. Then, by considering random series

$$\text{(S)} \qquad \sum_{n=0}^{\infty} a_n e^{2\pi i \omega_n} z^n, \qquad \overline{\lim_{n \to \infty}} \, a_n^{1/n} = 1,$$

it was not difficult to prove that noncontinuation holds almost surely (1923, 1929, [54,55]).

Paley and Zygmund proved the same result for Rademacher Taylor series

$$\text{(R)} \qquad \sum_{n=0}^{\infty} \pm a_n z^n$$

in 1932. They also considered random trigonometric series, and this was the beginning of the theory of random series of functions. In 1933, Wiener joined them and proposed to study Gaussian series, such as the Wiener Fourier series of Brownian motion, in parallel with Steinhaus or Rademacher trigonometric series. To make the notation simpler, I shall consider only Gaussian Taylor series

$$\text{(G)} \qquad \sum_{n=0}^{\infty} \zeta_n a_n z^n,$$

where the $\zeta_n \in \mathbb{C}$ are independent, normalized Gaussian random variables. When $z = e^{2\pi i t}$, (S), (R), and (G) appear as random trigonometric series with only positive frequencies. The history of these series, culminating with the work of Marcus and Pisier, can be found in the second edition of my book "Some Random Series of Functions" (1985) [24]. Roughly speaking, it was known in 1933 that the series (S), (R), and (G) represent a.s. functions in H^1, H^2, and H^p with $p < \infty$, if and only if $\sum |a_n|^2 < \infty$. Marcus and Pisier established that uniform convergence in the closed unit disc has the same probability, 0 or 1, for series

(S), (R), and (G), and in this way reduced the study of uniform convergence of Rademacher Taylor series to the same, but more tractable, problem involving Gaussian processes.

Nevertheless, not all results are the same for the series (S), (R), and (G). An interesting example is the problem of divergence everywhere at the boundary of the unit disc, which was considered by Dvoretzky and Erdős [23]. The condition

(*)
$$\overline{\lim_{n \to \infty}} \frac{1}{\log n} |a_1 + a_2 + \cdots + a_n| > 0$$

implies that each of the series (S), (R), and (G) diverges everywhere on the circle $|z| = 1$ [24]. But there are series (R) that converge somewhere, while the corresponding (G) diverges everywhere. It suffices to choose

$$(R) : \sum_{j=0}^{\infty} \pm \alpha_j z^{4^j} \quad \text{and} \quad (G) : \sum_{j=0}^{\infty} g_j \alpha_j z^{4^j},$$

where $g_j = \zeta_{4^j}$, $\alpha_j = o(1)$, $\lim_{j \to \infty} \alpha_j \sqrt{\log j} = \infty$, and $\overline{\lim}_{j \to \infty} \alpha_j |g_j| = \infty$ a.s. This example also shows that $\log n$ cannot be replaced by any $o(\log n)$ in condition (*). (See the Paley–M. Weiss theorem in Section 7.)

Divergence everywhere raises apparently difficult questions. Is it true that the probability of everywhere divergence increases when we increase the moduli of the coefficients? The corresponding questions for convergence everywhere, convergence almost everywhere, and divergence almost everywhere (decreasing or increasing the moduli of the coefficients, according to the case) have positive answers. The question can be considered for other series of functions, for example, for random trigonometric series or for random Walsh series.

Other problems arise when we consider the properties of analytic functions defined by the series (R), (S), or (G). Does their range cover the plane $\mathbb{C}$ and by how much? The answer is fairly precise for (G), but it is still incomplete for (R) and (S) (references can be found in [27], p. 267).

5. Generic trigonometric and power series

Let me turn to another interpretation of Borel's statement on the noncontinuation of Taylor series as a generic phenomenon. Let $H(D)$ denote the space of analytic functions defined on the unit disc $D = \{z \mid |z| < 1\}$ and endowed with the topology of uniform convergence on compact subsets of D. Let X be a complete metric space consisting of analytic functions defined on D such that the mappings $f \mapsto f^{(n)}(z)$ are continuous from X to $\mathbb{C}$ whenever

$n \in \mathbb{N}$ and $z \in D$. Given $a \in D, r > 0$, and $A \in \mathbb{N}$, let

$$G(a, r, A) = \left\{ f \in X \mid \sum_{n=0}^{\infty} \frac{1}{n!} |f^{(n)}(a)| r^n > A \right\},$$

$$J(a, r) = \bigcap_{A>0} G(a, r, A) = \left\{ f \in X \mid \sum_{n=0}^{\infty} \frac{1}{n!} |f^{(n)}(a)| r^n = \infty \right\}.$$

The $G(a, r, A)$ are open, and $J(a, r)$ is a G_δ set. Define

$$J = \bigcap_{a,r} J(a, r),$$

where the coordinates of a and r are rational and $r > 1 - |a|$. Then J is also a G_δ set, and the functions f belonging to J are not continuable across the circle $|z| = 1$.

It follows that quasi all f in X are noncontinuable as soon as, given any domain Ω strictly larger than D, the set of f that cannot be continued analytically on Ω is dense in X. In particular, functions represented by (R), (S), or (G), as well as functions belonging to $H(D)$, are quasi surely not continuable across the circle. This result is due to Kierst and Szpilrajn for $H(D)$ [39], and it can be extended in a number of ways, by replacing D with any other domain in $\mathbb{C}$ or $\mathbb{C}^d$, in the following form: Either all functions belonging to the space under consideration can be extended in the same larger domain, or quasi all functions are noncontinuable [28].

Article [28] contains many generic properties of trigonometric and Taylor series. Let me quote a few of them.

1. $X = C(\mathbb{T})$. The partial sums $S_n(f, t)$ of the Fourier series of quasi all f satisfy

$$\varlimsup_{n \to \infty} \frac{1}{\omega_n} S_n(f, t) = \infty \quad \text{quasi everywhere}$$

whenever $0 < \omega_n = o(\log n)$ as $n \to \infty$.

2. $X = L^1(\mathbb{T})$. The partial sums $S_n(f, t)$ of quasi all f satisfy

$$\varlimsup_{n \to \infty} \frac{1}{\lambda_n} S_n(f, t) = \infty \quad \text{everywhere}$$

whenever $0 < \lambda_n = o\left(\frac{\sqrt{\log n}}{\sqrt{\log \log n}} \right)$ as $n \to \infty$. (This is the generic Konyagin theorem [40].)

3. $X = H(D)$. Given any open subset of D, say Δ, such that the boundaries of D and Δ have at least one point in common, quasi all f satisfy

$$f(e^{i\theta}\Delta) = \mathbb{C} \quad \text{for all } \theta \in \mathbb{R}.$$

4. $X = H(D)$. The Taylor series of quasi all f satisfy the Nestoridis universality property: Given any compact set K in $\{ z \mid |z| \geqslant 1 \}$ such that $\mathbb{C} \setminus K$ is connected, and given

any function F that is continuous on K and analytic in its interior, there exists a sequence of partial sums that converge to F uniformly on K [47]. As a corollary, one can show that every continuous function on $\{z \mid |z| = 1\}$ is a pointwise limit of some sequence of partial sums.

5. $X = c_0(\mathbb{N})$. Quasi all Fourier Taylor series $\sum_{n=0}^{\infty} a_n e^{2\pi int}$, $(a_n) \in X$, have the Menchoff universality property: Every Lebesgue-measurable function $F(e^{2\pi it})$ is a limit of some sequence of partial sums almost everywhere [31].

6. Thin sets and function spaces

The last part of this paper concerns thin sets in $\mathbb{Z}$ and $\mathbb{T}$. "Thin" refers to properties related to trigonometric series. Furthermore, the thin sets we consider in $\mathbb{Z}$ are lacunary, and the thin sets we consider in $\mathbb{T}$ are closed and have zero Lebesgue measure.

Probability methods have proven to be quite efficient for the study of thin sets in $\mathbb{Z}$, however, so far, I know of no use of the Baire theory in this context. On the other hand, both probability methods and Baire's method are actively used for exhibiting properties of thin sets in $\mathbb{T}$. As one would expect, the results obtained by the two methods are quite different and often go in opposite directions. A way to relate them was discovered by Körner; it will be sketched at the end of the paper.

The first use of thin sets of integers was made by Hadamard for studying lacunary Taylor series. Assuming that (λ_n) is an increasing sequence of positive integers such that (*) $\lambda_{n+1}/\lambda_n > q > 1$ for all n, he showed that all series $\sum_{n=0}^{\infty} a_n z^{\lambda_n}$ with a finite, nonzero radius of convergence are noncontinuable. Condition (*) is far from being the best for this kind of result. However, (*) plays a role in a number of questions in harmonic analysis. It is called the Hadamard lacunary condition.

Let me proceed with the definitions. $C(\mathbb{T})$, $L^p(\mathbb{T})$ $(1 \leqslant p \leqslant \infty)$, $c_0(\mathbb{Z})$, and $l^p(\mathbb{Z})$ $(1 \leqslant p \leqslant \infty)$ have the usual meaning. $C_{as}(\mathbb{T})$ is the subspace of $L^2(\mathbb{T})$ consisting of functions $f \sim \sum_{n\in\mathbb{Z}} \hat{f}_n e^{2\pi int}$ such that $\sum_{n\in\mathbb{Z}} \pm \hat{f}_n e^{2\pi int}$ represents a.s. a continuous function. Then, according to Marcus and Pisier, the same is true for the Gaussian Fourier series $\sum_{n\in\mathbb{Z}} \zeta_n \hat{f}_n e^{2\pi int}$, where the ζ_n are i.i.d. normalized Gaussian variables, and $C_{as}(\mathbb{T})$ is a Banach space using either of the equivalent norms

$$E \left\| \sum_{n\in\mathbb{Z}} \pm \hat{f}_n e^{2\pi int} \right\|_{C(\mathbb{T})} \quad \text{or} \quad E \left\| \sum_{n\in\mathbb{Z}} \zeta_n \hat{f}_n e^{2\pi int} \right\|_{C(\mathbb{T})}.$$

The Pisier algebra is $C(\mathbb{T}) \cap C_{as}(\mathbb{T})$. $A(\mathbb{T}) = \mathcal{F}l^1(\mathbb{Z})$ is the Wiener algebra, which is the subspace of those functions in $C(\mathbb{T})$ whose Fourier series converge absolutely (see Section 1). The space of Radon measures on $\mathbb{T}$ is denoted by $M(\mathbb{T})$, and the cone of probability measures is denoted by $M_1^+(\mathbb{T})$. The pseudomeasures are linear forms on $A(\mathbb{T})$, in other words, Schwartz distributions on $\mathbb{T}$ whose Fourier coefficients are bounded. The space of pseudomeasures, $\mathcal{F}l^\infty(\mathbb{Z})$, is denoted by $PM(\mathbb{T})$. Pseudofunctions are pseudomeasures whose

Fourier coefficients tend to 0. The space of pseudofunctions, $\mathcal{F}c_0(\mathbb{Z})$, is denoted by $PF(\mathbb{T})$. The Rajchman measures are at the same time measures and pseudofunctions. We write

$$M_0(\mathbb{T}) = M(\mathbb{T}) \cap PF(\mathbb{T}) = \{\mu \in M(\mathbb{T}) \mid \hat{\mu}(n) = o(1), \ n \to \pm\infty\}.$$

We shall also be interested in the classes $M_\alpha(\mathbb{T})$, $0 \leqslant \alpha \leqslant 1$, defined as

$$M_\alpha(\mathbb{T}) = \{\mu \in M(\mathbb{T}) \mid \hat{\mu}(n) = o(|n|^{-\alpha/2}), \ n \to \pm\infty\}.$$

7. Sidon and Zygmund sets

Given any family $F(\mathbb{T})$ of functions or distributions defined on $\mathbb{T}$ and any subset Λ of $\mathbb{Z}$, we write F_Λ for the subfamily of $F(\mathbb{T})$ whose elements have their spectrum in Λ. Thus, $f \in F_\Lambda$ if and only if

$$f \sim \sum_{n \in \Lambda} \hat{f}_n e^{2\pi i n t}.$$

Here is the first and most important definition of a thin set of integers.

D1. Λ is a Sidon set means that $C_\Lambda = A_\Lambda$.

Equivalent definitions, known already in the 1930s, are $\widehat{M}|_\Lambda = \widehat{PM}|_\Lambda$ and $\widehat{L^1}|_\Lambda = \widehat{PF}|_\Lambda$. (Here we write $\widehat{F}|_\Lambda$ for the space of restrictions to Λ of Fourier transforms of elements of $F(\mathbb{T})$; then $\widehat{PM}|_\Lambda = l^\infty(\Lambda)$ and $\widehat{PF}|_\Lambda = c_0(\Lambda)$.) The structure of Sidon sets is not yet clarified. Probability methods have helped, and there have been two approaches: 1) the use of random functions with a given spectrum, and 2) the use of random sets of integers.

The first method was introduced in 1957 [18, 19] and gave the following result: If Λ is a Sidon set, there exists a $K > 0$ such that given positive integers n and s, every "net" of the form

$$N(a_1, a_2, \ldots, a_n; s) = \left\{ a_1 m_1 + a_2 m_2 + \cdots + a_n m_n \;\middle|\; m_j \in \mathbb{Z}, \sum_{j=1}^{n} |m_j| < 2^s \right\},$$

where the a_j are real, contains no more than Kns points of Λ. It is not known whether this necessary condition is also sufficient. The triumph of the method is an alternative definition of Sidon sets, from which the theorem of Drury—the union of two Sidon sets is a Sidon set—follows immediately:

D2. Λ is a Sidon set if and only if $C_{as\,\Lambda} = A_\Lambda$.

This is due to D. Rider [52]. The characterizations given by Pisier [49,50] and by Borgain [6] use the same approach.

The second method was introduced by Katznelson and Malliavin in 1966 [36] in connection with the "dichotomy problem": For a rather large class of random sets of integers, $\Lambda = \Lambda(\omega)$, either Λ is a Sidon set or only analytic functions operate on $\widehat{L^1}|_\Lambda$. Katznelson proved the following result in 1972 [35]: In the second case, Λ is dense in the Bohr group, that is, the Bohr compactification of $\mathbb{Z}$. Moreover, he indicated a way to construct Rudin's $\Lambda(p)$ sets by a random procedure. (Λ is $\Lambda(p)$ means that $\widehat{L^2}|\Lambda(= l^2(\Lambda)) = \widehat{L^p}|_\Lambda$.) This was later developed by Borgain in [7] to obtain $\Lambda(p)$ sets that are not $\Lambda(p + \varepsilon)$. Borgain's selector method consists in choosing Λ as the support of the random measure $\sum_{n \in \mathbb{Z}} X_n \delta_n$, where δ_n is the Dirac mass at n, and where the X_n are independent random variables with values 0 or 1 such that $EX_n = a_n$ for some given sequence a_n, $0 < a_n < 1$. When $a_n = O(1/\log|n|)$, $n \to \pm\infty$, Λ is a.s. a Sidon set. When $|a_n|\log|n|$ tends to ∞, there is a.s. a trigonometric series $\sum_{\lambda \in \Lambda} c_\lambda \sin \lambda t$ that is uniformly, but not absolutely, convergent. New developments are given in [45].

It is through definition D2 that Sidon sets are related to almost surely everywhere convergent (or equivalently, uniformly convergent) trigonometric series. The Zygmund sets (or sequences), which I shall define below, are related to almost surely somewhere convergent trigonometric series, which we have already considered. They originate from a theorem of Zygmund [57] and a theorem of Paley and Mary Weiss [56] on Hadamard real trigonometric series and Hadamard Taylor series with coefficients tending to zero. From now on, Λ will denote a set of positive integers, ordered increasingly.

D3. Λ is a Zygmund set whenever every real trigonometric series

$$\sum_{\lambda \in \Lambda} \text{Re}(a_\lambda e^{2\pi i \lambda t})$$

with $a_\lambda = o(1)$, $\lambda \to \infty$, converges at some point $t \in \mathbb{T}$.

D4. Λ is a Zygmund$^+$ set whenever every Taylor series $\sum_{\lambda \in \Lambda} a_\lambda z^\lambda$ with $a_\lambda = o(1)$, $\lambda \to \infty$, converges at some z, $|z| = 1$.

Every Zygmund$^+$ set is a Zygmund set. It is not known if the converse is true. Necessary conditions for Λ to be a Zygmund set were given by Erdős [13] and by me [21], and they are close to the necessary conditions known for Sidon sets. Is every Zygmund set a Sidon set? Is every Sidon set in $\mathbb{N}$ a Zygmund set? These are old questions, hardly considered.

8. Kronecker, Helson, M, and Salem sets

We are now going to consider closed subsets of $\mathbb{T}$. I shall restrict myself to Kronecker sets, Helson sets, M sets, M_α sets with $0 \leqslant \alpha < 1$, and Salem sets. Here are the definitions ([29,33]).

D5. E (a closed subset of $\mathbb{T}$) is a Kronecker set if each function of modulus 1 that is continuous on E is the uniform limit of some sequence of imaginary exponentials $\exp(2\pi i n_j t)$.

From now on I shall write $C(E)$ for the Banach space of continuous functions on E and $A(E)$ for the Banach space consisting of the restrictions to E of functions belonging to $A(\mathbb{T})$.

D6. E is a Helson set if $A(E) = C(E)$.

An equivalent definition is that the measure norm and the pseudomeasure norm are equivalent for all measures $\mu \in M(E)$, that is, measures supported by E. The first theorem about Helson sets, due to Helson, is that a Helson set carries no nonzero measure belonging to $M_0(\mathbb{T})$ [16]. A long standing question, whether a Helson set can carry a pseudofunction, was solved positively by T. Körner in 1972 [41]. The best proof uses a probability device of R. Kaufman [38].

D7. E is an M set if it carries a nonzero pseudofunction, or in our notation, if $PF(E) \neq \{0\}$.

An M set is also called a set of multiplicity because there are infinitely many trigonometric series that converge to 0 outside the set. The opposite is called a set of uniqueness. A Kronecker set is both a Helson set and a set of uniqueness.

D8. E is an M_0 set if it carries a nonzero measure μ such that $\hat{\mu}(n) = o(1)$ as $|n| \to \infty$.

Helson's theorem says that an M_0 set cannot be a Helson set; Körner's construction shows that an M set can be a Helson set.

D9. E is an M_α set if $M_\alpha(E) \neq \{0\}$, which means that E supports a probability measure μ such that $\hat{\mu}(n) = o(|n|^{-\alpha/2})$ as $|n| \to \infty$.

It follows from a theorem of Frostman that the Hausdorff dimension of an M_α set is $\geqslant \alpha$.

D10. E is a Salem set of dimension α if its Hausdorff dimension is α and if E is an M_β set for every $\beta < \alpha$.

9. Random thin sets

Probability methods have been used to obtain Salem sets in different ways. The problem of finding Salem sets of dimension α for each α between 0 and 1 was suggested by Beurling [33].

Salem's construction, apart from probability methods, uses Diophantine approximation, and it produces sets that have strong arithmetical properties, of the same type as the usual Cantor sets. This construction is described in [33].

Processes with independent increments, in particular Lévy processes and Brownian motion, provide Salem sets in a most natural way: The images of a fixed set of dimension $\alpha/2$ under Brownian motion is a.s. a Salem set [24]. There is an analogue of this result for Lévy flights [26, 30, 46]. The first result in this direction was suggested by B. Mandelbrot when he became interested in the "Lévy dust." Here is the result. Recall that any stationary increasing process with independent increments, a Lévy flight, is a mapping $X : \mathbb{R}^+ \times \Omega \to \mathbb{R}^+$ that satisfies the following properties: For almost all ω, $X(t, \omega)$ is an increasing function of t, it is continuous to the right, and $X(0, \omega) = 0$ for all ω. Furthermore, for each choice of $0 \leqslant t_1 \leqslant t_2 \leqslant \cdots \leqslant t_n$, the random variables $X(t_{j+1}, \cdot) - X(t_j, \cdot)$, $j = 1, 2, \ldots, n - 1$, are independent and their law depends only on $t_{j+1} - t_j$. The "ψ function" associated with this process is defined by the equation

$$Ee^{iuX(t)} = e^{-t\psi(u)}, \quad X(t) = X(t, \cdot),$$

and the law of the process depends only on ψ. Then the image of the Lebesgue measure on $[0, 1]$ under $X(t)$ is a random measure μ whose Fourier transform is

$$\hat{\mu}(u) = \int_0^1 e^{iuX(t)} \, dt.$$

Writing

$$h(u) = \inf_{|x| \geqslant u} \operatorname{Re} \psi(x),$$

one gets successively

$$E|\hat{\mu}(u)|^{2p} \leqslant \frac{(2p)! \, 2^p}{p! \, (h(u))^p},$$

and

$$\hat{\mu}(u) = O\left(\sqrt{\frac{\log u}{h(|u|)}}\right), \quad |u| \to \infty, \quad \text{a.s.}$$

The last steps are copied from Salem [33]. The definition of $h(u)$ as inf $|\psi(x)|$ given in [30] and [26] is incorrect. The correction is due to A. Benchérif–Madani (see [5], p. 87). In particular, for stable Lévy processes of index α, $\psi(u) = cu^\alpha$, where c is complex, and the support of μ, transported to $\mathbb{T}$, is a Salem set a.s.

It can be expected that most "naturally occurring" random sets are Salem sets. In particular, level sets of fractional Brownian motions should be Salem sets. This is true for ordinary Brownian motion because the level set starting from 0 coincides with the closure of the range of a Lévy process of index $1/2$. The measure to be considered is now $\delta(X(\cdot))$. The method is at hand in the last chapter of [24], but it was never carried out.

Examples of sets where spectral synthesis fails on the line $\mathbb{R}$ were first given by Malliavin. His idea was to construct a function $f \in A(\mathbb{T})$ and a pseudomeasure defined formally as $\delta'(f)$, where δ' is the derivative of the Dirac measure. When $\langle \delta'(f), f \rangle \neq 0$, the set $f^{-1}(0)$ does not permit spectral synthesis. Level sets of specially constructed Gaussian processes $X(t)$ have this property: $\delta'(X(\cdot) - x)$ is a pseudomeasure that cannot be approximated in the weak topology of $PM(\mathbb{T})$ by measures carried within the support of $\delta'(X(\cdot) - x)$, with a positive probability that depends on x. It was noticed that, for these special Gaussian processes, this pseudomeasure belongs to $\mathcal{F}l^p$ whenever $p > 2$ [20]. (The statement in [20] is: Whatever $p > 2$, there exists a process.... It is easy to obtain: There exists a process such that, whatever $p > 2$, ...) It is very likely that, for these Gaussian processes, the pseudomeasure belongs to $M_\alpha(\mathbb{T})$, whatever $0 < \alpha < 1/2$ (perhaps also for $\alpha = 1/2$), but it needs some work to prove this.

10. Generic thin sets

I shall be brief on Baire's methods—as introduced by Robert Kaufman [37], developed in a series of ways, and revived recently by Thomas Körner [42]—because the matter is treated in [28]. However, it is worth mentioning the power and versatility of Baire's methods for obtaining strange closed sets on $\mathbb{R}$ or $\mathbb{T}$. (Definitions D5 to D10 extend to $\mathbb{R}$, so I shall consider $\mathbb{R}$ instead of $\mathbb{T}$ from now on.)

The first idea is that, contrary to the smoothing effect of random processes, using Baire's theorem accentuates the wildest behavior of Fourier transforms: Probability provides M sets, and even M_α sets, as we have seen; Baire provides sets of uniqueness. Actually, Baire's theorem is a very good tool for obtaining Kronecker sets [22] or, in several dimensions, Helson curves and surfaces. That at least was my personal philosophy until 1993.

Then T. Körner discovered that his Helson–M sets (so strange!) and also Salem sets (so far from Kronecker!) are also generic if suitable Baire spaces are chosen [42]. His ideas for the construction of Salem sets are expounded in [28]. Probability does not disappear, but it is reduced to a technicality.

In [42], Körner called Baire's theorem "a profound triviality." Its proof is trivial, and its use by Kaufman and Körner is profound. The use of probability methods relies on long experience in analysis and on the stochastic processes that we have at hand. To apply Baire's theorem, the first point is to discover the right Baire space; the second step is to chose the right open sets. It is a good way both to render strange objects generic and to find new phenomena.

I wish to thank Robert Ryan for a careful reading of this paper.

References

[1] R. Baire. Sur les fonctions discontinues qui rattachent aux fonctions continues. *C.R. Acad. Sci. Paris*, 126:1621–1623, 1898.

[2] R. Baire. Sur la théorie des ensembles. *C.R. Acad. Sci. Paris*, 129:946–949, 1899.

[3] R. Baire. *Leçons sur les fonctions discontinues*. Gauthier–Villars, Paris, 1905.

[4] S. Banach. *Théorie des opérations linéaire*. Z subwencji Funduszu kultury narodowej (Monografjie matematyczne, vol. 1), Warsaw, 1932.

[5] A. Benchérif-Madani. *Sur quelques propriétés de mesure géométrique de fractales associées à l'image d'un subordinateur*. PhD thesis, University of Paris, Orsay, December 1997.

[6] J. Bougain. Sidon sets and Riesz products. *Ann. Inst. Fourier (Grenoble)*, 35(1):136–148, 1985.

[7] J. Borgain. Bounded orthonormal systems and the $\Lambda(p)$-set problem. *Acta. Math.*, 162:227–245, 1988.

[8] A. Denjoy. Calcul de la primitive de la fonction dérivée le plus générale. *C.R. Acad. Sci. Paris*, 154:1075–1078, 1912.

[9] A. Denjoy. Calcul des coefficients d'une série trigonométrique convergente quelconque dont la somme est donnée. *C.R. Acad. Sci. Paris*, 172:1218–1221, 1921.

[10] A. Denjoy. *Leçons sur le calcul des coefficients d'une série trigonométrique*. Gauthier–Villars, Paris, 1941–1949. 4 volumes.

[11] P. du Bois-Reymond. Ueber die Fourier'schen Reihen. *Nachrichten von der Königlichten Gesellschaft der Wissenschaften und der G. A. Univ. zu Göttingen*, 21:571–582, 1873.

[12] A. Dvoretzky and P. Erdős. Divergence of random power series. *Michigan Math. Journal*, 6:343–347, 1959.

[13] P. Erdős. Remarks on a theorem of Zygmund. *Proc. London Math. Soc.*, 14 A:81–85, 1965.

[14] P. Fatou. Séries trigonométriques et séries de Taylor. *Acta Mathematica*, 30:335–400, 1906.

[15] G. Freud. Über trigonometrische Approximation und Fouriersche Reihen. *Math. Zeitschrift*, 78:252–262, 1962.

[16] H. Helson. Fourier transforms on perfect sets. *Studia Math.*, 14:209–213, 1954.

[17] S. Jaffard. On the Frisch–Parisi conjecture. *J. Math. Pures et Appliquées*, 76(6):525–552, 2000.

[18] J.-P. Kahane. Généralisation d'un théorème de Bernstein. *Bull. Soc. Math. France*, 85:221–229, 1957.

[19] J.-P. Kahane. Sur les fonctions moyenne-périodiques bornées. *Ann. Inst. Fourier (Grenoble)*, 7:293–314, 1957.

[20] J.-P. Kahane. Sur la synthèse harmonique dans l^∞. *Anais de Academia Brasileira de Ciêncios*, 22(2):179–189, 1960.

[21] J.-P. Kahane. Remarks on a theorem of Erdős. *Proc. London Math. Soc.*, 17:315–318, 1967.

[22] J.-P. Kahane. *Séries de Fourier absolument convergentes*. Ergebnisse der Mathematik, Band 50. Springer, Berlin, 1970.

[23] J.-P. Kahane. Sur les séries de Dirichlet $\sum \pm n^{-s}$. *C.R. Acad. Sci. Paris*, 276:739–742, 1973.

[24] J.-P. Kahane. *Some Random Series of Functions, 2nd edition*. Cambridge Univ. Press, 1985.

[25] J.-P. Kahane. Geza Freud and lacunary Fourier series. *J. Approx. Theory*, 46:51–57, 1986.

[26] J.-P. Kahane. Definition of stable laws, infinitely divisible laws, and Lévy processes. In *Lévy flights and related topics in physics, Lecture Notes in Physics, Procedings Nice, France 1994*. Springer, 1995.

[27] J.-P. Kahane. A century of interplay between Taylor series, Fourier series and Brownian motion. *Bull. London Math. Soc.*, pages 257–279, 1997.

[28] J.-P. Kahane. Baire's category theorem and trigonometric series. *Journal d'Analyse Mathématiques*, 80:143–182, 2000.

[29] J.-P. Kahane and P.-G. Lemarié-Rieusset. *Séries de Fourier et ondelettes*. Cassini, Paris, 1998.

[30] J.-P. Kahane and B. Mandelbrot. Ensembles de multiplicité aléatoires. *C.R. Acad. Sci. Paris*, 261:3931–3933, 1965.

[31] J.-P. Kahane and N. Nestoridis. Séries de Taylor et séries trigonométriques universelles au sens de Menchoff. *J. Math. Pures et Appliquées*, 2000.

[32] J.-P. Kahane and H. Queffelec. Order, convergence et sommabilité des produits de séries de Dirichlet. *Ann. Inst. Fourier (Grenoble)*, 47:485–529, 1997.

[33] J.-P. Kahane and R. Salem. *Ensembles parfaits et séries trigonométriques, 2ème édition*. Hermann, Paris, 1994.

[34] Y. Katznelson. Sur les fonctions opérant sur l'algèbre des séries de Fourier absolument convergentes. *C.R. Acad. Sci. Paris*, 247:404–406, 1958.

[35] Y. Katznelson. Suites aléatoires d'entiers. In *L'Analyse mathématique dans le domaine complexe*, Lecture Notes in Mathematics, Vol. 336, pages 148–152. Springer-Verlag, 1973.

[36] Y. Katznelson and P. Malliavin. Vérification statistique de la conjecture de dichotomie sur une classe d'algèbres de restriction. *C.R. Acad. Sci. Paris*, 262:490–492, 1966.

[37] R. Kaufman. A functional method for linear sets. *Israël J. Math.*, 5:185–187, 1967.

[38] R. Kaufman. M-sets and distributions. *Astérisque*, 5:225–230, 1973.

[39] S. Kierst and E. Szpilrajn. Sur certaines singularités des fonctions analytiques uniformes. *Fund. Math.*, 21:267–294, 1933.

[40] S.V. Konyagin. On divergence of trigonometric Fourier series everywhere. *C.R. Acad. Sci. Paris*, 329:693–697, 1999.

[41] T. Körner. A pseudofunction on a Helson set, I and II. *Astérisque*, 5:3–224, 231–239, 1973.

[42] T. Körner. Kahane's Helson curve. *J. Fourier Analysis*, Special Issue, Orsay 1993:325–346, 1995.

[43] H. Lebesgue. Sur la définition de l'aire d'une surface. *C.R. Acad. Sci. Paris*, 129:870–873, 1899.

[44] H. Lebesgue. *Leçons sur les séries trigonométriques*. Gauthier–Villars, Paris, 1906.

[45] D. Li, H. Queffelec, and L. Rodriquez-Piazza. Some new thin sets of integers in harmonic analysis. *Preprint*, 2000.

[46] B. Mandelbrot. *Multifractals and* $1/f$ *Noise*. Springer, 1998.

[47] V. Nestoridis. Universal Taylor series. *Ann. Inst. Fourier (Grenoble)*, 46:1293–1306, 1996.

[48] I. Netuka and J. Veselý. Sto let Baireovy věty o kategorích. Preliminary version, University of Prague, July 2000.

[49] G. Pisier. Ensembles de Sidon et processus gaussiens. *C.R. Acad. Sci. Paris*, 286:671–674, 1978.

[50] G. Pisier. Arithmetic characterizations of Sidon sets. *Bull. Amer. Math. Soc.*, 8:87–89, 1983.

[51] H. Queffelec. Propriétés presque sûres et quasi-sûres des séries de Dirichlet et des produits d'Euler. *Canad. J. Math.*, 32:531–558, 1980.

[52] D. Rider. Randomly continuous functions and Sidon sets. *Duke Math. J.*, 42:759–764, 1975.

[53] B. Riemann. Ueber die Darstellbarkeit einer Function durch einer trigonometrische Reihe. In *Gesammelte Mathematische Werke*. Leipzig, 1892. Habilitation, Göttingen 1854.

[54] H. Steinhaus. Les probabilités dénombrables et leur rapport à la théorie de la mesure. *Fund. Math.*, 4:286–310, 1923.

[55] H. Steinhaus. Über die Wahrscheinlichkeit dafür, dass der Konvergenzkreis einer Potenzreihe ihre natürliche Grenze ist. *Math. Zeitschrift*, 31:408–416, 1929.

[56] M. Weiss. Concerning a theorem of Paley on lacunary series. *Acta Math.*, 102:225–238, 1959.

[57] A. Zygmund. On the convergence on lacunary trigonometric series. *Fund. Math.*, 16:90–107, 1930.

Representations of Gabor frame operators

A.J.E.M. Janssen

Philips Research Laboratories
WY-81 5656 AA Eindhoven
The Netherlands
A.J.E.M.Janssen@philips.com

ABSTRACT. Gabor theory is concerned with expanding signals f as linear combinations of elementary signals that are obtained from a single function g (the window) by shifting it in time and frequency over integer multiples of a time shift parameter a and a frequency shift parameter b. In these expansion problems a key role is played by the Gabor frame operator associated with the set of elementary signals used in the expansions. The Gabor frame operator determines whether stable expansions exist for any finite-energy signal f (that is, whether we have indeed a frame), and, if so, gives a recipe for computing the expansion coefficients by using the canonically associated dual frame. In this contribution we consider the Gabor frame operator and associated dual frames in the time domain, the frequency domain, the time-frequency domain, and, for rational values of the sampling factor $(ab)^{-1}$, the Zak transform domain. We thus have the opportunity to address the basic problems – whether we have a Gabor frame and how we can compute a dual frame – in any of these domains we find, depending on g and a, b, convenient. The representations in the time domain and the frequency domain are conveniently discussed in the more general context of shift-invariant systems, and for this we present certain parts of what is known as Ron-Shen theory, adapted to our needs with emphasis on computational aspects.

This contribution contains many examples, counter-intuitive and confusing results, statements that one would like to be true but that are not and vice versa, etc., to show that Gabor theory, despite its rapid development in the last ten years, is still far from being completed.

1. Introduction

To introduce the subject matter of this contribution conveniently, we start with a survey of the various origins and early developments of Gabor theory. We do not aim here at a complete historical account, certainly not for the later developments, but rather refer for this to some excellent recent and less recent papers and books containing such surveys. Gabor systems

J.S. Byrnes (ed.), Twentieth Century Harmonic Analysis - A Celebration, 73–101.

are systems of functions of a real variable t that are built from a single function g (called the window) by shifting it in time and frequency over integer multiples of a time shift parameter a and a frequency shift parameter b. That is, denoting for real x, y by $g_{x,y}$ the time-frequency shifted version

$$(1.1) \qquad g_{x,y}(t) = e^{2\pi i y t} g(t - x) , \quad t \in \mathbb{R}$$

of g, a Gabor system with shift parameters a, b and window g consists of the functions $g_{na,mb}$ with integer n, m. We denote this system by (g, a, b). These systems were considered by Gabor [1] in 1946, with the window g a Gaussian and $(ab)^{-1} = 1$, with the aim of constructing efficient, time-frequency localized, non-redundant expansions of finite-energy signals as linear combinations of the system's elements in which the coefficients "represent" the expanded signal. Gabor's choice of Gaussian elementary building blocks and densities a^{-1}, b^{-1}, with product equal to unity was motivated by his desire for non-redundant, unique expansions that should exist for any finite-energy signal. Indeed, Gaussians uniquely achieve equality in the uncertainty inequality $\Delta t \cdot \Delta f \geq 1/2$ (with the deltas referring to the standard deviations in the time and the frequency domain, respectively), whence they occupy in a sense the least amount of area in the time-frequency plane. Furthermore, the setting of Nyquist's theorem, saying that band-limited signals are uniquely determined by their sample values at regularly spaced sample points with the spacing determined by the bandwidth, can be recast into a limiting case of a Gabor expansion problem, and this suggests to take $(ab)^{-1} = 1$ as critical density when non-redundant expansions have to exist for all signals.

Already much earlier, in 1932, systems (g, a, b) with Gaussian g and $(ab)^{-1} = 1$ were considered by von Neumann [2] in a quantum mechanical context, and, apparently, he established the completeness of these systems in L^2. For that reason one also finds in the literature the name von Neumann lattice systems, and also Weyl-Heisenberg systems to emphasize the underlying continuous Weyl-Heisenberg group of translations in the phase plane, for what we have called Gabor systems.

Gabor's 1946 paper certainly did not go unnoticed by the engineering community, but it was not until 1980 that the attention to Gabor expansions was revived through the work of Portnoff [3], Bastiaans [4] and Janssen [5]. This revival coincided, not entirely by accident, with the increasing interest in the electrical engineering community in time-frequency tools, such as the Wigner-Ville distribution and the short-time Fourier transform. (It should, however, be noted that as early as 1961 Lerner [6] presented a theory of signal representations in which Gabor expansions play a dominant role; in [6] one can already find orthogonalization procedures reminiscent of what we presently would call the construction of canonically associated dual frames.) From the von Neumann lattice side, completeness results for the Gaussian window and $(ab)^{-1} = 1$ were already obtained by Perelomov [7] and Bargmann et al. [8] using the Bargmann transform (and, in [8], the Zak transform in disguised form) and by Bacry, Grossmann and Zak [9] using the Zak transform.

Although Perelomov, in [7], already presented some considerations on dual functions, it was Bastiaans in [4] who analytically computed a dual function for the case of Gaussian g and $(ab)^{-1} = 1$. These dual functions are important since they allow one to exhibit the expansion coefficients for a particular signal f as inner products of f with the dual function shifted in a similar way as the window g itself. The mathematical analysis given by Janssen in [5] and [10] of the convergence properties of Gabor expansions and of Bastiaans' dual function showed that Gabor systems with Gaussian g and $(ab)^{-1} = 1$ yield unstable expansions that do not properly reflect time-frequency localization of the signals to be expanded. This point was also observed by Davis and Heller in 1979 [11]; they suggested to consider Gabor systems with Gaussian window g and $(ab)^{-1} > 1$ (oversampling) and thus obtained expansions with much better convergence properties.

The interest of mathematicians in Gabor systems dates from around 1980 with Janssen's work [5], [10] on the connection between the Bargmann transform, Zak transform and Gabor expansions, and that of Feichtinger (joined later on by Gröchenig) focusing on the more functional analytic (modulation spaces) and group theoretic aspects of Gabor expansions [12], [13]. A major development in the mathematical theory of Gabor expansions was made in 1986 by Daubechies, Grossmann and Meyer [14] who placed the Gabor expansion problem in the context of frames for a Hilbert space. The latter concept was introduced by Duffin and Schaeffer [15] in 1951 for addressing completeness and expansion problems involving sets of exponentials in spaces of band-limited functions. For a Gabor system (g, a, b) one thus considers the frame operator S, defined for $f \in L^2$ by

$$(1.2) \qquad Sf = \sum_{n,m} (f, g_{na,mb}) \, g_{na,mb} \, .$$

By definition, the Gabor system (g, a, b) is a frame when the frame operator S is bounded and positive definite. In this case, the Gabor system $(°\gamma, a, b)$ with

$$(1.3) \qquad °\gamma = S^{-1}g$$

is also a frame, called the canonical dual frame, and for any $f \in L^2$ we have the L^2-convergent expansions

$$(1.4) \qquad f = \sum_{n,m} (f, °\gamma_{na,mb}) \, g_{na,mb} = \sum_{n,m} (f, g_{na,mb}) \, °\gamma_{na,mb} \, .$$

Many of the research efforts in Gabor theory after 1986 were directed at studying frame operators, finding criteria for when a Gabor system is a frame, identification of the tight Gabor frames (for which g and $°\gamma$ coincide, except for a factor), and how to efficiently compute the canonical dual and the expansion coefficients in (1.4). These problems and their solutions have attracted many scientists from quite diverse disciplines and fields, such as theoretical electrical engineering, mathematical physics, Fourier analysis, numerical analysis, complex function theory, functional analysis, where it should be noted that especially the last field has increased its share of practitioners considerably over the last few years.

We have now arrived at a point where we are able to describe the technical content of this contribution, and so we finish the historical survey by pointing at two basic references for the developments in Gabor theory (before and) after 1986. These are Ch. 4 of the book [16] by Daubechies (having an enormous influence on the more recent developments in time-frequency analysis, and, in particular, Gabor theory) and the book [17], edited by Feichtinger and Strohmer, which is entirely devoted to Gabor analysis and applications with an extensive and up-to-date bibliography. For a survey of Gabor theory until 1989 and an excellent tutorial for both Gabor theory and wavelet theory, one should also consult the survey paper [18] by Heil and Walnut.

In this contribution we focus on the various representations of the frame operator S in (1.2). To that end, we first present the basics of frame theory, specialized to Gabor systems and to the more general shift-invariant systems. For the latter type of systems we provide our version of certain parts of a theory developed by Ron and Shen [19], [20] where we pay special attention to the issue of how to compute canonical dual systems. This is applied in two ways to Gabor systems, yielding a description of the frame bound conditions, the Gabor frame operator, the duality relation (1.4) and a characterization of and a computation method for the canonical dual function, both in the frequency domain and the time domain. The representation of the frame operator in the time domain is well known as Walnut's representation [21]. We shall also consider Gabor systems in the time-frequency domain using spectrograms, and this yields the Tolimieri-Orr-Janssen representation [22], [23] of the Gabor frame operator with a corresponding description of the frame bound conditions, the Wexler-Raz biorthogonality condition [24] for the duality relation (1.4), and a characterization of the canonical dual function as the minimum-energy Wexler-Raz dual. For rational values of the sampling factor $(ab)^{-1}$ we can also consider Gabor systems in the Zak transform domain which yields the Zibulski-Zeevi description [25] of the frame bound conditions, frame operator, duality relation and characterization/computation of the canonical dual function in terms of Zak matrices. Each of the four representations just given is potentially useful as a tool for finding out whether a Gabor system (g, a, b) is indeed a frame, and, if so, offers a means for computation of (canonical) dual functions in the considered domain.

We conclude this contribution with various counter-intuitive and confusing results, statements that one would obviously like to be true but that are not and vice versa, comments on the basic and hard problem of when a particular triple (g, a, b) is a Gabor frame, etc. As examples of this we have the Balian-Low theorem (conflicting with the relaxed attitude of von Neumann, Gabor himself and Lerner towards completeness, existence and convergence issues for the Gaussian window Gabor system at critical density) and the beating of this same Balian-Low phenomenon by considering Wilson systems at critical density; the existence of a well-behaved, positive g with positive Fourier transform such that $(g, \frac{1}{2}, 1)$ is not a Gabor frame; the difficulties of deciding whether (g, a, b) is a frame with $ab < 1$ for windows g as elementary as a Gaussian or the characteristic function of an interval. All this shows that

Gabor theory, despite the great progress that has been made in recent years, is still far from being completed, with various basic questions still waiting to be answered.

Almost all results presented here are proved somewhere in the literature; we shall therefore omit all proofs and we shall give appropriate references instead. In Secs. 2–6 of this contribution we follow roughly the developments of Secs. 1.1–5 of [17], Ch. 1; however, the presentation of the results has been considerably enhanced by adopting a uniform organization per section, while some of the results have been worked out in more detail.

2. Basics from frame theory

In this section we present some basic facts from frame theory, with particular attention for Gabor systems and shift-invariant systems. A shift-invariant system consists of a collection of functions g_{nm}, $n, m \in \mathbb{Z}$, of the form

$$(2.1) \qquad g_{nm}(t) = g_m(t - na) \, , \; t \in \mathbb{R} \, ,$$

where $g_m \in L^2$, $m \in \mathbb{Z}$, and $a > 0$. We are interested in finding dual systems γ_{nm}, $n, m \in \mathbb{Z}$, with $\gamma_{nm}(t) = \gamma_m(t - na)$, $t \in \mathbb{R}$, by which we mean that any $f \in L^2$ has the L^2-convergent expansions

$$(2.2) \qquad f = \sum_{n,m} (f, \gamma_{nm}) \, g_{nm} = \sum_{n,m} (f, g_{nm}) \, \gamma_{nm} \, .$$

For this to be meaningful, we require the two systems to have a finite frame upper bound. A system g_{nm}, $n, m \in \mathbb{Z}$, as in (2.1), has a finite frame upper bound when there is a $B_g < \infty$ such that

$$(2.3) \qquad \sum_{n,m} |(f, g_{nm})|^2 \leqslant B_g \, \|f\|^2 \, , \; f \in L^2 \, ,$$

and any $B_g < \infty$ such that (2.3) holds, is called a frame upper bound.

When g_{nm}, $n, m \in \mathbb{Z}$, has a finite frame upper bound B_g, one can define the operators T_g (analysis operator) and T_g^* (synthesis operator) by

$$(2.4) \qquad T_g : f \in L^2 \to T_g f = ((f, g_{nm}))_{n,m \in \mathbb{Z}} \in l^2(\mathbb{Z}^2)$$

and

$$(2.5) \qquad T_g^* : \alpha \in l^2(\mathbb{Z}^2) \to T_g^* \alpha = \sum_{n,m} \alpha_{nm} g_{nm} \in L^2 \, ,$$

respectively. These T_g and T_g^* are bounded linear operators with operator norm $\leqslant B_g^{\frac{1}{2}}$, and they are indeed adjoint operators when the standard inner products for L^2 and $l^2(\mathbb{Z}^2)$ are taken. When the system γ_{nm}, $n, m \in \mathbb{Z}$, has a finite frame upper bound as well, the duality condition (2.2) can be written as

$$(2.6) \qquad T_g^* T_\gamma = T_\gamma^* T_g = I \, ,$$

where I denotes the identity operator of L^2.

When the system g_{nm}, $n,m \in \mathbb{Z}$, has a finite frame upper bound B_g, the frame operator S_g is defined by $S_g = T_g^* T_g$. Explicitly,

$$(2.7) \qquad S_g : f \in L^2 \to S_g f = \sum_{n,m} (f, g_{nm}) \, g_{nm} \in L^2 \, ,$$

and there holds $S_g \leqslant B_g I$. When there is, in addition, an $A_g > 0$ such that

$$(2.8) \qquad \sum_{n,m} |(f, g_{nm})|^2 \geqslant A_g \, \|f\|^2 \, , \ f \in L^2 \, ,$$

so that S_g is invertible with $S_g \geqslant A_g I$, we say that the system g_{nm}, $n,m \in \mathbb{Z}$, has a positive frame lower bound, and any $A_g > 0$ such that (2.8) holds is called a frame lower bound. A system g_{nm}, $n,m \in \mathbb{Z}$, having both a finite frame upper bound and a positive frame lower bound is called a frame. When we have $A_g = B_g$ in (2.3) and (2.8), we say that the frame is tight, and then we have $S_g = A_g I = B_g I$.

When the system g_{nm}, $n,m \in \mathbb{Z}$, is a frame, a dual system is given by

$$(2.9) \qquad {}^\circ\gamma_{nm} = S_g^{-1} g_{nm} \, , \ n,m \in \mathbb{Z} \, ,$$

and this system is also a frame with frame bounds $A_{{}^\circ\gamma} = B_g^{-1}$, $B_{{}^\circ\gamma} = A_g^{-1}$. Since S_g, and hence S_g^{-1}, commutes with all relevant time-shift operators $f \in L^2 \to f(\cdot - na) \in L^2$, $n \in \mathbb{Z}$, we have that

$$(2.10) \qquad {}^\circ\gamma_{nm} = S_g^{-1} g_{nm} = (S_g^{-1} g_m)(\cdot - na) \, , \ n,m \in \mathbb{Z} \, .$$

We have, furthermore, that $S_g S_{{}^\circ\gamma} = I$, whence $S_{{}^\circ\gamma}$ is the inverse of the frame operator S_g and vice versa. The system in (2.9) is called the canonical dual system.

When the system g_{nm}, $n,m \in \mathbb{Z}$, is a frame, there are, in general, other dual systems γ_{nm}, $n,m \in \mathbb{Z}$, than the canonical dual system in (2.9). When we have two systems g_{nm}, $n,m \in \mathbb{Z}$, and γ_{nm}, $n,m \in \mathbb{Z}$, both with a finite frame upper bound, such that the duality condition (2.2) with L^2-convergence for all $f \in L^2$ holds, then both systems are a frame. This can be a useful method for checking whether a particular system g_{nm}, $n,m \in \mathbb{Z}$, is indeed a frame, viz. in those cases that one can easily produce a dual system γ_{nm}, $n,m \in \mathbb{Z}$, that does not need to be the canonical dual frame in (2.9).

The dual system in (2.9) is special for several reasons. For any $f \in L^2$ and any $\alpha \in l^2(\mathbb{Z}^2)$ with

$$(2.11) \qquad f = \sum_{n,m} \alpha_{nm} g_{nm} \, ,$$

there holds

$$(2.12) \qquad \sum_{n,m} |(f, {}^\circ\gamma_{nm})|^2 \leqslant \sum_{n,m} |\alpha_{nm}|^2 \, ,$$

with equality if and only if $\alpha_{nm} = (f, {}^\circ\gamma_{nm})$ for all $n, m \in \mathbb{Z}$. By applying this to the trivial representation for $n, m \in \mathbb{Z}$

$$(2.13) \qquad g_{nm} = \sum_{n',m'} (g_{nm}, {}^\circ\gamma_{n'm'}) g_{n'm'} = g_{nm} + \sum_{(n',m') \neq (n,m)} 0 \cdot g_{n'm'} \,,$$

we find that

$$(2.14) \qquad |(g_{nm}, {}^\circ\gamma_{nm})|^2 \leqslant \sum_{n',m'} |(g_{nm}, {}^\circ\gamma_{n'm'})|^2 \leqslant 1 \,.$$

A different way to characterize the dual system in (2.9) is as follows. Assume that $\alpha \in l^2$ is given and that we consider the α_{nm} as noisy/distorted versions of the numbers (f, g_{nm}), $n, m \in \mathbb{Z}$, of some $f \in L^2$. Then an estimate of f can be obtained by minimizing

$$(2.15) \qquad J(f) = \sum_{n,m} |(f, g_{nm}) - \alpha_{nm}|^2 \,.$$

When the system g_{nm}, $n, m \in \mathbb{Z}$, is a frame, this yields for f the unique solution

$$(2.16) \qquad f = S_g^{-1} \left(\sum_{n,m} \alpha_{nm} g_{nm} \right) = \sum_{n,m} \alpha_{nm} {}^\circ\gamma_{nm} \,.$$

In particular, when

$$(2.17) \qquad \alpha_{nm} = \delta_{nn_0} \delta_{mm_0} \,, \quad n, m \in \mathbb{Z}$$

with some $n_0, m_0 \in \mathbb{Z}$ (the deltas denote Kronecker's delta), we obtain $f = {}^\circ\gamma_{n_0m_0}$. For more generalities about frames and shift-invariant systems we refer to [16], Sec. 3.2, [26], Sec. I.C and [19], Sec. 1.3.

A particular example of a shift-invariant system arises when we take for $m \in \mathbb{Z}$

$$(2.18) \qquad g_m(t) = e^{2\pi i m b t} g(t) \,, \quad t \in \mathbb{R} \,,$$

with $b > 0$ and $g \in L^2$. It is customary here to ignore the phase factors in g_{nm}, γ_{nm}, given by $\exp(-2\pi i n m a b)$ for $n, m \in \mathbb{Z}$, when studying duality questions, since these vanish anyway at the right-hand sides of (2.2). Thus one considers

$$(2.19) \qquad g_{na,mb}(t) = e^{2\pi i m b t} g(t - na) \text{ vs. } g_{nm}(t) = e^{2\pi i m b (t - na)} g(t - na)$$

for $n, m \in \mathbb{Z}$, and one arrives at a Gabor system (g, a, b) as in Sec. 1 with window g and shift parameters $a > 0$, $b > 0$.

In the case of a Gabor frame (g, a, b), the frame operator S_g commutes with all relevant time-frequency shift operators $f \in L^2 \to \exp(2\pi i m b \cdot) f(\cdot - na) \in L^2$, $n, m \in \mathbb{Z}$. As a consequence we have then that

$$(2.20) \qquad {}^\circ\gamma_{nm} = S_g^{-1} g_{nm} = ({}^\circ\gamma)_{na,mb}$$

with ${}^\circ\gamma = S_g^{-1} g$ the canonical dual window.

3. Shift-invariant systems

In this section we consider shift-invariant systems g_{nm}, $n, m \in \mathbb{Z}$, and γ_{nm}, $n, m \in \mathbb{Z}$, and we present, in the frequency domain an equivalent condition for a shift-invariant system to have a finite frame upper bound and to be a frame, a representation result for the frame operators, an equivalent condition for the two systems to be dual, and a characterization of and a computation method for the canonical dual system. Many of the results in this section can be found in [19] by Ron and Shen. However, the presentation of the results we give here is rather different from the one in [19], and, for instance, the results on frame operator representation as well as those on the characterization and computation of canonical dual systems cannot be found in [19], at least not in the form we present them here. For full details and proofs we refer to [27], Sec. 1.2.

We consider $L^2 = L^2(\mathbb{R})$ with the standard inner product and norm

(3.1)
$$(f, h) = \int_{-\infty}^{\infty} f(t)\, h^*(t)\, dt \; ; \; \|f\|^2 = (f, f) \, , \; f, h \in L^2 \, .$$

Furthermore, we denote for $f \in L^2$ by $\hat{f} = \mathcal{F}f$ the Fourier transform of f, given as

(3.2)
$$\hat{f}(\nu) = (\mathcal{F}f)(\nu) = \int_{-\infty}^{\infty} e^{-2\pi i \nu t} f(t)\, dt \, , \; \text{a.e. } \nu \in \mathbb{R} \, .$$

With $f_m \in L^2$, $m \in \mathbb{Z}$, we define the "matrices"

(3.3)
$$H_g(\nu) := (\hat{g}_m(\nu - k/a))_{k \in \mathbb{Z}, m \in \mathbb{Z}} \, , \; \text{a.e. } \nu \in \mathbb{R} \, ,$$

whose k^{th} "row" consists of the sample values $\hat{g}_m(\nu - k/a)$, $m \in \mathbb{Z}$.

THEOREM 1. *The system g_{nm}, $n, m \in \mathbb{Z}$, has a finite frame upper bound B_g if and only if $H_g(\nu)$ and $H_g^*(\nu)$ define for a.e. $\nu \in \mathbb{R}$ a bounded linear operator of $l^2(\mathbb{Z})$ with operator norm $\leqslant (a\,B_g)^{\frac{1}{2}}$. In particular, there then holds for $k, m \in \mathbb{Z}$*

(3.4)
$$\sum_m |\hat{g}_m(\nu - k/a)|^2 \leqslant a\,B_g \, , \; \sum_k |\hat{g}_m(\nu - k/a)|^2 \leqslant a\,B_g$$

for a.e. $\nu \in \mathbb{R}$.

THEOREM 2. *Let $A \geqslant 0$, $B < \infty$. Then we have*

(3.5)
$$A\,\|f\|^2 \leqslant \sum_{n,m} |(f, g_{nm})|^2 \leqslant B\,\|f\|^2 \, , \; f \in L^2 \, ,$$

if and only if

(3.6)
$$A\,I \leqslant \frac{1}{a}\,H_g(\nu)\,H_g^*(\nu) \leqslant B\,I \, , \; \text{a.e. } \nu \in \mathbb{R} \, ,$$

where the I in (3.6) denotes the identity operator of $l^2(\mathbb{Z})$.

We observe that $H_g(\nu) H_g^*(\nu)$ is given as the "matrix"

$$(3.7) \qquad H_g(\nu) H_g^*(\nu) = \left(\sum_m \hat{g}_m(\nu - k/a) \hat{g}_m^*(\nu - l/a) \right)_{k,l \in \mathbb{Z}}, \text{ a.e. } \nu \in \mathbb{R},$$

and that for all $j, k, l \in \mathbb{Z}$ there holds

$$(3.8) \qquad (H_g(\nu - l/a) H_g^*(\nu - l/a))_{jk} = (H_g(\nu) H_g^*(\nu))_{l+j,l+k}, \text{ a.e. } \nu \in \mathbb{R}.$$

Hence for checking (3.6) it is sufficient to consider ν in an interval of length $1/a$.

Also note that the system g_{nm}, $n, m \in \mathbb{Z}$, is a tight frame if and only if there is a constant c such that for a.e. $\nu \in \mathbb{R}$

$$(3.9) \qquad \sum_m \hat{g}_m(\nu - k/a) \hat{g}_m^*(\nu - l/a) = c \delta_{kl}, \ k, l \in \mathbb{Z}.$$

THEOREM 3. *Assume that the system g_{nm}, $n, m \in \mathbb{Z}$, has a finite frame upper bound B_g, and let $f \in L^2$. Then we have, with S_g the frame operator,*

$$(3.10) \qquad \widehat{S_g f}(\nu) = \frac{1}{a} \sum_k d_k(\nu) \hat{f}(\nu - k/a), \text{ a.e. } \nu \in \mathbb{R},$$

with absolute convergence of the right-hand series for a.e. $\nu \in \mathbb{R}$. Here

$$(3.11) \qquad \begin{aligned} d_k(\nu) &= (H_g(\nu) H_g^*(\nu))_{0k} \\ &= \sum_m \hat{g}_m(\nu) \hat{g}_m^*(\nu - k/a), \text{ a.e. } \nu \in \mathbb{R}, \ k \in \mathbb{Z}. \end{aligned}$$

Because of (3.8) there holds, more generally, for $f \in L^2$ and a.e. $\nu \in \mathbb{R}$

$$(3.12) \qquad \widehat{S_g f}(\nu - l/a) = \frac{1}{a} \sum_k (H_g(\nu) H_g^*(\nu))_{lk} \hat{f}(\nu - k/a), \ l \in \mathbb{Z}.$$

This gives a frame operator representation in the Fourier domain in terms of the matrices $H_g(\nu) H_g^*(\nu)$ where $L^2(\hat{\mathbb{R}})$ is identified with $L^2([0, 1/a) \times \mathbb{Z})$. The relation (3.12) can be extended as follows. Let A_g be a frame lower bound for the system g_{nm}, $n, m \in \mathbb{Z}$, and let φ be a function analytic in an open set containing the closed segment $[A_g, B_g]$. Then we have for $f \in L^2$ and a.e. $\nu \in \mathbb{R}$

$$(3.13) \qquad \widehat{\varphi(S_g) f}(\nu - l/a) = \sum_k \left(\varphi\left(\frac{1}{a} H_g(\nu) H_g^*(\nu) \right) \right)_{lk} \hat{f}(\nu - k/a), \ l \in \mathbb{Z}.$$

In particular, when $A_g > 0$, so that g_{nm}, $n, m \in \mathbb{Z}$, is a frame, the choice $\varphi(x) = x^{-1}$ yields a representation of the inverse of the frame operator S_g according to

$$(3.14) \quad \widehat{S_g^{-1}f}(\nu - l/a) = \widehat{S_{\alpha_\gamma}f}(\nu - l/a)$$
$$= \sum_k \left(\frac{1}{a} H_g(\nu) H_g^*(\nu) \right)_{lk}^{-1} \hat{f}(\nu - k/a), \quad l \in \mathbb{Z},$$

for $f \in L^2$ and a.e. $\nu \in \mathbb{R}$. Specialization of (3.15) to the case $f = g_m$, where $m \in \mathbb{Z}$, and $l = 0$ yields for a.e. $\nu \in \mathbb{R}$

$$(3.15) \quad \hat{\overset{\circ}{\gamma}}_m(\nu) = \widehat{S_g^{-1}g_m}(\nu) = \sum_k \left(\frac{1}{a} H_g(\nu) H_g^*(\nu) \right)_{0k}^{-1} \hat{g}_m(\nu - k/a).$$

THEOREM 4. *Assume that the systems g_{nm}, $n, m \in \mathbb{Z}$, and γ_{nm}, $n, m \in \mathbb{Z}$, have finite frame upper bounds. Then the two systems are dual in the sense of (2.2) if and only if*

$$(3.16) \quad H_g(\nu) H_\gamma^*(\nu) = H_\gamma(\nu) H_g^*(\nu) = aI, \quad a.e. \ \nu \in \mathbb{R},$$

if and only if

$$(3.17) \quad \sum_m \hat{g}_m(\nu - k/a) \hat{\gamma}_m^*(\nu) = \sum_m \hat{\gamma}_m(\nu - k/a) \hat{g}_m^*(\nu)$$
$$= a\delta_{k0}, \ k \in \mathbb{Z}, \quad a.e. \ \nu \in \mathbb{R}.$$

Note that (3.16) says that $a^{-1}H_\gamma^*(\nu)$ is a right-inverse of $H_g(\nu)$ and that $a^{-1}H_\gamma(\nu)$ is a left-inverse of $H_g^*(\nu)$ for a.e. $\nu \in \mathbb{R}$. The next theorem shows that the canonical dual system $\overset{\circ}{\gamma}_{nm}$, $n, m \in \mathbb{Z}$, is special in the sense that $a^{-1}H_{\alpha_\gamma}^*(\nu)$ is "the" generalized inverse of $H_g(\nu)$ for a.e. $\nu \in \mathbb{R}$.

THEOREM 5. *Assume that g_{nm}, $n, m \in \mathbb{Z}$, is a frame. Then*

$$(3.18) \quad H_{\alpha_\gamma}^*(\nu) = a H_g^*(\nu)(H_g(\nu) H_g^*(\nu))^{-1}, \quad a.e. \ \nu \in \mathbb{R},$$

and

$$(3.19) \quad \left(\frac{1}{a} H_{\alpha_\gamma}(\nu) H_{\alpha_\gamma}^*(\nu) \right)^{-1} = \left(\frac{1}{a} H_g(\nu) H_g^*(\nu) \right)^{-1}, \quad a.e. \ \nu \in \mathbb{R}.$$

Theorem 5 can be made more explicit for the purpose of calculating the canonical dual functions $\overset{\circ}{\gamma}_m$ as follows.

THEOREM 6. *Assume that the system g_{nm}, $n, m \in \mathbb{Z}$, is a frame, and denote by $\mathbf{c}(\nu) \in l^2(\mathbb{Z})$ for a.e. $\nu \in \mathbb{R}$ the least-norm solution $\mathbf{c} = (c_m)_{m \in \mathbb{Z}}$ of the linear system*

$$(3.20) \quad \sum_m \hat{g}_m(\nu - k/a) c_m = a\delta_{k0}, \ k \in \mathbb{Z}.$$

Then there holds

$$(3.21) \quad \hat{\overset{\circ}{\gamma}}_m(\nu) = c_m^*(\nu), \ m \in \mathbb{Z}, \ a.e. \ \nu \in \mathbb{R}.$$

As a consequence of Theorem 6 we have the following. Assume that g_{nm}, $n, m \in \mathbb{Z}$, and γ_{nm}, $n, m \in \mathbb{Z}$, are dual frames. Then we have

(3.22)
$$\sum_m |{}^{\circ}\hat{\gamma}_m(\nu)|^2 \leqslant \sum_m |\hat{\gamma}_m(\nu)|^2 , \ \text{a.e. } \nu \in \mathbb{R} ,$$

with equality if and only if ${}^{\circ}\hat{\gamma}_m = \hat{\gamma}_m$ a.e.

4. Gabor systems as shift-invariant systems

In this section we specialize the results of Sec. 3 to the case of a Gabor system (g, a, b), so that we have now a $g \in L^2$ and a $b > 0$ such that

(4.1)
$$g_m(t) = e^{2\pi i m b t} g(t) , \ m \in \mathbb{Z}, \ t \in \mathbb{R} ;$$

as already said it is customary in Gabor theory to consider $g_{na,mb}$ rather than

$$g_{nm} = \exp(2\pi i m b(\cdot - na))g(\cdot - na).$$

This amounts to dropping the phase factors $\exp(-2\pi i n m a b)$. Since (g, a, b) is a Gabor system if and only if $(\hat{g}, b, a)$ is a Gabor system, we have two ways of specialization of the results of Sec. 3, yielding a description of the various notions and conditions in the frequency domain and the time domain, respectively.

4.1. Frequency domain results

We first show what form the results of Sec. 3 take when we choose the g_m as in (4.1). This then yields a description of the frame bound conditions, the frame operator, the duality condition and a characterization of and a computation method for the canonical dual function ${}^{\circ}\gamma$. Since now

(4.2)
$$\hat{g}_m(\nu) = (\mathcal{F}g_m)(\nu) = \hat{g}(\nu - mb) , \ m \in \mathbb{Z}, \ \text{a.e. } \nu \in \mathbb{R} ,$$

the "matrix" $H_g(\nu)$ in (3.3) is given by

(4.3)
$$H_g(\nu) = (\hat{g}(\nu - mb - k/a))_{k \in \mathbb{Z}, m \in \mathbb{Z}} ,$$
$$\text{a.e. } \nu \in \mathbb{R} ,$$

and the "matrix" $H_g(\nu) H_g^*(\nu)$ in Theorem 2 is given by

(4.4)
$$H_g(\nu) H_g^*(\nu) = \left(\sum_{m=-\infty}^{\infty} \hat{g}(\nu - mb - k/a) \hat{g}^*(\nu - mb - l/a) \right)_{k,l \in \mathbb{Z}}$$
$$\text{a.e. } \nu \in \mathbb{R} .$$

Hence, Theorem 1 gives that (g, a, b) is a Gabor system with a finite frame upper bound B_g if and only if $H_g(\nu)$ in (4.4) and $H_g^*(\nu)$ define for a.e. $\nu \in \mathbb{R}$ a bounded linear operator of $l^2(\mathbb{Z})$

with operator norm $\leqslant (a\,B_g)^{\frac{1}{2}}$; in particular, we then have that

(4.5) $$\sum_m |\hat{g}(\nu - mb)|^2 \leqslant a\,B_g\,,\ \sum_k |\hat{g}(\nu - k/a)|^2 \leqslant a\,B_g\,,\ \text{a.e. } \nu \in \mathbb{R}\,.$$

Furthermore, from Theorem 2 we see that (g, a, b) is a Gabor frame with frame bounds $A > 0$, $B < \infty$ if and only if the matrices $H_g(\nu)\,H_g^*(\nu)$ in (4.5) satisfy

(4.6) $$AI \leqslant \frac{1}{a}\,H_g(\nu)\,H_g^*(\nu) \leqslant BI\,,\ \text{a.e. } \nu \in \mathbb{R}\,.$$

Moreover, by Theorem 3 the frame operator of the Gabor frame (g, a, b) has the representation

(4.7) $$\widehat{S_g f}(\nu) = \frac{1}{a} \sum_k d_k(\nu)\,\hat{f}(\nu - k/a)\,,\ \text{a.e. } \nu \in \mathbb{R}\,,$$

for $f \in L^2$ with absolute convergence of the right-hand side series for a.e. $\nu \in \mathbb{R}$, in which the $d_k(\nu)$ are given by

(4.8) $$d_k(\nu) = \sum_m \hat{g}(\nu - mb)\,\hat{g}^*(\nu - mb - k/a)\,,\ k \in \mathbb{Z}\,,\ \text{a.e. } \nu \in \mathbb{R}\,.$$

Also, tightness of the frame (g, a, b) is equivalent with

(4.9) $$\sum_m \hat{g}(\nu - mb - k/a)\,\hat{g}^*(\nu - mb - l/a) = c\,\delta_{kl}\,,\ k, l \in \mathbb{Z}$$

for a.e. $\nu \in \mathbb{R}$ with c some constant. And duality of two Gabor systems (g, a, b) and (γ, a, b) having a finite frame upper bound is equivalent with

$$\sum_m \hat{g}(\nu - mb - k/a)\,\gamma^*(\nu - mb) = \sum_m \hat{\gamma}(\nu - mb - k/a)\,g^*(\nu - mb)$$
$$= a\,\delta_{k0}\,,\ k \in \mathbb{Z}\,,\ \text{a.e. } \nu \in \mathbb{R}\,.$$

Finally, the canonical dual functions $^\circ\gamma_m = \exp(2\pi i m b\cdot)\,^\circ\gamma$ can be found by using that

(4.10) $$^\circ\gamma(\nu) = \sum_k \left(\frac{1}{a}\,H_g(\nu)\,H_g^*(\nu)\right)^{-1}_{0k}\,\hat{g}(\nu - k/a)\,,\ \text{a.e. } \nu \in \mathbb{R}\,.$$

This canonical dual function $^\circ\gamma$ is minimal in the sense that for any dual frame (γ, a, b) we have for a.e. $\nu \in \mathbb{R}$ that

(4.11) $$\sum_m |^\circ\hat{\gamma}(\nu - mb)|^2 \leqslant \sum_m |\hat{\gamma}(\nu - mb)|^2\,,$$

with equality if and only if $^\circ\hat{\gamma}(\nu - mb) = \hat{\gamma}(\nu - mb)$, $m \in \mathbb{Z}$. When we integrate (4.11) over an interval of length b and use Parseval's theorem, we obtain

(4.12) $$\|^\circ\gamma\|^2 \leqslant \|\gamma\|^2$$

with equality if and only if $^\circ\gamma = \gamma$ a.e. Hence the canonical dual $^\circ\gamma$ has the least L^2-norm among all dual functions.

4.2. Time-domain results

We next show how the results of Sec. 3 can be applied to yield a description of the frame bound conditions, the frame operator, the condition of duality, and a characterization of and a computation method for the canonical dual $^\circ\gamma$ in the time domain. To do so, we note that

(4.13) $\qquad \mathcal{F}^{-1}[g_{na,mb}](\nu) = e^{2\pi inmab}(\breve{g})_{-mb,na}(\nu), \; n, m \in \mathbb{Z}, \text{ a.e. } \nu \in \mathbb{R}.$

Here $\mathcal{F}^{-1}$ denotes the inverse Fourier transform and $\breve{g}$ is the inverse Fourier transform of g, so that

(4.14) $\qquad \breve{g}(\nu) = (\mathcal{F}^{-1}g)(\nu) = \int_{-\infty}^{\infty} e^{2\pi i\nu t} g(t)\,dt, \text{ a.e. } \nu \in \mathbb{R}.$

Furthermore, we have for $f \in L^2$

(4.15) $\qquad (f, g_{na,mb}) = e^{-2\pi inmab}(\breve{f}, (\breve{g})_{-mb,na}), \; n, m \in \mathbb{Z}.$

Consequently, when the system (g, a, b) has a finite frame upper bound or a positive frame lower bound, then, by Parseval's theorem, so has the system $(\breve{g}, b, a)$, and the respective frame bounds can be taken equal. And, in the case of finite frame upper bounds, the two frame operators are related according to

(4.16) $\qquad S_{(g,a,b)}f = \widehat{S_{(\breve{g},b,a)}\breve{f}}, \; f \in L^2.$

Furthermore, the two Gabor systems (g, a, b) and (γ, a, b) are dual if and only if the systems $(\breve{g}, b, a)$ and $(\breve{\gamma}, b, a)$ are dual. Also, when (g, a, b) is a Gabor frame with canonical dual Gabor frame $(^\circ\gamma, a, b)$, then $(\breve{g}, b, a)$ is a Gabor frame as well with canonical dual $((^\circ\gamma)\breve{}, b, a)$ (for the latter fact we have used (4.16) together with $(\hat{h})\breve{} = h$ for $h \in L^2$).

Accordingly, we consider now the "matrix"

(4.17) $\qquad M_g(t) := (g(t - na - i/b))_{i\in\mathbb{Z},n\in\mathbb{Z}}, \text{ a.e. } t \in \mathbb{R},$

instead of the "matrix" $H_g(\nu)$ in (3.3) and (4.4), and in Theorem 2 and (4.5) we have now the "matrix"

(4.18) $\qquad M_g(t)\, M_g^*(t) = \left(\sum_{n=-\infty}^{\infty} g(t - na - i/b)\, g^*(t - na - j/b) \right)_{i,j\in\mathbb{Z}},$

$\qquad\qquad$ a.e. $t \in \mathbb{R}$,

instead of the matrix $H_g(\nu)\, H_g^*(\nu)$. Theorem 1 gives that (g, a, b) has a finite frame upper bound B_g if and only if $M_g(t)$ in (4.17) and $M_g^*(t)$ define for a.e. $t \in \mathbb{R}$ a bounded linear operator of $l^2(\mathbb{Z})$ with operator norm $\leqslant (b\,B_g)^{\frac{1}{2}}$; in particular, we then have that

$$(4.19) \qquad \sum_n |g(t-na)|^2 \leqslant b\,B_g \,, \quad \sum_i |g(t-i/b)|^2 \leqslant b\,B_g \,, \quad \text{a.e. } t \in \mathbb{R} \,.$$

Furthermore, by Theorem 2 we see that (g, a, b) is a frame with frame bounds $A > 0, B < \infty$ if and only if

$$(4.20) \qquad AI \leqslant \frac{1}{b}\, M_g(t)\, M_g^*(t) \leqslant BI \,, \quad \text{a.e. } t \in \mathbb{R} \,.$$

Next, by Theorem 3 the frame operator of the Gabor frame (g, a, b) has the representation

$$(4.21) \qquad (S_{(g,a,b)}f)(t) = \frac{1}{b} \sum_i e_i(t)\, f(t-i/b) \,, \quad \text{a.e. } t \in \mathbb{R} \,,$$

for $f \in L^2$ with absolute convergence of the right-hand side series for a.e. $t \in \mathbb{R}$, in which the $e_i(t)$ are given by

$$(4.22) \qquad e_i(t) = \sum_n g(t-na)\, g^*(t-na-i/b) \,, \quad i \in \mathbb{Z}, \text{ a.e. } t \in \mathbb{R} \,.$$

Also, tightness of the frame (g, a, b) is equivalent with

$$(4.23) \qquad \sum_n g(t-na-i/b)\, g^*(t-na-j/b) = c\,\delta_{ij} \,, \quad i,j \in \mathbb{Z} \,,$$

for a.e. $t \in \mathbb{R}$ with c some constant. Moreover, the duality condition between two Gabor frames (g, a, b) and (γ, a, b) can be expressed as

$$\sum_n g(t-na-i/b)\, \gamma^*(t-na) \;=\; \sum_n \gamma(t-na-i/b)\, g^*(t-na)$$

$$(4.24) \qquad\qquad\qquad\qquad\qquad = \; b\,\delta_{i0} \,, \quad i \in \mathbb{Z}, \text{ a.e. } t \in \mathbb{R} \,.$$

Finally, the canonical dual function $^\circ\gamma$ can be computed as

$$(4.25) \qquad {}^\circ\gamma(t) = \sum_i \left(\frac{1}{b}\, M_g(t)\, M_g^*(t) \right)^{-1}_{0i} g(t-i/b) \,, \quad \text{a.e. } t \in \mathbb{R} \,,$$

and this $^\circ\gamma$ is minimal in the sense that for any other dual function γ we have that $\|^\circ\gamma\|^2 \leqslant \|\gamma\|^2$ with equality if and only if $^\circ\gamma = \gamma$ a.e.

We note that the representation (4.21–4.22) of the frame operator S_g (with shift parameters a, b) is called the Walnut representation [21] of the frame operator. Note that this representation holds for any $f \in L^2$ with absolute convergence for a.e. t. A detailed study of the convergence of the right-hand side of (4.21) as an operator of L^2 has been carried out in [28]. A sufficient condition that the Gabor system has a finite frame upper bound while the representation (4.21) converges unconditionally is that g satisfies the CC-condition: there is an $M < \infty$ such that

$$(4.26) \qquad \sum_i |e_i(t)| \leqslant M \,, \text{ a.e. } t \in \mathbb{R} \,,$$

with the e_i given in (4.22), see [28], Theorem 4.1 and 6.9.

5. Gabor systems in the time-frequency domain

In this section we consider Gabor systems (g, a, b) with $g \in L^2$ and $a > 0, b > 0$ in the time-frequency domain. We shall thus obtain a description in the time-frequency domain of the frame bound conditions and the frame operator, of the duality condition and of the canonical dual function $°\gamma$. We define time-frequency shift operators U_{kl} for $k, l \in \mathbb{Z}$ by

$$(5.1) \qquad U_{kl}\, h = h_{k/b,l/a} \,, \quad h \in L^2 \,.$$

The proofs of the main results in this section can be found in [23], [27], Sec. 1.4, while many of these main results can also be found in [20], [29]. It should be noted that the approaches used in [23], [27], Sec. 1.4 and in [20] and in [29] are quite different; indeed, [20], [23] and [29] were written independently of one another and more or less simultaneously. We follow here the the approach in [23], [27], Sec. 1.4 which is based upon what we call the Fundamental Identity. This identity can be traced back to the work of Tolimieri and Orr [22], and the sharp form that we present below is due to Janssen, [30], Proof of Prop. A, [23], Props. 2.3 and 2.4, [27], Subsec. 1.4.1.

THEOREM 7 (Fundamental Identity). *Let $f^{(1)}, f^{(2)}, f^{(3)}, f^{(4)} \in L^2$, and assume that at least one of the systems $(f^{(1)}, a, b)$, $(f^{(2)}, a, b)$ and at least one of the systems $(f^{(3)}, a, b)$, $(f^{(4)}, a, b)$ has a finite frame upper bound. Also assume that*

$$(5.2) \qquad \sum_{k,l} |(f^{(3)}, f^{(2)}_{k/b,l/a})|\,|(f^{(1)}_{k/b,l/a}, f^{(4)})| < \infty \,.$$

Then

$$(5.3) \qquad \sum_{n,m} (f^{(1)}, f^{(2)}_{na,mb})(f^{(3)}_{na,mb}, f^{(4)}) = \frac{1}{ab} \sum_{k,l} (f^{(3)}, f^{(2)}_{k/b,l/a})(f^{(1)}_{k/b,l/a}, f^{(4)}) \,.$$

The proof of this result consists of a careful inspection of the proof of the Poisson summation formula for functions of two variables and their $2D$-Fourier transforms where the $2D$-Fourier transform pair

$$(5.4) \qquad (x, y) \rightarrow (f^{(1)}, f^{(2)}_{x,y})(f^{(3)}_{x,y}, f^{(4)}) \,;$$
$$(v, w) \rightarrow (f^{(3)}, f^{(2)}_{w,-v})(f^{(1)}_{w,-v}, f^{(4)})$$

is taken.

Now let $g \in L^2$ and define the linear mapping U_g of L^2 by

(5.5) $$U_g f = ((f, g_{k/b,l/a}))_{k,l \in \mathbb{Z}} \, , \quad f \in L^2 \, .$$

THEOREM 8. *The Gabor system (g, a, b) has a finite frame upper bound B_g if and only if U_g and U_g^* are bounded linear mappings of L^2 into $l^2(\mathbb{Z}^2)$ and $l^2(\mathbb{Z}^2)$ into L^2, respectively, with operator norms $\leqslant (abB_g)^{\frac{1}{2}}$. In particular, the Gabor system $(g, 1/b, 1/a)$ then has the finite frame upper bound abB_g.*

Note that the mapping U_g^* is given by

(5.6) $$U_g^* c = \sum_{k,l} c_{kl} \, g_{k/b,l/a} \, , \quad c \in l^2(\mathbb{Z}^2) \, .$$

THEOREM 9. *Let $A \geqslant 0$, $B < \infty$. Then we have*

(5.7) $$A \|f\|^2 \leqslant \sum_{n,m} |(f, g_{na,mb})|^2 \leqslant B \|f\|^2 \, , \quad f \in L^2 \, ,$$

if and only if

(5.8) $$AI \leqslant \frac{1}{ab} U_g U_g^* \leqslant BI \, ,$$

where I is now the identity operator of $l^2(\mathbb{Z}^2)$. That is, (g, a, b) is a frame if and only if $(g, 1/a, 1/b)$ is a Riesz basis for its linear span.

We observe that $U_g U_g^*$ maps $l^2(\mathbb{Z}^2)$ into $l^2(\mathbb{Z}^2)$ (when (5.8) holds), with matrix elements given by

(5.9) $$(U_g U_g^*)_{k,l;k',l'} = (g_{k'/b,l'/a}, g_{k/b,l/a}) \, , \quad k,l \in \mathbb{Z}; \, k',l' \in \mathbb{Z} \, .$$

Hence, the frame upper bound conditions and tightness of the Gabor frame (g, a, b) can be read off from the operator $U_g U_g^*$ whose matrix elements are given in (5.9). In particular, (g, a, b) is a tight frame if and only if

(5.10) $$(g_{k'/b,l'/a}, g_{k/b,l/a}) = c \, \delta_{kk'} \, \delta_{ll'} \, , \quad k,l \in \mathbb{Z}; \, k',l' \in \mathbb{Z} \, ,$$

for some constant c.

We next give a result, Theorem 10 below, on frame operator representation. We first introduce a norm-preserving mapping of L^2 into $L^2(\mathbb{Z}^2 \times [0, b^{-1}) \times [0, a^{-1}))$. Let h be any member of Schwartz space $\mathcal{S}$ with $\|h\| = 1$. Then the mapping $STFT_h$, defined for $f \in L^2$ by

(5.11) $$(STFT_h f)(x, y) = (f, h_{x,y}) \, , \quad y \in \mathbb{R} \, ,$$

is a norm-preserving mapping of L^2 into $L^2(\mathbb{R}^2)$. Now there holds, see (5.1), for $f \in L^2$, $k,l \in \mathbb{Z}$ and $x,y \in \mathbb{R}$ that

(5.12) $$(U_{kl}f, h_{x,y}) = (f, h_{x-k/b,y-l/a}) \, e^{-2\pi iky/b + 2\pi ikl/ab} \, .$$

Hence the mapping V_h, defined for $f \in L^2$ by

$$(5.13) \qquad (V_h f)(k, l; x, y) = (U_{kl} f, h_{x,y}), \quad k, l \in \mathbb{Z};$$
$$x \in [0, b^{-1}), \; y \in [0, a^{-1}),$$

is a norm-preserving mapping from L^2 into $L^2(\mathbb{Z}^2 \times [0, b^{-1}) \times [0, a^{-1}))$.

THEOREM 10. *When the system (g, a, b) has a finite frame upper bound B_g, we have for* $f \in L^2$

$$(V_h S_g f)(\cdot, \cdot; x, y) = \frac{1}{ab} (U_g U_g^*)^T (V_h f)(\cdot, \cdot; x, y),$$
$$(5.14) \qquad\qquad x \in [0, b^{-1}), \; y \in [0, a^{-1}),$$

where $(U_g U_g^*)^T$ is the transpose of the "matrix" $U_g U_g^*$ in (5.9).

Note that in this representation of the frame operator S_g the matrix $(ab)^{-1} (U_g U_g^*)^T$ is independent of x, y. The relation (5.14) extends as follows. Assume that A_g is a frame lower bound of the Gabor frame (g, a, b) and that φ is analytic on an open set containing the closed segment $[A_g, B_g]$. Then (5.14) holds with S_g at the left-hand side replaced by $\varphi(S_g)$ and $(ab)^{-1} (U_g U_g^*)^T$ at the right-hand side replaced by $\varphi((ab)^{-1} (U_g U_g^*)^T)$.

The representation in Theorem 10 can be rephrased more loosely as follows. Assume that the Gabor system (g, a, b) has a finite frame upper bound B_g. Then the frame operator S_g has the representation

$$(5.15) \qquad S_g = \frac{1}{ab} \sum_{k,l} (g, g_{k/b, l/a}) U_{kl}$$

in the sense that for any $f, h \in L^2$ such that $((U_{kl} f, h))_{k,l \in \mathbb{Z}} \in l^2(\mathbb{Z}^2)$ there holds

$$(5.16) \qquad (S_g f, h) = \frac{1}{ab} \sum_{k,l} (g, g_{k/b, l/a})(U_{kl} f, h).$$

In the case that g satisfies

$$(5.17) \qquad \text{condition } A : \; E := \sum_{k,l} |(g, g_{k/b, l/a})| < \infty$$

of Tolimieri and Orr [22], Sec. 3, the system (g, a, b) has the finite frame upper bound E/ab, and the convergence in (5.15) is without any proviso.

THEOREM 11. *Assume that the systems (g, a, b) and (γ, a, b) have finite frame upper bounds. Then the two systems are dual if and only if*

$$(5.18) \qquad U_g U_\gamma^* = U_\gamma U_g^* = ab I.$$

Moreover, we have for the canonical dual $°\gamma$ that

$$(5.19) \qquad U_{°\gamma}^* = ab U_g^* (U_g U_g^*)^{-1},$$

and

(5.20)
$$\frac{1}{ab} U_{^\circ\gamma} U_{^\circ\gamma}^* = \left(\frac{1}{ab} U_g U_g^*\right)^{-1} ,$$

so that the inverse frame operator has the representation

(5.21)
$$S_g^{-1} = S_{^\circ\gamma} = \sum_{k,l} \left(\frac{1}{ab} U_g U_g^*\right)^{-1}_{kl;00} U_{kl}$$

with the same proviso as in (5.15).

The duality condition in (5.18) can be made more explicit as follows. We have that two Gabor systems (g, a, b) and (γ, a, b), both with a finite frame upper bound, are dual if and only if

(5.22)
$$(\gamma, g_{k/b, l/a}) = ab\, \delta_{k0} \delta_{l0} , \quad k, l \in \mathbb{Z} .$$

This is a rigorous form of the celebrated Wexler-Raz biorthogonality condition [24]. Also, (5.19) can be made more explicit as follows. We have that $^\circ\gamma$ is the unique element $\gamma \in L^2$ of minimum norm such that (5.22) holds. The latter result has become known as the "Wexler-Raz dual equals the frame dual"-result.

6. Gabor systems in the Zak transform domain

We consider in this section Gabor systems (g, a, b) for the special case that $(ab)^{-1} = q/p$ with integer q and p satisfying $\gcd(q, p) = 1$.

Let $\lambda > 0$. We define for $h \in L^2$ the Zak transform $Z_\lambda h$ of h by

(6.1)
$$(Z_\lambda h)(t, \nu) = \lambda^{\frac{1}{2}} \sum_{k=-\infty}^{\infty} h(\lambda(t - k))\, e^{2\pi i k \nu} , \quad \text{a.e. } t, \nu \in \mathbb{R} .$$

The following properties hold for the Zak transform. Let $f, h \in L^2$. Then $Zf, Zh \in L^2_{\text{loc}}(\mathbb{R}^2)$, they are quasi-periodic according to

(6.2)
$$F(t + 1, \nu) = e^{2\pi i \nu} F(t, \nu) ; \quad F(t, \nu + 1) = F(t, \nu) , \quad \text{a.e. } t, \nu \in \mathbb{R} ,$$

and there holds

(6.3)
$$(f, h) = (Zf, Zh) ,$$

where the inner product on the right-hand side involves any unit square in $\mathbb{R}^2$. Furthermore, any $F \in L^2_{\text{loc}}$ satisfying (6.2) is of the form $F = Zf$ with some unique $f \in L^2$. Some other properties are

(6.4)
$$\lambda^{\frac{1}{2}} f(\lambda t) = \int_0^1 (Z_\lambda f)(t, \nu) , \quad \text{a.e. } t \in \mathbb{R} ,$$

and

(6.5) $$(Z_\lambda \hat{f})(t, \nu) = e^{2\pi i \nu t}(Z_{1/\lambda} f)(-\nu, t), \text{ a.e. } t, \nu \in \mathbb{R}.$$

Finally, any continuous F satisfying (6.2) has a zero in any unit square in $\mathbb{R}^2$. See [31] for these and many more properties of the Zak transform.

The usefulness of the Zak transform for description of frame bound conditions, frame operator, the duality condition and characterization and computation of the canonical dual function was recognized and elaborated by Zibulski and Zeevi, see, for instance, [25]. Also see [26], pp. 978 and 981 and [32]. We make the choice $\lambda = b^{-1}$ and suppress the subscript λ in Z_λ so that

(6.6) $$(Zh)(t, \nu) = b^{-\frac{1}{2}} \sum_{k=-\infty}^{\infty} h\left(\frac{t-k}{b}\right) e^{2\pi i k \nu}, \text{ a.e. } t, \nu \in \mathbb{R},$$

for $h \in L^2$. See [27], Subsec. 1.5.7, where it is shown that the choice $\lambda = a$ yields equally useful results.

We set for $f, h \in L^2$ and a.e. $t, \nu \in \mathbb{R}$

(6.7) $$\Phi^f(t, \nu) = p^{-\frac{1}{2}}\left((Zf)\left(t - l\frac{p}{q}, \nu + \frac{k}{p}\right)\right) \begin{matrix} k = 0, \ldots, p-1 \\ l = 0, \ldots, q-1 \end{matrix},$$

(6.8) $$A^{fh}(t, \nu) = \Phi^f(t, \nu)(\Phi^h(t, \nu))^*.$$

THEOREM 12. *The Gabor system* (g, a, b) *has a finite frame upper bound* B_g *if and only if* $\Phi^g(t, \nu)$ *and* $(\Phi^g(t, \nu))^*$ *are for a.e.* $t, \nu \in [0, 1)$ *bounded linear mappings of* $\mathbb{C}^q$ *into* $\mathbb{C}^p$ *and* $\mathbb{C}^p$ *into* $\mathbb{C}^q$, *respectively, with norm* $\leqslant B_g^{\frac{1}{2}}$. *In particular, there holds*

(6.9) $$|(Zg)(t, \nu)|^2 \leqslant p B_g, \text{ a.e. } t, \nu \in \mathbb{R}.$$

We note here that it thus follows that (g, a, b) has a finite frame upper bound if and only if Zg is essentially bounded.

THEOREM 13. *Let* $A \geqslant 0$, $B < \infty$. *Then we have*

(6.10) $$A \|f\|^2 \leqslant \sum_{n,m} |(f, g_{na,mb})|^2 \leqslant B \|f\|^2, \ f \in L^2,$$

if and only if

(6.11) $$A I_{p \times p} \leqslant A^{gg}(t, \nu) \leqslant B I_{p \times p}, \text{ a.e. } t, \nu \in \mathbb{R},$$

where $I_{p \times p}$ *denotes the identity operator of* $\mathbb{C}^p$.

We observe that one can restrict oneself in checking the condition (6.11) to the set $(t, \nu) \in [0, q^{-1}) \times [0, p^{-1})$. We also note that the Gabor frame (g, a, b) is tight if and only if

(6.12) $$A^{gg}(t, \nu) = c\, I_{p \times p}, \quad \text{a.e. } t, \nu \in \mathbb{R},$$

for some constant c.

THEOREM 14. *Assume that the Gabor system (g, a, b) has a finite frame upper bound. With S_g the frame operator, there holds for $f \in L^2$*

(6.13) $$\Phi^{S_g f}(t, \nu) = A^{gg}(t, \nu)\, \Phi^f(t, \nu), \quad \text{a.e. } t, \nu \in \mathbb{R}.$$

Theorem 14 gives the representation of the frame operator S_g in the Zak transform domain via the matrices Φ^f in (6.7). More generally, when A_g is a frame lower bound for (g, a, b) and φ is analytic in an open set containing the closed segment $[A_g, B_g]$, we can replace S_g at the left-hand side of (6.13) by $\varphi(S_g)$ and $A^{gg}(t, \nu)$ at the right-hand side of (6.13) by $\varphi(A^{gg}(t, \nu))$.

THEOREM 15. *Assume that the systems (g, a, b) and (γ, a, b) have finite frame upper bounds. Then the two systems are dual if and only if*

(6.14) $$\Phi^g(t, \nu)(\Phi^\gamma(t, \nu))^* = \Phi^\gamma(t, \nu)(\Phi^g(t, \nu))^* = I_{p \times p}, \quad \text{a.e. } t, \nu \in \mathbb{R}.$$

Moreover, we have for the canonical dual that

(6.15) $$(\Phi^{\circ\gamma}(t, \nu))^* = (\Phi^g(t, \nu))^*\, (\Phi^g(t, \nu)\, (\Phi^g(t, \nu))^*)^{-1}, \quad \text{a.e. } t, \nu \in \mathbb{R},$$

and

(6.16) $$A^{\circ\gamma\circ\gamma}(t, \nu) = (A^{gg}(t, \nu))^{-1}, \quad \text{a.e. } t, \nu \in \mathbb{R}.$$

The condition of duality can be written more explicitly as follows. The systems (g, a, b) and (γ, a, b), both having a finite frame upper bound, are dual if and only if for a.e. $t, \nu \in \mathbb{R}$ we have

(6.17) $$\frac{1}{p} \sum_{l=0}^{q-1} (Zg)\left(t - l\frac{p}{q}, \nu + \frac{k}{p}\right)(Z\gamma)^*\left(t - l\frac{p}{q}, \nu\right) = \delta_{k0}, \quad k = 0, ..., p-1.$$

Furthermore, for any dual system (γ, a, b) and a.e. $t, \nu \in \mathbb{R}$ we have

(6.18) $$\sum_{k=0}^{q-1} \left|(Z^\circ\gamma)\left(t + \frac{k}{q}, \nu\right)\right|^2 \leq \sum_{k=0}^{q-1} \left|(Z\gamma)\left(t + \frac{k}{q}, \nu\right)\right|^2,$$

with equality if and only if $(Z^\circ\gamma)(t + k/q, \nu) = (Z\gamma)(t + k/q, \nu)$ for $k = 0, ..., q-1$.

7. When is (g, a, b) a Gabor frame?

In this section we present a collection of results, comments, observations, (counter)examples, open problems, etc., on the basic problem of deciding whether a triple (g, a, b) with $g \in L^2$ and $a > 0, b > 0$ is a frame. While the finite frame upper bound condition is reasonably easy to deal with by imposing rather mild smoothness and decay conditions, the positivity of the frame lower bound presents a much harder problem. The basic problem can be considered in each of the four domains of Secs. 4–6, and any one of these domains can come with a particular advantage. By nature, this section is not as well organized as the preceding sections. Other authors might well have chosen to include different specific topics.

7.1.

We start with the well-known result that when $g \in L^2$ and (g, a, b) is a frame then we must have $(ab)^{-1} \geqslant 1$. This result has a long and complicated history, see for this [16], Sec. 4.1, [26], p. 978, [30], Sec. 1. There is a stronger result, due to Howe and Steger using results of Rieffel in [33], that says that completeness of the system (g, a, b) in L^2 implies that $(ab)^{-1} \geqslant 1$. In [34], Sec. 2, Benedetto, Heil and Walnut present a somewhat unsettling example of a $g \in L^2$ and an irregular Gabor system (i.e. the time-frequency points involved in the shifts do not form a lattice) of arbitrarily low density such that the system is complete in L^2, but not a frame. Restricting again to frames, several proofs of the fact that $(ab)^{-1} \geqslant 1$ when (g, a, b) is a frame are known now. When $(ab)^{-1} = q/p$ is rational with integer q, p such that $\gcd(q, p) = 1$, a simple rank consideration of the matrices in (6.7) and (6.11) suffices to show that $p \leqslant q$, i.e. $(ab)^{-1} \geqslant 1$. For general values of $(ab)^{-1}$, a simple proof can be based upon the Wexler-Raz biorthogonality condition (5.22) together with the inequality $|(g, \mathring{\gamma})| \leqslant 1$ that follows from (2.14). Yet another proof follows upon integrating the duality condition (4.10) for the canonical dual $\mathring{\gamma}$ over an interval of length b and using (as in the previous proof) that $|(g, \mathring{\gamma})| \leqslant 1$. It is the latter approach that can be generalized to shift-invariant systems, the windows of which have certain frequency localization properties, see [35].

7.2.

We now consider the case that $(ab)^{-1} = 1$. It is particularly convenient to discuss this case in the Zak transform domain since now we have $p = q = 1$ in $(ab)^{-1} = q/p$, so that the matrices in (6.7–6.8) reduce to scalars. As already said in Sec. 6, when Zg is continuous it must have a zero in any unit square. Hence, when $ab = 1$ and (g, a, b) is a frame, it cannot be true that g is continuous and rapidly decaying (for then Zg is continuous and has a zero, whence the lower frame bound in (6.11) is zero). Another result for the case $(ab)^{-1} = 1$ is the Balian-Low theorem (for the complicated history of this result, see [34], Subsec. 1.1), according to which at least one of $g'(t)$ and $tg(t)$ is not square integrable as a function of $t \in \mathbb{R}$ when (g, a, b) is a frame. Somewhat surprisingly, there is a construction involving cosines and sines rather than exponentials where one does get a frame at critical density

$(ab)^{-1} = 1$ with a well-behaved window g (Wilson bases, see [16], Subsec. 4.2.2 for history and details of the construction).

In the remainder of this section we shall consider, with few exceptions, the case that $(ab)^{-1} > 1$.

7.3.

We shall first indicate a class of windows g such that (g, a, b) is a Gabor frame. To that end we choose a continuous g, positive on and vanishing outside an interval $(-\frac{1}{2}c, \frac{1}{2}c)$, where c is any number in the non-empty interval (a, b^{-1}). Now the "matrix" $M_g(t) M_g^*(g)$ in (4.19) is a diagonal matrix with strictly positive diagonal elements

$$(7.1) \qquad D(t) = \sum_n |g(t - na)|^2 \, , \; t \in \mathbb{R} \, ,$$

that are bounded away from 0 and ∞. Hence by (4.20) there are the frame bounds b^{-1} min D, b^{-1} max D. One easily sees, furthermore, that a tight frame (h, a, b) is obtained by choosing $h = g/D^{\frac{1}{2}}$.

7.4.

There are a few cases where one can show that a system (g, a, b) with a finite frame upper bound is a Gabor frame by displaying a dual function γ such that (γ, a, b) has a finite frame upper bound. This is so, for instance, for the case of the Gaussian window

$$(7.2) \qquad g_\alpha(t) = (2\alpha)^{\frac{1}{4}} \exp(-\pi\alpha t^2) \, , \; t \in \mathbb{R} \, ,$$

and for the one-sided exponential considered in 7.5 below. In [30], Sec. 3, the Wexler-Raz biorthogonality condition (5.22) for $g = g_\alpha$ is written out and elaborated to yield, for any $\varepsilon > 0$ with $\varepsilon < \alpha^{-1}(1 - ab)$, the biorthogonal function (apart from a constant factor)

$$(7.3) \qquad \gamma_{\varepsilon,\alpha}(t) = \int_0^t e^{-\pi\alpha cs^2} \sum_{k=-\infty}^\infty (-1)^k \exp\left(\frac{-\pi a}{\varepsilon bc}\left(k + \frac{1}{2} - bcs\right)^2\right) ds \, ,$$

$$t \in \mathbb{R} \, ,$$

where $c = (\alpha\varepsilon + ab)^{-1} > 1$. It can be shown that this $\gamma_{\varepsilon,\alpha}$ can be extended to an entire function of $t = x + iy \in \mathbb{C}$ satisfying

$$(7.4) \qquad \gamma_{\varepsilon,\alpha}(x + iy) = O(\exp(-\pi\alpha x^2 + \pi\varepsilon^{-1}y^2)) \, , \; x, y \in \mathbb{R} \, .$$

This implies that both $\gamma_{\varepsilon,\alpha}$ and $\hat{\gamma}_{\varepsilon,\alpha}$ have Gaussian decay, just like g_α and $\hat{g}_\alpha$. Such a $\gamma_{\varepsilon,\alpha}$ can also be constructed by using the Bargmann transform, see, for instance [7], [8]. Interestingly, when $\alpha = 1$ and we take $\varepsilon \downarrow 0$ we obtain a function $\gamma_{0,1}$ that would coincide with Bastiaans singular function in [4] when $ab = 1$. The latter singular function can be shown to be in any $L^\infty \backslash L^p$ with $1 \leqslant p < \infty$, see [10], Subsec. 4.4.

The result that (g, a, b) is a frame for Gaussian g and $(ab)^{-1} > 1$, and several generalizations of it, has a rich history for which we refer to [16], Subsec. 3.4.4.B, [26], pp. 980–982, [30], Sec. 1. In [36], Lyubarskii and Seip give a very careful and detailed analysis of what happens for Gaussian g and $(ab)^{-1} = 1$ when the lattice of the relevant time-frequency points is slightly disturbed.

It is unlikely that any of the $\gamma_{\varepsilon,\alpha}$ in (7.4) coincides with the canonical dual $^\circ\gamma$. For even values of $(ab)^{-1}$ and $\alpha = 1$ the canonical dual $^\circ\gamma_\alpha$ was computed in [37], Sec. 6 as

$$(7.5) \qquad {}^\circ\gamma_\alpha(t) = \frac{ab}{\vartheta_3(\pi t/a; \exp(-\pi/2a^2))} \sum_{k=-\infty}^{\infty} c_k \exp(-\pi(t - k/b)^2),$$
$$t \in \mathbb{R},$$

with ϑ_3 a theta function, see [37], (6.5), and c_k certain numbers decaying like $\exp(-\pi|k|/2b^2)$ as $|k| \to \infty$. This $^\circ\gamma_\alpha$ cannot be extended to an entire function and it decays only like $\exp(-\pi|t|/2b)$ as $t \in \mathbb{R}$, $|t| \to \infty$. For non-integral values of $(ab)^{-1}$ it does not seem easy to determine the canonical dual (even the case that $(ab)^{-1}$ is an odd integer presents serious problems).

7.5.

We consider next the one-sided exponential

$$(7.6) \qquad {}_\alpha g(t) = (2\alpha)^{\frac{1}{2}} e^{-\alpha t} \chi_{[0,\infty)}(t), \; t \in \mathbb{R},$$

with $\alpha > 0$. One can again guess a dual $_\alpha\gamma$ by looking at the Wexler-Raz biorthogonality condition (5.22), and one obtains as a dual function

$$(7.7) \qquad {}_\alpha\gamma(t) = \frac{b}{\sqrt{2\alpha}} e^{\alpha t}(\chi_{[0,a)}(t) - \chi_{[-a,0)}(t)), \; t \in \mathbb{R}.$$

We note that this $_\alpha\gamma$ works also for the case that $(ab)^{-1} = 1$, whence $(_\alpha g, a, b)$ is a Gabor frame for any $a > 0, b > 0$ with $(ab)^{-1} \geqslant 1$. When $(ab)^{-1}$ is an integer, one can compute the canonical dual $^\circ_\alpha\gamma$, see [37], (4.11), as

$$^\circ_\alpha\gamma(t) = \frac{b}{\sqrt{2\alpha}} \frac{1 - e^{-2\alpha a}}{1 - e^{-2\alpha/b}} e^{\alpha t - 2\alpha a \lfloor t/a \rfloor} \times$$
$$(7.8) \qquad \qquad \times (\chi_{[0,1/b)}(t) - e^{-2\alpha/b} \chi_{[-1/b,0)}(t)), \; t \in \mathbb{R},$$

and this $^\circ_\alpha\gamma$ differs from the $_\alpha\gamma$ in (7.7) (unless $(ab)^{-1} = 1$).

We note that in [37] there are some more specific examples, such as two-sided exponentials, hyperbolic secants, for which it is shown that they yield a Gabor frame with integer $(ab)^{-1} > 1$.

7.6.

While 1 is the lower bound for $(ab)^{-1}$ so that (g, a, b) can be a frame, it appears that the chances of having a frame increase with increasing value of $(ab)^{-1}$. For integer values N of $(ab)^{-1}$ this is apparent from Sec. 6, since now the matrix $A^{gg}(t, \nu)$ is a scalar $(p = 1, q = N)$, given by

$$(7.9) \qquad A^{gg}(t, \nu) = \sum_{l=0}^{N-1} \left| (Zg)\left(t - \frac{l}{q}, \nu \right) \right|^2 , \text{ a.e. } t, \nu \in \mathbb{R} ,$$

and this quantity is positive and bounded for many windows g, including certain smooth and rapidly decaying g's. More precisely, according to [26], Theorems 2.5–6, for any sufficiently well-behaved g there are $a_c > 0$, $b_c > 0$ such that (g, a, b) is a frame when $0 < a < a_c$, $0 < b < b_c$. Also see [13], Part 1, Theorem 6.1. Such a result can also be obtained from the frame operator representation (5.15) (holding under condition A in (5.17)) when $(g, g_{x,y})$ decays sufficiently rapidly when $x^2 + y^2 \to \infty$. Then in the right-hand side of (5.15) the terms with $(k, l) \neq (0, 0)$ are small compared to the term $(ab)^{-1}I$, corresponding to $(k, l) = (0, 0)$, as $a^{-2} + b^{-2} \to \infty$. When one allows g's that are not well-behaved, one gets problems with results of this type. In [38] there is constructed for any irrational α a smooth, bounded $g \in L^2$ such that for any rational $a > 0$, $b > 0$ the system (g, a, b) has a finite frame upper bound while for any $\beta > 0$ and any rational $c > 0$ the system $(g, c\alpha, \beta)$ has no such bound. Also in [38] there is an example of a g, bounded and supported by $[0, 1]$, such that the above a_c, b_c do not exist, and an example of a g such that 0 is accumulation point of points a such that (g, a, b) has frame lower bound 0 and, at the same time, accumulation point of points a such that (g, a, b) has frame lower bound $\geqslant 1$ (arbitrary $b \in (0, 1)$).

7.7.

It seems hard to find a general condition on windows $g \in L^2$ (with reasonable smoothness and decay properties) ensuring (g, a, b) to be a frame for all or even some specific values of $a > 0$, $b > 0$ with $(ab)^{-1} > 1$. In [39] two such classes of windows, for integer values of $(ab)^{-1}$, are found. The first class consists of all g supported, positive, strictly decreasing and continuous and integrable on $[0, \infty)$, for which $(ab)^{-1}$ is allowed to take any positive integer $\geqslant 1$ as its value. This class contains the one-sided exponentials in 7.5. The second class consists of all even, positive, continuous and integrable g such that g has on $[0, \infty)$ the form

$$(7.10) \qquad g(t) = b(t) + b(t + 1) , \; t \geqslant 0 ,$$

with b strictly convex and positive on $[0, \infty)$. In these classes, the respective conditions of strictness cannot be weakened yielding, for instance, a well-behaved rapidly decaying g such that both g and $\hat{g}$ are strictly positive while $(g, \frac{1}{2}, 1)$ is not a frame. We also observe that the Gaussians do not belong to either one of these two classes.

7.8.

An example of a g that just fails to belong to the first class in 7.7 is the characteristic function $\chi_{[0,c)}$ of the interval $[0, c)$. Assuming without restriction that $b = 1$ the author has obtained the following results for this g. The frame upper bound condition is always satisfied in this case, and we can restrict to $a < 1$, $c > 1$ for it is not difficult to show that

(a) $c < a$ or $a > 1 \Rightarrow$ no frame,
(b) $a \leqslant c \leqslant 1 \Rightarrow$ frame,
(c) $a = 1, c > 1 \Rightarrow$ no frame.

Assuming $a < 1$, $c > 1$ one can show furthermore that

(d) $a \notin \mathbb{Q}, c \in (1, 2) \Rightarrow$ frame,
(e) $a = p/q \in \mathbb{Q}$, $\gcd(p, q) = 1$, $2 - \frac{1}{q} < c < 2 \Rightarrow$ no frame,
(f) $a > \frac{3}{4}, c = L - 1 + L(1 - a)$ with integer $L \geqslant 3 \Rightarrow$ no frame,
(g) $|c - \lfloor c \rfloor - \frac{1}{2}| < \frac{1}{2} - a \Rightarrow$ frame.

While (f) shows that one may fail to have a frame for irrational a, it seems to hold that one does have a frame when a, c, a/c are irrational. And for rational $a = p/q$, $\gcd(p, q) = 1$, one can give an algorithm that determines whether one has a frame for any $c > 1$ (complexity determined by q, p). In particular one thus can find rational $a > \frac{1}{2}$ and $c > 1$ such that one has a frame.

This shows that the answer to the basic question is already rather bewildering for windows g as elementary as a characteristic function.

7.9.

We continue by giving some comments on inheritance of certain desirable properties of a $g \in L^2$ for which (g, a, b) is a frame by the canonical dual. We recall from Subsec. 4.1 the CC-condition of essential boundedness of $\sum_i |e_i(t)|$, with e_i given by (4.22), guaranteeing the Walnut representation (4.21) of the frame operator to converge unconditionally as an operator of L^2. It is an open problem whether ${}^{\circ}\gamma$ inherits the CC-condition from g. In [28], Theorem 4.14, it is shown that for rational values of $(ab)^{-1}$ a slightly stronger condition, viz. the uniform CC-condition, is inherited by ${}^{\circ}\gamma$ from g.

A similar situation occurs for condition A of Tolimieri and Orr, see (5.17). While it is an open problem whether this condition A is inherited for all values of $(ab)^{-1} > 1$, it has been shown in [32] that it does so for rational values of $(ab)^{-1}$.

7.10.

The next topic concerns the construction of a tight frame canonically associated to a Gabor frame. Given a Gabor frame (g, a, b) with frame operator S_g, the Gabor frame (h, a, b) with

(7.11)
$$h = S_g^{-\frac{1}{2}} g$$

is tight. It should be noted that this h can be computed using the functional calculus (with $\varphi(x) = x^{-\frac{1}{2}}$) for the frame operator in the various domains (see the comments after Theorems 3, 10 and 14). For instance, when $(ab)^{-1} = q/p$ is rational, the h in (7.11) is given in the Zak transform domain according to

$$(7.12) \qquad \Phi^h(t, \nu) = (A^{gg}(t, \nu))^{-\frac{1}{2}} \Phi^g(t, \nu) \,, \text{ a.e. } t, \nu \in \mathbb{R} \,.$$

In the particular case that $(ab)^{-1}$ is an integer N (so that $q = N, p = 1$) we get

$$(7.13) \qquad \Phi^h(t, \nu) = \frac{\Phi^g(t, \nu)}{\left(\sum_{l=0}^{N-1} \left| (Zg)\left(t + \frac{l}{N}, \nu\right) \right|^2 \right)^{\frac{1}{2}}} \,, \text{ a.e. } t, \nu \in \mathbb{R} \,,$$

i.e.

$$(7.14) \qquad (Zh)(t, \nu) = \frac{(Zg)(t, \nu)}{\left(\sum_{l=0}^{N-1} \left| (Zg)\left(t + \frac{l}{N}, \nu\right) \right|^2 \right)^{\frac{1}{2}}} \,, \text{ a.e. } t, \nu \in \mathbb{R} \,.$$

7.11.

A further interesting point is the problem of finding out whether certain decay and smoothness properties of a g generating a Gabor frame are inherited by the canonical dual $°\gamma$ or by the tight frame generating h of (7.11). In [23], see Properties 5.5–6, it is shown, by using Banach algebra methods, that when $g \in S$ generates a Gabor frame, then both $°\gamma$ and h of (7.11) are in S as well. A different proof of this fact, as pointed out by K.-H. Gröchenig, can be based on the frame operator representations of Sec. 4 together with a result of S. Jaffard [40] on inheritance of decay of the off-diagonal elements of an invertible operator of $l^2(\mathbb{Z})$ by the inverse and the inverse square root of the operator. For inheritance of exponential decay in the time and/or frequency domain the latter approach has been worked out in [41], Secs. 2 and 4.

There are some more results in [41] on inheritance of smoothness and decay by $°\gamma$ and h from g. For instance, it is shown that in the case of integer values of $(ab)^{-1}$, there is no inheritance of faster-than-exponential decay unless we are dealing with tight frames. Whether there do exist tight frames (g, a, b) with a g having faster-than-exponential decay in both the time domain and the frequency domain is an open problem.

7.12.

We finally make some comments on discrete-time Gabor systems. We now take $N, M \in \mathbb{N}$ and $g \in l^2(\mathbb{Z})$, and we consider the system of sequences $g_{nN,m/M}$ with $n \in \mathbb{Z}$, $m = 0, 1, ..., M - 1$, defined by

$$(7.15) \qquad g_{nN,m/M} = (e^{2\pi i mr/M} g(r - nN))_{r \in \mathbb{Z}} \,.$$

It turns out, see [27], Sec. 1.6, that the theory of discrete-time Gabor systems with respect to frame bound conditions, frame operator representation, the duality condition and characterization and computation of canonical dual systems can be developed in largely the same way as for continuous-time Gabor systems.

A basic question that arises here is whether one can obtain a discrete-time Gabor frame by sampling the windows of a continuous-time Gabor frame $(g, N, 1/M)$ at the integers, and how the respective frame operators, representations, canonical duals etc. are related. In [42] it is shown that (and how) this can be done under a decay condition and a regularity condition on the window g. These conditions are the condition A of Tolimieri and Orr, see (5.17), for the system $(g, N, 1/M)$ and the regularity condition

$$(7.16) \qquad R: \quad \lim_{\varepsilon \downarrow 0} \sum_{r=-\infty}^{\infty} \frac{1}{\varepsilon} \int_{-\frac{1}{2}\varepsilon}^{\frac{1}{2}\varepsilon} |g(r+u) - g(r)|^2 \, du = 0 \, .$$

It is shown in [42] that when $(g, N, 1/M)$ is a Gabor frame with g satisfying A and R, then the canonical dual function ${}^{\circ}\gamma$ satisfies A and R as well (so that it can be sampled at the integers, just like g). Furthermore, denoting the sampling operation by a superscript D, we have that $(g^D, N, 1/M)$ is a discrete-time Gabor frame with the same frame bounds as the continuous-time frame. Denoting the frame operator of $(g^D, N, 1/M)$ by S_g^D, we have for a large class of $h \in L^2$ (including g and ${}^{\circ}\gamma$) that

$$(7.17) \qquad (S_g h)^D = S_g^D h^D \, .$$

In particular we get that ${}^{\circ}\gamma^D$ is the canonical dual sequence for the discrete-time Gabor system $(g^D, N, 1/M)$.

Acknowledgement

The author wishes to thank H.G. Feichtinger and D. Walnut for a careful reading of an early version of Sec. 1 of this contribution. In addition, the author is most grateful to Jurgen Rusch for a superb hit-and-run action making the LaTeX-work of this contribution ready for integration in the book.

References

[1] D. Gabor, Theory of communication, J. Inst. Elec. Eng. (London), 93 (1946), pp. 429–457.

[2] J. von Neumann, "Mathematical Foundations of Quantum Mechanics", Princeton, 1955, p. 407.

[3] M.R. Portnoff, Time-frequency representation of digital signals and systems based on short-time Fourier analysis, IEEE Trans. ASSP, 28 (1980), pp. 55–69.

[4] M.J. Bastiaans, Gabor's expansion of a signal into Gaussian elementary signals, Proc. IEEE, 68 (1980), pp. 538–539.

[5] A.J.E.M. Janssen, Gabor representation of generalized functions, J. Math. Anal. Appl., 83 (1981), pp. 377–394.

[6] R.M. Lerner, Representations of signals, Ch. 10 in "Lectures in Communication System Theory", ed. E.J. Baghdady, McGraw-Hill, 1961.

[7] A.M. Perelomov, Remark on the completeness of the coherent states system, Teoret. Mat. Fiz., 6 (1971), pp. 213–224.

[8] V. Bargmann, P. Butera, L. Girardello and J.R. Klauder, On the completeness of the coherent states, Rep. Math. Phys., 2 (1971), pp. 221–228.

[9] H. Bacry, A. Grossmann and J. Zak, Proof of the completeness of lattice states in the kq-representation, Phys. Rev. B, 12 (1975), pp. 1118–1120.

[10] A.J.E.M. Janssen, Bargmann transform, Zak transform, and coherent states, J. Math. Phys., 23 (1982), pp. 720–731.

[11] M.J. Davis and E.J. Heller, Semiclassical Gaussian basis set method for molecular vibrational wave functions, J. Chem. Phys., 71 (1979), pp. 3383–3395.

[12] H.G. Feichtinger, On a new Segal algebra, Monatsh. Math., 92 (1981), pp. 269–289.

[13] H.G. Feichtinger and K.-H. Gröchenig, Banach spaces related to integrable group representations and their atomic decompositions, J. Funct. Anal., 86 (1989), pp. 307–340 (part 1) and Monatsh. Math., 108 (1989), pp. 129–148 (part 2).

[14] I. Daubechies, A. Grossmann and Y. Meyer, Painless non-orthogonal expansions, J. Math. Phys., 27 (1986), pp. 1271–1283.

[15] R.J. Duffin and A.C. Schaeffer, A class of nonharmonic Fourier series, Trans. Amer. Math. Soc., 72 (1952), pp. 341–366.

[16] I. Daubechies, "Ten Lectures on Wavelets", SIAM, Philadelphia, 1992.

[17] H.G. Feichtinger and T. Strohmer (eds.), "Gabor Analysis and Algorithms", Birkhäuser, Boston, 1998.

[18] C. Heil and D. Walnut, Continuous and discrete wavelet transforms, SIAM Review, 31 (1989), pp. 628–666.

[19] A. Ron and Z. Shen, Frames and stable bases for shift-invariant subspaces of $L^2(\mathbb{R}^d)$, Canadian J. Math., 47 (1995), pp. 1051–1094.

[20] A. Ron and Z. Shen, Weyl-Heisenberg systems and Riesz bases in $L^2(\mathbb{R}^d)$, Duke Math. J., 89 (1997), pp. 237–282.

[21] D. Walnut, Continuity properties of the Gabor frame operator, J. Math. Anal. Appl., 165 (1992), pp. 479–504.

[22] R. Tolimieri and R.S. Orr, Poisson summation, the ambiguity function and the theory of Weyl-Heisenberg frames, J. Fourier Anal. Appl., 1 (1995), pp. 233–247.

[23] A.J.E.M. Janssen, Duality and biorthogonality for Weyl-Heisenberg frames, J. Fourier Anal. Appl., 1 (1995), pp. 403–436.

[24] J. Wexler and S. Raz, Discrete Gabor expansions, Signal Processing, 21 (1990), pp. 207–220.

[25] M. Zibulski and Y.Y. Zeevi, Oversampling in the Gabor scheme, IEEE Trans. Signal Proc., 41 (1993), pp. 2679–2687.

[26] I. Daubechies, The wavelet transform, time-frequency localization and signal analysis, IEEE Trans. Inform. Theory, 36 (1990), pp. 961–1005.

[27] A.J.E.M. Janssen, The duality condition for Weyl-Heisenberg frames, Ch. 1 in "Gabor Analysis and Applications", eds. H.G. Feichtinger and T. Strohmer, Birkhäuser, Boston, 1998.

[28] P.G. Casazza, O. Christensen and A.J.E.M. Janssen, Weyl-Heisenberg frames, translation invariant systems and the Walnut representation, submitted to J. Funct. Anal., May 1999.

[29] I. Daubechies, H.J. Landau and Z. Landau, Gabor time-frequency lattices and the Wexler-Raz identity, J. Fourier Anal. Appl., 1 (1995), pp. 437–478.

[30] A.J.E.M. Janssen, Signal analytic proofs of two basic results on lattice expansions, Appl. Comp. Harm. Anal., 1 (1994), pp. 350–354.

[31] A.J.E.M. Janssen, The Zak transform: a signal transform for sampled, time-continuous signals, Philips J. Res., 43 (1988), pp. 23–69.

[32] A.J.E.M. Janssen, On rationally oversampled Weyl-Heisenberg frames, Signal Processing, 47 (1995), pp. 239–245.

[33] M.A. Rieffel, Von Neumann algebras associated with pairs of lattices in Lie groups, Math. Ann., 257 (1981), pp. 403–418.

[34] J.J. Benedetto, C. Heil and D. Walnut, Differentiation and the Balian-Low theorem, J. Fourier Anal. Appl., 1 (1995), pp. 344–402.

[35] A.J.E.M. Janssen, A density theorem for time-continuous filter banks, pp. 513–523 in "Signal and Image Representation in Combined Spaces", eds. Y. Zeevi and R. Coifman, Academic Press, San Diego, 1998.

[36] Y.I. Lyubarskii and K. Seip, Convergence and summability of Gabor expansions at the Nyquist density, J. Fourier Anal. Appl., 5 (1999), pp. 127–157.

[37] A.J.E.M. Janssen, Some Weyl-Heisenberg frame bound calculations, Indag. Mathem., 7 (1996), pp. 165–182.

[38] H.G. Feichtinger and A.J.E.M. Janssen, Validity of WH-frame bound conditions depends on lattice parameters, Appl. Comp. Harm. Anal., 8 (2000), pp. 104–112.

[39] A.J.E.M. Janssen, On Zak transforms with few zeros, preprint (1999).

[40] S. Jaffard, Propriétés des matrices bien localisées près de leur diagonale et quelques applications, Ann. Inst. Henri Poincaré, 7 (1990), pp. 461–476.

[41] H. Bölcskei and A.J.E.M. Janssen, Gabor frames, unimodularity and window decay, to appear in J. Fourier Anal. Appl. (2000).

[42] A.J.E.M. Janssen, From continuous to discrete Weyl-Heisenberg frames through sampling, J. Fourier Anal. Appl., 3 (1997), pp. 583–596.

Does Order Matter

T. W. Körner

DPMMS
16 Mill Lane
Cambridge
twk@pmms.cam.ac.uk

ABSTRACT. I describe the work of Olevskiĭ, Tao and others on the rearrangement of orthogonal series. It turns out that arbitrary rearrangements produce trouble for all orthogonal and wavelet methods, that decreasing rearrangements produce trouble for Fourier series, but that wavelet expansions continue to work well under decreasing rearrangement.

1. Introduction

In this paper we shall move without comment between the circle $\mathbb{T} = \mathbb{R}/\mathbb{Z}$, the closed interval $[0, 1]$ and the half closed interval $[0, 1)$ as seems most convenient. When our results deal with almost everywhere behaviour this does not create any problems, when we want everywhere convergence readers may have to produce their own argument to deal with the point 1.

What do we mean when we consider

$$\sum_{u=-\infty}^{\infty} \hat{f}(u) \exp 2\pi i u t \, ?$$

Traditionally we take

$$\lim_{N \to \infty} \sum_{u=-N}^{N} \hat{f}(u) \exp 2\pi i u t$$

and the only variations that we allow concern the mode of convergence (pointwise, L^p, etc). A sign that this may be too narrow an approach appears when we consider the two dimensional case of a function $f : \mathbb{T}^2 \to \mathbb{C}$. The obvious way to proceed is to take

$$\sum_{(u,v) \in \mathbb{Z}^2} \hat{f}(u,v) \exp 2\pi i (ut + vs) = \lim_{N \to \infty} \sum_{(u,v) \in \Gamma(N)} \hat{f}(u,v) \exp 2\pi i (ut + vs)$$

103

J.S. Byrnes (ed.), Twentieth Century Harmonic Analysis - A Celebration, 103–125.

with the $\Gamma(N)$ finite subsets of $\mathbb{Z}^2$ such that $\Gamma(1) \subseteq \Gamma(2) \subseteq \ldots$ and $\bigcup_{N=1}^{\infty} \Gamma(N) = \mathbb{Z}^2$. However, it is well known that, even when f is quite well behaved, different choices of the sequence $\Gamma(N)$ give rise to different behaviour.

The question of 'correct order' also arises, even in the one dimensional case, from signal processing. If we seek to store and reconstruct a function $f : \mathbb{T} \to \mathbb{C}$ by using its Fourier coefficients it is natural to to use them in decreasing order of magnitude and to consider

$$\sum_{|\hat{f}(u)| \geqslant \eta} \hat{f}(u) \exp 2\pi iut$$

rather than

$$\sum_{|u| \leqslant N} \hat{f}(u) \exp iut.$$

Any easy optimism about rearrangements is quenched by the following result.

THEOREM 1. *There exists a real $f \in L^2(\mathbb{T})$ and a bijection $\sigma : \mathbb{Z} \to \mathbb{Z}$ such that*

$$\limsup_{N \to \infty} \left| \sum_{u=-N}^{N} \hat{f}(\sigma(u)) \exp i\sigma(u)t \right| = \infty.$$

for almost all $t \in \mathbb{T}$.

This theorem was first stated by Kolmogorov. A proof of Kolmogorov's statement was sketched by Zahorskiĭ, given in detail by Ulyanov and much simplified by Olevskiĭ. The reader who consults [7] will find an excellent bibliography.

Pólya says that sometimes the easiest way to prove a result is to generalise it and prove the generalisation. By the time Kolmogorov's theorem reached Ulyanov and Olevskiĭ it had taken the following form.

THEOREM 2. *Let ϕ_1, ϕ_2, ϕ_3, ... form a complete orthonormal system in $L^2([0,1))$. Then there exists a real $f \in L^2(\mathbb{T})$ and a bijection $\sigma : \mathbb{N} \to \mathbb{N}$ such that*

$$\limsup_{N \to \infty} \left| \sum_{u=0}^{N} \hat{f}(\sigma(u)) \phi_{\sigma(u)}(t) \right| = \infty.$$

for almost all $t \in [0,1)$.

Here as usual we write

$$\hat{f}(\phi_r) = \langle f, \phi_r \rangle = \int_0^1 f(t) \phi_r(t) \, dt.$$

In fact Olevskiĭ improved Theorem 2 by replacing $f \in L^2(\mathbb{T})$ by f continuous. We shall prove a result which includes this as Theorem 5.

In the years since Olevskiĭ published his result, more general systems than those of complete orthonormal type have assumed practical importance.

DEFINITION 3. *We say that*
$$\phi_1, \phi_2, \phi_3, \ldots$$
form a Riesz basis for $L^2([0,1))$ if the linear span of the ϕ_n is dense in L^2 and there exists an A with $A \geqslant 1$ such that

$$A^{-1} \sum_{n=0}^{\infty} |a_n|^2 \leqslant \left\| \sum_{n=0}^{\infty} a_n \phi_n \right\|_2^2 \leqslant A \sum_{n=0}^{\infty} |a_n|^2.$$

We call *A* the Riesz constant of the system. Easy functional analysis reveals the following lemma.

LEMMA 4. *Let $\phi_1, \phi_2, \phi_3, \ldots$ form a Riesz basis for $L^2([0,1))$. Then there exists a unique sequence $\psi_1, \psi_2, \psi_3, \ldots$ of bounded L^2 norm such that $\sum_{n=0}^{\infty} |\langle \psi_n, f \rangle|^2 < \infty$ and*

$$f = \sum_{n=0}^{\infty} \langle \psi_n, f \rangle \phi_n$$

for every $f \in L^2$. If $f = \sum_{n=0}^{\infty} a_n \phi_n$ then $a_n = \langle \psi_n, f \rangle$ for all n.

We write $\hat{f}(n) = \langle \psi_n, f \rangle$.

As Olevskiĭ indicates very clearly his method can be extended to Riesz bases.

THEOREM 5. *Let $\phi_1, \phi_2, \phi_3, \ldots$ form a Riesz basis for $L^2([0,1])$. Then there exists a real continuous f and a bijection $\sigma : \mathbb{N} \to \mathbb{N}$ such that*

$$\limsup_{N \to \infty} \left| \sum_{u=0}^{N} \hat{f}(\sigma(u)) \phi_{\sigma(u)}(t) \right| = \infty.$$

for almost all $t \in [0,1]$.

If we ask for which complete orthonormal system Theorem 2 is least plausible one answer would be the Haar system. Set

$$\Lambda(N) = \{(r,s) \in \mathbb{Z}^2 : 0 \leqslant s \leqslant 2^r - 1 \text{ and } N \geqslant r \geqslant 0\}$$

and $\Lambda = \bigcup_{N \geqslant 0} \Lambda(N)$. Consider intervals of the form

$$E(r,s) = [s2^{-r}, (s+1)2^{-r})$$

where $(r,s) \in \Lambda$. We define

$$\begin{aligned}
\chi_{r,s}(t) &= 1 && \text{if } t \in E(r+1, 2s) \\
\chi_{r,s}(t) &= -1 && \text{if } t \in E(r+1, 2s+1) \\
\chi_{r,s}(t) &= 0 && \text{otherwise}
\end{aligned}$$

where, again, $(r,s) \in \Lambda$. We call the $\chi_{r,s}$ together with the function $1 = \chi_{-1,0}$ the Haar system. We call the $T_{r,s} = 2^{r/2} \chi_{r,s}$ together with the function 1 the normalised Haar system.

It is well known that the normalised Haar system is a complete orthonormal system. If we put the standard order

$$(r, s) \ll (r', s') \text{ if } r' > r \text{ or } r = r' \text{ and } s' \geqslant s$$

on the Haar system it enjoys remarkable convergence properties. For example if f is continuous

$$\sum_{(u,v) \ll (r,s)} \hat{f}(u,v) T_{u,v} \to f$$

uniformly as we allow (r, s) to increase so as to exhaust the system.

If we can prove Theorem 2 for the Haar system, one is tempted to say, we can surely prove it for any orthonormal system. Olevskiĭ showed that this is indeed the case and we shall see that once we have Theorem 2 for the Haar system we can obtain the rest of the theorem from this particular case. As we might expect on general grounds the key to the Haar system case lies in a finite version of the theorem.

LEMMA 6. *Let $\epsilon > 0$ and $K \geqslant 1$ be given. Then we can find a bijection*

$$\sigma : \{1, 2, 3, \ldots, 2^{N+1} - 1\} \to \Lambda(N)$$

and real $a_{r,s} [(r, s) \in \Lambda(N)]$ such that

$$\left| \sum_{(r,s) \in \Lambda(N)} a_{r,s} \chi_{r,s}(t) \right| \leqslant 1$$

for all $t \in [0, 1)$ but

$$\max_{1 \leqslant k \leqslant 2^{N+1}-1} \left| \sum_{j=1}^{k} a_{\sigma(j)} \chi_{\sigma(j)}(t) \right| \geqslant K$$

for all $t \notin E$ where E is a set of measure at most ϵ.

The next section is devoted to the proof of this key lemma. Once it is understood the rest is relatively routine.

2. Olevskiĭ's Lemma

We shall need two simple results, the first combinatorial and the second probabilistic.

LEMMA 7. *If $a_1, a_2, \ldots, a_m \geqslant 0$ and $a_{m+1}, a_{m+2}, \ldots, a_N \leqslant 0$ then*

$$\max_{1 \leqslant j \leqslant N} \left| \sum_{r=1}^{j} a_r \right| \geqslant \frac{1}{3} \sum_{r=1}^{N} |a_r|.$$

PROOF. Observe that

$$\sum_{r=1}^{N} |a_r| = 2 \sum_{r=1}^{m} a_r - \sum_{r=1}^{N} a_r \leqslant 3 \max_{1 \leqslant j \leqslant N} \left| \sum_{r=1}^{j} a_r \right|.$$

□

LEMMA 8. *Let $X_1, X_2, \ldots, X_N$ be independent identically distributed random variables with $\Pr(X_r = 1) = \Pr(X_r = -1) = 1/2$ and let K be an integer with $K \geqslant 1$. Then*

(i) $\Pr \left(\max_{1 \leqslant j \leqslant N} \sum_{r=1}^{j} X_r \geqslant K \right) \leqslant 2 \Pr \left(\sum_{r=1}^{N} X_r \geqslant K \right).$

(ii) $\Pr \left(\max_{1 \leqslant j \leqslant N} \left| \sum_{r=1}^{j} X_r \right| \geqslant K \right) \leqslant 4 \Pr \left(\left| \sum_{r=1}^{N} X_r \right| \geqslant K \right).$

(iii) $\Pr \left(\max_{1 \leqslant j \leqslant N} \left| \sum_{r=1}^{j} X_r \right| \geqslant K \right) \leqslant 2N K^{-2}.$

PROOF. (i) This is the simplest form of the reflection principle. (See e.g. [2] Chapter 3.)

(ii) Use symmetry.

(iii) Use Chebychev's inequality to bound $\Pr \left(\left| \sum_{r=1}^{N} X_r \right| \geqslant K \right).$

□

Of course, we can get much better estimates in (iii) but the only use I can find for such estimates is to study the Hausdorff dimension of the exceptional set of convergence in Theorem 2 when applied to Haar functions.

By using the standard interpretation of Haar functions in terms of coin tossing, Lemma 8 gives the following result.

LEMMA 9. *If N and K are strictly positive integers then we can find $a_{p,q}$ taking the values 0 or 1 $[(p, q) \in \Lambda(N)]$ and a set E of measure at most $4N K^{-2}$ such that*

$$\left| \sum_{(p,q) \in \Lambda(N)} a_{p,q} \chi_{p,q}(t) \right| \leqslant K$$

for all $t \in [0, 1)$, but

$$\sum_{(p,q) \in \Lambda(N)} |a_{p,q} \chi_{p,q}(t)| = N$$

for all $t \notin E$.

PROOF. Consider $[0, 1)$ with Lebesgue measure as a probability space. Let

$$X_j(t) = (-1)^{[2^j t]}$$

(where $[2^j t]$ is the integer part of $2^j t$). The X_j satisfy the conditions of Lemma 8. It follows by Lemma 8 (iii) that the set

$$E = \left\{ t : \max_{1 \leqslant j \leqslant N} \left| \sum_{r=1}^{j} X_r \right| \geqslant K \right\}$$

has measure at most $4NK^{-2}$.

To define $a_{p,q}$ we look at the interval $E(p,q) = [p2^{-q}, (p+1)2^{-q})$ and observe that $\sum_{r=1}^{j} X_r(t)$ is constant on $E(p,q)$ for all $1 \leqslant j \leqslant q-1$. We may thus define $a_{p,q} = 1$ if

$$\max_{1 \leqslant j \leqslant N} \left| \sum_{r=1}^{j} X_r(t) \right| \leqslant K - 1$$

for $t \in E(p,q)$, and $a_{p,q} = 0$ otherwise.

If we now set

$$Y(t) = \sum_{(p,q) \in \Lambda(N)} a_{p,q} \chi_{p,q}(t)$$

then Y is the random variable defined by

$$Y(t) = \sum_{r=1}^{N} X_r(t)$$

if $| \max_{1 \leqslant j \leqslant N} \sum_{r=1}^{j} X_r(t) | \leqslant K$ but

$$Y(t) = \sum_{r=1}^{j(t)} X_r(t)$$

where $j(t)$ is the smallest j with $| \sum_{r=1}^{j(t)} X_r(t) | = K$. In more vivid terms, toss a fair coin keeping track of the difference between the number of heads and tails thrown. If this ever takes the value K or $-K$ stop and record the value as Y. If, after N throws this has not happened, record the value after N throws as Y. By definition $|Y(t)| \leqslant K$ and if $t \notin E$ (so we complete the N throws)

$$\sum_{(p,q) \in \Lambda(N)} |a_{p,q} \chi_{p,q}(t)| = N$$

as required.   □

Now, instead of taking the sequence of heads and tails as chance presents them, we wish to take all the heads first and then all the tails. To do this we introduce the Olevskiĭ order on Λ

$$(p', q') \succeq (p, q) \text{ if } (2p' + 1)2^{-q'} \geqslant (2p + 1)2^{-q}.$$

The Olevskiĭ order is more intricate than may at first appear and the reader should try ordering the (p, q) with $(p, q) \in \Lambda(N)$ for $N = 5$ or $N = 6$. Observe that $(p', q') \succeq \chi(p, q)$ if the mid-point of supp $\chi_{p',q'}$ is to the right of (i.e. greater than or equal to) the mid-point of supp $\chi_{p,q}$. We observe that $\chi_{p,q}(t) \geq 0$ if t is strictly to the left of (i.e. strictly less than) the mid-point of supp $\chi_{p,q}$ and $\chi_{p,q}(t) \leq 0$ if t is to the right of (i.e. greater than or equal to) the mid-point of supp $\chi_{p,q}$. It follows that for each $t \in [0, 1)$ there is a $\lambda(t) \in \Lambda(N)$ such that

$$\chi_{p,q}(t) \leq 0 \text{ for} \lambda \succeq (p, q), \ \chi_{p,q}(t) \geq 0 \text{ for} (p, q) \succeq \chi_\lambda.$$

Using this observation we easily arrive at the result we require.

LEMMA 10. *Suppose that N and K are strictly positive integers and $a_{p,q}$ and E are as in Lemma 9. Then*

$$\max_{\lambda \in \Lambda(N)} \left| \sum_{\lambda \succeq (p,q)} a_{p,q} \chi_{p,q}(t) \right| \geq N/3$$

for all $t \notin E$.

PROOF. By Lemma 7 and the last sentence of the previous paragraph

$$\max_{\lambda \in \Lambda(N)} \left| \sum \lambda \succeq (p, q) a_{p,q} \chi_{p,q}(t) \right| \geq \frac{1}{3} \sum_{(p,q) \in \Lambda(N)} |a_{p,q} \chi_{p,q}(t)| = N/3$$

for all $t \notin E$. $\square$

THEOREM 11. *Let $\epsilon > 0$ and $\kappa > 1$ be given. Then we can find an $N \geq 1, 1 \geq b > 0$, $b_{p,q}$ taking the values 0 or b [$(p, q) \in \Lambda(N)$] and a set E such that*

(i) $|\sum_{(p,q) \in \Lambda(N)} b_{p,q} \chi_{p,q}(t)| \leq 1$ for all $t \in [0, 1)$,
(ii) $\max_{\lambda \in \Lambda(N)} |\sum_{\lambda \succeq (p,q)} a_{p,q} \chi_{p,q}(t)| \geq \kappa$ for all $t \notin E$
(iii) $|E| < \epsilon$.

PROOF. Choose an integer M with $M > 4\epsilon^{-1}$ and $M^2 \geq 3\kappa$. Set $N = M^5$, $K = M^3$ and choose $a_{p,q}$ and E as in Lemma 9. If we now put $b_{p,q} = a_{p,q}/K$, $b = 1/K$ then all the conclusions of the lemma with the exception of condition (ii) follow at once from Lemma 9. Condition (ii) itself follows from Lemma 10. $\square$

Standard 'rolling hump' (or 'condensation of singularities') methods now give the following result.

EXERCISE 12. *Let χ_n [$n \in \mathbb{N}$] be the Haar system enumerated in some way. Then there exists a real $f \in L^\infty([0, 1))$ and a bijection $\sigma : \mathbb{N} \to \mathbb{N}$ such that*

$$\limsup_{N \to \infty} \left| \sum_{u=0}^{N} \hat{f}(\sigma(u)) \chi_{\sigma(u)}(t) \right| = \infty.$$

for almost all $t \in [0, 1)$.

We leave it as an exercise for the reader because we shall prove stronger results. However these results will require extra complications in their proofs and the reader may find the more complex proofs easier to follow if he or she has worked through an easier case.

3. Extension To General Spaces

We have not yet exhausted the strength of Pólya's dictum. What happens if we seek to extend Theorem 2 to more general measure spaces? A moment's reflection brings to mind the the fact that, so far as measure theory is concerned, all nice measure spaces which are not obviously different are the same. Recall that a measure space $(X, \mathcal{F}, \mu)$ with positive measure μ is *non-atomic* if given $E \in \mathcal{F}$ with $\mu(E) > 0$ we can find $F \in \mathcal{F}$ with $F \subseteq E$ and $\mu(E) > \mu(F) > 0$. The following result is typical (see e.g. theorem 9, page 327 of [8]).

THEOREM 13. *Let X be a complete separable metric space equipped with $\mathcal{B}_X$ its σ-algebra of Borel sets and μ a non-atomic measure on $\mathcal{B}_X$. If $I = [0, 1]$ is the unit interval with its usual metric, $\mathcal{B}_I$ its σ-algebra of Borel sets, and m is the usual Lebesgue measure then there exists a bijective map $F : I \to X$ such that F and F^{-1} carry Borel sets to Borel sets of the same measure.*

We shall not use Theorem 13 but we shall use the lemma which underlies its proof and the proof of results like it.

LEMMA 14. *Let $(X, \mathcal{F}, \mu)$ be a non-atomic probability space. If $E \in \mathcal{F}$ and $1 \geqslant \alpha \geqslant 0$ then we can find $F \in \mathcal{F}$ with $E \supseteq F$ and $\mu(F) = \alpha\mu(E)$.*

Lemma 17 follows easily from a lemma of Saks given as Lemma 7 of section IV.9.8 of Dunford and Schwartz [1]. There is a discussion of isomorphism theorems in Chapter VIII of Halmos's *Measure Theory* [3]. However before readers rush off to inspect the wilder shores of measure theory they should note that all they will learn is that they could have stayed at home, since $[0, 1]$ with Lebesgue measure is the type of all non-atomic measure spaces.

In view of the preceding discussion, it is natural to aim at the following generalisation of Theorem 2. (The extension of Definition 3 to general probability spaces is obvious.)

THEOREM 15. *Let $(X, \mathcal{F}, \mu)$ be a probability space with μ non-atomic. Let $\phi_1, \phi_2, \ldots$ be a Riesz basis in $L^2(X)$. Then there exists a real $f \in L^2(X)$ and a bijection $\sigma : \mathbb{N} \to \mathbb{N}$ such that*

$$\limsup_{N \to \infty} \left| \sum_{u=0}^{N} \hat{f}(\sigma(u))\phi_{\sigma(u)}(t) \right| = \infty.$$

for almost all $t \in X$.

The result is clearly false if μ is not non-atomic. Suppose $E \in \mathcal{F}$ is an atom, that is $\mu(E) > 0$ and if $F \in \mathcal{F}$ with $F \subseteq E$ then $\mu(F) = \mu(E)$ or $\mu(F) = 0$. Then if $g_n, g \in L^2(X)$ and $\|g_n - g\|_2 \to 0$ it follows that $g_n(t) \to g(t)$ for μ-almost all $t \in E$.

It is not hard to obtain Theorem 15 from Theorem 2 by the standard argument used to prove results like Theorem 13. However, it is more interesting to ask how we might tackle Theorem 15 directly and, in particular, what we are to make of Theorem 11 in the general context. If we do so we are rewarded with a key insight — Theorem 11 is a combinatorial theorem.

THEOREM 16. *Let $(X, \mathcal{F}, \mu)$ be a probability space. Let $\epsilon > 0$ and $\kappa > 1$ be given. Then we can find an $N \geqslant 1$, $1 \geqslant b > 0$ and $b_{p,q}$ taking the values 0 or b $[(p, q) \in \Lambda(N)]$ with the following property.*

Suppose that we have a collection of sets $E_{n,r} \in \mathcal{F}$ such that

(A) $E(2r - 1, n + 1) \cap E(2r, n + 1) = \emptyset$, $E(2r - 1, n + 1) \cup E(2r, n + 1) = E(r, n)$ *for all $(r, n) \in \Lambda_N$.*

(B) $|E(r, n)| = 2^{-n}$ *for all $(r, n) \in \Lambda_{N+1}$.*

Let us set

$$
\begin{aligned}
H_{r,n}(t) &= 1 && \text{for } t \in E(2r - 1, n + 1), \\
H_{r,n}(t) &= -1 && \text{for } t \in E(2r, n + 1), \\
H_{r,n}(t) &= 0 && \text{otherwise.}
\end{aligned}
$$

whenever $(r, n) \in \Lambda_N$. Then

(i) $|\sum_{(p,q)\in\Lambda(N)} b_{p,q} H_{p,q}(t)| \leqslant 1$ *for all $t \in [0, 1)$,*

and there is a set $E \in \mathcal{F}$ such that

(ii) $\max_{\lambda\in\Lambda(N)} |\sum_{\lambda\succeq(p,q)} b_{p,q} H_{p,q}(t)| \geqslant \kappa$ *for all $t \notin E$*

(iii) $|E| < \epsilon$.

PROOF. This is just Theorem 11.   □

To see why this is indeed an insight let us return to the concrete case of Lebesgue measure on $[0, 1)$ and consider the most important Riesz basis of all, the exponentials $e_n(t) = \exp(2\pi i n t)$. If we try to convert Theorem 11 into a result on Fourier series we run into the problem that different Haar functions do not 'occupy different parts of the frequency spectrum'. (Indeed the Fourier coefficients of the Haar functions $\chi_{p,q}$ with fixed p are all of same amplitude differing only in phase.) However, if we 'shuffle $[0, 1)$' the $E_{p,q}$ can be chosen so that the $H_{p,q}$ are, for practical purposes, in 'different parts of the frequency spectrum'. That is, although we do not have the ideal outcome in which at most one of the $\hat{H}_{p,q}(j)$ is non-zero for each j, we can arrange that at most one of the $\hat{H}_{p,q}(j)$ is large for each j.

Since there seems no further advantage in considering general measure spaces we shall stay within $[0, 1)$ and $[0, 1]$. However, readers who wish to work more generally will find that they only need the following simple consequence of Lemma 14.

LEMMA 17. *Let $(X, \mathcal{F}, \mu)$ be a non-atomic probability space. If $F \in \mathcal{F}$ and $\mu(F) > 0$ we can find a sequence e_j of orthogonal functions and sets $F_j \in \mathcal{F}$ with $F_j \subseteq F$ such that*

(i) $e_j(t) = 1$ for $t \in F_j$, $e_j(t) = -1$ for $t \in F \setminus F_j$, $e_j(t) = 0$ otherwise,
(ii) $\mu(F_j) = \mu(F)/2$.

PROOF. Set $E(0,0) = F$. By repeated use of Lemma 14 with $\alpha = 1/2$ we can find $E_{n,r} \in \mathcal{F}$ such that

(A) $E(2r - 1, n + 1) \cap E(2r, n + 1) = \emptyset$, $E(2r - 1, n + 1) \cup E(2r, n + 1) = E(r, n)$ for all $1 \leqslant r \leqslant 2^n$,

(B) $|E(r, n)| = 2^{-n}|F|$ for all $1 \leqslant r \leqslant 2^n$.

Now set $F_j = \bigcup_{r=1}^{2^j} E(2r - 1, j + 1)$ and define e_j as in condition (i). $\qquad \square$

4. Extension to general Riesz bases

In this section we work in $[0, 1]$ with Lebesgue measure. In order to put into effect the programme sketched at the end of the last section we need a sequence of easy lemmas. It may be helpful for the reader to keep in mind as examples both the 'well behaved' orthonormal system of exponentials and some other system ϕ_n where $\phi_n \notin L^\infty$.

LEMMA 18. Let ϕ_1, ϕ_2, ... be a Riesz basis. If e_1, e_2, ... form an orthonormal sequence then

$$\hat{e}_k(j) \to 0$$

as $k \to \infty$ for each fixed j.

PROOF. Referring back to Lemma 4 we see that $\hat{e}_k(j) = \langle \psi_j, e_k \rangle$ and the result is obvious. $\qquad \square$

LEMMA 19. Let ϕ_1, ϕ_2, ... be a Riesz basis. Suppose that $\delta > 0$, $M \geqslant 0$ and that F is a set of strictly positive measure. Then we can find E a measurable subset of F with $|E| = |F|/2$ and $M' > M$ such that if we set $H(t) = 1$ for $t \in E$, $H(t) = -1$ for $t \in F \setminus E$ and $H(t) = 0$ otherwise then

$$\sum_{j=1}^{M} |\hat{H}(j)|^2 + \sum_{j=M'}^{\infty} |\hat{H}(j)|^2 < \delta.$$

PROOF. Combining the results of Lemma 17 and Lemma 18 we see that we can find E a measurable subset of F with $|E| = |F|/2$ such that, if H is defined as stated,

$$\sum_{j=1}^{M} |\hat{H}(j)|^2 < \delta/2.$$

Since $H \in L^2$ there exists an $M' > M$ such that $\sum_{j=M'}^{\infty} |\hat{H}(j)|^2 < \delta/2$, so we are done. $\qquad \square$

We now have our basic construction.

THEOREM 20. *Let ϕ_1, ϕ_2, ... be a Riesz basis. Suppose that $\delta > 0$ and $N, M'(0,0) \geqslant 1$ are given. Then we can find integers $M(r,n) < M'(r,n)$ with $M(r,n) > M'(0,0)$—together with a collection of measurable sets $E(n,r)$ such that*

(A) $E(2r-1, n+1) \cap E(2r, n+1) = \emptyset$, $E(2r-1, n+1) \cup E(2r, n+1) = E(r,n)$ *for all* $(r,n) \in \Lambda_N$,
(B) $|E(r,n)| = 2^{-n}$ *for all* $(r,n) \in \Lambda_{N+1}$,
(C) *if we write* $\Delta(r,n) = \{k : M(r,n) \leqslant k \leqslant M'(r,n)\}$ *then* $\Delta(r,n) \cap \Delta(s,m) = \emptyset$ *whenever* $(r,n) \neq (s,m)$,

and such that, if we set

$$
\begin{aligned}
H_{r,n}(t) &= 1 && \text{for } t \in E(2r-1, n+1), \\
H_{r,n}(t) &= -1 && \text{for } t \in E(2r, n+1), \\
H_{r,n}(t) &= 0 && \text{otherwise.}
\end{aligned}
$$

whenever $(r,n) \in \Lambda_N$ *then*

$$
\sum_{j \notin \Delta(r,n)} |\hat{H}_{r,n}(j)|^2 < \delta.
$$

PROOF. Use Lemma 19 repeatedly. $\qquad\square$

THEOREM 21. *Let $\epsilon > 0$ and $\kappa > 1$ be given and take $N \geqslant 1$, b and $b_{p,q}$ $[(p,q) \in \Lambda(N)]$ be as in Theorem 16. Suppose that ϕ_1, ϕ_2, ... is a Riesz basis, $\delta > 0$ and $M'(0,0) \geqslant 1$ are given and that $E(n,r)$, $\Delta(r,n)$ and $H_{r,n}$ are constructed as in Theorem 20. Then if we set*

$$
f(t) = \sum_{(p,q) \in \Lambda(N)} b_{p,q} H_{p,q}(t)
$$

we have

(i) $\|f\|_\infty \leqslant 1$.

Further, provided only that δ is small enough, there is a measurable set F such that

(ii) $\max_{\lambda \in \Lambda(N)} |\sum_{\lambda \succeq (p,q)} \sum_{j \in \Delta(p,q)} |\hat{f}(j)\phi_j(t)| \geqslant \kappa - 1$ *for all* $t \notin F$,
(iii) $|F| < 2\epsilon$.

PROOF. Conclusion (i) is just conclusion (i) of Theorem 20. To obtain (ii) and (iii) we proceed as follows. Observe that

$$\left\| \sum_{\lambda \succeq (p,q)} \sum_{j \in \Delta(p,q)} \hat{f}(j)\phi_j - \sum_{\lambda \succeq (p,q)} b_{p,q} H_{p,q} \right\|_2$$

$$\leq \sum_{\lambda \succeq (p,q)} |b_{p,q}| \sum_{j \in \Delta(p,q)} \left\| H_{p,q} - \sum_{j \in \Delta(p,q)} \hat{f}(j)\phi_j \right\|_2$$

$$\leq \sum_{\lambda \succeq (p,q)} \sum_{j \in \Delta(p,q)} \left\| H_{p,q} - \sum_{j \in \Delta(p,q)} \hat{f}(j)\phi_j \right\|_2.$$

But, for each $(p,q) \in \Lambda(N)$,

$$\left\| H_{p,q} - \sum_{j \in \Delta(p,q)} \hat{f}(j)\phi_j \right\|_2$$

$$\leq \left\| \sum_{j \notin \Delta(p,q)} \hat{H}_{p,q}(j)\phi_j \right\|_2 + \sum_{(r,s) \neq (p,q)} \left\| \sum_{j \in \Delta(p,q)} \hat{H}_{r,s}(j)\phi_j \right\|_2$$

$$\leq \left\| \sum_{j \notin \Delta(p,q)} \hat{H}_{p,q}(j)\phi_j \right\|_2 + \sum_{(r,s) \neq (p,q)} \left\| \sum_{j \notin \Delta(r,s)} \hat{H}_{r,s}(j)\phi_j \right\|_2$$

$$= \sum_{(r,s) \in \Lambda(N)} \left\| \sum_{j \notin \Delta(r,s)} \hat{H}_{r,s}(j)\phi_j \right\|_2.$$

But, writing A for the Riesz constant of the basis, Theorem 20 tells us that

$$\left\| \sum_{j \notin \Delta(r,s)} \hat{H}_{r,s}(j)\phi_j \right\|_2^2 \leq A \sum_{j \notin \Delta(r,s)} |\hat{H}_{r,s}(j)|^2 < A\delta.$$

Thus, retracing our steps,

$$\left\| H_{p,q} - \sum_{j \in \Delta(p,q)} \hat{f}(j)\phi_j \right\|_2 < 2^N (A\delta)^{1/2}$$

and

$$\left\| \sum_{\lambda \succeq (p,q)} \sum_{j \in \Delta(p,q)} \hat{f}(j)\phi_j - \sum_{\lambda \succeq (p,q)} b_{p,q} H_{p,q} \right\|_2 < 2^{2N} (A\delta)^{1/2}$$

for all $\lambda \in \Lambda(N)$.

If we now write

$$F_\lambda = \left\{ t : \left| \sum_{\lambda \succeq (p,q)} \sum_{j \in \Delta(p,q)} \hat{f}(j)\phi_j - \sum_{\lambda \succeq (p,q)} b_{p,q} H_{p,q} \right| \geqslant 1 \right\},$$

then Chebychev's inequality and the last inequality of the preceding paragraph tell us that

$$|F_\lambda| < 2^{4N} A\delta.$$

Thus if we set $F = E \cup \bigcup_{\lambda \in \Lambda(N)} F_\lambda$ conclusion (ii) follows from conclusion (ii) Theorem 20 whilst conclusion (iii) Theorem 20 tells us that

$$|F| < \epsilon + 2^{5N} A\delta,$$

and (iii) holds provided only that δ is small enough. $\qquad\square$

Theorems 20 and 21 give us what we want. However, we have accumulated a fair amount of notation in the course of the construction which we can now jettison to provide a simpler conclusion.

THEOREM 22. *Let* ϕ_1, ϕ_2, ... *be a Riesz basis. Given* $K \geqslant 1$ *we can find an integer* $M(K) \geqslant 1$ *with the following properties. Given any integer* $m \geqslant 1$ *and any* $\eta > 0$ *we can find a function* $f \in L^\infty([0,1])$, *an integer* $m' > m$, *a bijection* $\sigma : \mathbb{N} \to \mathbb{N}$ *with* $\sigma(r) = r$ *for* $1 \leqslant r \leqslant m$ *and for* $r \geqslant m'$, *integers* $m \leqslant p(1) \leqslant p(2) \leqslant p(3) \dots p(M) \leqslant m'$ *and a measurable set* E *such that*

(i) $\|f\|_\infty \leqslant 1$,
(ii) $\sum_{j=1}^m |\hat{f}(j)|^2 \leqslant \eta$,
(iii) $\max_{1 \leqslant k \leqslant M} |\sum_{j=1}^{p(k)} \hat{f}(\sigma(j))\phi_{\sigma(j)}| \geqslant K$ *for all* $t \in E$,
(iv) $|E| > 1 - K^{-1}$.

The distance from $L^\infty([0,1])$ to $C([0,1])$ is usually not very great. The present case is no exception.

THEOREM 23. *In Theorem 22 we may take f continuous.*

PROOF. This is entirely routine. Let the $M(K)$ of our new Theorem 23 be chosen to be the $M(2K+2)$ of the old Theorem 22. Then, by Theorem 22, given any $m \geqslant 1$ and any $\eta > 0$ we can find a function $f \in L^\infty([0,1))$, an integer $m' > m$, a bijection $\sigma : \mathbb{N} \to \mathbb{N}$ with $\sigma(r) = r$ for $1 \leqslant r \leqslant m$ and for $r \geqslant m'$, integers $m \leqslant p(1) \leqslant p(2) \leqslant p(3) \dots p(M) \leqslant m'$ and a measurable set E' such that

(i) $' \|f\|_\infty \leqslant 1$,

(ii) $' \sum_{j=1}^{m} |\hat{f}(j)|^2 \leqslant \eta$,

(iii) $' \max_{1 \leqslant k \leqslant M} |\sum_{j=1}^{p(k)} \hat{f}(\sigma(j)) \phi_{\sigma(j)}(t)| \geqslant K + 1$ for all $t \in E$,

(iv) $' |E'| > 1 - K^{-1}/2$.

Let $\delta > 0$ be a small number to be determined and choose $f \in C([0, 1))$ such that f satisfies condition (i) and $\|f - g\|_2 \leqslant \delta/A^2$. Automatically

$$\left(\sum_{j=1}^{\infty} |\hat{f}(j) - \hat{g}(j)|^2 \right)^{1/2} \leqslant \delta/A$$

so condition (ii) is satisfied provided only that we choose δ small enough.

Next we observe that

$$\left\| \sum_{j=1}^{p(k)} \hat{f}(j) \phi_j - \sum_{j=1}^{p(k)} \hat{g}(j) \phi_j \right\|_2 \leqslant A \left(\sum_{j=1}^{p(k)} |\hat{f}(j) - \hat{g}(j)|^2 \right)^{1/2}$$

$$\leqslant A \left(\sum_{j=1}^{\infty} |\hat{f}(j) - \hat{g}(j)|^2 \right)^{1/2} \leqslant \delta.$$

Thus writing

$$E_k = \left\{ t \in [0, 1) : \left\| \sum_{j=1}^{p(k)} \hat{f}(j) \phi_j - \sum_{j=1}^{p(k)} \hat{g}(j) \phi_j \right\| \geqslant 1 \right\},$$

we have, by Chebychev's inequality,

$$|E_k| \leqslant \left\| \sum_{j=1}^{p(k)} \hat{f}(j) \phi_j - \sum_{j=1}^{p(k)} \hat{g}(j) \phi_j \right\|_2 = \delta.$$

Thus, setting $E = E' \setminus \bigcup_{k=1}^{N}$, we see that (iii) holds automatically and

$$|E| \geqslant |E'| - \sum_{k=1}^{N} |E_k| \geqslant (1 - \eta/2) - N\delta$$

so that (iv) holds provided only that we choose δ small enough.   □

The remainder of the construction follows a standard rolling hump (condensation of singularities) pattern using Chebychev's inequality in the same way as in the two previous proofs. In view of the amount of notation involved readers will probably prefer to do it themselves rather than follow my proof.

It is easy to put the bricks of Theorem 23 together.

LEMMA 24. *Let ϕ_1, ϕ_2, ... be a Riesz basis with constant A. Let $m'(0) = 1$ We can construct inductively positive integers $M(n)$, $m(n)$, $m'(n)$ with $m'(n-1) < m(n)$, continuous functions f_n, bijections $\sigma_n : \mathbb{N} \to \mathbb{N}$ with $\sigma_n(r) = r$ for $1 \leqslant r \leqslant m'(n-1)$ and for $m'(n) \leqslant r$, integers $m \leqslant p(1,n) \leqslant p(2,n) \leqslant p(3,n) \ldots p(n, M(n)) \leqslant m'$ and a measurable set E such that*

(i) $_n$ $\|f_n\|_\infty \leqslant 2^{-n}$,
(ii) $_n$ $\sum_{r=1}^{m(n)} |\hat{f}_n(r)|^2 \leqslant A^{-1} 2^{-4n-4} m(n)^2$,
(iii) $_n$ $\max_{1 \leqslant k \leqslant M(n)} \left| \sum_{j=1}^{p(k,n)} \hat{f}_n(\sigma(j)) \phi_{\sigma_n(j)}(t) \right| \geqslant 2^n$ *for all* $t \in E$,
(iv) $_n$ $|E_n| > 1 - 2^{-n}$,
(v) $_n$ $\sum_{j=1}^{n} \sum_{r=m'(n)}^{\infty} |\hat{f}_j(r)|^2 \leqslant A^{-1} 2^{-2n-4} M(n+1)^{-2}$.

PROOF. Conditions (i)$_n$ to (iv)$_n$ come directly from Theorem 23. The key point is that Theorem 23 allows us to define $M(n+1)$ before we define $m'(n)$ in such a way as to satisfy condition (v)$_n$. □

PROOF OF THEOREM 5. As I said above, this is routine. It is easy to check that $\sigma(r) = \sigma_n(r)$ for $m'(n-1) \leqslant r \leqslant m'(n)$ gives a well defined bijection $\sigma : \mathbb{N} \to \mathbb{N}$. By the conditions (i)$_n$ $g_n = \sum_{r=1}^{n} f_r$ converges uniformly to a continuous function f as $n \to \infty$. Trivially

$$\left| \sum_{j=1}^{p(k,n)} \hat{f}(\sigma(j)) \phi_{\sigma(j)}(t) - \sum_{j=1}^{p(k,n)} \hat{f}_n(\sigma(j)) \phi_{\sigma(j)}(t) \right| \leqslant \|g_{n-1}\|_\infty + |G_{k,n}(t)|$$

$$\leqslant 1 + |G_{k,n}(t)|,$$

where

$$G_{k,n}(t) = \sum_{r=1}^{n-1} \sum_{j=p(k,n)+1}^{\infty} \hat{f}_r(\sigma(j)) \phi_{\sigma(j)}(t) + \sum_{r=n+1}^{\infty} \sum_{j=1}^{p(k,n)} \hat{f}_r(\sigma(j)) \phi_{\sigma(j)}(t).$$

Using the properties of Riesz bases together with conditions (ii)$_{(r)}$ and (v)$_{(r)}$ we have

$$\|G_{k,n}\|_2$$

$$\leqslant \sum_{r=1}^{n-1} \left\| \sum_{j=p(k,n)+1}^{\infty} \hat{f}_r(\sigma(j))\phi_{\sigma(j)} \right\| + \sum_{\frac{}{}=n+1}^{\infty} \left\| \sum_{j=1}^{p(k,n)} \hat{f}_r(\sigma(j))\phi_{\sigma(j)} \right\|_2$$

$$\leqslant A^{1/2} \sum_{r=1}^{n-1} \left(\sum_{j=p(k,n)+1}^{\infty} |\hat{f}_r(\sigma(j))|^2 \right)^{1/2} + A^{1/2} \sum_{r=n+1}^{\infty} \left(\sum_{j=1}^{p(k,n)} |\hat{f}_r(\sigma(j))|^2 \right)^{1/2}$$

$$\leqslant A^{1/2} \sum_{r=1}^{n-1} \left(\sum_{j=m'(n-1)}^{\infty} |\hat{f}_r(\sigma(j))|^2 \right)^{1/2} + A^{1/2} \sum_{r=n+1}^{\infty} \left(\sum_{j=1}^{m(n)} |\hat{f}_r(\sigma(j))|^2 \right)^{1/2}$$

$$\leqslant A^{1/2} \sum_{r=1}^{n-1} \left(\sum_{j=m'(n-1)}^{\infty} |\hat{f}_r(\sigma(j))|^2 \right)^{1/2} + A^{1/2} \sum_{r=n+1}^{\infty} \left(\sum_{j=1}^{m(r-1)} |\hat{f}_r(\sigma(j))|^2 \right)^{1/2}$$

$$\leqslant 2^{-n-1} M(n)^{-1} + \sum_{r=n+1}^{\infty} 2^{-2r-4} m(r)^{-1} \leqslant 2^{-n} M(n)^{-1}.$$

Thus writing

$$E(k,n) = \{t : |G_{k,n}(t)| \geqslant 1\},$$

we have, by Chebychev's theorem, $|E(k,n)| \leqslant 2^{-n} M(n)^{-1}$ and, if we set

$$E'(n) = E(n) \bigcup_{k=1}^{M(n)} E(k,n),$$

we have, by $(iv)_n$,

$$|E'(n)| \leqslant 2^{-n+1}$$

and by (iii)$_n$

$$\max_{1 \leqslant k \leqslant M(n)} \left| \sum_{j=1}^{p(k,n)} \hat{f}(\sigma(j))\phi_{\sigma(j)}(t) \right| \geqslant 2^n - 2$$

for all $t \in E'(n)$. The theorem follows.   □

The reader may readily check that Theorem 15 may be obtained by very similar arguments as can the following direct generalisation of Theorem 5

THEOREM 25. *Let (X, τ) be a compact Hausdorff space. Let $(X, \mathcal{F}, \mu)$ be a regular non-atomic probability space. Let $\phi_1, \phi_2, \ldots$ form a Riesz basis for $L^2(\mu)$. Then there exists* *a real continuous f and a bijection σ : N → N such that*

$$\limsup_{N\to\infty}\left|\sum_{u=0}^{N}\hat{f}(\sigma(u))\phi_{\sigma(u)}(t)\right|=\infty.$$

for almost all t ∈ X.

5. Further Questions

It is natural to ask if our results can be improved so as to replace 'divergence almost everywhere' by 'divergence everywhere'. The obvious answer is no, since if we take a complete orthonormal system ϕ_n on $[0,1)$ and define $\tilde{\phi}_n$ by $\tilde{\phi}_n(t) = \phi_n(t)$ for $t \neq 0$, $\tilde{\phi}_n(0) = 0$ the result remains a complete orthonormal system but any linear combination of a finite set of $\tilde{\phi}_n$ will take the value 0 at zero.

The "unfair" example just given means that we must reformulate our question and suggests that, at least initially, we should consider particular complete orthonormal systems. In the case of Fourier series we recall the remarkable theorem of Kahane and Katznelson that every set of measure zero is a set on which the Fourier sum of a continuous function diverges (see e.g. [4], Chapter II, Section 3) and consider the following sequence of lemmas.

LEMMA 26. *Given any $\epsilon > 0$, any $K > 1$ and any integer $N \geqslant 1$ we can find a trigonometric polynomial P, a set E which is the union of a finite set of intervals and a bijection $\sigma : \mathbf{N} \to \mathbf{Z}$ such that*

(i) $\|P\|_\infty \leqslant 1$,
(ii) $\hat{P}(n) = 0$ *for all $n \leqslant N$*,
(iii) $\max_{k \geqslant 0} |\sum_{j \leqslant k} \hat{P}(\sigma(j)) \exp(i\sigma(j)t)| \geqslant K$ *for all $t \notin E$*.
(iv) $|E| < \epsilon$.

PROOF. Apply de la Vallée Poussin summation to Theorem 23 (with the exponentials $\exp(2\pi int)$ as the Riesz basis) to obtain a P satisfying all the conditions except possibly (ii). To obtain (ii) it suffices to replace $P(t)$ by $\exp(iMt)P(t)$ with M a suitable large positive integer. □

LEMMA 27 (Kahane and Katznelson). *Given any $K > 1$ there exists an $\epsilon(K)$ with the following property. Given any set E which is the union of a finite set of intervals and has $|E| < \epsilon(K)$ and any integer $N \geqslant 1$ we can find a trigonometric polynomial P such that*

(i) $\|P\|_\infty \leqslant 1$,
(ii) $\hat{P}(n) = 0$ *for all $n \leqslant N$*,
(iii) $\max_{k \geqslant 0} |\sum_{j \leqslant k} \hat{P}(j) \exp(ijt)| \geqslant K$ *for all $t \notin E$*.

PROOF. See [4], Chapter II, Section 3. □

Combining the last two lemmas we obtain the following result.

LEMMA 28. *Given any $\epsilon > 0$, any $K > 1$ and any integer $N \geqslant 1$ we can find a trigonometric polynomial P and a bijection $\sigma : \mathbb{N} \to \mathbb{Z}$ such that*

(i) $\|P\|_\infty \leqslant 1$,

(ii) $\hat{P}(n) = 0$ *for all* $n \leqslant N$,

(iii) $\max_{k \geqslant 0} |\sum_{j \leqslant k} \hat{P}(\sigma(j)) \exp(2\pi i \sigma(j)t)| \geqslant K$ *for all* $t \in \mathbb{T}$.

It is now easy to prove the desired result.

THEOREM 29. *We can find a continuous function $f : \mathbb{T} \to \mathbb{C}$ with $\hat{f}(n) = 0$ for $n < 0$ and a bijection $\sigma : \mathbb{N} \to \mathbb{N}$ such that*

$$\sup_{k \geqslant 0} |\sum_{j \leqslant k} \hat{f}(\sigma(j)) \exp(2\pi i j t)| = \infty$$

for all $t \in \mathbb{T}$.

If instead of considering complex valued functions and the orthonormal system $\exp 2\pi i n t$, we wish to consider real valued functions and the orthonormal system formed by $\sin 2\pi n t$ and $\cos 2\pi n t$ then we can replace the P of Lemma 28 by $\Re P + \Im P$ and first sum the $\cos 2\pi \sigma(j)t$ terms and then the $\sin 2\pi \sigma(j)t$ terms. The existence of a continuous function whose rearranged Fourier series diverges everywhere was first proved by L. V. Taĭkov [10] using Olevskiĭ's theorem in a different way.

So far as I know the general question remains open. In particular we may ask the following question.

QUESTION 30. *Consider the Haar system on $[0, 1)$ ordered in some way. Does there exist a continuous function $f : [0, 1] \to \mathbb{C}$ and a bijection $\sigma : \mathbb{N} \to \mathbb{N}$ such that*

$$\sup_{k \geqslant 0} \left| \sum_{j=0}^{k} a_{\sigma(j)} \chi_{\sigma(j)}(t) \right| = \infty$$

for all $t \in [0, 1)$?

We remark that a similar proof to the one above (replacing the non-trivial lemma of Kahane and Katznelson by a trivial parallel lemma) gives the following.

LEMMA 31. *Consider the Haar system on $[0, 1)$ ordered in some way. There exists a real $f \in L^2$ and a bijection*

$$\sigma : \mathbb{N} \to \mathbb{N}$$

such that

$$\sup_{k \geqslant 1} \left| \sum_{j=1}^{k} \hat{f}(\chi_{\sigma(j)}) \chi_{\sigma(j)}(t) \right| = \infty$$

for all $t \in [0, 1)$.

The method of proof applies to any 'well behaved' system (though the 'unfair example' with which we began the section shows that it can not apply to all Riesz bases).

6. Hard summation

The reader with more practical interests will observe that striking as Kolmogorov's theorem and its generalisations may be, they do not answer the question in the case of any particular rearrangement. Let us return to the question we started with but apply it to a general othonormal system $\phi_1, \phi_2, \ldots$. If $f \in L^2$ we consider

$$S_\delta(t) = \sum_{|\hat{f}(n)| \geqslant \delta} \hat{f}(n)\phi_n(t)$$

where $\hat{f}(n) = \langle f, \phi_n \rangle$ is the usual Fourier coefficient for the given system. This method of forming a sum is called 'hard summation' and, as we said at the beginning of the talk, is a very natural one to use.

(The reader may well ask what 'soft summation' is. Here we consider something like

$$\sigma_\delta \sum_{|\hat{f}(n)| \geqslant \delta} \hat{f}(n)\phi_n(t) + \sum_{\delta \geqslant |\hat{f}(n)| \geqslant \delta/2} \frac{|\hat{f}(n)| - \delta/2}{\delta/2} \hat{f}(n)\phi_n(t).$$

I suspect that the idea of soft summation derives from the use of Cesàro sums and similar filtering techniques to improve the behaviour of Fourier sums. I also suspect that the analogy is false since I know of no reason why soft summation should behave better than hard summation. However, this is merely opinion, and like all opinion liable to be disproved by fact.)

In [5] I constructed the following example.

THEOREM 32. *There exists a $f \in L^2(\mathbb{T})$ such that*

$$\limsup_{\eta \to 0+} \left| \sum_{|\hat{f}(u)| \geqslant \eta} \hat{f}(u) \exp 2\pi i u t \right| = \infty.$$

for almost all $t \in \mathbb{T}$.

Later in [6] I showed that we could take f continuous.

It therefore came as a great surprise to me to learn that Tao had proved the following theorem

THEOREM 33. *hard summation works for well behaved (rapidly decreasing) wavelets.*

Still more surprising was the simple nature of his proof. In order to show how it works I will prove it in a special case where the argument is particularly clean.

THEOREM 34. *Consider the Haar system on* $\mathbb{T}$. *If* $f : \mathbb{T} \to \mathbb{C}$ *is continuous then*

$$\sum_{|\hat{f}(\chi)| \geqslant \eta} \hat{f}(\chi)\chi \to f$$

uniformly on $\mathbb{T}$.

[The exceptionally good behaviour of the Haar system means that just as in some sense it is the *hardest* to prove a Kolmogorov type theorem for, so it is in some sense the *easiest* to prove a Tao type theorem for. However, I think all the essential ideas can be seen even in this simple special case.]

The reader should notice that we now use a different normalisation for the Haar functions since we want an *orthonormal* system in which $\langle \chi, \chi \rangle = 1$. We shall return to this point when we consider Lemma 37.

Tao's idea is to use one of the key characters in 20th century harmonic analysis — the maximal function. If we write

$$P_N f(t) = \sum_{\text{rank}(\chi) \leqslant N} \hat{f}(\chi)\chi(t)$$

$$S_\eta f(t) = \sum_{|\hat{f}(\chi)| \geqslant \eta} \hat{f}(\chi)\chi(t)$$

then corresponding maximal functions are

$$P^* f(t) = \sup_{N \geqslant 0} |P_N(f)(t)|$$

$$S^* f(t) = \sup_{\eta > 0} |S_\eta(f)(t)|.$$

Since the Haar system is so well behaved P^* is easy to bound.

LEMMA 35. $|P^* f(t)| \leqslant \|f\|_\infty$.

PROOF. Observe that $P_N f(t)$ is piecewise constant with $P_N f(t)$ taking the average value of f on the interval of the form $[q2^{-N}, (q+1)2^{-N})$ to which t belongs. Thus $|P_N f(t)| \leqslant \|f\|_\infty$ for all N and so $|P^* f(t)| \leqslant \|f\|_\infty$. $\qquad\qquad\square$

In general maximal functions are not so easy to bound and the main work of the proof of Tao's result lies in obtaining a bound for S^*.

THEOREM 36. *There exists a constant* K *such that if* $f : \mathbb{T} \to \mathbb{C}$ *is continuous then*

$$|S^* f(t)| \leqslant K\|f\|_\infty.$$

PROOF OF THEOREM 34 FROM THEOREM 36. Let $\epsilon > 0$. Then, since $P_N f(t) \to f$ uniformly we can find an M such that $\|P_M f - f\|_\infty < \epsilon$. Set $g = f - P_M(f)$. We observe that $\hat{g}(\chi) = 0$ if rank $\chi \leqslant N$ and $P_M \hat{(f)}(\chi) = 0$ if rank $\chi \geqslant N + 1$.

Let $\delta_0 = 1$ if $P_M(f) = 0$ and set

$$\delta_0 = \min\{|P_M\hat{}(f)(\chi)|;\, P_M\hat{}(f)(\chi) \neq 0\}$$

otherwise. If $\delta_0 > \delta > 0$ we have

$$S_\delta(f) = P_M(f) + S_\delta(g)$$

and so

$$\|S_\delta(f) - f\|_\infty \leqslant \|P_M(f) - f\|_\infty + \|S_\delta(g)\|_\infty \leqslant \epsilon + S^*(g) \leqslant (K+1)\epsilon$$

using Theorem 36. Since ϵ was arbrary the result follows. $\qquad\square$

In the proof of Theorem 36 we look out for ways in which (well behaved) wavelets differ from their Fourier counterparts. The first is that although the Riemann Lebesgue lemma represents the strongest statement we can make about the decrease of Fourier coefficients, much stronger results hold for wavelets.

LEMMA 37. *If $f : [0,1] \to \mathbb{C}$ is continuous and χ has rank n then*

$$|\hat{f}(\chi)| \leqslant \|f\|_\infty 2^{-n/2}$$

PROOF. Direct calculation gives

$$
\begin{aligned}
|\hat{f}(\chi)| &= \left|\int_{\operatorname{supp}\chi} f(t)\chi(t)\,dt\right| \leqslant \|f\|_\infty \|\chi\|_\infty |\operatorname{supp}\chi| \\
&= \|f\|_\infty 2^{n/2} 2^{-n} = \|f\|_\infty 2^{-n/2}.
\end{aligned}
$$

Note how the normalisation of χ comes into the calculation. $\qquad\square$

The second and best known difference is the localisation property of wavelets. This will play an important role in our estimate of the maximal function $S^* f$.

PROOF OF THEOREM 36. . It is suficicient to show that if $\|f\|_\infty = 1$, and $1 \geqslant \delta > 0$ then $|S_\delta(f)(t)| \leqslant K$ for some fixed K. To this end, choose n so that $2^{-n/2} \geqslant \delta > 2^{-(n+1)}/2$ and observe that by Lemma 37 this means that $|\hat{f}(\chi)| < \delta$ for $rank\chi > n$.

Thus

$$|S_\delta(f)(t)| = \left| P_n(f)(t) - \sum_{\text{rank}\,\chi \leqslant n,\,|\hat{f}(\chi)|<\delta} \hat{f}(\chi)\chi(t) \right|$$

$$\leqslant |P_n(f)(t)| + \left| \sum_{\text{rank}\,\chi \leqslant n,\,|\hat{f}(\chi)|<\delta} \hat{f}(\chi)\chi(t) \right|$$

$$\leqslant P^*(f)(t) + \delta \sum_{\text{rank}\,\chi \leqslant n} |\chi(t)|$$

$$\leqslant \|f\|_\infty + \delta \sum_{\text{rank}\,\chi \leqslant n} |\chi(t)|,\, \chi(t) \neq 0$$

$$\leqslant 1 + \delta \sum_0^n 2^{r/2}$$

since exactly one χ of each rank is non-zero at t (this is an extreme version of the localisation property of wavelets).

Doing some simple calculations we have

$$|S_\delta(f)(t)| \leqslant 1 + \delta \sum_0^n 2^{r/2} \leqslant 1 + \frac{\delta 2^{(n+1)/2}}{2^{1/2}-1} \leqslant 1 + \frac{2^{1/2}}{2^{1/2}-1},$$

which is an inequality of the required form (with $K = 1 + 2^{1/2}/(2^{1/2}-1)$). $\qquad\square$

To extend beyond continuous functions we need to use the Hardy-Littlewood maximal operator

$$Mf(t) = \sup_{r>0} \frac{1}{2r} \int_{t-r}^{t+r} |f(x)|\,dx.$$

If we do so, we get pointwise convergence almost everywhere for $f \in L^p$ $[1 \leqslant p \leqslant \infty]$ for the Haar system.

We end by stating Tao's general theorem from [9]. He works on $\mathbb{R}$ which is a more useful space than $\mathbb{T}$.

THEOREM 38 (Tao). *We work on $\mathbb{R}$. Suppose ϕ has integral zero, and is bounded and rapidly decreasing. Suppose that the set of functions (wavelets) given by*

$$\phi_{j,k}(x) = 2^{j/2}\phi(2^j(x-k))$$

form an orthonormal system. Then, if $f \in L^p(\mathbb{R})$ $[1 < p < \infty]$, we have

$$\sum_{|\hat{f}(\phi_{j,k})|>\lambda} \hat{f}(\phi_{j,k})\phi_{j,k}(x) \to f(x)$$

almost everywhere as $\lambda \to 0+$.

References

[1] N. Dunford and J. T. Schwartz, *Linear Operators (Part I)* (2nd Ed), Wiley, 1957.

[2] W. Feller, *An Introduction to Probability Theory and its Applications*, (3rd Ed) Wiley, 1968.

[3] P. R. Halmos *Measure Theory*, Van Nostrand, 1950.

[4] Y. Katznelson *An Introduction to Harmonic Analysis* Wiley, 1968.

[5] T. W. Körner, *Divergence of decreasing rearranged Fourier series* Ann. of Math. **144**, no. 1, 167–180 (1996).

[6] T. W. Körner *Decreasing rearranged Fourier series* J. Fourier Anal. Appl. **5** , no. 1, 1–19 (1999).

[7] A. M. Olevskiĭ, *Fourier Series with Respect to General Orthogonal Systems* Springer 1975.

[8] H. L. Royden, *Real Analysis* (2nd Ed), MacMillan, New York, 1968.

[9] T. Tao *On the almost everywhere convergence of wavelet summation methods* Appl. Comput. Harmon. Anal. **3**, no. 4, 384–387 (1996).

[10] L. V. Taĭkov, *The divergence of Fourier series of continuous functions with respect to a rearranged trigonometric system*, Dokl. Akad. Nauk SSSR, **150**, 262–265 (1963).

Wavelet expansions, function spaces and multifractal analysis

S. Jaffard

Département de Mathématiques
Université Paris XII
Faculté des Sciences et Technologie
61 Av. du Gal. de Gaulle
94010 Créteil Cedex, France

ABSTRACT. The purpose of this tutorial is to describe the interplay between three subjects: function spaces, wavelet expansions, and multifractal analysis. Some relationships are now classical. Wavelet bases were immediately considered as remarkable by analysts because they are unconditional bases of 'most' function spaces. This property is a key feature of the denoising algorithms of Donoho, for instance. multifractal analysis tries to derive the Hausdorff dimensions of the Holder singularities. Wavelet techniques proved the most efficient tool in the numerical computation of the spectra of singularities of turbulent flows.

Our purpose is first to present these points, and then to show how ideas have developed in the recent interplay between these three fields.

Refinements of the numerical techniques introduced to compute turbulence spectra have led to the introduction of new function spaces, which turn out to be the right setting to determine the fractal dimensions of graphs, and offer natural extensions of the Besov spaces to negative p's.

The 'function space setting' allows one to derive Baire-type results for the value of spectra.

Keeping the histograms of wavelet coefficients gives richer information than just keeping the moments of these histograms (which corresponds to keeping only the knowledge of the function spaces to which the function belongs). We compare the probabilistic results (obtained from histograms) with the above Baire-type results.

1. Introduction

We will describe the interplay between three subjects: function spaces, wavelet expansions, and multifractal analysis. Some relationships are now well established:

J.S. Byrnes (ed.), Twentieth Century Harmonic Analysis - A Celebration, 127–144.

- Wavelet bases were immediately considered as remarkable because they are unconditional bases of 'most' function spaces. This property has very practical implications; for instance, it is a key feature of the signal denoising algorithms introduced by David Donoho and his collaborators, see [11].
- The purpose of the *multifractal formalism* introduced by Uriel Frisch and Georgio Parisi is to derive the Hausdorff dimensions of the Hölder singularities of a function (the so-called 'spectrum of singulgularities') from the knowledge of the Besov spaces to which this function belongs, see e.g. [17].
- Among the several variants of the multifractal formalism that have been introduced, those based on wavelet techniques proved the most efficient for the numerical computation of the spectra of singularities of turbulent flows, see [2].

We will present these topics, and show how ideas which developed in these three fields interplayed recently; we will particularly focus on the following points:

Refinements of the numerical techniques introduced to compute turbulence spectra have led to the introduction of new function spaces, which recently found unexpected applications: they turn out to be the right setting to determine the fractal dimensions of graphs, and they offer natural extensions of the Besov spaces $B_p^{s,q}$ when the exponent p takes negative values.

The knowledge of the Besov spaces to which a collection of functions belongs allows one to derive quasi-sure results (in the sense of Baire's categories) for the value of the spectra of singularities of these functions.

In image or signal processing, Besov regularity is usually a by-product, deduced from the knowledge of the histogram of the wavelet coefficients at each scale j; so that more information is actually available. We will determine the maximal information which can be derived from the wavelet histograms and is independent of the wavelet basis chosen. Note that working on histograms of wavelet coefficients is not new; for instance cascade-type models for the evolution of the p.d.f. (probability density function) of the wavelet coefficients through the scales have been proposed to model the velocity in the context of fully developed turbulence, see [3]. We will compare probabilistic results (obtained from drawing at random wavelet coefficients at each scale inside such a preassigned sequence of histograms) with the Baire-type results.

2. Wavelets as unconditional bases of Besov spaces

The most important basis in analysis has certainly been the trigonometric system. This is so because the resolution of several key problems in physics is particularly simple when formulated in this setting. Unfortunately, the convergence of the corresponding series posed important mathematical problems since Du Bois-Reymond showed in 1873 that the Fourier series of a continuous function may diverge. (See [23] where the fine properties of Fourier series and the development of ideas that led to wavelet analysis are described.) Is this phenomenon inherent to any orthogonal decomposition? Hilbert posed this problem to his student

Alfred Haar, who gave a negative answer in his thesis by constructing in 1909 the following orthonormal basis of $L^2([0,1])$. It is composed of the function 1, and of the $\psi_{j,k}$ defined by

$$(2.1) \qquad \psi_{j,k}(x) = 2^{j/2}\psi(2^j x - k)$$

where $j \geqslant 0$, $k = 0, \ldots 2^{j-1}$, $\psi(x) = 1_{[0,1/2]} - 1_{[1/2,1]}$, and 1_A denotes the characteristic function of the set A. One can also omit the function 1, but use all positive and negative integer values of j and k and thus obtain an orthonormal basis of $L^2(\mathbb{R})$. Haar showed that the partial sums of the decomposition of a continuous function in this basis are uniformly convergent. The comparison with the trigonometric system is striking: A basis composed of discontinuous functions is more adapted to the analysis and reconstruction of continuous functions than the trigonometric system, though this system is composed of C^∞ functions. The Haar basis has another important property which the trigonometric system lacks: Marcinkiewicz showed in 1937 that it is an unconditional basis of the spaces L^p when $1 < p < \infty$; this means that any function of L^p can be written in only one way as $\sum c_{j,k}\psi_{j,k}$ and the convergence is *unconditional*, i.e. does not depend on the order of summation. This result still has important implications in current research. In 1999, Bourgain, Brezis and Mironescu used the characterization of L^p on the Haar basis as a key tool in the *lifting problem*, which consists in determining if any function f which belongs to a given function space and satisfies $|f| = 1$ can be written $f(x) = e^{i\theta(x)}$ where θ belongs to the same function space. (When the answer is positive, this is a key-step in the linearization of some nonlinear PDEs where the constraint $|f| = 1$ is imposed by the physics, such as in the Ginzburg-Landau model, see [8].)

Of course, since the Haar basis is not composed of continuous functions, it cannot be a basis for spaces of continuous functions. This last remark motivated researches to 'smooth' the Haar basis. The goal was to construct bases of the similar algorithmic type, and which would be unconditional bases of as many function spaces as possible. In 1910, Faber considered on $[0,1]$ the basis composed of 1, x and the primitives of the Haar basis. This *Schauder basis* (so-called because it was rediscovered by Schauder in 1927) has the same algorithmic form as the Haar basis: it is still of the type (2.1) where ψ is the primitive of the Haar wavelet. Faber showed that this system is a basis of the continuous functions on $[0,1]$. The price to be paid is that it is no longer a basis of L^2. Should one necessarily lose on one hand what has been obtained by the other? In 1928, Franklin showed that you can have your cake and eat it by applying the Gram-Schmidt orthonormalization procedure to the Schauder basis, thus obtaining a basis which is simultaneously unconditional for all spaces $L^p([0,1])$ ($1 < p < \infty$), for $C([0,1])$ and for the Sobolev spaces of low regularity. One can go on and iterate one step of integration (which regularizes) and one step of Gram-Schmidt orthonormalization; Ciesielski thus constructed in 1972 bases which are unconditional for a wider and wider range of function spaces on $[0,1]$. Of course, applying the Gram-Schmidt orthonormalization procedure iteratively makes these bases essentially impossible to compute numerically: Something has been lost in the end: algorithmic simplicity. It is therefore no wonder that the Ciesielski bases were never used in practical applications. (Note that this is in sharp contrast to the Haar basis which, despite its lack of regularity, has been widely used in image processing.)

However, algorithmic simplicity and regularity can go together. In 1981, Strömberg had the idea of applying the Gram-Schmidt orthonormalization on the whole line instead of on $[0, 1]$ (loosely speaking, one starts the orthonormalization at $-\infty$). Because of the dilation and translation invariance of the real line, this substitute of the Shauder basis now has the exact algorithmic form (2.1). Starting the orthonormalization with B-splines of arbitrary high degree, Strömberg thus constructed orthonormal wavelet bases of arbitrarily large regularity. These bases are unconditional for a wider and wider range of Sobolev or Besov spaces. The ultimate perfection was found by Yves Meyer and Pierre-Gilles Lemarié who constructed, in 1986, C^∞ wavelets $(\psi^{(i)})_{i=1,\ldots,2^d-1}$ such that the functions

(2.2) $$2^{dj/2}\psi^{(i)}(2^j x - k), \quad j \in \mathbb{Z}, \ k \in \mathbb{Z}^d$$

form an orthonormal basis of $L^2(\mathbb{R}^d)$; this basis allows one to characterize functions of arbitrary regularity (or distributions of arbitrary irregularity by duality), see [27]. In order to be more specific, we start by introducing some notation.

Wavelets and wavelet coefficients will be indexed by dyadic cubes: λ will denote the cube $\lambda_{j,k} = k2^{-j} + [0, 2^{-j}]^d$, ψ_λ will denote the wavelet $\psi^{(i)}(2^j x - k)$ (note that we 'forget' to write the index i of the wavelet, which is of no consequence). Thus

(2.3) $$f(x) = \sum_\lambda c_\lambda \psi_\lambda(x),$$

where the wavelet coefficients of f are given by

$$c_\lambda = \int_{\mathbb{R}^d} 2^{dj} \psi_\lambda(t) f(t) dt.$$

(Note that we do not use the usual L^2 normalization; the natural normalization for the problems we will consider is the L^∞ normalization.)

Let $p \in (1, +\infty)$ and $s \geqslant 0$; by definition, the Sobolev space $L^{p,s}(\mathbb{R}^d)$ is composed of the functions of $L^p(\mathbb{R}^d)$ whose fractional derivatives of order s also belong $L^p(\mathbb{R}^d)$. A function f belongs to $L^{p,s}(\mathbb{R}^d)$ if and only if its wavelet coefficients c_λ satisfy the following condition, see [29]

(2.4) $$f \in L^{p,s}(\mathbb{R}^d) \Leftrightarrow \left(\sum_{\lambda,i} |c_\lambda|^2 (1 + 2^{2sj}) 1_\lambda(x) \right)^{1/2} \in L^p.$$

(Note the sharp contrast with Fourier series: When $p \neq 2$, there exists no characterization of $L^{p,s}$ by conditions on the moduli of the Fourier coefficients.)

These characterizations are quite difficult to handle and, in the context of wavelet analysis, Besov spaces are preferred for the two following reasons:

- They are very close to the Sobolev spaces, as shown by the following embeddings

$$\forall \epsilon > 0, \quad \forall p \geqslant 1, \quad \forall q, \quad L^{p,s+\epsilon} \hookrightarrow B_q^{s,p} \hookrightarrow L^{p,s-\epsilon}.$$

- They have a very simple wavelet characterization, see [7], [26] and [29],

(2.5) $\qquad f \in B_p^{s,q}(\mathbb{R}^d) \iff \left(\sum_{i,k} |c_\lambda 2^{(s-\frac{d}{p})j}|^p \right)^{1/p} = \epsilon_j \quad \text{with} \quad (\epsilon_j)_{j \in \mathbb{Z}} \in l^q.$

Note that in all such characterizations, wavelets are assumed to be smooth enough, say, with at least derivatives up to order $[s]+1$ having fast decay (see [7] for optimal regularity assumptions on the wavelets). In sharp contrast with the Sobolev case, Besov spaces are defined for any $p > 0$.

The global regularity information about a function f is given by its *Besov domain* B_f, which is the set of (q, s) such that f belongs to $B_{1/q,loc}^{s,1/q}$. By interpolation, the Besov domain has to be a convex subset of $\mathbb{R}^2$, and the Besov embeddings imply that, if (q, s) belongs to B_f, then the segment joining (q, s) and $(0, s - dq)$ also belongs to B_f, see [31]. It follows that the boundary of the Besov domain is the graph of a function $s(q)$ which is concave and satisfies

(2.6) $\qquad\qquad\qquad\qquad 0 \leqslant s'(q) \leqslant d.$

The following proposition of [19] shows that (2.6) characterizes the possible functions $s(q)$.

PROPOSITION 1. *Any concave function that* $s(q)$ *satisfies (2.6) defines the boundary of the Besov domain of a distribution* f.

One of the reasons for the success of wavelet decompositions in applications is that they often lead to very sparse representations of signals. This sparsity can be characterized by determining to which Besov spaces $B_p^{s,q}$ the function considered belongs when p is close to 0. Let us illustrate this assertion by an example. Consider the function

$$H(x) = 1 \quad \text{if} \quad |x| \leqslant 1,$$
$$= 0 \quad \text{elsewhere},$$

and suppose that the wavelet used is compactly supported, say on $[-A, A]$. For each j, there are less than $4A$ non-vanishing wavelet coefficients, so that the wavelet expansion of f is extremely sparse. Since $H(x)$ is bounded, $|c_\lambda| \leqslant C \; \forall \lambda$. Using (2.5), it follows that $H(x)$ belongs to $B_p^{s,q}(\mathbb{R})$ as soon as $s < 1/p$. Let us check that, conversely, this property is a way to express that the wavelet expansion of f is sparce. We suppose that a bounded function f satisfies

$$\forall p, q > 0, \quad \forall s < \frac{1}{p}, \quad f \in B_p^{s,q}(\mathbb{R}).$$

We claim that $\forall A > 0$, $\forall \epsilon > 0$, at each scale j there are less than $C(\epsilon, A)2^{\epsilon j}$ coefficients of size larger than 2^{-Aj}. Indeed, if it were not the case, taking $p = \epsilon/(2A)$, we get $\sum_k |c_\lambda|^p \to +\infty$ when $j \to +\infty$, hence a contradiction.

Here is another illustration of the relationship between sparsity of the wavelet expansion and Besov regularity. Suppose that f belongs to

$$\bigcap_{p>0} B_p^{d/p,p}(\mathbb{R}^d).$$

Going back to (2.5), this condition exactly means that the sequence c_λ belongs to l^p for all $p > 0$, which is also equivalent to the fact that the decreasing rearrangement of the sequence $|c_\lambda|$ has fast decay, which, again, is a way to express sparsity, see [21].

Besov spaces when $p < 1$ are no longer locally convex, which partly explains the difficulties met when using them. Before the introduction of wavelets, these spaces were either characterized by the order of approximation of f by rational functions whose numerator and denominator have a given degree, or equivalently by the order of approximation by splines with 'free nodes' (which means that the points where the piecewise polynomials are connected are left free, and can thus be fitted to the function considered), see [10] and [21]. However such characterizations were much more difficult to handle, and of hardly any use in numerical applications.

3. Beyond Besov spaces: oscillation spaces

One important drawback when using Besov spaces is that any information concerning possible correlations on the position of large wavelet coefficients is lost, since the wavelet norm (2.5) is clearly invariant under permutations of the wavelet coefficients at the same scale; this can be a drawback. For instance, piecewise smooth functions clearly have their large wavelet coefficients located at the singularities, so that such functions exhibit very strong correlation between the positions of large wavelet coefficients. Let us show another occurrence of this problem, concerning the computation of the fractal dimensions of graphs.

DEFINITION 1. *Let K be a bounded subset of $\mathbb{R}^{d+1}$, $N(K, j)$ will denote the number of dyadic cubes of size 2^{-j} necessary to cover K. The fractal dimension of K (also called upper box dimension) is*

$$\overline{dim_b}(K) = \limsup_{j\to\infty} \frac{\log N(K, j)}{j \log 2}.$$

If K is the graph of a compactly supported continuous function, $N(K, j)$ is related to the oscillation of f.

DEFINITION 2. *Let λ be a dyadic cube included in $\mathbb{R}^d$; if f is a continuous, real valued function defined on $\mathbb{R}^d$, let*

(3.1) $$osc(f, \lambda) = \sup_{x\in\lambda} f(x) - \inf_{x\in\lambda} f(x)$$

denote the oscillation of f on the cube λ. The p-oscillation of f at scale j is defined by

$$Osc_p(f, j) = \sum_k (osc(f, \lambda_{j,k}))^p$$

where $\lambda_{j,k}$ is the dyadic cube $k2^{-j} + 2^{-j}[0,1]^d$ (when $p = 1$ one uses the term 'oscillation' instead of '1-oscillation'). The p-oscillation exponent is

$$\omega_p(f) = \liminf_{j \to \infty} \frac{\log (Osc_p(f, j))}{\log(2^{-j})}.$$

(The p-oscillation is a variant of the p-variation, see [18].) It follows easily that there exist two positive constants C and C' such that

(3.2) $C \left(2^{dj} + 2^j Osc_1(f, j)\right) N(Graph(f), j) \leqslant C' \left(2^{dj} + 2^j Osc_1(f, j)\right).$

Let us explain by an example why the box dimension of a graph cannot be deduced from the Besov domain of a function; we use essentially a construction due to Anna Kamont and Barbara Wolnick, cf [25].

Consider the following function: Let $j_1 >> j_0 >> 0$; all wavelet coefficients of f, defined on $[0, 1]$, vanish for $j \leqslant j_1$; at the scale j_1, there are $2^{j_1 - j_0}$ nonvanishing wavelet coefficients of size $2^{-\alpha j_1}$. Let us now consider the two extreme possibilities:

- All nonvanishing wavelet coefficients are packed in the $2^{j_1 - j_0}$ first locations ($k = 1, \ldots, 2^{j_1 - j_0}$). In this case, the oscillation at the scale 2^{-j_0} vanishes, except for the two first dyadic intervals, for which it is $\sim 2^{-\alpha j_1}$.
- If the nonvanishing wavelet coefficients are equidistributed, the oscillation at the scale 2^{-j_0} is $\sim 2^{-\alpha j_1}$ on each dyadic interval of length 2^{-j_0}.

The total oscillation at scale 2^{-j_0} is thus $\sim 2^{-\alpha j_1}$ in the first case, and $\sim 2^{j_0 - \alpha j_1}$ in the second. We can pick j_0 and j_1 such that the oscillations in both cases are not of the same order of magnitude. By piling up this construction on an infinite number of scales, it is easy to construct two functions with the same histograms of wavelet coefficients at each scale, and different box dimensions of graphs. The problem here is that the box dimension of graphs is clearly altered by clustering or spreading the large wavelet coefficients. Therefore, it can be measured only by norms that are able to take such phenomena into account. It is the purpose of the *oscillation spaces* which allow one to derive p-oscillation exponents from the wavelet coefficients. In the following, λ' denotes the dyadic cube $k'2^{-j'} + [0, 2^{-j'}]^d$.

DEFINITION 3. *Let $p > 0$, and $s, s' \in \mathbb{R}$; then a function f belongs to $\mathcal{O}_p^{s,s'}(\mathbb{R}^d)$ if its wavelet coefficients satisfy*

(3.3) $$\sup_{j \in \mathbb{Z}} 2^{sj} \left(\sum_k \sup_{\lambda' \subset \lambda} |c_{\lambda'} 2^{s'j'}|^p \right)^{1/p} < \infty.$$

The left hand-side defines the $\mathcal{O}_p^{s,s'}(\mathbb{R}^d)$-seminorm.

Proposition 2 will imply that this definition is independent of the wavelet basis chosen. Note that this definition exhibits the property we were looking for, and that Besov spaces were lacking. Namely, because of the $\sup_{\lambda' \subset \lambda}$ that appears in the definition, two functions that share the same histograms of wavelet coefficients at all scales may have very different $\mathcal{O}_p^{s,s'}$ norms, depending upon whether the large wavelet coefficients are more or less clustered. oscillation spaces take into account the geometric disposition of the wavelet coefficients.

THEOREM 1. *If f belongs to $C^\epsilon(\mathbb{R}^d)$ for an $\epsilon > 0$, the p-oscillation exponent of f is given by*

$$\omega_p(f) = \sup\{s : f \in \mathcal{O}_p^{s/p,0}\} = \limsup_{j \to +\infty} \frac{\log\left(\displaystyle\sum_{\lambda \in \Lambda_j} \sup_{\lambda' \subset \lambda} |c_{\lambda'}|^p\right)}{j \log 2}.$$

The spaces $\mathcal{O}_p^{s,s'}$ are defined by conditions on the wavelet coefficients, therefore one first has to check that their definition is *intrinseque*, i.e., independent of the wavelet basis chosen. One way to do it, following [29], is to check that condition (3.3) is invariant under the action of the "infinite matrices" which belong to the algebras $\mathcal{M}^\gamma$ for γ large enough; these algebras are defined as follows, see [29]: $A(\lambda, \lambda')$ (indexed by the dyadic cubes) belongs to $\mathcal{M}^\gamma$ if

$$|A(\lambda, \lambda')| \leqslant \frac{C \, 2^{-(\frac{d}{2}+\gamma)(j-j')}}{(1 + (j - j')^2)(1 + 2^{\inf(j,j')} dist(\lambda, \lambda'))^{d+\gamma}}.$$

Matrices of operators which map a wavelet basis onto another belong to these algebras, and more generally matrices (on wavelet bases) of pseudodifferential operators of order 0, such as the Hilbert transform in dimension 1, or the Riesz transforms in higher dimensions, see [29]. We denote by $\mathcal{O}p(\mathcal{M}^\gamma)$ the space of operators whose matrix on a wavelet basis belongs to $\mathcal{M}^\gamma$. The following proposition is proved in [18].

PROPOSITION 2. *If $\gamma \geqslant \sup(|s|, |s'|)$, then the operators which belong to $\mathcal{O}p(\mathcal{M}^\gamma)$ are continuous on $\mathcal{O}_p^{s,s'}(\mathbb{R}^d)$.*

The following corollary is an immediate consequence of (3.2) and Theorem 1.

COROLLARY 1. *Let $f : \mathbb{R}^d \to \mathbb{R}$ be a compactly supported function which belongs to $C^\epsilon(\mathbb{R}^d)$ for an $\epsilon > 0$. Then*

$$\overline{dim_b}(Graph(f)) = \sup\left(d, 1 - \sup\{s : f \in \mathcal{O}_1^{s,0}\}\right).$$

Similarly, Kamont recently proved the following wavelet characterization of the *lower box dimension* of a graph in terms of the wavelet expansion of the function, see [24]. The lower box dimension of a set K is by definition

$$\underline{dim_b}(K) = \liminf_{j \to \infty} \frac{\log N(K, j)}{j \log 2}.$$

THEOREM 2. *Let* $f : \mathbb{R}^d \to \mathbb{R}$ *be a compactly supported function which belongs to* $C^\epsilon(\mathbb{R}^d)$ *for an* $\epsilon > 0$; *let*

$$\Xi_j = \sum_k \sup_{\lambda' \subset \lambda} 2^{dj'/2}|c_{\lambda'}|, \quad and \quad Y_j = 2^{j(d-1)} \sup_{0 \leqslant j' \leqslant j} \left(2^{-j'(\frac{d}{2}-1)} \sum_{k'} |c_{\lambda'}| \right),$$

then

$$\underline{dim_b}(Graph(f)) = 1 + \liminf_{j \to \infty} \frac{\log(\Xi_j + Y_j + 2^{j(d-1)})}{j \log 2}.$$

Another (much more difficult) problem concerning the fractal nature of graphs is to determine their Hausdorff dimension. Let us recall the definition of the Hausdorff dimension of a subset $A \subset \mathbb{R}^d$. For $\varepsilon > 0$, let

$$M_\varepsilon^d = \inf_R \sum_i \varepsilon_i^d,$$

where R is a generic covering of the set A by balls B_i of diameter $\varepsilon_i \leqslant \varepsilon$, Then

$$dim_H(A) = \sup\{d : \lim_{\varepsilon \to 0} M_\varepsilon^d = +\infty\} = \inf\{d : \lim_{\varepsilon \to 0} M_\varepsilon^d = 0\}.$$

Extending the previously mentioned techniques, François Roueff proved that, if $f \in C^\epsilon$ for an $\epsilon > 0$, the Hausdorff dimenson of the graph of f is bounded by $d + 1 - \sup\{s : f \in B_1^{s,\infty}\}$ (actually, some sharper estimates can be found in [32]).

The wavelet characterization given by Corollary 1 has implications in rugosimetry. Indeed the fractal dimension of a surface has been shown to be a pertinent way to model the notion of *rugosity*, see [12]. (In [33] other generalizations of Sobolev spaces are introduced following this motivation.) Therefore, finding a numerically stable algorithm to measure this fractal dimension became an important issue. Numerical algorithms based on the oscillation are discussed in [12]; alternative algorithms based on the wavelet characterization of Corollary 1 should prove numerically more stable, since they wouldn't be based directly on the pointwise value of the function, but on wavelet coefficients, which are averaged quantities, and, as such, are less sensitive to noise.

Let us end this section by mentioning a remarkable property of oscillation spaces. In the multifractal formalism that we will present in the next section, the spectrum of singularities of a function f is deduced from its Besov domain by a Legendre transform. One drawback of this approach is that Besov spaces are defined only for positive p's. Thus, at best, this method allows one to recover the increasing part of the concave hull of the spectrum; obtaining the decreasing part would involve the extension of the Besov domain for negative p's, which is clearly absurd when starting from any of the usual definitions of Besov spaces. On the contrary, oscillation spaces have a natural extension to negative p's. Let us sketch how this extension can be derived. First, we remark that Definition 3 can be rewritten

(3.4) $$\forall j \in \mathbb{Z} \qquad \sum_k \left(\sup_{\lambda' \subset \lambda} |c_{\lambda'} 2^{s'j'}| \right)^p \leqslant C 2^{-spj}.$$

Note that, in particular, applying this condition for $j = 0$, we obtain that, for any $j' \geqslant 0$, $|c_{\lambda'}| \leqslant C2^{-s'j'}$, or, in other words, $f \in C^{s'}(\mathbb{R}^d)$. Conversely, the condition $f \in C^{s'}(\mathbb{R}^d)$ is necessary to make sure that the suprema in (3.4) are finite. Therefore, we adopt the following

DEFINITION 4. *Let $p < 0$, and $s, s' \in \mathbb{R}$; then a function f belongs to $\mathcal{O}_p^{s,s'}(\mathbb{R}^d)$ if f belongs to $C^{s'}(\mathbb{R}^d)$ and if its wavelet coefficients satisfy*

$$\forall j \in \mathbb{Z} \quad \sum_k \left(\sup_{\lambda' \subset \lambda} |c_{\lambda'} 2^{s'j'}| \right)^p \leqslant C2^{-spj}.$$

The remarkable property of this definition is that it is 'almost' independent of the wavelet basis (in the Schwartz class) which is chosen. More precisely, the spaces $\bigcap_{s''>s} \mathcal{O}_p^{s'',s'}$ are 'intrinseque', i.e., are independent of the wavelet basis, and more generally are invariant under the action of an operator whose matrix A satisfies

$$A \in \bigcap_{\gamma>0} \mathcal{M}^\gamma \quad \text{and} \quad A^{-1} \in \bigcap_{\gamma>0} \mathcal{M}^\gamma.$$

4. Multifractal analysis and the Frisch-Parisi formula

Large classes of signals exhibit a very irregular behavior. In the wildest situations, this irregularity may follow different regimes, and can switch from one regime to another almost instantaneously. This is the case for recordings of speech signals; precise recordings of turbulence data (which became available at the beginning of the 80s) showed that turbulence also falls in this category. Such signals cannot be modeled by standard stationary increments processes, such as fractional brownian motions. The techniques of multifractal signal analysis have been specifically designed to analyze such behavior. Initially developed in the mid 80's in the context of turbulence analysis, they were applied successfully to a large range of signals, including traffic data (cars *and* internet), stock market prices, speech signals, texture analysis, DNA sequences...(see [1] and [28] for instance).

We start by introducing the definitions related to pointwise regularity. Let α be a positive real number and $x_0 \in \mathbb{R}^m$; a function $f : \mathbb{R}^m \to \mathbb{R}$ is $C^\alpha(x_0)$ if there exists a polynomial P of degree less than α such that

(4.1) $|f(x) - P(x - x_0)| \leqslant C|x - x_0|^\alpha.$

The *Hölder exponent* $h_f(x_0)$ is the supremum of all the values of α such that (4.1) holds. We are interested in analyzing signals whose Hölder exponent may widely change from point to point. This instability usually makes the task of determining the Hölder exponent $h_f(x)$ very difficult numerically. This is the case for multifractal functions, where the Hölder exponent jumps from point to point. In that case, points with a given Hölder exponent form fractal sets, and one is not interested in determining the exact value of the Hölder exponent at every point but rather in extracting some relevant information concerning the size and geometry of the Hölder singularities. The relevant quantity is the *spectrum of singulgularities*.

DEFINITION 5. *Let A_H be the set of the points x where $h_f(x) = H$. The domain of definition of the spectrum of singulgularities $d(H)$ is the set of H's such that A_H is not empty. If this is the case, $d(H)$ is the Hausdorff dimension of A_H. Otherwise, if H is not a value taken by h_f, $d(H) = -\infty$.*

It is clearly impossible to estimate numerically the spectrum of singulgularities of a signal since it involves the successive determination of several intricate limits, and a blind application of the formula giving the definition of the Hausdorff dimension would yield enormous, totally unstable calculations. The only method is to find some 'reasonable' assumptions under which the spectrum could be derived using only averaged quantities (which should be numerically stable) extracted from the signal. Such formulas, called *multifractal formalisms*, were inferred first by physicists.

The initial formulation asserts that the spectrum of Hölder singularities of a function can be recovered from the scaling function $\eta_f(p)$, defined for $p > 0$ by

$$\eta_f(p) = \sup\{s : f \in B_p^{s/p,p}\} = d + \liminf_{j \to +\infty} \left(\log \left(\sum_k |c_\lambda|^p \right) \right) / (\log 2^{-j}).$$

It follows immediately from the definition of $s(q)$ that

(4.2) $\eta_f(p) = ps(1/p).$

The multifractal formalism may be surprising at first sight because it relates pointwise behavior (Hölder exponents) to global estimates (Besov regularity). Before studying its mathematical validity, it may be enlightening to give the heuristic argument from which it is derived. Though this argument cannot be transformed into a correct mathematical proof, it shows at least why these formulas can be expected to hold, and a careful study of its implicit asumptions shows its limitations. This argument can be decomposed into four steps, each involving specific assumptions that we will state explicitly in order to make clear the conditions under which the formalism can be expected to hold. It is initially based on the following characterization of the Hölder exponent based on decay estimates of the wavelet coefficients, see [14] (or [20] for the sharpest results).

PROPOSITION 3. *If f belongs to $C^\epsilon(\mathbb{R}^d)$ for an $\epsilon > 0$ (i.e. if $|c_\lambda| \leqslant C2^{-\epsilon j}$), the Hölder exponent of f at each point x_0 is given by*

(4.3) $h_f(x_0) = \liminf_{j \to \infty} \inf_k \dfrac{\log(|c_\lambda|)}{\log(2^{-j} + |k2^{-j} - x_0|)}.$

Step 1: The first asumption in the derivation of the multifractal formalism is that the Hölder exponent of f at every point x_0 is given by the rate of decay of the wavelet coefficients of f in a cone $|k2^{-j} - x_0| \leqslant C2^{-j}$. Coming back to (4.3), if these coefficients decay like 2^{-Hj}, we expect that $h_f(x_0) = H$. This statement is wrong in full generality but is true under the hypothesis that f has only *cusp-type* singularities (see [30]).

Step 2: We estimate, for each H, the contribution of the Hölder singularities of exponent H to the quantity

$$(4.4) \qquad\qquad \sum_k |c_\lambda|^p.$$

Each such singularity brings a contribution of $C2^{-Hpj}$. We need about $2^{d(H)j}$ cubes of width 2^{-j} to cover these singularities; the total contribution of the Hölder singularities of exponent H to (4.4) is thus

$$(4.5) \qquad\qquad 2^{d(H)j}2^{-Hpj} = 2^{-(Hp-d(H))j}.$$

This is clearly a critical step in the argument; it involves an inversion of limits which supposes that all Hölder singularities start to have coefficients $\sim 2^{-Hj}$ at a certain scale J, and that the Hausdorff dimension is estimated as a box dimension. It is remarkable that the multifractal formalism is valid in many situations where these two hypotheses do not hold.

Step 3: This consists of a steepest descent argument. When $j \to +\infty$, among the terms (4.5), the one which yields the main contribution to (4.4) is obtained for the exponent H realizing the infimum of $Hp - d(H)$; hence

$$\eta(p) - d = \inf_H \ (Hp - d(H)).$$

Step 4: If $d(H)$ is a concave function, $-d(H)$ and $-\eta(p) + d$ are convex conjugates, and each can be recovered from the other by a Legendre transform; it follows that

$$(4.6) \qquad\qquad d(H) = \inf_{p>0} \ (Hp - \eta(p) + d).$$

The hypothesis that $d(H)$ is a concave function is often wrong; there are three ways to counter this difficulty:

- Stop at Step 3, and check that $\eta(p)$ is the Legendre transform of $d(H)$; however this weak form of the multifractal formalism is of little interest since $d(H)$ is the mathematical object of interest, and $\eta(p)$ is the only computable quantity in practice.
- Assert that we thus obtain only the convex hull of the spectrum. This is fine when the function obtained is strictly concave, but it yields ambiguous information when it contains straight segments, which is often the case. Do these segments correspond to effective points of the spectrum, or are they just the convex hull in a non-concave region?
- Do not use the partition function (4.4, but instead deal directly with histograms of wavelet coefficients; we will discuss this approach in the next section.

There exist several mathematical examples where the Hölder exponent can be analytically determined, and the validity of the multifractal formalism has been successfully tested (including selfsimilar functions (see [17] and [6]), specific historical functions, (see [16], and references therein) and Lévy processes, see [15]). These examples give some insight about sufficient conditions for the validity of the multifractal formalism. In each case, the function (or its wavelet transform) exhibits some selfsimilarity (deterministic or statistical).

5. Multifractal formalism: Mathematical results

To simplify some arguments, we suppose from now on that the functions we consider are defined on $\mathbb{R}$, are 1-periodic, belong to $C^\epsilon(\mathbb{R})$, and that we use one-dimensional periodized wavelets.

5.1. *Upper bounds for spectra*

We start by describing bounds on spectra of singularities which hold in full generality. These bounds are based on histograms of wavelet coefficients, so that we start by defining relevant quantities derived from these histograms. For each j, let

$$(5.1) \qquad N_j(\alpha) = \# \left\{ |C_{jk}| \geqslant 2^{-\alpha j} \right\}.$$

If $\rho(\alpha, \epsilon) = \limsup\limits_{j \to \infty} j^{-1} \log_2(N_j(\alpha + \epsilon) - N_j(\alpha - \epsilon))$, we note $\tilde\rho(\alpha) = \inf\limits_{\epsilon > 0} \rho(\alpha, \epsilon)$. (There are about $2^{\tilde\rho(\alpha)j}$ coefficients of size of order $2^{-\alpha j}$.) The scaling function $\eta_f(p)$ can be derived from the wavelet histograms by

$$(5.2) \qquad \eta_f(p) = \liminf_{j \to +\infty} \left(\frac{-1}{j} \log_2 (2^{-j} \int 2^{-\alpha p j} N_j(\alpha) d\alpha) \right).$$

Indeed, by definition of N_j, $\sum_k |c_\lambda|^p = \int 2^{-\alpha p j} dN_j(\alpha)$. The following result of [19] gives the relationship between the "Besov approach" and the "wavelet histograms approach". It is a direct consequence of (5.2) and shows that $\eta_f(p)$ can be deduced from $\tilde\rho(\alpha)$ by a Legendre transform. Note that $\tilde\rho(\alpha)$ cannot be recovered from $\eta_f(p)$; only its convex hull can. Therefore $\tilde\rho(\alpha)$ contains more information on f than $\eta_f(p)$.

PROPOSITION 4. *For any function f,*

$$(5.3) \qquad \eta_f(p) = \inf_{\alpha \geqslant 0} (\alpha p - \tilde\rho(\alpha) + 1).$$

The following proposition gives the optimal upper bound on $d(H)$ that can be derived from the wavelet histograms.

PROPOSITION 5. *If $f \in C^\epsilon(\mathbb{R})$ for an $\epsilon > 0$, then*

$$(5.4) \qquad d(H) \leqslant H \sup_{\alpha \in [0,H]} \frac{\tilde\rho(\alpha)}{\alpha}.$$

If $f \in C^\epsilon(\mathbb{R})$ for an $\epsilon > 0$, there exists a unique critical exponent p_c such that $\eta_f(p_c) = 1$; (5.4) easily implies the classical bound

$$(5.5) \qquad d(H) \leqslant \inf_{p \geqslant p_c} (pH - \eta_f(p) + 1),$$

of [17]. Nonetheless (5.4) clearly yields a sharper estimate if $\tilde{\rho}(\alpha)$ is not concave. Thus, strictly more information can be deduced from the histograms of wavelet coefficients than from the scaling function. Note that, though (5.4) can be sharper than (5.5), nonetheless (5.5) is optimal in the sense that no better upper bound can be deduced in full generality, as we will see.

The optimal bounds (5.4) and (5.5) allow one to propose alternative formulas for the multifractal formalism. The *almost-sure multifractal formalism* holds if (5.4) is saturated, i.e. if

$$(5.6) \qquad d(H) = H \sup_{\alpha \in [0,H]} \frac{\tilde{\rho}(\alpha)}{\alpha};$$

and the *quasi-sure multifractal formalism* holds if (5.4) is saturated, i.e. if

$$(5.7) \qquad d(H) = \inf_{p \geqslant p_c} (pH - \eta_f(p) + 1).$$

We will see in the next subsections general results concerning the validity of these formulas.

We conclude this section with a general remark on histograms of wavelet coefficients. The precise values taken by the sequence $N_j(\alpha)$ clearly depend on the wavelet basis chosen. One may wonder what is the maximal information that can be extracted from wavelet histograms, and doesn't depend on the particular wavelet basis chosen; or, more generally, what is the maximal information that can be extracted from wavelet histograms, and is *intrinseque*, meaning here that it remains unchanged when applying to the sequence of wavelet coefficients an operator A such that

$$(5.8) \qquad A \in \bigcap_{\gamma > 0} \mathcal{M}^\gamma \quad \text{and} \quad A^{-1} \in \bigcap_{\gamma > 0} \mathcal{M}^\gamma.$$

One can check that $\rho(\alpha)$ is not intrinseque, but that

$$\lambda(\alpha) = \lim_{\epsilon \to 0} \left(\sup_{\alpha' \in (-\infty, \alpha + \epsilon]} \rho(\alpha') \right)$$

is intrinseque, and that it is 'maximal', in the sense that any intrinseque quantity deduced from $\rho(\alpha)$ can be deduced from $\lambda(\alpha)$.

5.2. Quasi-sure results

This section describes results from [19]. Formula (5.7) certainly does not hold in full generality, and it is extremely easy to construct counterexamples. On the other hand, each time it has been shown to hold, it was the consequence of a functional equation satisfied by the function under study (usually a selfaffinity property, either exact, approximate, or stochastic). Therefore, the general consensus among mathematicians and physicists was that the validity of the multifractal formalism must be a consequence of the precise inner structure of the function considered. Actually, the opposite is true; (5.7) holds for quasi-all functions, i.e., outside a set of the first class of Baire. Let us explain more precisely what we mean.

The multifractal formalism, reformulated as above, states that if f belongs to the topological vector space

(5.9) $$V = \bigcap_{\epsilon>0, p>0} B_{p,loc}^{(\eta(p)-\epsilon)/p, p}$$

then its spectrum of singulgularities satisfies (5.7).

The space V $(= V_\eta)$ is a Baire's space, i.e., any countable intersection of everywhere dense open sets is everywhere dense; we will see that, in V, the set of functions that satisfy (5.7) contains a countable intersection of everywhere dense open sets of V, i.e., contains a dense G_δ set (we say that quasi-all functions of V satisfy (5.7)). In order to precisely state our result, we first have to determine what are the conditions satisfied by a function $\eta(p)$ so that it is a scaling function. The following definition follows directly from Proposition 1 and (4.2); it characterizes scaling functions.

DEFINITION 6. *A function* $\eta(p) : \mathbb{R}^+ \to \mathbb{R}$ *is strongly admissible if* $s(0) > 0$ *and if* $s(q) = q\eta(1/q)$ *is concave and satisfies* $0 \leqslant s'(q) \leqslant 1$.

One immediately sees that if $\eta(p)$ is strongly admissible, it is concave.

THEOREM 3. *Let* $\eta(p)$ *be a strongly admissible function and* V *be the function space defined by* (5.9). *The domain of definition of the spectrum of singularities of quasi-all functions of* V *is the interval* $[s(0), 1/p_c]$ *where*

(5.10) $$d(H) = \inf_{p \geqslant p_c} (Hp - \eta(p) + 1).$$

Remarks: Formula (5.10) states that the spectrum of quasi-all functions is composed of two parts:

- A part defined by $H < \eta'(p_c)$ where the infimum in (5.10) is attained for $p > p_c$, and the spectrum can be computed as the 'usual' Legendre transform of $\eta(p)$

$$d(H) = \inf_{p>0} (Hp - \eta(p) + 1).$$

- A part defined by $\eta'(p_c) \leqslant H \leqslant 1/p_c$ where the infimum in (5.10) is attained for $p = p_c$, and the spectrum is a straight segment $d(H) = Hp_c$.

This second case shows that the initial formulation of Frisch and Parisi (where the Legendre transform is taken on all positive p's) fails in this part of the spectrum. Comparing (5) and (5.10) we see that quasi-all functions of V strive to have their Hölder singularities on a set as large as possible.

The study of the properties of quasi-all functions with a given *a priori* regularity goes back to the famous paper of Banach [5], which gives differentiability properties of quasi-all continuous functions. Recently Z. Buczolich and J. Nagy proved in [9] that quasi-all monotone continuous functions on $[0, 1]$ are multifractal with spectrum $d(H) = H$ for $H \in [0, 1]$.

5.3. Almost-sure results from histograms

This section describes results from [4]. We start by describing the processes we will study. We suppose that, at each scale j, the wavelet coefficients of the process are picked independently from a given histogram. We denote by ρ_j the common probability measure of the 2^j random variables $X_{j,k} = -(\log_2 |c_\lambda|)/j$ (the signs of the wavelet coefficients have no consequence for Hölder regularity; therefore, we do not need to make any assumption on them). The measure ρ_j thus satisfies

$$\mathbb{P}\left(\log(|c_\lambda|) \leqslant 2^{-\alpha j}\right) = \rho_j((-\infty, a]).$$

We need to make two assumptions on the ρ_j. The first one is

$$\exists \epsilon > 0: \quad Supp\,(\rho_j) \subset [\epsilon, +\infty);$$

This assumption means that the sample paths belong to C^ϵ.

Let us now define some quantities that will be pertinent in our study. For each j, let $N_j(\alpha) = \# \{k: |C_{jk}| \geqslant 2^{-\alpha j}\}$ obtained after these 2^j draws have been performed. Therefore, $\mathbb{E}(N_j(\alpha)) = 2^j \rho_j([0, a])$. Note that the word *empirical* will be used in the following in relation to random quantities that are measured on the sample paths (as opposed to deterministic quantities that are derived from the ρ_j). We note

$$\rho(\alpha) = \lim_{\epsilon \to 0} \limsup_{j \to +\infty} \frac{\log_2 \left(2^j \rho_j([\alpha - \epsilon, \alpha + \epsilon])\right)}{j},$$

and $\tilde{\rho}(\alpha)$ is the corresponding empirical quantity already defined. The purpose of the following hypothesis is to make sure that some sizes of wavelet coefficients do not appear with small but nonvanishing probability. If it holds, quantities deduced from histograms and sample paths will coincide.

$$(\mathcal{H}) \quad \left\{ \begin{array}{l} \text{Either } \rho(\alpha) = -\infty \text{ or } \rho(\alpha) \geqslant 0. \text{ If } \rho(\alpha) = 0, \text{ there exists} \\ \text{a subsequence } j_n \text{ and a sequence } \epsilon_n \to 0 \text{ such that} \\ 2^{j_n} \rho_{j_n}([\alpha - \epsilon_n, \alpha + \epsilon_n]) \geqslant 2 j_n^2. \end{array} \right.$$

PROPOSITION 6. *If Hypothesis $(\mathcal{H})$ holds, with probability one, $\forall \alpha > 0\ \rho(\alpha) = \tilde{\rho}(\alpha)$.*

From now on, we suppose that $\rho(\alpha) > 0$ for at least one α. Let

$$H_{max} = \left(\sup_{\alpha > 0} \left(\frac{\rho(\alpha)}{\alpha}\right)\right)^{-1}.$$

THEOREM 4. *Let f be a random wavelet series satisfying* (H). *The spectrum of almost every sample path of f has support included in* $[\epsilon, H_{max}]$, *where*

$$(5.11) \qquad\qquad d(H) = H \sup_{\alpha \in [0,H]} \frac{\rho(\alpha)}{\alpha}.$$

Remarks: By inspecting (5.11), it is clear that $d(H)$ need not be concave, which shows another possible occurrence of the failure of the standard multifractal formalism.

Comparing Proposition 5 and Theorem 4, we see that the spectrum $d(H)$ of a random wavelet series takes the largest possible values compatible with the bounds (5.4), which shows that these bounds are optimal.

References

[1] A. ARNEODO, F. ARGOUL, E. BACRY, J. ELEZGARAY J.-F. MUZY *Ondelettes, multifractales et turbulences,* Diderot, Paris (1995).

[2] A. ARNEODO, E. BACRY J.-F. MUZY *The thermodynamics of fractals revisited with wavelets,* Physica A Vol.213 p.232-275 (1995).

[3] A. ARNEODO, E. BACRY J.-F. MUZY *Random cascades on wavelet dyadic trees,* J. Math. Phys. Vol.39 n.8 p.4142-4164 (1998).

[4] J.-M. AUBRY AND S. JAFFARD *Random wavelet series* Preprint (2000).

[5] S. BANACH *Über die Baire'sche Kategorie gewisser Funktionenmengen,* Studia Math. 3 pp. 174-179 (1931).

[6] M. BEN SLIMANE *multifractal formalisms and anisotropic selfsimilar functions.* Math. Proc. Camb. Philos. Soc. 124, No.2, 329-363 (1998).

[7] G. BOURDAUD *Ondelettes et espaces de Sobolev* Rev. Mat. Iberoam. Vol. 11 n.3 p.477-512 (1995)

[8] J. BOURGAIN, H. BREZIS AND P. MIRONESCU *Lifting in Sobolev spaces* Preprint, 1999

[9] Z. BUCZOLICH AND J. NAGY *Hölder spectrum of typical monotone continuous functions,* Preprint, Eötvös Loránd University, Budapest (1999).

[10] R. DEVORE *Nonlinear Approximation* Acta Numerica. p. 1-99 (1998)

[11] D. DONOHO I. M. JOHNSTONE, G.KERKIACHARIAN AND D.PICARD *Wavelet shrinkage: Asymptopia* J.Roy.Stat. Soc. B 57 (1995) 301-369

[12] B. DUBUC, S. W. ZUCKER, C. TRICOT, J.-F. QUINIOU AND D. WEHBI, Evaluating the fractal dimension of surfaces, Proc. R. Soc. Lond. A 425, 113-127 (1989)

[13] U. FRISCH AND G. PARISI *Fully developed turbulence and intermittency.* Proc. Int. Summer school Phys. Enrico Ferrmi pp.84-88 North Holland (1985).

[14] S. JAFFARD *Pointwise smoothness, two-microlocalization and wavelet coefficients* Publicacions Matematiques Vol.35 p.155-168 (1991).

[15] S.JAFFARD *The multifractal nature of Lévy processes.* Prob. Th. Rel. Fiel., V. 114 N.2 p.207-227 (1999).

[16] S.JAFFARD *Old friends revisited. The multifractal nature of some classical functions.* J. Four. Anal. App., V.3 N.1 p.1-22 (1997).

[17] S. JAFFARD *Multifractal formalism for functions,* S.I.A.M. Journal of Mathematical Analysis V.28 N.4 pp.944-998 (1997).

[18] S. JAFFARD *Oscillation spaces: Properties and applications to fractal and multifractal functions* To appear in J. Math. Phys.

[19] S. JAFFARD *On the Frisch-Parisi conjecture* To appear in J. Math. Pure Appl. (June 2000).

[20] S. JAFFARD AND Y. MEYER *Wavelet methods for pointwise regularity and local oscillations of functions,* Memoirs of the A.M.S. Vol.123 n.587 (1996)

[21] S. JAFFARD, Y. MEYER AND R. RYAN *Wavelet Tools for Science and Technology,* S.I.A.M., to appear (2000)

[22] J.-P. KAHANE *Some random series of functions.* Cambridge University Press (1968).

[23] J.-P. KAHANE AND P.-G. LEMARIÉ *Fourier series and Wavelets,* Gordon Breach, New-York (1996).

[24] A. KAMONT *East J. on Approx.,* Vol 4 n.4 p. 541-564 (1998)

[25] A. KAMONT AND B. WOLNIK *wavelet expansions and Fractal dimension,* 15, No.1, 97-108 (1999).

[26] G. KYRIAZIS *wavelet coefficients measuring smoothness in* $H_p(\mathbb{R}^d)$ Appl. Comp. Har. Anal. 3 p.100-119 (1996).

[27] P.-G. LEMARIÉ AND Y. MEYER. *Ondelettes et bases hilbertiennes.* Revista Math.Iberoamericana Vol.1 (1986).

[28] B. MANDELBROT *Fractals and Scalings in Finance* Springer (1997).

[29] Y. MEYER *Ondelettes et opérateurs* Hermann (1990).

[30] Y. MEYER Wavelets, Vibrations and Scalings, CRM Ser. AMS Vol. 9, *Presses de l'Université de Montréal* (1998).

[31] J. PEETRE New thoughts on Besov spaces, Duke Univ. Math. Ser. I (1976).

[32] F. ROUEFF *Thèse de l'Ecole Nationale Supérieure des Télécommunications,* (2000).

[33] C. TRICOT, *Function norms and fractal dimension,* SIAM J. Math. Anal., 28, 189-212 (1997)

Some Plots of Bessel Functions of Two Variables

F.A. Grünbaum

Department of Mathematics
University of California
Berkeley, CA 94720

ABSTRACT. For a finite reflection group G there is a rich theory developed by Dunkl, Heckman and Opdam leading to the notion of a commuting set of Bessel differential operators. These systems play an important role in the study of Calogero–Moser systems and other problems of physical interest. When G acts on the real line one recovers the usual Bessel function with a well known power series expansion at the origin. We obtain some such expansions in the case of $G = A_2$ acting in the plane and we use these to produce plots of some of these functions.

1. Introduction

Special functions such as Gauss' hypergeometric series and the Bessel functions have played a crucial role in math-physics for a very long time, and many branches of engineering and technology are good examples of such use. The appearance of symbolic/numerical/graphics packages such as Macsyma, Maple, Mathematica, and others has put plots of these functions at the fingertips of many users in applied fields.

The pathbreaking work of Dunkl, Heckman and Opdam, de Jeu and others has produced a rich theory of Bessel functions of several variables which extends in a natural way the one dimensional situation. There are also several versions of Gauss' function in the case of several variables, and a large program dealing with polynomials has been developed by Macdonald, Koornwinder and Cherednik, as well as others.

The author was supported in part by NSF Grant # DMS94-00097 and by AFOSR under Contract FDF49620-96-1-0127.

J.S. Byrnes (ed.), Twentieth Century Harmonic Analysis - A Celebration, 145–149.

These functions have arisen at times in connection with integrable systems or as spherical functions for appropriate symmetric spaces. One can wonder if these relatively recent mathematical objects will eventually have an impact on technology comparable to the one variable case, and how much they will penetrate the engineering literature.

There are at least three interrelated answers to the question above:

a) this will happen only if interesting "down-to-earth" problems ever get solved in terms of these new functions.

b) the amount of heavy going mathematical sophistication needed to go through this new material makes it unlikely that enough "applied math" people will take the time to learn about these new tools and test their applicability.

c) maybe some few plots will not hurt. After all, the best way to see the usefulness of sines and cosines—as well as Bessel functions, elliptic functions, and many other such gems—is to view some few graphs of them. Keep in mind that whether we like it or not, more and more engineering students will be trained in front of a screen with increasing graphical capabilities.

The goal of this note is rather modest. I want to show some of these plots and hopefully provoke someone into doing a better job. Given the quality of the graphs that I show this should not be too much of a challenge. I have not seen any such graphs, or even some of the power series expressions that I will use to compute with, in the literature. It is clear that for computational purposes eventually one would like something smarter than just power series, like piecewise rational approximation or similar things. This can wait until we see if these new functions get used enough so as to warrant such an effort.

2. The eigenvalue problem

On the plane with coordinates (a, b) consider the operators op_1 and op_2 given by

$$op_1(f) = \frac{d^2 f}{db^2} + \frac{d^2 f}{da^2}$$
$$+ k \left(\frac{1}{(\sqrt{3}b + 3a)^2} + \frac{1}{(\sqrt{3}b - 3a)^2} + \frac{1}{12b^2} \right) f$$

$$op_2(f) = \frac{27k \left(\frac{1}{(\sqrt{3}b-3a)^2} - \frac{1}{(\sqrt{3}b+3a)^2} \right) \frac{df}{db}}{2\sqrt{3}}$$
$$+ \frac{d^3 f}{da^3} - 3\frac{d^3 f}{dadb^2}$$
$$+ \frac{3k \left(\frac{1}{(\sqrt{3}b+3a)^2} + \frac{1}{(\sqrt{3}b-3a)^2} - \frac{1}{6b^2} \right) \frac{df}{da}}{2}.$$

Here k is an arbitray parameter.

These operators are invariant under the operations 1) b goes into $-b$ as well as 2) (a, b) goes into $(-a/2 + \sqrt{3}/2b, \sqrt{3}/2a + b/2)$. The first one is a reflection across the a axis, the second one a reflection across the axis given by the vector $(-\sqrt{3}, 1)$. The same is true if one considers one more reflection, across the axis making an angle of -60 degrees with the a axis. There is then a three-fold symmetry, and since these reflections generate a group of six elements (the symmetric group on three symbols) the full group of symmetries has order six.

The operators commute and we define the Bessel function as the function made symmetric by adding six terms of the form

$$f(a, b, s_1, s_2) = e^{s_1 a + s_2 b}(1 + \text{small at infinity})$$

that solves the system

$$op_1(f) = \lambda_1 f$$

and

$$op_2(f) = \lambda_2 f.$$

This clearly requires $\lambda_1 = s_1^2 + s_2^2$ and $\lambda_2 = s_1^3 - 3s_1 s_2^2$.

Put $a = r \cos t$ and $b = r \sin t$ and put $x = \cos 3t$ and make the analoguous change of variables on the spectral side $s_1 = \beta \cos s$, $s_2 = \beta \sin s$, $y = \cos 3s$.

For later use, introduce p as any root of the equation

$$k = -\frac{4p^2 - 12p}{3}.$$

Conjugate the resulting operators by the factor

$$r^p(1 - x^2)^{p/6}$$

i.e. define opp_i by

$$r^p(1 - x^2)^{p/6} opp_i = op_i \, r^p(1 - x^2)^{p/6}$$

to get new operators opp_i $(i = 1, 2)$ given by

$$-\frac{9\frac{d^2 f}{dx^2}(x - 1)(x + 1)}{r^2} - \frac{3\frac{df}{dx}(2p + 3)x}{r^2} + \frac{\frac{df}{dr}(2p + 1)}{r} + \frac{d^2 f}{dr^2}$$

and

$$-\frac{3\frac{df}{dx}(35x^2 - 4p^2 + 6p - 17)}{r^3} + \frac{9\frac{d^2 f}{dr\,dx}(6x^2 + 2p - 3)}{r^2}$$

$$-\frac{27\frac{d^3 f}{dx^3}(x - 1)^2(x + 1)^2}{r^3} + \frac{27\frac{d^3 f}{dr\,dx^2}(x - 1)x(x + 1)}{r^2}$$

$$-\frac{135\frac{d^2 f}{dx^2}(x - 1)x(x + 1)}{r^3} - \frac{9\frac{d^3 f}{dr^2\,dx}(x - 1)(x + 1)}{r}$$

$$-\frac{3\frac{d^2 f}{dr^2}x}{r} + \frac{3\frac{df}{dr}x}{r^2} + \frac{d^3 f}{dr^3}x$$

respectively.

Consider now the resulting equations

$$opp_1 \, f = \beta^2 f$$

$$opp_2 \, f = \beta^3 y f.$$

It is easy to see that they admit solutions of the form

$$1 + \sum c_{ijkl} r^{2i} (r^3 x)^j \beta^{2k} (\beta^3 y)^l$$

with $c_{ijkl} = c_{klij}$.

In particular we get the well known fact that the Bessel functions are symmetric in the interchange between "spatial" and "spectral" variables. This is then a "trivial or basic" instance of the bispectral property discussed in [DG] but for systems of partial differential operators. In this connection it is pleasing to see that the relations

$$ad^3(opp_i)(r^2) = 0 \quad i = 1, 2$$

and

$$ad^4(opp_i)(r^3 x) = 0 \quad i = 1, 2$$

hold in this case.

If we are in the simpler case when $\beta = y = 1$ we have solutions of the form

$$1 + \sum c_{ij} r^{2i+3j} x^j$$

to the equations

$$opp_1 \, f = f, \; opp_2 \, f = f.$$

For this case we have a simple expression for the coefficients c_{ij}, namely

$$c_{ij} = \Gamma(p)/\Gamma(p/3)3^i\Gamma(p/3 + i + j)/(2^{2j+2i}i!j!\Gamma(p + 2i + 3j))$$

and this will be used in the plots below. These plots, made for different values of p, give an indication of f for $x = \cos 3\theta$ in $(-1, 1)$ and r in $(0, 1)$.

To get the true Bessel functions we still need to multiply these expressions by the conjugating factor

$$r^p(1 - x^2)^{p/6}$$

used above.

E. Opdam [O2] mentioned to me that M. de Jeu had made some relevant computations. I have received some notes from de Jeu [dJ2] who has obtained other expressions which could be used in producing plots of the Bessel functions.

There are also expressions for those Bessel functions in [O] that could be used for plotting purposes.

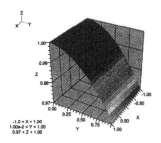

FIGURE 1. p=-19/2

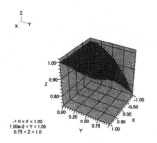

FIGURE 2. p=-5/2

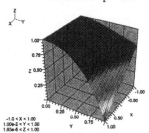

FIGURE 3. p=-3/2

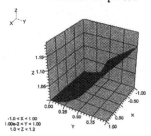

FIGURE 4. p=1/2

References

[dJ1] de Jeu, M. F. E., *The Dunkl transform*, Invent. Math. **113**, No. 1 (1993), 147–162.

[dJ2] _____, Private communication.

[DG] Duistermaat, J., and Grünbaum, F. A., *Differential equations in the spectral parameter*, CMP **103** (1986), No. 2, 177–240.

[D] Dunkl, C., *Hankel transforms associated to finite reflection groups, hypergeometric functions in domains of positivity, Jack polynomials, and applications*, (Tampa, Florida, 1991), Contemp. Math. **138** (1992), 123–138.

[H] Heckman, G., *Integrable systems and reflection groups*, Unpublished lecture notes, Fall School on Hamiltonian Geometry, Woudschoten, 1992.

[O] Okounkov, A., and Olshanski, G., *Shifted Jack polynomials, binomial formula, and applications*, Math. Res. Letters **4** (1997), 69–78.

[O1] Opdam, E., *Dunkl operators, Bessel functions and the discriminant of a finite Coxeter group*, Compositio Math. **85** (1993), 333–373.

[O2] _____, Private communication.

Lesser Known FFT Algorithms

R. Tolimieri and M. An

ABSTRACT. The introduction of the Cooley-Tukey Fast Fourier transform (C-T FFT) algorithm in 1968 was a critical step in advancing the widespread use of digital computers in scientific and technological applications. Initial efforts focused on realizing the potential of the immense reduction in arithmetic complexity afforded by the FFT for computing the finite Fourier transform and convolution. On existing serial butterfly architectures, this limited implementations of the FFT to transform sizes a power of two.

The original C-T FFT algorithms and its many extensions rested mainly on additive structures of the data indexing set and divide-and-conquer strategies for reducing complexity. The relative cost of multiplications as compared with additions on these early machines motivated the study of new algorithms for reducing multiplications usually at the cost of increasing additions. These new algorithms were based on multiplicative structures of the data indexing set. They resulted in greater flexibility as compared to earlier efforts by allowing for fast small prime size computations which could be embedded in a wide collection of large transform sizes.

During the 1980s, the increasing importance of RISC and parallel computers removed some of the initial motivation for these multiplicative FFTs. Many RISC architectures featured a hardwired multiply and accumulate which permitted multiplications to be nested in additions. The goal was to arrange the computations so that most multiplications were followed by an addition. Parallel architectures placed the major algorithmic burden on controlling the data flow. The exotic data flow of the multiplicative FFTs were often incompatible or at least required an immense coding effort for efficient parallel computation. Only the need for flexible transform sizes justified continued efforts.

The recent importance of FPGAs and reconfigurable hardware renews the need to reduce multiplication counts. In this talk, we will review some of the work on multiplicative approaches introduced mainly at IBM in the 1970s and 1980s and place these results within the realm of harmonic analysis.

1. Introduction

The fast Fourier transform (FFT) algorithm has a long history dating from Gauss but has over the years been rediscovered in a variety of contexts. It is closely related to the Poisson

J.S. Byrnes (ed.), Twentieth Century Harmonic Analysis - A Celebration, 151–162.

summation formula and the Chinese remainder theorem for polynomials. In the latter it reflects the semi-simplicity of the complex group algebra of $\mathbf{Z}/N$, $N \geqslant 2$.

In 1965 the FFT was introduced to IBM by James Cooley [7] and was the joint work of James Cooley and John Tukey. Although other FFT algorithms such as the Good-Thomas prime factor algorithm had been in use, especially in geophysics [10, 16], the great reduction in computation time for large size problems, flexibility, ease of coding, and widespread applicability of the Cooley-Tukey (CT) FFT algorithm was quickly recognized by IBM management. They saw it as the key tool for the rapidly expanding use of digital computers.

The success of the CT FFT algorithm often hid several of its disadvantages in scientific and engineering applications. In fact, from the very beginning, especially at IBM, other algorithms for computing the finite Fourier transform had been discovered and sporadically found use. Typically these algorithms increase the number of additions and decrease the number of multiplications as compared with the CT FFT.

During the 1980s, with the introduction of reduced instruction set computer (RISC) processors which are capable of nesting multiplications inside additions (pipelined dual ops), these approaches were ignored.

We will discuss two historically neglected approaches which, driven by recent technological advances, have become increasingly important, the Rader-Winograd multiplicative FFT algorithms and the reduced-polynomial transform algorithms.

Much of this work was carried out during the late 1970s and early 1980s by S. Winograd [19–22] and H. J. Nussbaumer [12, 13] and placed in a mathematical framework by several of their collaborators including L. Auslander, E. Feig [3] and at a later time by C. S. Burrus [5, 6], I. Gertner [9], and M. Rofheart [15]. Additional results and references can be found in texts [4, 8, 11, 17, 18].

Proofs of theorems stated in this presentation are in [18].

2. Multiplicative Theory

2.1. *Introduction*

Standard divide-and-conquer FFT algorithms are based on the existence of nontrivial subgroups of the natural additive group structure of the underlying indexing set. The specification of a subgroup decomposes the indexing set into cosets and the computation is decomposed relative to the coset partitioning. The importance of two-to-a-power size indexings in early programming efforts is due to the plentitude of subgroups and the simplicity of implementing the two-point Fourier transform (butterfly). Data readdressing is especially regular (striding) and was eventually hardwired into many architectures (LOAD-STORE).

The immense speed-up in run time of two-to-a-power FFT codes as compared with direct computation resulted in significant compromises in many applications. The need to zero-pad from a natural computation size to the nearest two-to-a-power size introduced increased memory requirements and less accuracy, especially in multidimensional problems. In some applications, such as X-ray crystallography, symmetry relations on data which had played an essential role before the FFT became difficult to fully exploit inside standard FFT codes.

Moreover, the regularity of these codes caused memory call conflicts on many large vector and parallel processors. The need for fast Fourier transform codes acting on more flexible data set sizes became, by the end of the 1970s, an increasingly important goal. The most important step in achieving this goal was quickly seen to be the designing of algorithms for small prime size Fourier transforms which could then be nested in standard FFT codes.

C. Rader [14] and independently S. Winograd had developed such algorithms in the early 1970s, but for the most part these algorithms were viewed as mathematical oddities. These algorithms suffered from several disadvantages which at the time precluded their widespread adoption. Coding is difficult, with each size requiring a machine dependent special, time-consuming coding effort to implement complex data readdressing. On reduced instruction set computer (RISC) architectures the instruction set for these codes can use significant memory.

From the late 1980s codes for small prime size Fourier transforms have increasingly become part of standard programming packages. In this section we will first describe the original derivations of C. Rader and S. Winograd and then place them in a harmonic analysis framework.

2.2. Rader algorithm

The additive group $\mathbf{Z}/p$, p a prime, has no nontrivial subgroups, but the multiplicative group $U(p)$ of nonzero elements is cyclic. The main idea underlying the p-point FFT as described by C. Rader is that the p-point Fourier transform relative to input and output data, ordered by powers of any generator of $U(p)$, has an especially simple form: the p-point Fourier transform becomes a $(p-1)$-point cyclic convolution.

For odd prime p, the group of units $U(p^m)$ of $\mathbf{Z}/p^m$ is again a cyclic group. Using this result S. Winograd independently extended Rader's result to odd prime power sizes. Fourier transform algorithms over finite fields can be developed in a similar manner. These last FFT algorithms are important in number theoretic transforms and in error-correcting coding [1,2].

Suppose p is an odd prime and $t = p - 1$. The *unit group* $U(p)$ of $\mathbf{Z}/p$ is a cyclic group of order t. Choosing a generator g of $U(p)$, we have

$$U(p) = \{1, g, \ldots, g^{t-1}\}.$$

Denote by π_g the permutation of $\mathbf{Z}/p$ defined by

$$\pi_g = \begin{pmatrix} 0 & 1 & g & \cdots & g^{t-1} \end{pmatrix},$$

and by P_g the $p \times p$ permutation matrix defined by

$$P_g = [\mathbf{e}_0 \ \mathbf{e}_1 \ \mathbf{e}_g \ \cdots \ \mathbf{e}_{g^{t-1}}]$$

where $\mathbf{e}_n$ is the p-tuple having 1 in the n-th position and 0 otherwise, $0 \leqslant n < p$. The main result of C. Rader is contained in the following theorem.

THEOREM 1.

$$P_g^{-1} F(p) P_g = \begin{bmatrix} 1 & 1 & \cdots & 1 \\ 1 & & & \\ \vdots & & C(p) & \\ 1 & & & \end{bmatrix},$$

where $C(p)$ is the $(p-1) \times (p-1)$ skew-circulant matrix

$$C(p) = \left[v^{g^{j+k}} \right] = \begin{bmatrix} v & v^g & \cdots \\ v^g & v^{g^2} & \\ \vdots & & \\ v^{g^{t-1}} & & \end{bmatrix}, \qquad v = e^{2\pi i \frac{1}{p}}.$$

EXAMPLE 1. For $p = 5$ and $g = 2$,

$$P_2 = \begin{bmatrix} 1 & 0 & 0 & 0 & 0 \\ 0 & 1 & 0 & 0 & 0 \\ 0 & 0 & 1 & 0 & 0 \\ 0 & 0 & 0 & 0 & 1 \\ 0 & 0 & 0 & 1 & 0 \end{bmatrix}$$

and

$$P_2^{-1} F(5) P_2 = \begin{bmatrix} 1 & 1 & 1 & 1 & 1 \\ 1 & v & v^2 & v^4 & v^3 \\ 1 & v^2 & v^4 & v^3 & v \\ 1 & v^4 & v^3 & v & v^2 \\ 1 & v^3 & v & v^2 & v^4 \end{bmatrix}.$$

Set

$$\mathbf{H} = F(t) \begin{bmatrix} v \\ v^g \\ \vdots \\ v^{g^{t-1}} \end{bmatrix}.$$

THEOREM 2.

$$\left(1 \otimes F(t)^{-1}\right) P_g^{-1} F(p) P_g \left(1 \oplus F(t)\right) = \begin{bmatrix} 1 & t & & & & \\ & & & & \Large 0 & \\ 1 & -1 & & & & \\ & & & & & & H_{t-1} \\ & & & & & & & \\ \Large 0 & & & & & \ddots & \\ & & & & & \cdot & \\ & & H_1 & & & \end{bmatrix}.$$

Consider the functions χ_n, $0 \leqslant n < t$, on $\mathbf{Z}/p$ defined by the following table.

	χ_0	χ_1	$\cdots$	χ_{t-1}
0	0	0		0
1	1	1		1
g	1	w		w^{t-1}
g^2	1	w^2		$w^{2(t-1)}$
$\vdots$				
g^{t-1}	1	w^{t-1}		$w^{(t-1)(t-1)}$

$w = e^{2\pi i \frac{1}{t}}$.

TABLE 1. Multiplicative characters

Since H_n, $0 \leqslant n < t$, can be written as

$$H_n = \begin{bmatrix} 1 & v & v^g & \cdots & v^{g^{t-1}} \end{bmatrix} \begin{bmatrix} 0 \\ 1 \\ w^n \\ \vdots \\ w^{(t-1)n} \end{bmatrix},$$

we have the following result.

THEOREM 3.

$$H_n = F(p)(\chi_n)(1), \quad 0 \leqslant n < t.$$

2.3. *Multiplicative characters*

Table 1 describes the multiplicative characters of $\mathbf{Z}/p$. Such characters were introduced by Gauss and studied in his work on reciprocity laws. The complex constants H_n, $0 \leqslant n < t$, are called the Gauss sums of multiplicative characters.

The general definition is given as follows. A mapping $\chi : U(p) \to \mathbf{C}$ is called a multiplicative character mod p if χ is a homomorphism from the multiplicative group $U(p)$ into the multiplicative group of nonzero complex numbers,

$$\chi(ab) = \chi(a)\chi(b), \quad a, b \in U(p).$$

A multiplicative character mod p is completely determined by its value on the generator g of $U(p)$ by

$$\chi(g^m) = \chi(g)^m, \quad 0 \leqslant m < t.$$

Since $g^t = 1$, $\chi(g)$ is a t-th root of unity. For $0 \leqslant n < t$, denote by χ_n the multiplicative character defined by

(2.1) $$\chi_n(g) = w^n$$

and observe that the multiplicative characters χ_n, $0 \leqslant n < t$, defined by (2.1) coincide with the functions in Table 1.

Denote by $L^2(\mathbf{Z}/p)$, the inner product space of all complex valued functions on $\mathbf{Z}/p$ with inner product

$$\langle f, g \rangle = \sum_{n=0}^{p-1} f(n)g^*(n), \quad f, g \in L(\mathbf{Z}/p),$$

where $*$ denotes complex conjugation.

Set $r = \frac{t}{2}$.

THEOREM 4.

$$\chi_r(g) = -1$$

and

$$\chi_n^* = \chi_{t-n}, \quad 0 \leqslant n < t.$$

THEOREM 5. *For any two multiplicative characters χ and χ', we have*

$$\langle \chi, \chi' \rangle = \begin{cases} t, & \chi = \chi' \\ 0, & \chi \neq \chi'. \end{cases}$$

Extend the domain of definition of each multiplicative characters χ to $\mathbf{Z}/p$ by setting $\chi(0) = 0$. Set $\mu_n = t^{-\frac{1}{2}}\chi_n$, $0 \leqslant n < t$.

THEOREM 6. *The set*

$$e_0, \mu_0, \mu_1, \ldots, \mu_{t-1}$$

is an orthonormal basis of $L^2(\mathbf{Z}/p)$.

Since $F(n)^2 f(x) = nf(-x)$ we have

$$F(p)^2\chi(x) = p\chi(-x) = p\chi(-1)\chi(x),$$

implying that the multiplicative characters are eigenvectors of $F(p)^2$,

$$F(p)^2\chi = p\chi(-1)\chi$$

and that the space spanned by

$$\{\chi, F(p)\chi\}$$

is invariant under the action of $F(p)$. The action of $F(p)$ on the multiplicative characters is described in the next two theorems.

THEOREM 7.

$$\begin{aligned} F(p)e_0 &= e_0 + \chi_0. \\ F(p)\chi_0 &= te_0 - \chi_0. \end{aligned}$$

For any nontrivial multiplicative character χ, we have

THEOREM 8.

$$F(p)\chi = \sqrt{p}G_p(\chi)\chi^*,$$

where

$$F(p)\chi(1) = \sqrt{p}G_p(\chi).$$

Proof

$$\begin{aligned} F(p)\chi(x) &= \sum_{y \in U(p)} \chi(y)e^{2\pi i\frac{xy}{p}} \\ &= \sum_{y \in U(p)} \chi(x^{-1}y)e^{2\pi i\frac{y}{p}} \\ &= \chi(x^{-1})\sum_{y \in U(p)} \chi(y)e^{2\pi i\frac{y}{p}} \\ &= \chi(x^{-1})F(p)\chi(1). \end{aligned}$$

By $\|F(p)f\|^2 = p\|f\|^2$, we have

$$|G_p(\chi)| = 1.$$

Relative to the orthonormal basis $e_0, \mu_0, \mu_1, \ldots, \mu_t$ of $L^2(\mathbf{Z}/p)$, we have

$$\begin{aligned} F(p)e_0 &= e_0 + t^{\frac{1}{2}}\mu_0, \\ F(p)\mu_0 &= t^{\frac{1}{2}}e_0 - \mu_0 \end{aligned}$$

and

$$F(p)\mu_n = \sqrt{p}G_p(\chi)u_{t-n}, \quad 1 \leqslant n < t.$$

The matrix of $F(p)$ relative to this basis is given by

$$
\begin{bmatrix}
1 & t^{\frac{1}{2}} & & & & \\
& & & & \quad 0 & \\
t^{\frac{1}{2}} & -1 & & & & \\
& & & & & \\
& & & & \sqrt{\bar{p}}G_p(\chi_{t-1}) & \\
& & & \ddots & & \\
\quad 0 & & & & \ddots & \\
& & \sqrt{\bar{p}}G_p(\chi_1) & & &
\end{bmatrix}.
$$

This matrix representation of $F(p)$ is essentially the same as that in theorem 2. It is the first step in several studies into the p-point Fourier transform [3]. Other studies include:

- construction of an orthonormal basis diagonalizing $F(p)$,
- determining rational subspaces of $F(p)$.

3. Multidimensional Algorithm

3.1. Introduction

The reduced transform algorithms (RTA) [9, 15, 17] compute a multidimensional Fourier transform by first projecting the multidimensional input data onto lines (more generally, lower dimensional planes) and then computing the one-dimensional Fourier transform of the projected data. By the periodic version of the projection slice theorem from tomography, these one-dimensional Fourier transforms compute the multidimensional Fourier transform. The RTA can be viewed as periodic Radon transforms.

Polynomial rings play many roles in algorithm design. Linear convolution can be defined as polynomial product and cyclic convolution mod N can be defined as polynomial product mod $x^N - 1$. Fourier transform can be defined over certain quotient polynomial rings. The Chinese remainder theorem (CRT) for polynomial rings is especially important in the Cook-Toom and Winograd convolution algorithms [18, 22]. The convolution theorem which diagonalizes cyclic convolution mod N relative to the N-point Fourier transform basis can be viewed as the CRT applied to the quotient polynomial ring $\mathbf{C}[x]/x^N - 1$,

$$
\mathbf{C}[x]/x^N - 1 \cong \mathbf{C}[x]/x - 1 \oplus \mathbf{C}[x]/x - v \oplus \cdots \oplus \mathbf{C}[x]/x - v^{N-1},
$$

where $v = e^{2\pi i \frac{1}{N}}$.

Polynomial ring theory is central to many of the results in the works of H. J. Nussbaumer [12, 13], both for multidimensional convolution and Fourier transform (polynomial transform). One goal in these works is to rely as much as possible on shifts and cyclic shifts to carry out these computations. Polynomial transforms compute multidimensional convolutions by first using the CRT to decompose the computation into subcomputations, some of which can be computed by Fourier transform computations over rings and then completed by using the inverse CRT isomorphism to combine the subcomputations.

The RTA can be adapted to computing multidimensional convolution and provides a geometric setting for the Nussbaumer polynomial transform. A multidimensional convolution can be computed by RTA which first projects the multidimensional data onto lines, computes the convolution of the one-dimensional data and then combines the subcomputations. It differs from the RTA for multidimensional Fourier transform in that the last step does not rely on the projection slice theorem. We will address this last problem by defining an inverse projection algorithm which reconstructs multidimensional data from its one-dimensional order projections.

In some ways this is a curious result since, in the continuous case, the inversion requires either Fourier transform computations or convolutions and back projections and can be extremely costly. However a close inspection of these stages in the Nussbaumer polynomial transform algorithm shows that this step can be accomplished using only additions. We will show, using the constructions underlying the RTA, how this last step can be implemented using integer matrix multiplication.

3.2. *Periodic back projection*

Consider $\mathbf{Z}/3 \times \mathbf{Z}/3$. By a line in $\mathbf{Z}/3 \times \mathbf{Z}/3$, we mean a maximal cyclic subgroup of $\mathbf{Z}/3 \times \mathbf{Z}/3$. There are 4 distinct lines in $\mathbf{Z}/3 \times \mathbf{Z}/3$,

$$
\begin{aligned}
L(1,j) &= \{(0,0),\ (1,j),\ (2,j)\}, \quad 0 \leqslant j < 3. \\
L(0,1) &= \{(0,0),\ (0,1),\ (0,2)\}.
\end{aligned}
$$

Suppose $f \in L(\mathbf{Z}/3 \times \mathbf{Z}/3)$. The periodizations of f orthogonal to the 4 lines can be constructed as follows. Denote by $\mathbf{f}$ the lexicographically ordered two dimensional array f

$$
\mathbf{f} = \begin{bmatrix} f(0,0) \\ f(1,0) \\ f(2,0) \\ f(0,1) \\ f(1,1) \\ f(2,1) \\ f(0,2) \\ f(1,2) \\ f(2,2) \end{bmatrix}.
$$

The periodization orthogonal to $L(0,1)$ consists of the sums

$$
f(0,0) + f(1,0) + f(2,0),
$$

$$f(0,1) + f(1,1) + f(2,1),$$
$$f(0,2) + f(1,2) + f(2,2),$$

which we can write as

$$(I_3 \otimes 1_3^t)\mathbf{f}.$$

The periodization orthogonal to $L(1,0)$ consists of the sums

$$f(0,0) + f(0,1) + f(0,2),$$
$$f(1,0) + f(1,1) + f(1,2),$$
$$f(2,0) + f(2,1) + f(2,2),$$

which we can write as

$$(I_3 \otimes 1_3^t)P(9,3)\mathbf{f} = (1_3^t \otimes I_3)\mathbf{f}.$$

Set

$$S_3 = \begin{bmatrix} 0 & 0 & 1 \\ 1 & 0 & 0 \\ 0 & 1 & 0 \end{bmatrix}$$

and $Z_3 = I_3 \oplus S_3 \oplus S_3^2$. The periodization orthogonal to $L(1,1)$ is given by

$$(I_3 \otimes 1_3^t)P(9,3)Z_3\mathbf{f} = (1_3^t \otimes I_3)Z_3\mathbf{f}$$

and the periodization orthogonal to $L(1,2)$ is given by

$$(I_3 \otimes 1_3^t)P(9,3)Z_3^2\mathbf{f} = (1_3^t \otimes I_3)Z_3^2\mathbf{f}.$$

We assume that we know

$$\begin{bmatrix} I_3 \otimes 1_3^t \\ 1_3^t \otimes I_3 \\ (1_3^t \otimes I_3)Z_3 \\ (1_3^t \otimes Z_3^2) \end{bmatrix}\mathbf{f} = A\mathbf{f}$$

and we want to compute $\mathbf{f}$. We will show

$$A^T A\mathbf{f} = (9I_3 + E_3 \otimes E_3)\mathbf{f},$$

where E_3 is the 3×3 matrix of all ones.

Since $(E_3 \otimes E_3)\mathbf{f}$ is known from the sum

$$\sum_{j=0}^{2}\sum_{k=0}^{2} f(j,k),$$

We can compute $\mathbf{f}$ from $A^T A\mathbf{f}$.

The following table lists the matrices in A, their transposes and the products of the matrices with their transposes. $A^T A$ is the sum of the matrices in the last column.

X	X^T	$X^T X$
$I_3 \otimes 1_3^t$	$I_3 \otimes 1_3$	$I_3 \otimes E_3$
$1_3^t \otimes I_3$	$1_3 \otimes I_3$	$E_3 \otimes I_3$
$(1_3^t \otimes I_3)Z_3$	$Z_3^2(1_3 \otimes I_3)$	$\begin{bmatrix} I_3 & S_3 & S_3^2 \\ S_3^2 & I_3 & S_3 \\ S_3 & S_3^2 & I_3 \end{bmatrix}$
$(1_3^t \otimes I_3)Z_3^2$	$Z_3(1_3 \otimes I_3)$	$\begin{bmatrix} I_3 & S_3^2 & S_3 \\ S_3 & I_3 & S_3^2 \\ S_3^2 & S_3 & I_3 \end{bmatrix}$

Since $I + S_3 + S_3^2 = E_3$, $A^T A$ is as claimed.

References

[1] R. C. Agarwal and C. S. Burrus, "Fast convolution using Fermat number transforms with applications to digital filtering," *IEEE Trans. ASSP* **ASSP22**, 87-97, 1974.

[2] R. C. Agarwal and C. S. Burrus, "Number theoretic transforms to implement fast digital convolution," *Proc. IEEE* **63**, 550-560, 1975.

[3] L. Auslander, E. Feig and S. Winograd, "The multiplicative complexity of the discrete Fourier transform," *Adv. in Appl. Math.* (5), 31-55, 1984.

[4] R. E. Blahut, *Fast Algorithms for Digital Signal Processing*, Addison-Wesley, Reading, MA 1985.

[5] C. S. Burrus, "An in-place, in-order prime factor FFT algorithm," *IEEE Trans. ASSP*, **ASSP29**, 806-817, 1979.

[6] C. S. Burrus and T. W. Park, *DFT/FFT and Convolution Algorithms*, Wiley and Sons, New York, 1985.

[7] J. Cooley and J. Tukey, "An algorithm for the machine calculation of complex Fourier series," *Math. Comp.*, **19** 297-301, 1965.

[8] Hari Krishna Garg, *Digital Signal Processing: Number theory, convolution, fast Fourier Transforms, and applications*, CRC Press, London, 1998.

[9] I. Gertner, "A new efficient algorithm to compute the two-dimensional discrete Fourier transform," *IEEE Trans. ASSP* **ASSP36**(7), 1036-1050, 1988.

[10] I. J. Good, "The interaction algorithm and practical Fourier analysis," *J. Royal Statistics*, **B**(2), 361-375, 1958.

[11] J. H. McClellan and C. M. Rader, *Number theory in Digital Signal Processing*, Prentice Hall, 1979.

[12] H. J. Nussbaumer, "Fast polynomial transform algorithms for digital convolution," *IEEE Trans. ASSP* **ASSP28**, 205-215, 1980.

[13] H. J. Nussbaumer, *Fast Fourier Transform and Convolution Algorithms*, Springer-Verlag, Berlin, Heidelberg and New York, 1981.

[14] C. Rader, "Discrete Fourier transform when the number of data samples is prime," *Proc. IEEE*, **56**, 1107-1108, 1968.

[15] M. Rofheart, *Algorithms and Methods for Multidimensional Digital Signal Processing*, PhD Thesis, the City University of New York, 1991.

[16] L. H. Thomas, "Using computers to solve problems in physics," *Applications of Digital Computers*, Ginn and Co., Boston, 1963.

[17] R. Tolimieri, M. An and C. Lu, *Mathematics of Multidimensional Fourier Transform Algorithms 2nd ed.*, Springer, New York 1997.

[18] R. Tolimieri, M. An and C. Lu, *Algorithms for Discrete Fourier Transform and Convolution, 2nd ed.*, Springer, New York 1997.

[19] S. Winograd, "On computing the discrete Fourier transform," *Math. of Computation*, **32**(141) 175-199, 1978.

[20] S. Winograd, "On computing the discrete Fourier transform," *Proc. Nat. Acad. Science, USA*, **73**(4), 1005-1006, 1976.

[21] S. Winograd, "Complexity of computations," *CBMS Regional Conf. Ser. in Math.*, **33** Philadelphia, 1980.

[22] S. Winograd, "On multiplication of polynomials modulo a polynomial," *SIAM J. on Computing* **9**(2), 225-229, 1980.

The Phase Problem of X-ray Crystallography

H.A. Hauptman

Hauptman-Woodward Medical Research Institute, Inc.
73 High Street
Buffalo, NY, USA
hauptman@hwi.buffalo.edu

ABSTRACT. The intensities of a sufficient number of X-ray diffraction maxima determine the structure of a crystal, that is, the positions of the atoms in the unit cell of the crystal. The available intensities usually exceed the number of parameters needed to describe the structure. From these intensities a set of numbers $|E_\mathbf{H}|$ can be derived, one corresponding to each intensity. However, the elucidation of the crystal structure also requires a knowledge of the complex numbers $E_\mathbf{H} = |E_\mathbf{H}| \exp(i\varphi_\mathbf{H})$, the normalized structure factors, of which only the magnitudes $|E_\mathbf{H}|$ can be determined from experiment. Thus, a "phase" $\varphi_\mathbf{H}$, unobtainable from the diffraction experiment, must be assigned to each $|E_\mathbf{H}|$, and the problem of determining the phases when only the magnitudes $|E_\mathbf{H}|$ are known is called the "phase problem". Owing to the known atomicity of crystal structures and the redundancy of observed magnitudes $|E_\mathbf{H}|$, the phase problem is solvable in principle.

Probabilistic methods have traditionally played a key role in the solution of this problem. They have led, in particular, to the so-called tangent formula which, in turn, has played the central role in the development of methods for the solution of the phase problem.

Finally, the phase problem may be formulated as one in constrained global optimization. A method for avoiding the countless local minima in order to arrive at the constrained global minimum leads to the *Shake-and-Bake* algorithm, a completely automatic solution of the phase problem for structures containing as many as 1000 atoms when data are available to atomic resolution.

In the case that single wavelength anomalous scattering (SAS) data are available, the probabilistic machinery leads to estimates of special linear combinations of the phases, the so-called structure invariants. A method of going from estimates of the structure invariants to the values of the individual phases is described.

J.S. Byrnes (ed.), Twentieth Century Harmonic Analysis - A Celebration, 163–171.

1. Introduction

When a crystal is irradiated with a beam of X-rays the resulting interference effect gives rise to the so-called diffraction pattern which is uniquely determined by the crystal structure. Only the intensities of the scattered rays can be measured; the phases, which are also needed in order to work backwards, from diffraction pattern to the atomic positions, are lost in the diffraction experiment. However, owing to the known atomicity of real structures and the large number of observable intensities, the lost phase information is in fact contained in the measured intensities. The problem of recovering the missing phases, when only the intensities are available, is known as the phase problem. Alternatively, since the magnitudes $|E|$ of the normalized structure factors $E = |E| \exp(i\varphi)$ are readily determined from the measured diffraction intensities, the phase problem may be defined as the problem of determining the phases φ when the magnitudes $|E|$ are given.

Due to the redundancy of known magnitudes $|E|$, the phase problem is an over determined one and is therefore solvable in principle. This over determination implies the existence of relationships among the Es and, therefore, since the magnitudes $|E|$ are presumed to be known, the existence of identities among the phases φ alone, dependent on the known magnitudes $|E|$, which must of necessity be satisfied. The so-called direct methods are those which exploit these relationships in order to go directly from known magnitudes $|E|$ to desired phases φ. They do not depend on the presence of heavy atoms or atoms having other special scattering properties, for example anomalous scatterers, or prior structural knowledge.

The techniques of modern probability theory lead to the joint probability distributions of arbitrary collections of normalized structure factors from which the conditional probability distributions of selected sets of phases, given the values of suitably chosen magnitudes $|E|$, may be inferred. These distributions, dependent on known magnitudes $|E|$, constitute the foundation on which direct methods are based. They have provided the unifying thread from the beginning, circa 1950, until the present time. In particular, they have led to the recent formulation of the phase problem as one of constrained global optimization [1].

In the case that the structure consists of N identical atoms in the unit cell the relationship between diffraction pattern and crystal structure is given by the pair of equations

$$(1.1) \qquad E_{\mathbf{H}} = |E_{\mathbf{H}}| \exp(i\varphi_{\mathbf{H}}) = N^{-1/2} \sum_{j=1}^{N} \exp(2\pi i \mathbf{H} \cdot r_j)$$

$$(1.2) \qquad \langle E_{\mathbf{H}} \exp(-2\pi i \mathbf{H} \cdot r) \rangle_{\mathbf{H}} \approx \begin{cases} N^{-1/2} & \text{if } r = r_j \\ 0 & \text{if } r \neq r_j \end{cases}$$

where r_j is the position vector of the atom labeled j and $|E_H|$ is obtained from the intensity of the scattered beam in the direction labeled by the reciprocal lattice vector E_H. Clearly if one is to determine the crystal structure from Eq.(1.2) it is necessary to know not only the magnitudes $|E_H|$, obtainable from the diffraction experiment, but also the phases φ_H, lost in the diffraction experiment.

2. Method

2.1. The Structure Invariants

Equation (1.2) implies that the normalized structure factors E_H determine the crystal structure. However, Eq.(1.1) does not imply that, conversely, the crystal structure determines the values of the normalized structure factors E_H since the position vectors r_j depend not only on the structure but on the choice of origin as well. It turns out, nevertheless, that the magnitudes $|E_H|$ of the normalized structure factors are in fact uniquely determined by the crystal structure and are independent of the choice of origin, but that the values of the phases φ_H depend also on the choice of origin. Although the values of the individual phases depend on the structure and the choice of origin, there exist certain linear combinations of the phases, the so-called structure invariants, whose values are determined by the structure alone and are independent of the choice of origin. The most important class of structure invariants, and the only one to be considered here, consists of the three-phase structure invariants (triplets),

(2.1) $$\varphi_{HK} = \varphi_H + \varphi_K + \varphi_{-H-K},$$

where H and K are arbitrary reciprocal lattice vectors.

2.2. The Probabilistic Background

It is assumed that the atomic position vectors r_j are the primitive random variables, uniformly and independently distributed in the unit cell. Then the normalized structure factors E_H, as functions of the r_j's, are themselves random variables. The structure invariants φ_{HK} in turn, as functions of the individual phases φ (Eq.(2.1)), are therefore also random variables.

2.3. The Conditional Probability Distribution of φ_{HK}, Given $|E_H|, |E_K|, |E_{H+K}|$

Under the conditions set forth in § 2.2 the conditional probability distribution of the triplet φ_{HK} (Eq.(2.1)), given the presumed known values of $|E_H|, |E_K|, |E_{H+K}|$, is known to be

(2.2) $$P(\Phi|A_{HK}) = [2\pi I_0(A_{HK})]^{-1} \exp(A_{HK} \cos \Phi)$$

where Φ represents the triplets φ_{HK}, the parameter A_{HK} is defined by

(2.3) $$A_{HK} = 2N^{-1/2}|E_H E_K E_{H+K}|,$$

and I_0 is the Modified Bessel Function. Equation (2.3) implies that the mode of φ_{HK} is zero, and the conditional expected value (or average) of $\cos \varphi_{HK}$, given A_{HK}, is

(2.4) $$\varepsilon(\cos \varphi_{HK}|A_{HK}) = I_1(A_{HK})/I_0(A_{HK}) > 0,$$

where I_1 is the Modified Bessel Function. It is also readily confirmed that the larger the value of $A_{\mathbf{HK}}$ the smaller is the conditional variance of $\cos \varphi_{\mathbf{HK}}$, given $A_{\mathbf{HK}}$. It is to be stressed that the conditional expected value of the cosine, Eq.(2.4), is always positive since $A_{\mathbf{HK}} > 0$.

2.4. The Minimal Principle

In view of Eq.(2.4) one obtains the following estimate of $\cos \varphi_{\mathbf{HK}}$:

$$(2.5) \qquad \cos \varphi_{\mathbf{HK}} \approx I_1(A_{\mathbf{HK}})/I_0(A_{\mathbf{HK}})$$

and expects that the smaller the variance, that is the larger $A_{\mathbf{HK}}$, the more reliable this estimate will be. Hence one is led to construct the function (the so-called minimal function), determined by the known magnitudes $|E|$,

$$(2.6) \qquad R = R(\varphi) = \frac{1}{\sum_{\mathbf{H,K}} A_{\mathbf{HK}}} \sum_{\mathbf{H,K}} A_{\mathbf{HK}} \left(\cos \varphi_{\mathbf{HK}} - \frac{I_1(A_{\mathbf{HK}})}{I_0(A_{\mathbf{HK}})} \right)^2$$

which, in view of Eq.(2.1), is seen to be a function of phases φ alone. Equation (2.4) then implies that the global minimum of the minimal function $R(\varphi)$, where the phases are constrained to satisfy the identities known to exist (§ 1), yields the desired phases (the minimal principle). Thus the phase problem is formulated as a problem in constrained global minimization [2]. There remains only the problem of avoiding the myriad local minima of $R(\varphi)$ in order to arrive at the constrained global minimum. The next section shows how this minimum is reached *via* the computer program *Shake-and-Bake*.

2.5. The Computer Program "Shake-and-Bake"

[3]
The six-part *Shake-and-Bake* phase determination procedure, shown by the flow diagram in Figure 1, combines minimal-function phase refinement and real-space filtering. It is an iterative process that is repeated until a solution is achieved or a designated number of cycles have been performed. With reference to Figure 1, the major steps of the algorithm are described next.

2.5.1. *Generate invariants.* Normalized structure-factor magnitudes ($|E|$'s) are generated by standard scaling methods and the triplet invariants that involve the largest corresponding $|E|$'s are generated. Parameter choices that must be made at this stage include the numbers of phases and triplets to be used. The total number of invariants is ordinarily chosen to be at least 100 times the number of atoms whose positions are to be determined.

2.5.2. *Generate trial structure.* A trial structure or model is generated that is comprised of a number of randomly positioned atoms equal to the number of atoms in the unit cell. The starting coordinate sets are subject to the restrictions that no two atoms are closer than a specified distance (normally 1.2Å) and that no atom is within bonding distance of more than four atoms.

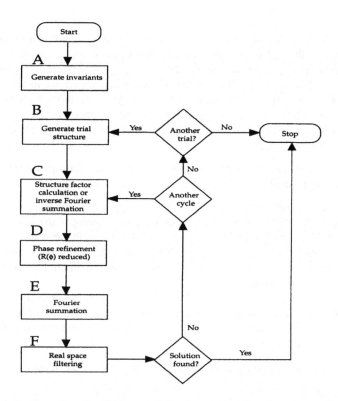

FIGURE 1. Flow chart for *Shake-and-Bake*, the minimal-function phase refinement and real-space filtering procedure.

2.5.3. Structure-factor calculation. A normalized structure-factor calculation (see Eq.(1.1)) based on the trial coordinates is used to compute initial values for all the desired phases simultaneously. In subsequent cycles, peaks selected from the most recent Fourier series are used as atoms to generate new phase values.

2.5.4. Phase refinement. The values of the phases are perturbed by a *parameter-shift* method in which $R(\varphi)$, which measures the mean-square difference between estimated and calculated structure invariants, is reduced in value. $R(\varphi)$ is initially computed on the basis of the set of phase values obtained from the structure-factor calculation in step C. The phase set is ordered in decreasing magnitude of the associated $|E|$'s. The value of the first phase is incremented by a preset amount and $R(\varphi)$ is recalculated. If the new calculated value of $R(\varphi)$ is lower than the previous one, the value of the first phase is incremented again by the preset amount. This is continued until $R(\varphi)$ no longer decreases or until a predetermined number of increments has been applied to the first phase. A completely analogous course is taken if, on the initial incrementation, $R(\varphi)$ increases, except that the value of the first phase is decremented until $R(\varphi)$ no longer decreases or until the predetermined number of decrements has been applied. The remaining phase values are varied in sequence as just described. Note that, when the ith phase value is varied, the new values determined for the previous $i - 1$ phases are used immediately in the calculation of $R(\varphi)$. The step size and number of steps are variables whose values must be chosen.

2.5.5. *Fourier summation.* Fourier summation is used to transform phase information into an electron-density map (Refer to Eq.(1.2)).

2.5.6. *Real-space filtering (Identities among phases imposed).* Image enhancement is accomplished by a discrete electron-density modification consisting of the selection of a specified number of the largest peaks on the Fourier map for use in the next structure-factor calculation. The simple choice, in each cycle, of a number of the largest peaks corresponding to the number of expected atoms has given satisfactory results. No minimum-interpeak-distance criterion is applied at this stage.

   Steps C, D, E, and F are repeated until a pre-assigned number of cycles has been completed or until the process converges. The smallest of the final values of the minimal function (one for each trial) reveals the constrained global minimum of the minimal function $R(\varphi)$ and the true values of the phases.

3. Single Wavelength Anomalous Scattering (SAS) Data are Available

3.1. *Introduction*

   In this case the normalized structure factors $E_{\mathbf{H}}$ (compare Eq.(1.1)) are defined by

$$(3.1) \qquad E_{\mathbf{H}} = \frac{1}{\alpha_2} \sum_{j=1}^{N} f_j \exp(2\pi i \mathbf{H} \cdot r_j)$$

$$(3.2) \qquad \alpha_2 = \sum_{j=1}^{N} |f_j|^2$$

where N is the number of atoms in the unit cell, r_j is the position vector of the atom labeled j and f_j is the (complex-valued) atomic scattering factor, presumed to be known.

3.2. *The Probabilistic Background*

   With the assumption that SAS diffraction data are available, the conditional probability distribution $P(\Phi)$ of the triplet

$$(3.3) \qquad \varphi_{\mathbf{HK}} = \varphi_{\mathbf{H}} + \varphi_{\mathbf{K}} + \varphi_{-\mathbf{H}-\mathbf{K}},$$

given the six magnitudes

(3.4) $$|E_{\mathbf{H}}|, |E_{-\mathbf{H}}|, |E_{\mathbf{K}}|, |E_{-\mathbf{K}}|, |E_{\mathbf{H}+\mathbf{K}}|, |E_{-\mathbf{H}-\mathbf{K}}|$$

is known to be [4]

(3.5) $$P(\Phi) = [2\pi I_0(A_{\mathbf{HK}})]^{-1} \exp\{A_{\mathbf{HK}} \cos(\Phi - \omega_{\mathbf{HK}})\}$$

in which I_0 is the Modified Bessel Function and $A_{\mathbf{HK}}$ and $\omega_{\mathbf{HK}}$ are expressed in terms of the six magnitudes (3.4) and the (presumed known) complex-valued atomic scattering factors f. Compare Eq.(3.5) with Eq.(2.2) and note that the $A_{\mathbf{HK}}$ of Eq.(3.5) is no longer defined by Eq.(2.3) but is instead a much more complicated function of the six magnitudes (3.4) and the atomic scattering factors f. Hence, $A_{\mathbf{HK}}(> 0)$ and $\omega_{\mathbf{HK}}$ are here assumed to be known for every pair $(\mathbf{H}, \mathbf{K})$. Note that, owing to the anomalous scattering, the six magnitudes (3.4) are, in general, distinct in contrast to the normal case when $|E_{-\mathbf{H}}| = |E_{\mathbf{H}}|$, etc.

In view of (3.5), the most probable value of $\varphi_{\mathbf{HK}}$ is $\omega_{\mathbf{HK}}$, and the larger the value of $A_{\mathbf{HK}}$ the better is this estimate of $\varphi_{\mathbf{HK}}$:

(3.6) $$\varphi_{\mathbf{HK}} = \varphi_{\mathbf{H}} + \varphi_{\mathbf{K}} + \varphi_{-\mathbf{H}-\mathbf{K}} \approx \omega_{\mathbf{HK}}$$

3.3. *The System of SAS Tangent Equations*

Fix the reciprocal lattice vector $\mathbf{H}$. From Eq.(3.6)

(3.7) $$\varphi_{\mathbf{H}} \approx \omega_{\mathbf{HK}} - \varphi_{\mathbf{K}} - \varphi_{-\mathbf{H}-\mathbf{K}}$$

(3.8) $$\sin \varphi_{\mathbf{H}} \approx \sin(\omega_{\mathbf{HK}} - \varphi_{\mathbf{K}} - \varphi_{-\mathbf{H}-\mathbf{K}})$$

which has approximate validity for each fixed value of $\mathbf{K}$. Averaging the right-hand side of (3.8) over $\mathbf{K}$, naturally using weights $A_{\mathbf{HK}}$, one obtains

(3.9) $$\sin \varphi_{\mathbf{H}} \approx \frac{1}{\sum_{\mathbf{K}} A_{\mathbf{HK}}} \sum_{\mathbf{K}} A_{\mathbf{HK}} \sin(\omega_{\mathbf{HK}} - \varphi_{\mathbf{K}} - \varphi_{-\mathbf{H}-\mathbf{K}})$$

Similarly

(3.10) $$\cos \varphi_{\mathbf{H}} \approx \frac{1}{\sum_{\mathbf{K}} A_{\mathbf{HK}}} \sum_{\mathbf{K}} A_{\mathbf{HK}} \cos(\omega_{\mathbf{HK}} - \varphi_{\mathbf{K}} - \varphi_{-\mathbf{H}-\mathbf{K}})$$

Eqs.(3.9) and (3.10) imply

(3.11) $$\tan \varphi_{\mathbf{H}} \approx \frac{\sum_{\mathbf{K}} A_{\mathbf{HK}} \sin(\omega_{\mathbf{HK}} - \varphi_{\mathbf{K}} - \varphi_{-\mathbf{H}-\mathbf{K}})}{\sum_{\mathbf{K}} A_{\mathbf{HK}} \cos(\omega_{\mathbf{HK}} - \varphi_{\mathbf{K}} - \varphi_{-\mathbf{H}-\mathbf{K}})}$$

the system of SAS tangent equations. For each fixed $\mathbf{H}$, Eq.(3.11) yields two values for $\varphi_{\mathbf{H}}$ differing by π. Eqs.(3.9) and (3.10) serve to fix the quadrant.

3.4. The Maximal Function $M(\varphi)$

One defines the maximal function $M(\varphi)$, a function of the phases φ, by means of

$$(3.12) \qquad M(\varphi) = \frac{1}{\sum_{\mathbf{H},\mathbf{K}} A_{\mathbf{HK}}} \sum_{\mathbf{H},\mathbf{K}} A_{\mathbf{HK}} \cos(\varphi_{\mathbf{HK}} - \omega_{\mathbf{HK}})$$

and infers that $M(\varphi)$ has a global maximum when all the phases appearing in Eq.(3.12) are set equal to their true values.

3.5. The Maximal Property of the System of SAS Tangent Equations

Fundamental maximal property. Fix $\mathbf{H}$. Assume that the values of all phases other than $\varphi_{\mathbf{H}}$ are specified arbitrarily. Then the maximal function $M(\varphi)$ becomes a function, $M(\varphi_{\mathbf{H}}|\varphi)$, of the single phase $\varphi_{\mathbf{H}}$. As a function of $\varphi_{\mathbf{H}}$, $M(\varphi_{\mathbf{H}}|\varphi)$ has a unique maximum in the whole interval $(0, 2\pi)$ and the value of $\varphi_{\mathbf{H}}$ that maximizes $M(\varphi_{\mathbf{H}}|\varphi)$ is given by the SAS tangent equation (3.11).

3.6. Solving the System of SAS Tangent Equations

Specify arbitrarily initial values for all the phases φ. Fix $\mathbf{H}$. Calculate a new value for the phase $\varphi_{\mathbf{H}}$ by means of the SAS tangent equations (3.9) to (3.11), in this way, in view of § 3.5, increasing the initial value of the maximal function $M(\varphi)$. Fix $\mathbf{H}' \neq \mathbf{H}$. Calculate a new value for $\varphi_{\mathbf{H}'}$, again using (3.9) to (3.11), the new value for $\varphi_{\mathbf{H}}$, and initial values for the remaining phases, thus increasing still further the value of $M(\varphi)$. Continue in this way to obtain new values for all the phases, thus completing the first iteration and, in the process, continuously increasing the value of $M(\varphi)$. Complete as many iterations as necessary in order to secure convergence. Convergence is assured since the iterative process yields a monotonically increasing sequence of numbers, the values of $M(\varphi)$, bounded above by unity. Evidently also, the process leads to a local maximum of $M(\varphi)$ and a corresponding set of values for all the phases φ which depends on the values of the phases chosen initially. By choosing different starting values for the phases one obtains different solutions for the system of SAS tangent equations and different local maxima of $M(\varphi)$. That solution yielding the global maximum of $M(\varphi)$ is the one we seek.

3.7. The Linear Congruence Connection

The problem of going from the estimated values $\omega_{\mathbf{HK}}$ of the three-phase structure invariants $\varphi_{\mathbf{HK}}$ to the values of the individual phases φ may be formulated as the problem of solving the redundant system of linear congruences

$$(3.13) \qquad \varphi_{\mathbf{H}} + \varphi_{\mathbf{K}} + \varphi_{-\mathbf{H}-\mathbf{K}} \equiv \omega_{\mathbf{HK}} \; (\text{modulo } 2\pi)$$

each with weight $A_{\mathbf{HK}}$. Our solution of the system of SAS tangent equations also yields the solution of the redundant system of linear congruences (3.13).

This research is supported by National Institutes of Health Grant No. GM-46733.

References

[1] Hauptman, H.A. (1991). *Crystallographic Computing 5: From Chemistry to Biology*, edited by D. Moras, A.D. Podnarny & J.C. Thierry, 324–332. IUCr/Oxford University Press.

[2] DeTitta, G.T., Weeks, C.M., Thuman, P., Miller, R. & Hauptman, H.A. (1994). *Structure Solution by Minimal Function Phase Refinement and Fourier Filtering: I. Theoretical Basis*. Acta Cryst. **A50**, 203–210.

[3] Weeks, C.M., DeTitta, G.T., Hauptman, H.A., Thuman, P. & Miller, R. (1994). *Structure Solution by Minimal Function Phase Refinement and Fourier Filtering: II. Implementation and Applications*. Acta Cryst. **A50**, 210–220.

[4] Hauptman, H. A. (1982) *On Integrating the Techniques of Direct Methods with Anomalous Dispersion: I. The Theoretical Basis*. Acta Cryst. **A38**, 632–641.

Multiwindow Gabor-type Representations and Signal Representation by Partial Information

Y.Y. Zeevi

Department of Electrical Engineering
Technion—Israel Institute of Technology
Haifa 32000, Israel

ABSTRACT. Harmonic analysis has been the longest lasting and most powerful tool for dealing with signals and systems which involve both periodic and transient phenomena. Further impact on signal processing was facilitated by what we refer to as representations in combined spaces. These representations, motivated by Gabor's concept of time-frequency information, cells have evolved in recent years into a rich repertoire of Gabor-type windowed Fourier transforms, relating Fourier analysis to the Heisenberg group. Such localized bases or frames are useful in the representation, processing, compression and transmission of speech, images and other natural signals that, by their very nature, are nonstationary. To incorporate scale that lends itself to multiresolution analysis as is the case with wavelets, the Gabor scheme is generalized to multiwindow Gabor frames. The properties of such sequences of functions are characterized by an approach that combines the concept of frames and the Zak Transform. Results on signal representation and reconstruction from partial information in the frequency domain are related to and derived, using relevant results obtained in the time or positional information domain. Some results concerning the representation of Fourier-transformed discrete time (finite) sequences by partial information are rederived by exploiting the duality of the Fourier-Stieltjes transform and its inverse. Results related to discrete signals are extended to continuous one-dimensional signals. Signal and image representation by phase only information is considered also in the context of localized (Gabor) phase, where restoration of magnitude by iterative techniques is much more efficient than in the case of global (Fourier) phase.

173

J.S. Byrnes (ed.), Twentieth Century Harmonic Analysis - A Celebration, 173–199.
© 2001 *Kluwer Academic Publishers. Printed in the Netherlands.*

1. Introduction

The subject of harmonic analysis is extremely broad, as can be concluded, for example, from the viewpoint of Katzenelson [16], who considers it as the study of objects (functions, measures, etc.) defined on topological groups. Indeed, groups are relevant to Gabor-type representations and wavelet-type transforms [25]. Consider the quasi-regular representation

$$(1.1) \qquad T(g)h(\overline{x}) = kh(g^{-1}\overline{x}),$$

where k is a constant that may depend on g, and the independent variable $\overline{x}$ is an n-dimensional vector.

This representation that satisfies the functional homomorphism

$$T(g_1 g_2) = T(g_1)T(g_2)$$

underlines the decomposition of a function h into a package of waves, i.e., wavelets

$$\{T(g_0)h, \cdots, T(g_N)h, \cdots\}.$$

We say that $T(g)$ is a unitary representation with respect to a measure μ if

$$(1.2) \qquad \langle T(g)f, T(g)h \rangle_\mu = \langle f, h \rangle_\mu.$$

A wavelet-type transform is defined, according to the group theoretic approach, as a cross correlation between a signal $f(\overline{x})$ and the wavelets

$$\{T(g)h(\overline{x})\}$$

that is defined as follows:

$$
\begin{aligned}
C_f(g) &= c \int_M f(\overline{x}) T(g) h^*(\overline{x}) \mu(\overline{x}) \\
(1.3) \qquad &= c \langle f, T(g)h \rangle_\mu,
\end{aligned}
$$

where h^* stands for conjugate, $c > 0$ is a normalization constant, h is a template function (or a mother wavelet), and $\mu(\overline{x}) = \rho(\overline{x})d\overline{x}$ is an appropriate invariance measure (if one exists) under the action of the group G.

Given the equation for the analysis, (1.3), we are looking for a way to recover $f(\overline{x})$ from $C_f(g)$, i.e. the equation of the synthesis and the conditions under which it exists.

Most studies of wavelets have been devoted to the special case of the affine group $G = G(A^{-1}, \overline{b})$ of planar scaling and translation. A^{-1} is in this case a diagonal matrix with entries a_i^{-1} and $\overline{b}$ is the translation vector. The quasi-regular unitary representation over $L^2(\mathbb{R}^n)$ is defined in this case by

$$(1.4) \qquad T(g)f(\overline{x}) = kf(A(\overline{x} - \overline{b})),$$

where $k = \sqrt{J}$ and $J = |\det(A)|$. Gabor-type representations originate from the Heisenberg group and the associated Weyl operational calculus. As such this representation lacks the scaling that is characteristic of wavelets. Scaling is, of course, important in the analysis of natural signals. However, the explicit harmonic, or spectral nature of Gabor-type representations is just as important. Thus, it is desirable to incorporate scaling into the Gabor-type representation. This is accomplished by the generalized Gabor scheme [21], [35]. In this paper we focus on signal and image representation by multi-window Gabor-type schemes, where by proper choice of the set of windows we incorporate scaling [35]. We then proceed to present various aspects of signal and image representation by partial information, mostly in the context of harmonic analysis. These two subjects are of utmost importance in signal processing, and have proven to be useful in a wide range of applications (not discussed in this paper). The two interrelated subjects are also instrumental in gaining insight into the structure of natural signals, and of images as such. In both cases we discuss the analysis as well as the synthesis.

2. Multiwindow Gabor-type frames

2.1. *Generalized Gabor-type schemes and the Zak transform*

Many problems in physics and engineering involve the representation and analysis of nonstationary signals and processes. Such problems call for the development and application of sets of functions that are localized in the sense that they rapidly decay in both time (or position in the case of images) and frequency. Such are the wavelets, Gabor functions and the Wigner distribution.

The two widely-studied sequences of the so-called wavelets and Gabor functions are special cases of the following generalized sequence:

$$(2.1) \qquad s_{r,m,n}(x) = g_r(x - na_r)\phi_{r,m}(x),$$

where $\{g_r(x)\}$ is a sequence of window functions, $\{a_r\}$ is a set of real numbers and $\{\phi_{r,m}(x)\}$ is a set of kernel functions. In the classical Gabor case [9], there is a single-window function $g(x)$, with $a_r = a$ and b some positive constant, $\phi_{r,m}(x) = e^{j2\pi mbx}$, and $m, n \in \mathbb{Z}$. In the wavelet-type Gabor case, $\phi_{r,m}(x) = 1$, $g_r(x) = b^{-r/2}g(x/b^r)$, $a_r = ab^r$, $r, n \in \mathbb{Z}$, and $g(x)$ is a "mother wavelet" function.

Zibulski and Zeevi [35] analyzed such sequences of functions by developing a matrix algebra approach based on the concept of frames and the Zak transform (ZT). The basic results are presented here without the proofs. The idea of using frames in order to examine the sequence of Gabor functions (and affine wavelets as well) was first proposed and implemented by Daubechies et al. [4]; the so-called Weyl-Heisenberg frames. The ZT of a signal $f(x)$ is defined as follows [14]:

$$(2.2) \qquad (\mathcal{Z}f)(x, u) \triangleq \alpha^{1/2} \sum_{k \in \mathbb{Z}} f\big[\alpha(x + k)\big] e^{-j2\pi uk}, \quad -\infty < x, u < \infty,$$

with a fixed parameter $\alpha > 0$. The ZT satisfies the following periodic and quasiperiodic relations:

(2.3) $(\mathcal{Z}f)(x, u + 1) = (\mathcal{Z}f)(x, u),\ (\mathcal{Z}f)(x + 1, u) = e^{j2\pi u}(\mathcal{Z}f)(x, u).$

As a consequence of these two relations, the ZT is completely determined by its values over the unit square $(x, u) \in ([0, 1)^2)$. This is the essence of this unitary mapping.

Based on the ZT defined by (2.2), we define the Piecewise Zak Transform (PZT) as a vector-valued function $F(x, u)$ of size p [34]:

(2.4) $$F(x, u) = \Big[F_0(x, u), F_1(x, u) \ldots, F_{p-1}(x, u) \Big]^T,$$

where

(2.5) $$F_i(x, u) = (\mathcal{Z}f)\left(x, u + \frac{i}{p}\right),\quad 0 \leqslant i \leqslant p - 1,\ i \in \mathbb{Z}.$$

The vector-valued function $F(x, u)$ belongs to $L^2([0, 1) \times [0, 1/p); \mathbb{C}^p)$, which is a Hilbert space with the inner-product:

$$\langle F, G \rangle = \int_0^1 dx \int_0^{1/p} du \sum_{i=0}^{p-1} F_i(x, u)\overline{G_i(x, u)}.$$

Since the ZT is a unitary mapping from $L^2(\mathbb{R})$ to $L^2([0, 1)^2)$, the PZT is a unitary mapping from $L^2(\mathbb{R})$ to $L^2([0, 1) \times [0, 1/p); \mathbb{C}^p)$. As a consequence we obtain the following inner-product preserving property:

$$\int_{-\infty}^{\infty} f(x)\overline{g(x)}dx = \int_0^1 \int_0^1 (\mathcal{Z}f)(x, u)\overline{(\mathcal{Z}g)(x, u)}dxdu$$

$$= \int_0^1 dx \int_0^{1/p} du \sum_{i=0}^{p-1} F_i(x, u)\overline{G_i(x, u)}.$$

This unitary property of the PZT allows us the transformation from $L^2(\mathbb{R})$ to $L^2([0, 1) \times [0, 1/p); \mathbb{C}^p)$, where issues regarding Gabor-type representations are often easier to deal with and understand.

2.2. Multi-window Gabor-type expansions

Generalizing the Gabor scheme, by using several window functions instead of a single-one, the representation of a given signal $f(x) \in L^2(\mathbb{R})$ is given by:

(2.6) $$f(x) = \sum_{r=0}^{R-1} \sum_{m,n} c_{r,m,n} g_{r,m,n}(x),$$

where

(2.7) $$g_{r,m,n}(x) = g_r(x - na)e^{j2\pi mbx},$$

and $\{g_r(x)\}$ is a set of R distinct window functions. Such a set can incorporate, for example, Gaussian windows of various widths. In this case, with proper oversampling, one can overcome in a way the uncertainty constraint expressed by the limitations of having either a high temporal (spatial) resolution and low frequency resolution or vice versa. This type of a richer Gabor-type representation can be instrumental in applications such as detection of transient signals [8], or in identification and recognition of various prototypic textures and other features in images [22]. Clearly, if $R = 1$ we obtain the single-window Gabor representation.

The characterization of the sequence $\{g_{r,m,n}\}$ can be divided into three categories according to the sampling density of the combined space (the so-called phase space density) defined by $d \triangleq R(ab)^{-1}$: undersampling - $d < 1$, critical sampling - $d = 1$ and oversampling - $d > 1$.

2.3. *Matrix algebra approach for the analysis of frames' properties*

In order to examine the properties of the sequence $\{g_{r,m,n}\}$, we consider the operator:

(2.8) $$Sf = \sum_{r=0}^{R-1} \sum_{m,n} \langle f, g_{r,m,n} \rangle g_{r,m,n},$$

which is a frame operator if $\{g_{r,m,n}\}$ constitutes a frame. For the single-window Gabor scheme, this operator was examined by straightforward methods [5, 13], where application of the ZT was restricted to the case $ab = 1$. In [35], Zibulski and Zeevi show that, if the product ab is a rational number, it might be advantageous to examine this operator in $L^2([0,1) \times [0, 1/p); \mathbb{C}^p)$ by using the PZT. A major result of analysis in the PZT domain is the representation of the frame operator as a finite order matrix-valued function, as formulated in the following theorem.

THEOREM 1 (Zibulski and Zeevi [35]). *Let $ab = p/q$, $p, q \in \mathbb{N}$, and let S_z be the frame operator of the sequence, which is the PZT of $\{g_{r,m,n}\}$. The action of S_z in $L^2([0,1) \times [0, 1/p); \mathbb{C}^p)$ is given by the following matrix algebra:*

(2.9) $$(S_z F)(x, u) = \mathbf{S}(x, u)F(x, u),$$

where $\mathbf{S}(x, u)$ is a $p \times p$ matrix-valued function whose entries are given by:

(2.10)
$$S_{i,k}(x, u) = \frac{1}{p} \sum_{r=0}^{R-1} \sum_{l=0}^{q-1} \mathcal{Z}g_r\left(x - l\frac{p}{q}, u + \frac{i}{p}\right) \overline{\mathcal{Z}g_r\left(x - l\frac{p}{q}, u + \frac{k}{p}\right)};$$
$$i, k = 0, \cdots, p - 1,$$

and the vector-valued function $F(x, u)$ is given by (2.4) and (2.5).

Since the PZT is a unitary transform, (2.9) is an isometrically isomorphic representation of S (2.8).

Using the PZT and the matrix representation of the operator S, Zibulski and Zeevi [35] examined the properties of the sequence $\{g_{r,m,n}\}$ for a rational ab. Since, in the case of undersampling the sequence $\{g_{r,m,n}\}$ is not complete, the results presented next are relevant in the cases of critical sampling and oversampling.

The next theorem examines the completeness of the sequence $\{g_{r,m,n}\}$ in relation to the structure of the matrix-valued function $S(x, u)$.

THEOREM 2 (Zibulski and Zeevi [35]). *Given* $g_r \in L^2(\mathbb{R})$, $0 \leqslant r \leqslant R - 1$, *and a matrix-valued function* $S(x, u)$, $(x, u) \in ([0, 1) \times [0, 1/p))$ *as in (2.10), the sequence* $\{g_{r,m,n}\}$ *associated with* $\{g_r\}$, $ab = p/q$, $p, q \in \mathbb{N}$ *is complete if and only if* $\det(S)(x, u) \neq 0$ *a.e. on* $[0, 1) \times [0, 1/p)$.

The frame bounds of the sequence $\{g_{r,m,n}\}$ are determined by the eigenvalues of the matrix-valued function $S(x, u)$. Let

$$(2.11) \qquad \lambda_{max}(S) \triangleq \text{ess sup}_{(x,u) \in ([0,1) \times [0,1/p))} \max_{1 \leqslant i \leqslant p} \lambda_i(S)(x, u)$$

$$(2.12) \qquad \lambda_{min}(S) \triangleq \text{ess inf}_{(x,u) \in ([0,1) \times [0,1/p))} \min_{1 \leqslant i \leqslant p} \lambda_i(S)(x, u),$$

where $\lambda_i(S)(x, u)$ are the eigenvalues of the matrix $S(x, u)$. Then, the upper frame-bound $B = \lambda_{max}(S)$, and the lower frame-bound $A = \lambda_{min}(S)$. This result yields the following theorem.

THEOREM 3 (Zibulski and Zeevi [35]).
The sequence $\{g_{r,m,n}\}$ *associated with* $\{g_r\}$, $g_r \in L^2(\mathbb{R})$, $0 \leqslant r \leqslant R - 1$, *and* $ab = p/q$, $p, q \in \mathbb{N}$ *constitutes a frame if and only if* $0 < \lambda_{min}(S) \leqslant \lambda_{max}(S) < \infty$.

An alternative approach to determining whether $\{g_{r,m,n}\}$ constitutes a frame is as follows. The following Lemma formulates a necessary and sufficient condition for the existence of an upper frame bound $B < \infty$.

LEMMA 1. *The sequence* $\{g_{r,m,n}\}$ *associated with* $\{g_r\}$, $g_r \in L^2(\mathbb{R})$, $0 \leqslant r \leqslant R - 1$ *and* $ab = p/q$, $p, q \in \mathbb{N}$, *has an upper frame bound* $B < \infty$ *if and only if* $(\mathcal{Z}g_r)(x, u)$ *are all bounded a.e. on* $(0, 1]^2$ $(\mathcal{Z}g_r \in L^\infty((0, 1]^2))$.

Theorem 4 below determines whether the sequence $\{g_{r,m,n}\}$ constitutes a frame, when an upper frame bound exists, and which does not necessitate calculation of the eigenvalues of $S(x, u)$.

THEOREM 4 (Zibulski and Zeevi [35]). *Given* $g_r \in L^2(\mathbb{R})$, $0 \leqslant r \leqslant R - 1$, *such that there exists an upper frame bound* $B < \infty$ *for the sequence* $\{g_{r,m,n}\}$ *associated with* $\{g_r\}$, *and* $ab = p/q$, $p, q \in \mathbb{N}$. *The sequence* $\{g_{r,m,n}\}$ *constitutes a frame if and only if*

$0 < K \leqslant \det(\mathbf{S})(x, u)$ *a.e. on* $[0, 1) \times [0, 1/p)$, *where the matrix-valued function* $\mathbf{S}(x, u)$ *is as in (2.10).*

The following theorem concerns tight frames and the matrix representation of the frame operator. Recall that a set of functions $\{\psi_n\}$ in a Hilbert space H constitutes a *tight frame* if

$$\sum_n |\langle f, \psi_n \rangle|^2 = A\|f\|^2$$

THEOREM 5 (Zibulski and Zeevi [35]). *Given* $g_r \in L^2(\mathbb{R})$, $0 \leqslant r \leqslant R - 1$, *and a matrix-valued function* $\mathbf{S}(x, u)$ *as in (2.10), the set of functions* $\{g_{r,m,n}\}$ *associated with* $\{g_r\}$, $ab = p/q$, $p, q \in \mathbb{N}$ *constitutes a tight frame if and only if* $\mathbf{S}(x, u) = A\mathbf{I}$ *a.e., where* $\mathbf{I}$ *is the identity matrix, and* $A = \frac{q}{p} \sum_{r=0}^{R-1} \|g_r\|^2$.

In order to complete the formalism of representation, one has to obtain the dual frame. This can be done by using operator techniques [5, 7]. For the single-window scheme the dual frame $\{\mathcal{S}^{-1}g_{m,n}\}$ is generated by a single dual frame window function [5, 13]. This is, indeed, also the case for the multi-window scheme. Let $\{\gamma_{r,m,n}\}$ denote the dual frame of $\{g_{r,m,n}\}$. Then $\{\gamma_{r,m,n}\}$ is generated by a finite set of R dual frame window functions $\{\gamma_r\}$:

$$\gamma_{r,m,n}(x) = \gamma_r(x - na)e^{j2\pi mbx}, \quad 0 \leqslant r \leqslant R - 1,$$

where $\gamma_r = \mathcal{S}^{-1}g_r$. Using the matrix representation (2.9) of the frame operator, the PZT of γ_r, is:

(2.13) $$\Gamma_r(x, u) = \mathbf{S}^{-1}(x, u)G_r(x, u),$$

that is, $\Gamma_r(x, u), G_r(x, u)$ are vector-valued functions in

$$L^2([0, 1) \times [0, 1/p); \mathbb{C}^p)$$

and $\mathbf{S}^{-1}(x, u)$ is the inverse of the matrix $\mathbf{S}(x, u)$; for example,

$$\mathbf{S}^{-1}(x, u) = [\det(\mathbf{S})(x, u)]^{-1}\text{adj}(\mathbf{S})(x, u).$$

2.4. *The Balian-Low theorem in the case of multi-windows*

In the case of a single-window Gabor scheme with critical sampling, i.e. $ab = 1$, by choosing an appropriate $g(x)$, the sequence $\{g_{r,m,n}\}$ can be complete and constitute a frame, notwithstanding the problems of stability [5]. The sequence $\{g_{r,m,n}\}$ can, however, be complete and not constitute a frame, in which case the representation is unstable. A classical example of such an unstable scheme is the one with a Gaussian window function.

In the case of a single window and critical sampling, the following theorem of Balian and Low indicates that a wide range of well behaved – rapidly decaying and smooth – functions $g(x)$ are excluded from being proper candidates for generators of frames.

THEOREM 6 (Balian-Low Theorem [2, 5, 6]). . *Given* $g \in L^2(\mathbb{R})$, $a > 0$ *and* $ab = 1$, *if the sequence* $\{g_{m,n}\}$ *constitutes a frame, then either* $xg(x) \notin L^2(\mathbb{R})$ *or* $g'(x) \notin L^2(\mathbb{R})$.

Note that $g'(x) \in L^2(\mathbb{R}) \Leftrightarrow \omega \hat{g}(\omega) \in L^2(\mathbb{R})$, where $\hat{g}$ is the Fourier transform of g.

One of the solutions for this problem is oversampling. In fact, it was proven that in the case of a Gaussian the $\{g_{m,n}\}$ constitutes a frame for all $ab < 1$ [15, 26].

In the case of critical sampling of the multi-window scheme, an interesting question is whether one can overcome the constraint imposed by the Balian-Low Theorem by utilizing several windows. According to the following theorem, if all the windows in the set $\{g_r\}$ are well behaved functions, this is not possible:

THEOREM 7 (Zibulski and Zeevi [35]). *Given* $g_r \in L^2(\mathbb{R})$, $0 \leqslant r \leqslant R - 1$, $a > 0$ *and* $R(ab)^{-1} = 1$, *if the sequence* $\{g_{r,m,n}\}$, *as in (2.7), constitutes a frame, then either* $xg_r(x) \notin L^2(\mathbb{R})$ *or* $g_r'(x) \notin L^2(\mathbb{R})$ *for some* $0 \leqslant r \leqslant R - 1$.

One of the advantages of using more than one window is the possibility of overcoming the constraint imposed by the Balian-Low Theorem on the choice of window functions, by adding an extra window function of proper nature such that the resultant scheme of critical sampling constitutes a frame. Whether one can find a non-well-behaved window function, complementary to a set of well-behaved window functions, such that the inclusive set will generate a frame for critical sampling, depends on the nature of the set of the well-behaved window functions as indicated by the following proposition.

PROPOSITION 1. *Let a set,* $\{g_r\}$, $0 \leqslant r \leqslant R - 2$, *of* $R - 1$ *window functions be given. Denote by* $\mathbf{G}^0(x, u)$ *the* $R - 1 \times R$ *matrix-valued function with entries* $G^0_{r,k}(x, u) = \overline{\mathcal{Z}g_r(x, u + \frac{k}{R})}$, *and* $\mathbf{P}(x, u) = \mathbf{G}^0(x, u)\mathbf{G}^{0*}(x, u)$. *There exists a window function* $g_{R-1}(x)$ *such that the inclusive set* $\{g_r\}$, $0 \leqslant r \leqslant R - 1$ *generates a frame for the critical sampling case, if and only if* $0 < K \leqslant \det(\mathbf{P})(x, u)$ *a.e. on* $[0, 1) \times [0, 1/R)$.

An example of $R - 1$ well-behaved window functions satisfying $0 < K \leqslant \det(\mathbf{P})(x, u)$ a.e. on $[0, 1) \times [0, 1/R)$, can be constructed in the following manner. Take a window function $g(x)$ such that the sequence $\{g(x - n/b)e^{j2\pi mx/a}\}$ constitutes a frame for $ab = R/(R - 1)$. Note, that this is an oversampling scheme $(1/(ab) = (R - 1)/R < 1)$ and that there exist, therefore, well-behaved window functions $g(x)$ such that $\{g(x - n/b)e^{j2\pi mx/a}\}$ constitutes a frame (for example the Gaussian function). Construct the following $R - 1$ window functions:

$$g_r(x) = g\left(x - \frac{rR}{b(R - 1)}\right).$$

Clearly these are well-behaved window functions. Moreover, we obtain

$$(\mathcal{Z}g_r)(x, u) = (\mathcal{Z}g)(x - r\frac{R}{R - 1}, u)$$

and the matrix-valued function $\mathbf{G}^0(x, u)$ equals the matrix-valued function $\mathbf{G}(x, u)$ which corresponds to the sequence $\{g(x - na)e^{j2\pi mbx}\}$ (which we denote by $\{g_{r,m,n}\}$). In this case $\{g_{r,m,n}\}$ corresponds to an undersampling scheme. By the duality principle, since

$\{g(x - n/b)e^{j2\pi mx/a}\}$ constitutes a frame, $\{g_{r,m,n}\}$ constitutes a Riesz basis for a sub-space of $L^2(\mathbb{R})$. It can therefore be shown that $0 < K \leqslant \det(\mathbf{P})(x, u)$ a.e. on $[0, 1) \times [0, 1/R)$.

2.5. A wavelet-type Gabor scheme

Consider a generalization of the sequence $\{g_{r,m,n}\}$, where for each window function $g_r(x)$ there is a different set of parameters a_r, b_r:

(2.14) $$g_{r,m,n}(x) = g_r(x - na_r)e^{j2\pi mb_r x},$$

and the sampling density of the combined space is:

$$d \triangleq \sum_{r=0}^{R-1}(a_r b_r)^{-1}.$$

In the case of the single window, the characterization of the sequence $\{g_{r,m,n}\}$ can be divided into the three categories of undersampling, critical sampling and oversampling according to $d < 1$, $d = 1$, $d > 1$, respectively.

In order to analyze this kind of a scheme, we consider an equivalent one with $a_r = a$, $b_r = b$ for all r [35] and utilize the tools presented in the previous section for the analysis of the sequence properties.

Utilizing the degrees of freedom of choosing a different set of parameters a_r, b_r for each window function g_r, we construct a wavelet-type scheme. Let α, β be positive, real numbers. Given a window function $g(x)$ let

$$g_r(x) = \alpha^{-r/2}g(\alpha^{-r}x).$$

Also, let $a_r = \beta\alpha^r$, and $a_r b_r = R/d$ for all r, where d is the sampling density of the combined space. We then have

$$g_{r,m,n}(x) = \alpha^{-r/2}g(\alpha^{-r}x - n\beta)e^{j2\pi\frac{mxR}{\beta\alpha^r d}}.$$

In this scheme the width of the window is proportional to the translation step a_r, whereas the product $a_r b_r$ is constant. This scheme incorporates scaling characteristics of wavelets and of the Gabor scheme with the logarithmically-distorted frequency axis [21]. However, in contrast with wavelets, this scheme has a finite number of window functions, i.e., resolution levels, and each of the windows is modulated by the infinite set of functions defined by the kernel. Each subset for fixed m can be considered as a finite (incomplete) wavelet-type set, in that it obeys the properties of scaling and translation of a complex prototypic "mother wavelet". For example, if R/d is an integer, the mother wavelet is defined by $g(x)e^{j2\pi\frac{mxR}{\beta d}}$; if not, this definition holds within a complex phase. Thus, all the functions corresponding to each of these mother wavelets are self-similar.

Scaling has been realized as an important property of sets of representation functions which are used in the analysis of natural signals and images. Hence the importance of this type of generalization of the Gabor scheme.

3. Signal representation by partial information

3.1. *Fourier and related transforms*

Let $f(t)$ be a complex function which belongs to $L_1(-\infty, \infty)$ or $L_2(-\infty, \infty)$. The ordinary Fourier transform denoted by $\hat{f}(\omega)$ and its inverse are defined by

$$(3.1) \qquad \hat{f}(\omega) \;=\; \int_{-\infty}^{\infty} f(t)e^{-j\omega t}dt$$

$$(3.2) \qquad f(t) \;=\; \frac{1}{2\pi}\int_{-\infty}^{\infty} \hat{f}(\omega)e^{j\omega t}d\omega\,.$$

If $\hat{f}(\omega)$ is zero outside the interval $[a, b]$ then

$$(3.3) \qquad f(t) = \frac{1}{2\pi}\int_{a}^{b} \hat{f}(\omega)e^{j\omega t}d\omega \quad |a|, |b| < \infty\,,$$

is called a band-limited function, i.e., $f(t) \in \mathbf{B}^{\Omega}$, where $\mathbf{B}^{\Omega}$ is the space of bandlimited functions.

It is well known from the Paley-Wiener theoremm [23] that such a $f(t)$, with t being interpreted as a complex variable, is an entire function of exponential type (EFET). Due to the duality of the Fourier transform and its inverse, if $f(t)$ is time-limited to $[\alpha, \beta]$, then

$$(3.4) \qquad \hat{f}(\omega) = \int_{\alpha}^{\beta} f(t)e^{-j\omega t}dt\,.$$

In this case, with ω being viewed as a complex variable, $\hat{f}(\omega)$ is an EFET. This duality is complete if we change the argument sign and multiply (divide) by 2π. If we define a dual Fourier transform pair $(\hat{f}'(\omega), f'(t))$ by

$$(3.5) \qquad \hat{f}'(\omega) \;=\; \frac{1}{2\pi}\int_{-\infty}^{\infty} f'(t)e^{j\omega t}dt$$

$$(3.6) \qquad f'(t) \;=\; \int_{-\infty}^{\infty} \hat{f}(\omega)e^{-j\omega t}d\omega$$

we have perfect duality between time-limited signals and band-limited signals. If $f(t) = f'(t)$, then $\hat{f}(\omega) = 2\pi\hat{f}'(-\omega)$. Thus, if $\hat{f}(\omega)$ is given by

$$(3.7) \qquad \hat{f}(\omega) = P(\omega) + jQ(\omega) = R(\omega)\exp(i\vartheta(\omega)), \quad R(\omega) > 0\,,$$

where $P(\omega)$ is the real and $Q(\omega)$ the imaginary part, $\hat{f}'(\omega)$ is given by

$$(3.8) \qquad 2\pi\hat{f}'(\omega) = \hat{f}(-\omega) = P(\omega) - jQ(\omega)\,.$$

Alternatively, $\hat{f}'(\omega)$ can be expressed in polar coordinates

$$(3.9) \qquad 2\pi\hat{f}'(\omega) = R(\omega)\exp(-j\vartheta(\omega))\,,$$

where $R(\omega)$ is the magnitude and $\vartheta(\omega)$ the phase. All subsequent results relating to time-domain (when the signals are band-limited in frequency) will be used with reference to the frequency-domain (when the signals are of finite duration in time or finite extent in space), with proper interpretation of the duality that exists between frequency and time domain representations.

In general, the signals in the time or frequency domain will not belong to $L_1(-\infty, \infty)$ or $L_2(-\infty, \infty)$. We will use delta functions to describe discrete time signals or discrete Fourier spectra. The application of these functions can be mathematically justified by introducing the Fourier-Stieltjes transform [3], [23],

$$(3.10) \qquad \hat{\theta}(\omega) - \hat{\theta}(0) = \int_{-\infty}^{\infty} \frac{e^{-j\omega t} - 1}{-j\omega} d\hat{f}(t)$$

$$(3.11) \qquad \hat{f}(T) - \hat{f}(0) = \frac{1}{2\pi} \int_{-\infty}^{\infty} \frac{e^{+j\omega t} - 1}{+j\omega} d\hat{\theta}(\omega),$$

where $\hat{\theta}(\omega)$ is the cumulative spectrum and $\hat{f}(t)$ the cumulative time function. The Fourier-Stieltjes integrals are properly defined even when the time or frequency functions contain discrete components, while $\hat{\theta}(\omega)$ and $\hat{f}(t)$ are functions of bounded variation in $(-\infty, \infty)$. The duality of the Fourier-Stieltjes transform and its inverse is self-evident by their definitions (3.10), (3.11). $\hat{f}(t)$ is bandlimited in the Fourier-Stieltjes sense if $\hat{\theta}(\omega)$ is constant outside an interval $[a, b]$. In a dual manner we define time-limited signals. In cases where $\hat{\theta}(\omega)$ is unbounded in its variation, more general results should be used, such as Zakai's definition of band- (or time-) limited signals [32], or generalized functions [10]. An important result due to Polya's theorem [17] or Paley-Wiener-Schwartz [10] ensures that if a signal is bandlimited in the Fourier-Stieltjes sense in one domain, its counterpart is an EFET.

A discrete complex time signal is defined by

$$(3.12) \qquad x(t) = \sum_n x_n \delta(t - t_n),$$

and its Fourier transform by

$$(3.13) \qquad \hat{x}(\omega) = \sum_n x_n e^{-j\omega t_n}.$$

Using a transformation of the whole complex plane into itself (ω and z being viewed as complex variables),

$$(3.14) \qquad z = e^{j\omega}$$

the generalized z-transform is defined by

$$(3.15) \qquad \hat{x}(z) = \sum_n x_n z^{-t_n}.$$

The ordinary Fourier sequence and z-transforms are derived through uniformly spaced (normalized) sampling $t_n = n$. The same (within the duality) definitions apply to cases where a discrete signal describes the spectral function.

We use properties of a class of EFET called B-functions. These include all functions of finite bandwidth (according to various definitions of bandwidth). The properties of B-functions are well established by classical theory of entire functions, and a complete characterization of their zeros is known. Furthermore, for each B-function there corresponds a unique expansion into the product of its zeros, determined by the Hadamard factorization [23]:

$$(3.16) \qquad f(s) = cs^m e^{jks} \lim_{r \to \infty} \prod_{|s_n| \leqslant r} (1 - s/s_n) \,,$$

where $f(s)$ is a B-function, c a complex constant, m an integer, k a real constant, and s_n are the complex roots of $f(s)$. Convergence of (3.16) is conditional on correct order of the zeros [23]. This form of expansion shows that a band-limited function is uniquely determined (within some parameters) by all zeros.

3.2. *Time and position (i.e. 2D) domains – zero crossings*

Logan's work [17] addresses the issue of whether a one-dimensional bandpass signal can be uniquely determined by its real zero crossings. His work characterizes analytically the class of these interesting signals, and proves that a bandpass signal can be uniquely determined by its real zero crossings (within a multiplicative constant) even if it has complex zeros. Logan's existence theorem defines the conditions that must be satisfied in order to have a representation: 1. The bandpass function and its Hilbert transform should have no zeros in common other than real simple zeros, and 2. The bandwidth of the signal must be less than an octave. These common zeros of the bandpass signal and its Hilbert transform (free zeros in Logan's terminology) are those that may be removed or moved around without destroying the bandpass property of the signal.

The formal conditions for a real bandpass signal representation by its zero crossings are stated in Theorem 8, and for real periodic function, as a special case, in Theorem 9.

THEOREM 8 (Real Bandpass Signal Version (Logan [17])).
Let $h(t) = Re(f(t)e^{j\omega_0 t})$ be a real bandpass function, where $f(t)$ is bandlimited and bounded. $h(t)$ is uniquely determined by its real zero-crossings within a multiplicative constant, if $h(t)$ has no zeros in common with its Hilbert transform $\tilde{h}(t) = I_m(f(t)e^{j\omega_0 t})$ (free zeros) other than real simple zeros, and if the band is less than one octave in width.

THEOREM 9 (Real Periodic Functions (Poggio et al. [20])).
Let $h_s(t)$ be a real periodic bandpass function

$$(3.17) \qquad h_s(t) = \sum_{n=-N}^{N} c_n e^{jnt} \qquad c_n = c_{-n}^* \,,$$

where c_n is a discrete complex sequence, and $*$ denotes the conjugate operation. The function $h_s(t)$ is uniquely determined by its real zeros within a multiplicative constant provided $h_s(t)$ and its Hilbert transform $\tilde{h}_s(t)$, given by

(3.18)
$$\tilde{h}_s(t) = \sum_{n=-N}^{N} \tilde{c}_n e^{jnt} \quad \tilde{c}_n = -j\,sign(n)c_n\,,$$

have no zeros in common except real simple zeros, and $c_n = 0$ for $|n| \leqslant N/2$ (or $(N+1)/2$ when N is odd).

It should be noted that these are existence theorems and as such they define the necessary and sufficient conditions. However, they do not provide an algorithm for signal reconstruction from its zero crossings, nor do they address the important issue of stability. Algorithms for one-dimensional signals were put forward by Voelecker and Requicha [31]. These algorithms were modified by Rotem and Zeevi [24], and applied in reconstruction of images from their zero crossings.

THEOREM 10 (Bandpass Signals in Two Dimensions).
(**Rotem and Zeevi [24]**) *Let $f(x, y)$ be a real function that satisfies the following conditions:*

(i) *$\int \int_{-\infty}^{\infty} f^2(x, y) dx\, dy < \infty$.*
(ii) *$\int_{-\infty}^{\infty} f^2(x, y_0) dx < \infty$ for a certain y_0 (such y_0 must exist due to (i)).*
(iii) *$\int_{\infty}^{\infty} f^2(x_k, y) dy < \infty$ for $\{x_n\}$ defined below (see (vii)).*
(iv) *The two-dimensional Fourier transform of $f(x, y)$*

(3.19)
$$\hat{f}(u, v) = \int\!\!\int_{-\infty}^{\infty} f(x, y)e^{-j(ux+vy)} dx\, dy$$

satisfies the bandwidth condition:

(3.20)
$$\hat{f}(u, \nu) = 0 \begin{cases} u < -b \cup -a < u < a \cup b < u \\ \nu < -d \cup -c < \nu < c \cup d < u \end{cases}$$

where

(3.21)
$$0 < a < b < 2a; \qquad 0 < c < d < 2c.$$

(v) *For y_0 in (ii), the one-dimensional function of x, $f(x, y_0)$, and its Hilbert transform $\tilde{f}(x, y_0)$ do not have zeros in common except real zeros of degree one.*
(vi) *For $\{x_n\}$ in (iii), the one-dimensional function in y, $f(x_n, y)$, and its Hilbert transform $\tilde{f}(x_n, y)$ do not have zeros in common except real zeros of degree one.*
(vii) *The set of points $\{x_n\}$ constitutes a sampling set for $\mathbf{B}(a, b)$ by having density greater than $(b - a)/\pi$. More precisely: if $N(L)$ is the number of sampling points on the*

interval $(0, L)$, then the sufficient condition is

(3.22)
$$\lim_{L \to \infty} \sup \frac{N(L)}{L} > \frac{b-a}{\pi}.$$

Then, if there exists another function $g(x, y)$ which also satisfies conditions (i)–(vii), and if

(3.23)
$$\text{sign}(g(x, y)) = \text{sign}(f(x, y)),$$

we have

(3.24)
$$g(x, y) \equiv \alpha f(x, y).$$

That is, f and g are identical within a multiple constant. The proof is a direct extension of Logan's theorem [17]. Since the one-dimensional functions $f(x_n, y)$ and $f(x, y_0)$ satisfy Logan's conditions, we can construct the functions along vertical lines $x = x_n$, scale them according to the function on the horizontal line y_0, and then, having a sampling set $\{x_n\}$, the value of $f(x, y)$ is determined (up to a constant) at every point in the plane. Obviously, x and y can be interchanged in conditions (ii)–(vii) and in the proof [24].

THEOREM 11 (Bandpass in One Dimension, Lowpass in Other).
(Rotem and Zeevi [24]) *Let $f(x, y)$ be a real function that satisfies the following conditions:*

(i) $\int \int_{-\infty}^{\infty} f^2(x, y) dx \, dy < \infty.$
(ii) $\int \int_{-\infty}^{\infty} f^2(x, y_n) dx < \infty$ *for $\{y_n\}$ defined below.*
(iii) *The two-dimensional Fourier transform of $f(x, y)$ satisfies*

(3.25)
$$\hat{f}(u, \nu) = 0 \begin{cases} u < -b \cup -a < u < a \cup b < u \\ \nu < -c \cup \nu > c \end{cases}$$

where

(3.26)
$$0 < a < b < 2a; \quad 0 < c < \infty.$$

(iv) *For $\{y_n\}$ in (ii), the one-dimensional function in x, $f(x, y_n)$ and its Hilbert transform $\tilde{f}(x, y_n)$ do not have zeros in common except real zeros of degree one.*
(v) *There exists a straight line ℓ of angle θ to the x axis, where*

(3.27)
$$\tan \theta < \frac{2a - b}{3c}$$

(by (iii), $b < 2a$, so $\theta > 0$) such that the one-dimensional function $f(s)$ satisfies along ℓ,

(3.28)
$$\int_{-\infty}^{\infty} f(s) ds < \infty.$$

(vi) *The function $f(s)$ and its Hilbert transform $\tilde{f}(s)$ do not have zeros in common except real zeros of degree one.*

(vii) *The set of points $\{y_n\}$ constitutes a sampling set for $\mathbf{B}(c)$.*

Then, if there exists another function $g(x,y)$ which also satisfies conditions (i)–(vii), and if

(3.29) $$\text{sign}(g(x,y)) = \text{sign}(f(x,y)),$$

we have

(3.30) $$g(x,y) \equiv \alpha f(x,y).$$

In this case, the function is determined on horizontal lines y_n and scaled on the diagonal ℓ. Having $\{y_n\}$, a sampling set for $\mathbf{B}(c)$, we can determine the value for every point between the lines.

The conditions can be interchanged so that the function is determined on multiple diagonals and one horizontal line, or another diagonal angle less than θ relative to the x axis.

The importance of formulating the conditions of free zeros in the case of a two-dimensional signal (e.g. an image), is in the observation that these conditions have to be satisfied only on a sparse set of lines in the plane. Furthermore, if the signal is periodic (and every finite image can be considered as one period of a periodic, infinite signal, thus enabling bandpass conditions to exist) – only a finite number of lines, out of the infinite number of lines in the plane, have to satisfy the free-zero conditions. Thus, even if the strict condition of having no free zero is not realized in an image which satisfies one of the bandpass conditions, the image is still likely to be represented by its zero crossings.

3.2.1. *Reconstruction of images from their zero crossings.* The proposed algorithms directly follow the proofs of Theorems 10 and 11. If the image satisfies the conditions of Theorem 10, the one-dimensional functions on several rows (or columns) are reconstructed, using an iterative algorithm described below. Then, another one-dimensional function is reconstructed along a column (or a row), and used for scaling the previous functions. If the image satisfies the conditions of Theorem 11, the scaling is done along a diagonal at the proper angle. The rest of the image is reconstructed by interpolation.

The reconstruction algorithm is as follows: Given the clipped signal S, pass it through an ideal bandpass filter to yield S_0. Designating the bandpass operation as $B_p\{\cdot\}$, obtain the initial signal

(3.31) $$S_0 = B_p\{\text{sign } S\}$$

and then continue iteratively

(3.32) $$S_{j+1} = S_n - c[B_p\{\text{sign } S_n\} - S_0].$$

The algorithm converges when $\text{sign}(S_i) = \text{sign}(S)$. It can be shown that the process of these successive approximations is contracting. If Logan's conditions for uniqueness are met, then $S_n \equiv \alpha S$, where α is some constant. This algorithm yields good results for most bandpass signals, with the number of iterations varying from a few up to 100. If $\text{sign}(S_i(nT)) = \text{sign}(S(nt))$ for every n, there can still exist another continuous signal $S'(t)$ where $S'(nT) =$

$S_i(nT)$ but sign$(S'(t)) \neq$ sign$(S(t))$, so $S'(t) \neq S(t)$ and also $S_i(nT) \neq \alpha S(nT)$. This ambiguity can be viewed as a result of real-zeros jitter that may also change the location of the complex zeros [24]. It should be noted, however, that the dependence of the deviation of the complex zeros on that of the reals zeros is closely related to the question of stability. This issue is dealt with elsewhere [33].

The error analysis indicates some simple ways to reduce the errors in the scaling factors. For example, select a column along which the reconstructed signal converged; locate the maxima of the signal. (On these points the normalized errors are minimal.) Reconstruct the signals along some rows intersecting the selected column at the peak points and scale them according to the peak values. Then reconstruct the signals in the other columns. The intersection points of each column with the rows form two vectors: that of the (as yet unscaled) values of the signal along the column and that of the scaled values of the signals along the rows. Denote the unscaled values along the column by $c(k)$ and the values of the rows by $r(k)$. If there are M rows, then the optimal scaling factor for the column is

$$(3.33) \qquad\qquad a = \frac{\sum_{k=1}^{M} c(k)r(k)}{\sum_{k=1}^{M} c^2(k)}.$$

In this way the effect of reconstruction errors in the columns and the rows tends to cancel out [24].

The same method can be applied to images which are bandpass in one dimension and lowpass in the other (Theorem 11). The scaling is carried out in this case by reconstruction on several diagonals intersecting one column at its maxima, scaling the diagonals, and then scaling the other columns by values of the points of intersection with the diagonals.

In order to have an ideal bandpass (or lowpass) image, the signal should be looked upon as periodic, the actual image being one period (in the two-dimensional sense). The conditions of finite Fourier integrals should in this case be modified for periodic signals.

The bandpass operation can be carried out by calculating the FFT of each column and simply deleting all components out of the passband, and then transposing the image and repeating the operation. This is equivalent to convolving the image with an infinite, periodic bandpass filter.

Prior to performing the sign operation, it is recommended to interpolate the signal so as to reduce the zero-crossings jitter. Interpolation is also needed in creating the diagonals, so that they can be inclined at an angle of less than $45°$. In order to create a bandpass signal along the diagonal, the signal should also be periodic. This is done by continuing the diagonal along some periods of the image until it reaches the starting point. Because of the assumed periodicity of the image, continuing the diagonal means "folding" from bottom to top each time it reaches the last row.

3.3. *Frequency domain – discrete signals*

We now present theorems concerning one-dimensional discrete signal representation by partial information in the frequency domain. These include the cases of phase only, sampled phase, one bit of phase, magnitude, and signed-magnitude. The theorems are quoted as stated in the original papers.

THEOREM 12 (Hayes, Lim and Oppenheim [12]). *Let $x(n)$, $y(n)$ be two finite length real sequences whose z-transforms have no zeros in reciprocal conjugate pairs or on the unit circle. If $\vartheta_x(\omega) = \vartheta_y(\omega)$ for all ω, then $x(n) = \beta y(n)$ for some positive constant β. If $\tan \vartheta(\omega) = \tan \vartheta_y(\omega)$ for all ω, then $x(n) = \beta y(n)$ for some real constant β. $\vartheta_x(\omega)$ and $\vartheta_y(\omega)$ are the respective phases of the Fourier sequences of $x(n)$ and $y(n)$.*

The results have been extended to complex sequences and the condition prohibiting simple zeros on the unit circle relaxed.

THEOREM 13 (Hayes [11]). *Let $x(n)$, $y(n)$ be complex sequences which are zero outside the interval $[0, N-1]$, with z-transforms having no zeros in conjugate reciprocal pairs or on the unit circle. If $M \geqslant 2N - 1$ and $\vartheta_x(\omega) = \vartheta_y(\omega)$ at M distinct frequencies in the interval $[0, 2\pi]$, then $x(n) = \beta y(n)$ for some positive real number β. If $M \geqslant 2N - 1$ and $\tan(\vartheta_z(\omega)) = \tan(\vartheta_y(\omega))$ at M distinct frequencies in the interval $[0, 2\pi]$, then $x(n) = \beta y(n)$ for some real number β.*

THEOREM 14 (Hayes, Lim and Oppenheim [12]). *Let $x(n)$ be a real sequence which is zero outside the interval $[0, N-1]$ with $x(0) \neq 0$ and which has a z-transform with no zeros in conjugate reciprocal pairs or on the unit circle. Let $y(n)$ be any real sequence which is also zero outside the interval $[0, N-1]$. If $\vartheta_x(\omega) = \vartheta_y(\omega)$ at $N-1$ distinct frequencies in the interval $(0, \pi)$, then $y(n) = \beta x(n)$ for some positive constant β. If $\tan(\vartheta_x(\omega)) = \tan(\vartheta_y(\omega))$ for $N-1$ distinct frequencies in the interval $(0, \pi)$, then $y(n) = \beta x(n)$ for some real constant β.*

THEOREM 15 (Hayes, Lim and Oppenheim [12]). *Let $x(n)$ and $y(n)$ be two real sequences whose z-transform contain no reciprocal pole-zero pairs and which have all poles, not at $z = \infty$, inside the unit circle. If the magnitude of the Fourier transforms of $x(n)$ and $y(n)$ are equal, then $x(n) = \pm y(n + m)$ for some integer m.*

THEOREM 16 (Thm. 2 of Van Hove, Hayes, Lim, Oppenheim).
[29] *Let $x(n)$ and $y(n)$ be two real, causal (or anticausal), and finite extent sequences with z-transforms which have no zeros on the unit circle. If the signed Fourier magnitudes of the sequences (denoted by $G_x^\alpha(\omega)$ and $G_y^\alpha(\omega)$, respectively, and defined by (3.34) and (3.35)) are equal for all ω, $G_x^\alpha(\omega) = G_y^\alpha(\omega)$ $0 < \alpha < \pi$, then $x(n) = y(n)$. When $\alpha = \pi$ (or 0), and if $G_x^*(\omega) = G_y^*(\omega)$ for all ω and $x(0) = y(0)$, then $x(n) = y(n)$.*

The signed Fourier magnitude $G_x^\alpha(\omega)$ and Fourier magnitude $R_x(\omega)$ are related through

(3.34)
$$G_x^\alpha(w) = S_x^\alpha(\omega)|R_x(\omega)|,$$

where the bipolar function $S_x^\alpha(\omega)$, defined by

(3.35) $$S_x^\alpha(\omega) = \left\{ \begin{array}{ll} +1 & \alpha - \pi \leqslant \vartheta_x(\omega) < \alpha \\ -1 & \text{otherwise} \end{array} \right\}$$

incorporates one bit of phase information at each frequency [19], [29].

THEOREM 17 (Oppenheim, Lim, and Curtis [19]). *Let* $x(n), y(n)$ *be two real, finite extent, causal (or anticausal) sequences with all zeros outside (inside) the unit circle. If*

$$S_x^\alpha(\omega) = S_y^\alpha(\omega), \text{for all } \omega,$$

$\alpha = 0, \pi$ *then* $x(n) = \beta y(n)$, *where* β *is a scale factor. (For* $\alpha = \pi/2, \beta$ *is a positive factor.)*

4. Duality of results in time and frequency domains

Based on the duality of the Fourier-Stieltjes transform and its inverse, relevant well-known results for time domain problems, concerned with unique representation of bandlimited functions by partial information, were applied by Shitz and Zeevi [27] in order to derive results about unique representation of a Fourier transformed discrete or continuous time signal by partial information in the Fourier domain. This approach not only enables the derivation of new results for the continuous signals, but more importantly, also highlights the interrelationship of the problems in the two domains.

Shitz and Zeevi [27] have derived, and in various cases extended the results concerning signal representation by partial information in the frequency domain by application of Logan's results (Theorems 8 and 9), and those of Voelecker [30]. In the sequel we only outline their approach.

Let $x(t)$ be a complex, finite time signal; $x(t)$ equal to zero outside the interval $[0, T]$. Let $\hat{x}(\omega)$ be the Fourier transform of $x(t)$, assumed to exist at least in the Fourier-Stieltjes sense. Then

(4.1) $$\hat{x}(\omega) = \int_{-\infty}^{\infty} x(t)e^{-j\omega t}dt = \int_0^T x(t)e^{-j\omega t}dt$$

(4.2) $$x(t) = \frac{1}{2\pi} \int_{-\infty}^{\infty} \hat{x}(\omega)e^{j\omega t}d\omega$$

(4.3) $$\begin{aligned} \hat{x}(\omega) &= P_x(\omega) + jQ_x(\omega) \\ &= R_x(\omega)\exp(i\vartheta_x(\omega)) \quad R_x(\omega) > 0, \end{aligned}$$

where $P_x(\omega), Q_x(\omega), R_x(\omega)$, and $\vartheta_x(\omega)$ are, respectively, the real part, imaginary part, magnitude, and phase of $\hat{x}(\omega)$.

4.1. *Signal representation by its Fourier phase*

THEOREM 18 (Shitz and Zeevi [27]). *Let $x(t)$ be a complex finite-time signal the Fourier transform of which $\hat{x}(\nu)$ has no complex conjugate zeros. The signal $x(t)$ is uniquely determined (within a positive multiplicative constant) by its Fourier phase $\vartheta_x(\omega), \forall \omega$. If $\hat{x}(\nu)$ has no conjugate zeros at all (neither complex nor real), then $x(t)$ is uniquely defined (within a real constant) by $\tan(\vartheta_x(\omega))$.*

This version of Theorem 12 is valid for continuous signals. Its discrete version, obtained by using the discrete model

$$(4.4) \qquad x(n) = \sum_n x_n \delta(t - t_n) \quad 0 \leqslant t_n < T$$

and by interpreting the free zeros (see Theorem 8) as reciprocal conjugate zeros (in the z-plane), turns out to be an extended version of Theorem 12. In fact, $x(n)$ may under this generalization be a countable sequence provided $0 < t_n < T$ for all n (t_n may, for example, be a convergent sequence) and $\Sigma_n |x_n| < \infty$. This result is evident due to the entireness of $\hat{x}(\nu)$ which contains a countable but not necessarily finite number of zeros (3.16). (Theorem 12 is obtained by choosing $T = N, t_n = N$ and $0 \leqslant n < N$).

Theorem 2 of Hayes, Lim, and Oppenheim [12], which is the no-poles-in-reciprocal-pairs version of Theorem 12, follows the latter and can be derived by using an anticausal signal $x(t) \neq 0$ for $-T < t \leqslant 0$, or alternatively by using the transformation $z = e^{-j\nu}$ instead of $z = e^{j\nu}$.

An alternative and much simpler derivation of Theorem 12 is possible by application of Theorem 9 which already deals with discrete signals (trigonometric sequences). However, under this approach the obtained theorem is restricted to discrete signals.

4.2. *Signal representation by its sampled Fourier phase*

Consider a finite extent complex sequence d_n, which equals zero outside the interval $[0, N - 1]$. The discrete Fourier transform $\hat{d}(\omega)$ of d_n is given by

$$\hat{d}(\omega) = \sum_{n=0}^{N-1} d_n e^{-jn\omega}$$

$$(4.5) \qquad \hat{d}(\omega) = P_{\hat{d}}(\omega) + jQ_{\hat{d}}(\omega) = R_{\hat{d}}(\omega) \exp(j\vartheta_{\hat{d}}(\omega))$$

where $P_{\hat{d}}(\omega), Q_{\hat{d}}(\omega), R_{\hat{d}}(\omega)$, and $\vartheta_{\hat{d}}(\omega)$ are the real part, imaginary part, magnitude, and phase of $\hat{d}(\omega)$. The transform is extended to the whole complex plane by replacing ω with $\nu = \omega + ju$.

Similarly to our consideration of the continuous case, we define a new sequence α_n which is time constrained, or "bandpass in time".

$$(4.6) \qquad \alpha_n = 0.5(d_{n-M} + d^*_{-n-M}) \quad M > 0.$$

Its Fourier sequence

(4.7)
$$\hat{\alpha}(\nu) = \sum_{n=-(M+N-1)}^{M+N-1} \alpha_n e^{-jn\nu}$$

is alternatively written as the sum of two sequences

(4.8)
$$\hat{\alpha}(\nu) = M_{\hat{a}}(\nu) + N_{\hat{a}}(\nu)$$

defined by

(4.9)
$$M_a(\nu) = \sum_{n=-M-(N-1)}^{-M} \alpha_n e^{-jn\nu} = 0.5\hat{d}^*(\nu)e^{j\nu M}$$

(4.10)
$$N_a(\nu) = \sum_{n=M}^{M+(N-1)} \alpha_n e^{jn\nu} = 0.5\hat{d}(\nu)e^{-j\nu M}$$

(4.11)
$$\hat{\alpha}(\nu) = Re(\hat{d}(\nu)e^{-j\nu M}) = R_{\hat{d}}(\nu)\cos(M\nu - \vartheta_{\hat{d}}(\nu)).$$

$\hat{\alpha}(\omega)$ is real since $d_n = d_{-n}^*$.

The Hilbert transform $\hat{\underset{\sim}{\alpha}}(\nu)$ of $\hat{\alpha}(\nu)$ is defined by

(4.12)
$$\hat{\underset{\sim}{\alpha}}(\nu) = \sum_{n=-M-(N-1)}^{M+(N-1)} \underset{\sim}{\alpha_n} e^{-jn\nu}, \ \underset{\sim}{\alpha_n} = i\,sign(n)\alpha_n$$

(4.13)
$$\hat{\underset{\sim}{\alpha}}(\nu) = R_{\hat{d}}(\nu)\sin(M\nu - \vartheta_{\hat{d}}(\nu)).$$

Now we can apply Theorem 9 [20]. The one octave bandwidth requirement is satisfied by choosing $M > (N-1)$. We note that $\hat{\alpha}(-\nu)$ is equivalent to $h(t)$ in the notation of [20]. Interpreting this work in the context of our own notation, it was shown that if ν_0 is a free zero of $\hat{\alpha}(-\nu)$, then both this zero and its conjugate ν_0^* are zeros of $N_{\hat{a}}(\nu)$ [20, eq. (6)]:

(4.14)
$$N_{\hat{a}}(\nu_0) = N_{\hat{a}}(\nu_0^*) = 0.$$

Derivation of Theorem 13 is straightforward, by application of the formalism provided in [20]. Accordingly $\hat{\alpha}1(\nu)$ is the Fourier sequence of another time bandpass sequence $\alpha1_n$ which is zero for $|n| \notin [M, M+(N-1)]$ and defined similarly to (4.6), where d_n is replaced by $d1_n$. It was shown in [20] that

$$\hat{\alpha}(\omega)\hat{\underset{\sim}{\alpha}}(\nu)1(\omega) - \hat{\underset{\sim}{\alpha}}(\nu)1(\omega)\hat{\alpha}1(\omega) = R_{\hat{d}}(\omega)R_{\hat{d}1}(\omega)\sin(\vartheta_{\hat{d}}(\omega) - \vartheta_{\hat{d}1}(\omega))$$

(4.15)
$$= \sum_{-(N-1)}^{N-1} e_n e^{-jn\omega}.$$

This equation can be interpreted as an appropriate convolution of two complex sequences of length N, one of them being causal and the other anticausal.

It was shown in [20] that the Vandermonde determinant associated with the real roots of (4.15) is not equal to zero if Logan's conditions are satisfied, so if $\vartheta_{\hat{a}}(\nu) = \vartheta_{\hat{a}1}(\nu)$ for at least $2N - 1$ distinct frequencies in the interval $(0, 2\pi)$ then e_n in (4.15) all equal zero. Therefore $\sin(\vartheta_{\hat{a}}(\omega) - \vartheta_{\hat{a}1}(\omega)) = 0$ for all ω or, equivalently, $\tan \vartheta_{\hat{a}}(\omega) = \tan \vartheta_{\hat{a}1}(\omega)$ for all ω.

Logan's condition prohibiting complex free zeros, when expressed in the context of z, dictates a no reciprocal conjugate zeros condition, but still allows zeros on the unit circle. By the fact that $\tan \vartheta_{\hat{a}}(\omega) = \tan \vartheta_{\hat{a}1}(\omega)$ and through Theorem 12 we have $d_n = \beta d1_n$ for some real β, and Theorem 13 follows. It is emphasized, though, that in this case zeros on the unit circle are not allowed except for the case when the points ω_i at which $\vartheta_{\hat{a}}(\omega_i)$ is specified are all the real zeros of $\hat{a}(\omega)$, as discussed in the preceding subsection.

Theorem 5 of Hayes, Lim, and Oppenheim [12] is a special case of Theorem 13 considering a real sequence d_n. The phase $\vartheta_{\hat{a}}(0)$ at $\omega = 0$ is always zero. It is defined in this case for at least N distinct frequencies over the interval $[0, \pi)$, provided $M \geqslant N - 1$. Theorem 4 of [12], which is equivalent to Theorem 3 of [12], can be derived in the same way, using anticausal sequences or the complex transformation $z = e^{-j\nu}$.

4.3. Derivation of Theorem 14

Theorem 14 in its complex version is derived by directly applying Theorem 9. Through (4.15) we reach the following equality: if $\vartheta_{\hat{a}}(\omega) = \vartheta_{\hat{a}1}(\omega)$, at least for $2N - 1$ distinct frequencies in the interval $(0, 2\pi)$, then

$$(4.16) \qquad \frac{\hat{a}(\omega)}{\underset{\sim}{\hat{a}}(\omega)} = \frac{\hat{a}1(\omega)}{\underset{\sim}{\hat{a}}1(\omega)}$$

$$(4.17) \qquad \tan \vartheta_{\hat{a}}(\omega) = \tan \vartheta_{\hat{a}1}(\omega) \forall \omega .$$

Since $\hat{a}(\nu)$ satisfies Logan's conditions, it can be uniquely determined within a real constant. If $\hat{a}1(\nu)$ also does not possess forbidden free zeros (in Logan's sense), the theorem is proved immediately. If, however, it does possess such free zeros, complex, or real, multiple zeros, then we can identify $\hat{a}1(\omega)$ within a real function $\hat{b}(\omega)$ which is positive or negative for all ω, i.e.,

$$(4.18) \qquad \hat{a}1(\omega) = \hat{b}(\omega)\hat{a}(\omega) .$$

¿From (4.18) we thus realize that $a1_n$ is a convolution between a_n and some conjugate symmetric sequence b_n, so that $a1_n$ cannot be time-bandpass in the same band as a_n, $|n| \in [M, M + (N - 1)]$. This is possible only when $\hat{b}(\omega) = \beta$, β being a real constant. Again, if the equality refers to $\vartheta_{\hat{a}}(\omega) = \vartheta_{\hat{a}1}(\omega)$ for distinct $2N - 1$ frequencies in $(0, 2\pi)$, the signum function of $\hat{a}1(\omega)$ and $\hat{a}(\omega)$ is uniquely defined, so that $\hat{b}(\omega)$ in (4.18) can be only a positive function and the multiplication factor β is therefore positive. It should be emphasized that free zeros on the real axis of $\hat{\alpha}(\omega)$ are not permitted according to this theorem, except for the case when the instants ω_i, at which $\vartheta_{\hat{d}}(\omega_i)$ is specified, are all real zero crossing instants of $\hat{\alpha}(\omega)$ (because for zero crossing Logan permits simple free zeros to occur). Theorem 14 is merely the real version of the theorem derived here, restricted to real sequences.

Theorems 6 and 7 (in Hayes et al. [12]) are dual of Theorem 14 here for pole-only sequences. They can be derived using the methods implemented in the case of the dual Theorems 12 and 13 for pole-only sequences.

4.4. *Signal representation by Fourier magnitude*

We turn back to a continuous causal case where $x(t)$ is defined by (4.2) and $\hat{x}(\omega)$ is an analytic signal expressed by

$$
\begin{aligned}
\hat{x}(\omega) &= P_x(\omega) + jQ_x(\omega) = P_x(\omega) - j\tilde{P}_x(\omega) \\
&= R_x(\omega)\exp(j\vartheta_x(\omega)), \quad R_x(\omega) > 0.
\end{aligned}
$$

(4.19)

This is an analytic representation of a lower sideband signal in the frequency domain, with $P_x(\omega)$ being real. Equivalently, $(2\pi)^{-1}\hat{x}(-\omega)$ is an analytic representation of an upper sideband [30], [18].

The question in which cases $(2\pi)^{-1}\hat{x}(-\omega)$ [or equivalently, $\hat{x}(\omega)$] is determined uniquely by its magnitude is equivalent to the problem of determining when a single-side-band (SSB) signal can be unambiguously demodulated (within a sign factor) by an amplitude modulation (AM) detector. We apply Voelecker's [30] results directly to $(2\pi)^{-1}\hat{x}(-\omega)$, and extend Theorem 15 to the continuous case. According to Voelecker, $\hat{x}(-\nu)$ [or $\hat{x}(\nu)$] cannot have zeros in either the upper half plane (UHP) or lower half plane (LHP). Substituting $\nu = \omega + ju$, we have that $u > 0$ and $u < 0$ [or vice versa with respect to $\hat{x}(\nu)$]. Accordingly, we have for continuous signals the following theorem.

THEOREM 19 (Shamai and Zeevi [27]). *Let $x(t)$ be a complex finite-time signal whose Fourier transform $\hat{x}(\nu)$ possesses no zeros either in the lower half-plane (LHP) or in the upper half-plane (UHP). Then $x(t)$ is uniquely determined by the Fourier magnitude $R_x(\omega)$ within a constant time shift and a complex phase factor e^{ja}.*

Representing a discrete time signal by (4.4) and applying the complex transformation $z = e^{j\nu}$, we derive the all zero version of Theorem 15 for complex sequences. The zeros at the origin of z^{-m}, m being an integer, are permitted since they form pure delays. Because these delays cannot be identified by the magnitude, we have the ambiguity $x(n) = \pm y(n + m)$. The formal derivation which includes terms of z^{-m} is obtained directly by defining $x(t)$ to be zero outside an interval (T_0, T), where $T > T_0 > 0$. The pole-zero version of Theorem 15 is derived by resorting to the same reasoning. However, $x(t)$ should in this case be defined as a convolution between two appropriate causal (all-zero) sequences one of which has only the zero terms of Theorem 15 and the other zeros which are the reciprocals of the poles defined in that theorem. The new signal is also an analytic signal, and Theorem 15 follows.

Since Theorem 8 of Hayes, Lim, and Oppenheim [12] is just a mirror version of Theorem 15, it can be derived by the same arguments using the lower half-plane instead of the upper one or vice versa.

4.5. *Signal representation by Fourier signed-magnitude*

Applying well-known results, obtained in the time domain for continuous band-limited functions, Shamai and Zeevi [27] derived the following extended version of Theorem 16.

THEOREM 20 (Shamai and Zeevi [27]).
Let $x(t)$ be a complex signal that vanishes outside the interval $[0,T)$ with $x(0)$ being real. The signal $x(t)$ is uniquely determined (within $x(0)$ for $\alpha = 0, \pi$) by the Fourier signed-magnitude for any $0 \leqslant \alpha \leqslant \pi$, provided the Fourier transform $\hat{x}(\nu)$ contains no multiple real zeros.

4.6. *Signal representation by one bit of Fourier phase*

By direct application of Logan's results [17], a continuous version of Theorem 17, generalized to also incorporate complex signals, was derived.

THEOREM 21 (Shamai and Zeevi [27]).
Let $x(t)$ be a complex signal that vanishes outside the interval $[0,T)$, with $x(0)$ being real. The signal $x(t)$ is uniquely determined (within a positive multiplicative constant for $\alpha \neq 0, \pi$, or within a real constant and within $x(0)$ for $\alpha = 0, \pi$) by the one bit of Fourier phase, provided the Fourier transform $\hat{x}(\nu)$ is zero-free in the closed upper half-plane.

Using the analytic signal

$$\hat{x}(\nu)e^{-j\alpha} = P_x^\alpha(\nu) + jQ_x^\alpha(\nu)$$
(4.20)
$$= R_x(\nu)\exp(j(\vartheta_x(\nu) - \alpha)),$$

$f(t)$ in Logan's notation becomes equivalent to $(2\pi)^{-1}\hat{x}(-\nu)e^{-j\alpha}e^{-jT\nu/2}$, in our case, because the Fourier transform (not the inverse) of this function is zero outside the interval $(-T/2, T/2)$. According to Logan's theorem in the context of our notations

$$Re\{(2\pi)^{-1}\hat{x}(-\nu)e^{-j\alpha-jT\nu/2} \cdot e^{j\mu\nu}\}$$

has only real simple zeros if $(2\pi)^{-1}\hat{x}(-\nu)e^{-j(jT\nu/2+\alpha)}$ is zero-free in the closed lower half-plane $(\nu = \omega + ju, u \leqslant 0)$.
$P_x^\alpha(\nu)$ is an EFET which is real on the axis $\nu = \omega$ (in fact it is a B-function) and it determines $x(t)$ except perhaps for $x(0)$. If the zeros of $\hat{x}(\nu)$ are all in the closed upper half-plane, $P_x^{\alpha+\pi/2}(\nu)$ has only simple real zeros and it is of course determined within a multiplicative constant by $S_x^\alpha(\nu)$. The latter, in turn, determines the real zero crossings of $Q_x^\alpha(\nu)$ (utilized in the preceding subsection). $S_x^{\alpha+\pi/2}(\nu)$ determines eventually the real zero crossings of $P_x^\alpha(\nu)$. In fact, the multiplicative constant is positive unless $\alpha = 0, \pi$ since, as

was mentioned before, $S_x^\alpha(\nu)$ determines not only the zero crossing instances of $P_x^{\alpha+\pi/2}(\nu)$ but also its sign. In the case of $\alpha = \pi/2$, $P_x^\alpha(\nu)$ also determines $x(0)$ which is assumed in our theorem to be true.

Interpretation of these results for the discrete version is straightforward, by defining $x(t)$ according to (4.4). The conditions with respect to the z variable are found by using $z = e^{j\nu}$. Since $\nu = \omega + ju$ and $u \leqslant 0$, all the zeros in z-notations have to be outside or on the unit circle; those on the unit circle must be simple zeros.

The extended version of Theorem 21 applies to complex sequences, and permits simple zeros on the unit circle; it is valid also for the generalized z-transform, since the results are valid for complex [except for $x(0)$], continuous causal, and finite extent signals. The anticausal case can be derived by the same procedure, using anticausal finite extent complex and continuous signals instead of $x(t)$, or alternatively, in the case of discrete signals, by applying the transformation $z = e^{j\nu}$.

4.7. Signal and image reconstruction from localized phase

The subject of one-dimensional signal and image representation by phase information was extended to localized phase. Porat and Zeevi [21] represented images by Gabor phase and showed that such representations by partial information preserve edge information of the original image. Their work was further extended to restoration of magnitude by constrained iterative techniques [1]. Images (and other signals) can be reconstructed from localized (Gabor-type) phase more efficiently than from global (Fourier) phase. One straightforward saving is in the number of computational operations (i.e. computational complexity). The other reason is the much faster convergence rate of the iterative algorithm in the case of localized phase [1]. The reason for a faster convergence rate remained obscured until recently, when Urieli, Porat and Cohen [28] proved the following theorem:

THEOREM 22 (Urieli, Porat and Cohen [28]).
Let L and K be two convex sets whose intersection contains the solution of the POCS (projection onto convex sets) algorithm of iterative orthogonal projections between the two convex sets.

Given a discrete signal $x(n) = [x_0, x_1, \cdots, x_{n-1}]^T$, the optimal convergence angle $\alpha = \pi/4$ is obtained if and only if the $x(n)$ is a geometric sequence, i.e.,

$$x(n) = c[1, q, q^2, \cdots q^{n-1}]^T,$$

where c and q are scalars. For all other signals $\alpha < \pi/4$ (i.e. slower convergence).

How does Theorem 22 relate to the convergence rate reported in [1]? As Urieli et al. [28] observed, if a signal x is sequentially bisected into smaller segments, the ratio of monotonic to nonmonotic sequences increases. Since smaller segments are more likely to be monotonic than larger segments, they are likely to more closely correspond to geometric sequences.

This finding provides an additional good reason for further investigation of natural signals' representation in combined spaces, such as the Gabor-type and wavelet representations.

Acknowledgment

Research support in part by the Ollendorff Center, by the Fund for promotion of Research at the Technion, and by the Israeli Ministry of Science.

References

[1] J. Behar, M. Porat and Y.Y. Zeevi, "Image reconstruction from localized phase", *IEEE Trans. on SP*, vol. 40, no. 4, pp. 736–743, April 1992

[2] J.J. Benedetto, C. Heil and D.F. Walnut, Differentiation and the Balian-Low theorem, *Journal of Fourier Analysis and Applications* vol. 1, 355–402, 1995.

[3] S. Bochner, *Lectures on Fourier Integrals*, Princeton, NJ: Princeton University, 1959.

[4] I. Daubechies, A. Grossman and Y. Meyer, "Painless non-orthogonal expansions", *J. Math. Phys.*, vol. 27, pp. 1271–1283, 1986.

[5] I. Daubechies, "The wavelet transform, time-frequency localization and signal analysis", *IEEE Trans. Inform. Theory* vol. 36, 961–1005, 1990.

[6] I. Daubechies and A.J.E.M. Janssen, "Two theorems on lattice expansions", *IEEE Trans. Inform. Theory* vol. 39, 3–6, 1993.

[7] R.J. Duffin, and A.C. Schaeffer, A class of nonharmonic Fourier series. *Trans. Amer. Math. Soc.* vol. 72, 341–366, 1952.

[8] B. Friedlander and B. Porat, "Detection of transient signals by the Gabor representation," *IEEE Trans. Acoust., Speech, Signal Processing* vol. 37, pp. 169–180, 1989.

[9] D. Gabor, "Theory of communication", *J. Inst. Elect. Eng. (London)* vol. 93(III) pp. 429–457, 1946.

[10] I.M. Gel'fand and G.E. Shilov, *Generalized Functions*. New York: Academic, 1968.

[11] M.N. Hayes, "Signal reconstruction from phase or magnitude," D.Sc. Dissertation, Dep. Elec. Eng. Comput. Sci., MIT, June 1981.

[12] M.H. Hayes, J.S. Lim, and A.V. Oppenheim, "Signal reconstruction from phase or magnitude," *IEEE Trans. Acoust., Speech, Signal Processing*, vol. ASSP-28, pp. 672–680, Dec. 1980.

[13] C. Heil and D. Walnut, "Continuous and discrete wavelet transforms", *SIAM Rev.* vol. 31, 628–666, 1989.

[14] A.J.E.M. Janssen, "The Zak transform: A signal transform for sampled time-continuous signals", *Philips J. Res.* vol. 43, pp. 23–69, 1988.

[15] A.J.E.M. Janssen, "Signal analytic proofs of two basic results on lattice expansions," *Applied and Computational Harmonic Analysis*, 350–354, 1994.

[16] Y. Katzenelson, *An Introduction to Harmonic Analysis*, Dover Pub. Inc., NY 1968.

[17] B.F. Logan, Jr. "Information in the zero crossings of bandpass signals," *Bell Syst. Tech. J.*, vol. 56, pp. 487–510, Apr. 1977.

[18] B.F. Logan, Jr. "Theory of analytic modulation systems," *Bell Syst. Tech. J.*, vol. 57, pp. 491–576, Mar. 1978.

[19] A.V. Oppenheim, J.S. Lim and S.R. Curtis, "Signal synthesis and reconstruction from partial Fourier domain information," *J. Opt. Soc. Amer.*, vol. 73, no. 11, pp. 1413–1420, Nov. 1983.

[20] T. Poggio, H.K. Nishihara, and K.R.K. Nielsen, "Zero-crossings and spatiotemporal interpolation in vision: Aliasing and coupling between sensor," M.I.T.A.I. memo G75, pp. 1–27, Apr. 1982.

[21] M. Porat and Y.Y. Zeevi, "The generalized Gabor scheme of image representation in biological and machine vision" *IEEE Trans. PAMI*, vol. 10, pp. 452–468, 1988.

[22] M. Porat, and Y.Y. Zeevi, "Localized texture processing in vision: Analysis and synthesis in the Gaborian space", *IEEE Trans. on Biomedical Engineering* vol. 36, pp. 115–129, 1989.

[23] A.A.G. Requicha, "The zeros of entire functions: Theory and engineering applications," *Proc. IEEE*, vol. 68, pp. 308–328, March 1980.

[24] D. Rotem and Y.Y. Zeevi, "Image reconstruction from zero crossings," *IEEE Trans. ASSP*, vol. ASSP-34, pp. 1269-1277, Oct. 1986.

[25] J. Segman and Y.Y. Zeevi, "Image analysis by wavelet-type transform: Group theoretic approach", *J. Mathematical Imaging and Vision*, Vol. 3, pp. 51–77, 1993.

[26] K. Seip, and R. Wallstén, "Density theorems for sampling and interpolation in the Bargmann-Fock space II", *J. Reine Angew. Math* vol. 429, 107–113, 1992.

[27] S. Shitz and Y.Y. Zeevi, "On the duality of time and frequency domain signal reconstruction from partial information," *IEEE Trans. ASSP*, vol. ASSP-33, pp. 1486-1498, Dec. 1985.

[28] S. Urieli, M. Porat and N. Cohen, "Optimal reconstruction of images from localized phase", *IEEE Trans. on IP*, vol. 7, no. 6, pp. 838–859. June 1998.

[29] P.H. Van Hove, M.H. Hayes, J.S. Lim and A.V. Oppenheim, "Signal reconstruction from signal Fourier transform magnitude," *IEEE Trans. Acoust., Speech, Signal Processing*, vol. ASSP-31, pp. 1286–1293, Oct. 1983.

[30] H.B. Voelecker, "Demodulation of single-side band signals via envelope detection," *IEEE Trans. Commun.*, vol. COM–14, pp. 22–30, Feb. 1966.

[31] H.B. Voelecker and A.A.G. Requicha, "Clipping and signal determinism: Two algorithms requiring validation," *IEEE Trans. Commun.*, vol. COM-21, pp. 738–744, June 1973.

[32] M. Zakai, "Band-limited functions and the sampling theorem," *Inform. Contr.* vol. 8, pp. 143–158, Apr. 1965.

[33] Y.Y. Zeevi, A. Gavrieli and S. Shitz, "Image representation by zero and sinewave crossings," *J. Opt. Soc. Am. A*. vol. 4, pp. 2045–2060, Nov. 1987.

[34] M. Zibulski and Y.Y. Zeevi, "Gabor representation with oversampling", in *Proc. SPIE Conf. on Visual Communications and Image Processing'92*, P Maragos (ed.), Boston, MA, vol. 1818, pp. 976–984, Nov. 1992,

[35] M. Zibulski and Y.Y. Zeevi, "Analysis of multiwindow Gabor-type schemes by frame methods", *Applied and Computational Harmonic Analysis*, vol. 4, pp. 188–221, 1997.

Some polynomial extremal problems which emerged in the twentieth century

Bahman Saffari

Université de Paris-Sud
Bahman.Saffari@math.u-psud.fr

ABSTRACT. Most of the "extremal problems" of Harmonic (or Fourier) Analysis which emerged before the year 2000 were actually born in the twentieth century, and their emergences were scattered throughout that century, including the two world war periods. A great many of these problems pertain to polynomials, trigonometric polynomials and (finite) exponential sums. Writing a reasonably complete monograph on this huge subject (even if we choose to restrict it to polynomials only) would be a monumental task, although the literature does indeed contain some valuable monographs on various aspects of the subject. The present text just *touches upon* a number of extremal problems on polynomials and trigonometric polynomials, with the hope of expanding this same text in the near future to a much larger version, and ultimately to a "reasonably complete" monograph (but only with the help of other mathematicians.)

The theory of polynomials on the unit circle is, of course, part of classical Fourier Analysis, studied with the tools of real and complex analysis. But it also leads to studying polynomials on the (cyclic) finite subgroups of the unit circle, and this is part of Fourier Analysis on finite groups. In many ways this leads to *cyclotomy*, which is part of Number Theory and Algebra. Also, some *combinatorial designs* (cyclic difference sets) show up in connection with this study. Thus the analysis of polynomials and trigonometric polynomials, even in one single variable, is at the crossroad of many important areas of contemporary mathematics. It is also much connected with some areas of engineering, such as signal processing.

J.S. Byrnes (ed.), Twentieth Century Harmonic Analysis - A Celebration, 201–233.

1. Introduction

Whatever is worth doing is worth doing badly
Gilbert Keith Chesterton

Everybody writes, nobody reads
Paul Erdös

I have borrowed from Z. A. Melzak's excellent book "Companion to Concrete Mathematics" (volume II) [26] the first of the above two epigraphs: *"Whatever is worth doing is worth doing badly"*. Indeed this sentence is a most relevant epigraph for the present paper, and *even more so* for any attempt to write a fairly complete monograph on the topics touched upon in this paper: I shall explain this in some detail in Section 1.2 (on the "peculiarities and aims of this paper"). The second epigraph: *"Everybody writes, nobody reads"* is a frightening truth so concisely enunciated by the great Paul Erdös (1913–1996), and seems to be an explanation for a good many of the evils that are infecting current science research, whether pure or applied. More about this later.

1.1. *A little history*

Let me start with brief historical remarks concerning (only the origins of) the subject: "extremal problems on polynomials and trigonometric polynomials: a century of progress", as this was the (much too ambitious) initial title I had given the organizers of this ASI, before toning it down to the present title.

To the best of my knowledge (at the time this paper is being written, *i.e.*, in the first week of November 2000), the subject of *"extremal problems on polynomials and trigonometric polynomials"*, or at least the *analytic* theory of this subject, seems to find its main roots in two (somewhat distinct) fertile grounds: on one hand the nineteenth century theory of *trigonometric and Fourier series*, and on the other hand the nineteenth century theory of *approximation and interpolation*. The *discrete* aspects of the subject are much related to number theory and can even be tracked down (somewhat loosely) to the time of Gauss and Lagrange. But the century-old *golden age* of the subject seems to have really started around 1900 (*precisely* in 1889 and in 1911) with two *totally independent* major results: first *Markov's inequality*, proved in 1887 in a very special case by the chemist Mendeleiev [27] and in 1889 by his mathematician friend A. A. Markov [25] in the general case, and then *Bernstein's inequality* (proved in 1911 by S. N. Bernstein [3] *in a somewhat weaker form* than what is presently known as "Bernstein's inequality".)

Markov's inequality (in its modern formulation) says that

$$(1.1) \qquad \qquad \|P'\|_{[-1,1]} \leqslant n^2 \|P\|_{[-1,1]}$$

whenever $P(X) = \sum_{k=0}^{n} a_k X^k \in \mathbb{C}[X]$ is a polynomial with complex coefficients, $P'(X)$ is its derivative, and

(1.2) $$\|P\|_{[-1,1]} := \max_{-1 \leqslant x \leqslant 1} |P(x)|.$$

(1.1) is an equality if we take $P(X) = T_n(X)$ where the Chebishev (or Tchebyshev) polynomial $T_n(X)$ is defined by $\cos nu = T_n(\cos u)$. Markov's inequality had been proved in the special case $n = 2$ as early as 1887 by the chemist Mendeleiev who studied it in connection with a problem on substances dissolved in a liquid [27]. He then mentioned it to A. A. Markov who became very interested and proved [25] the general case of (1.1) shortly after Mendeleiev's work [27]. For more historical details, see *for example* [31] or [32].

Unlike Markov's inequality (which, as we saw, originated from Mendeleiev's research in chemistry), *Bernstein's inequality* originated from pure mathematics (approximation theory). Indeed, at the beginning of the twentieth century, the Belgian mathematician De la Vallée Poussin [11] asked whether any piecewise linear continuous function defined on a compact interval of $\mathbb{R}$ could be approximated by polynomials of degree n with a (uniform) error $o(1/n)$ as $n \to \infty$. With the less drastic error $O(1/n)$, the (affirmative) answer had been given by De la Vallée Poussin himself. In a celebrated memoir (which was awarded a prize by the Royal Academy of Belgium on 15 December 1911), S. N. Bernstein [3] answered De la Vallée Poussin's question *negatively*: He proved that the best (uniform) approximation of the function $|x|$ on $[-1, 1]$ by a polynomial of degree $2n$, $(n \geqslant 1)$, lies between $\frac{\sqrt{2}-1}{4} \cdot \frac{1}{2n-1}$ and $\frac{2}{\pi} \cdot \frac{1}{2n+1}$. Bernstein's proof of this theorem heavily uses the following inequality which he proves in the same memoir [3]: Whenever $P(X) = \sum_{k=0}^{n} a_k X^k \in \mathbb{C}[X]$,

(1.3) $$|P'(x)| \leqslant \frac{n}{\sqrt{1-x^2}} \|P\|_{[-1,1]} \quad (-1 < x < 1).$$

Actually, Bernstein [3] proved (1.3) only for $P(X) \in \mathbb{R}[X]$, but the extension of (1.3) to $P(X) \in \mathbb{C}[X]$ is straightforward: Indeed, if $P(X) \in \mathbb{C}[X]$, put $Q(x) := \Re(P(x))$ for $-1 \leqslant x \leqslant 1$. Choose $\alpha \in \mathbb{R}$ so that $e^{i\alpha} P'(x)\sqrt{1-x^2}$ attains the maximum $\|P'(x)\sqrt{1-x^2}\|_{[-1,1]}$ at, say, $x = x_0$. Then

$$\|P'(x)\sqrt{1-x^2}\|_{[-1,1]} = e^{i\alpha} P'(x_0)\sqrt{1-x_0^2}$$
$$= Q'(x_0)\sqrt{1-x_0^2} \leqslant n\|Q\|_{[-1,1]} \leqslant n\|P\|_{[-1,1]},$$

as claimed.

Putting $\|f\|_\infty := \max_{t \in \mathbb{R}} |f(t)|$ where $f(t)$ is a real trigonometric polynomial

(1.4) $$f(t) = A_0 + \sum_{k=1}^{n} (A_k \cos kt + B_k \sin kt),$$

the inequality

$$\|f'\|_\infty \leqslant n\|f\|_\infty \tag{1.5}$$

(where equality holds if and only if $f(t) = A\cos nt + B\sin nt$) is nowadays known as *"Bernstein's inequality"* although Bernstein never proved this result. Let us explain this. The change of variable $x = \cos t$ shows that any real cosine polynomial

$$f(t) = \sum_{k=0}^{n} A_k \cos kt \tag{1.6}$$

can be written in the form $f(t) = P(\cos t)$ with $P(X) \in \mathbb{R}[X]$ as in (1.3), and vice versa. Thus, apart from the equality case, (1.5) is indeed equivalent to (1.3) *in the case when $f(t)$ is a cosine polynomial* (1.6). So Bernstein [3] did state and prove (1.5) for all cosine polynomials. Then by a quite complicated (although very interesting) argument he showed that from the truth of (1.5) for all *cosine* polynomials one can deduce the truth of (1.5) for all *sine* polynomials of the form

$$f(t) = \sum_{k=1}^{n} B_k \sin kt. \tag{1.7}$$

Actually, his argument, as presented in [3], had a gap that Bernstein did correct in his famous book [4] published fourteen years later, in 1926.

Now, having proved (1.5) for all cosine and sine polynomials, Bernstein could only conclude that in the general case of "mixed" real-valued trigonometric polynomials (1.4) one has

$$\|f'\|_\infty \leqslant 2n\|f\|_\infty. \tag{1.8}$$

This is not as good as (1.5), and equality in (1.8) *never* holds unless $f(t) \equiv 0$.

Who supplied the first proof of "Bernstein's inequality" (1.5) for "mixed" trigonometric polynomials (1.4)? In his 1926 book [4], Bernstein stated that the first proof was supplied by E. Landau [20] in a personal letter he sent Bernstein shortly after the publication of Bernstein's original memoir [3]. In the book [4] Bernstein did give Landau's proof, which consists of a simple and elegant argument showing that the truth of (1.5) for all *sine* polynomials (1.7) in fact implies the truth of (1.5) for all "mixed" polynomials (1.4). Thus Bernstein had done most of the hard work by proving (1.5) for all *sine* polynomials, but had missed Landau's simple argument leading to the general case! (Such things happen quite often.)

So Landau's 1912 proof of "Bernstein's inequality" (1.5) remained unpublished until 1926, when Bernstein's book [4] appeared. However, in a 1914 paper devoted to conjugate trigonometric series, Fejér [15] states the truth of Bernstein's inequality (1.5) in the general case, and gives a very important application of it. Fejér's proof of Bernstein's inequality appeared two years later, in 1916, in a paper of Fekete [16] who indeed does attribute the proof to Fejér. It is thus plausible that the first proofs of "Bernstein's inequality" (1.5) for all real "mixed" trigonometric polynomials (1.4) were found *independently* by Landau [20] and by Fejér [15], at about the same time, and that both of them communicated their proofs privately to fellow mathematicians (Bernstein and Fekete, respectively.) However, note that in the same year 1914 when Fejér's paper [15] appeared, the brothers M. Riesz [35] and F. Riesz Riesz, F. [34] published two new proofs of Bernstein's inequality (1.5), entirely different from each other and entirely different from those of Landau [20] and of Fejér [15].

Nowadays there are numerous generalizations, extensions and refinements of (1.5) (some really profound, others less), but most of which are simply called "Bernstein's inequality" or "Bernstein type inequality". A big treatise would not suffice to present all of them. I recall that about twenty years ago, around 1981, the late S. K. Pichorides (1940–1992) and myself thought of writing a fairly complete monograph *just on the sup-norm versions* of the Markov-Bernstein type inequalities, and in view of the enormity of the literature, we gave up.

This being said, what is the "most classical" statement (or, if we prefer, the "most widely known" statement of "Bernstein's inequality"? In other words, if we ask a random mathematical analyst to tell us quickly what is meant by "Bernstein's inequality", what is he/she likely to reply off the top of his/her head? In my opinion there are three such "likely" statements. The first is the above inequality (1.5) for all *real* trigonometric polynomials (1.4). The second is the extension of this same (1.5) to all *complex-valued* trigonometric polynomials of the form

$$(1.9) \qquad f(t) = \sum_{k=-n}^{n} a_k e^{ikt} \quad (a_k \in \mathbb{C} \text{ for all } k = -n, \dots, n)$$

(with equality in (1.5) now reached if and only if $f(t) = ae^{int} + be^{-int}$ where a and b are complex constants.) The third is the following: If

$$P(X) = \sum_{k=0}^{n} a_k X^k \in \mathbb{C}[X] \quad \text{and} \quad \|P\|_\infty := \max_{t \in \mathbb{R}} |P(e^{it})|,$$

then

$$(1.10) \qquad \|P'\|_\infty \leqslant n\|P\|_\infty$$

(with equality in (1.10) if and only if $P(X) = aX^n$.)

Now, are the above three statements of "the classical Bernstein inequality" *equivalent*? (By "equivalent" I mean, in the present context, that each can be deduced from the others *in a quite easy manner*. Otherwise I am aware that any two true mathematical statements are equivalent!) The answer is that the first two are indeed equivalent, but that (1.10) is a weaker result. Indeed, the first statement is an obvious special case of the second, but the second is also a straightforward consequence of the first. (Here is a simple proof from G. G. Lorentz's book [23], by the same argument as the one presented right after (1.3) in this introduction, but I do not know who gave it first: With $f(t)$ of the form (1.9), select $\alpha \in \mathbb{R}$ so that $e^{i\alpha} f'(t)$ attains the value $\|f'\|_\infty$, say for $t = t_0$. Now $g(t) := \Re(e^{i\alpha} f(t))$ is of the form (1.4), so

$\|f'\|_\infty = e^{i\alpha} f'(t_0) = g'(t_0) \leqslant \|g\|_\infty \leqslant \|f\|_\infty$. The equality case is ignored in [23], but this is not hard either.) Now the third statement (1.10) is obviously a consequence of the second, and it is also (easily) derived *directly* from the first on page 45 of Bernstein's book [4]. Yet I do not know of any simple way of deriving the first or second statement from (1.10). On the other hand (1.10), as well as its extensions to all L^p norms with $p \geqslant 1$, have direct proofs much simpler than any of those available for the first two statements. The reader can work them out herself or look them up, *for example*, in [32]. Here is yet another way of seeing that the first two statements (1.5) are stronger than (1.10). Indeed they imply

$$(1.11) \qquad\qquad |P'(e^{it})| \leqslant \frac{n}{2} \left(\|P\|_\infty + |P(e^{it})| \right)$$

(as easily checked, or see for example [17]), and (1.11) is obviously a deeper form of (1.10).

Does Markov's inequality (1.1) compare, depth-wise, to any ("classical" form of) Bernstein's inequality, although it was discovered utterly *independently* and twenty-two years earlier, as we saw? Using Bernstein's original theorem (1.3) and a nice theorem of Schur (Theorem 6, Chapter 3 of Lorentz's book [23]), we see that Markov's inequality is a corollary to (1.3) and thus to (1.5) for all *cosine* polynomials, but modulo Schur's theorem which is not trivial (although not very difficult either.) Altogether we may rate Markov's inequality (1.1) at about the same level of depth as Bernstein's inequality (1.5) for all trigonometric polynomials of the form (1.4) or (1.9).

The above inequalities of Markov and Bernstein (together with their equality cases) are typical examples of extremal problems, as they are equivalent to the problems of finding the maximum of $\|P'\|_{[-1,1]}/\|P\|_{[-1,1]}$ and of $\|f'\|_\infty/\|f\|_\infty$ where $\deg P \leqslant n$ and $\deg f \leqslant n$, $(P \not\equiv 0$ and $f \not\equiv 0)$. I chose (the earliest versions of) the Markov-Bernstein inequalities as typical illustrations of extremal problems for two main reasons: 1) They were among the first (but certainly not *the very* first) that marked the beginning of the century-old "golden age" (twentieth century.) 2) They have been the starting point of an enormous literature concerning their generalizations, extensions, refinements and applications. Their usefulness in numerous areas of mathematics, physics and engineering has been extraordinary, ever since they were discovered. See *for example* [1], [8], [32] and many other good references.

This being said, there are at least two different meanings to the expression "extremal problem":

1. One meaning is obviously the *search* for (attained or unattained, global or local) suprema and/or infima of real-valued functions defined on a set (of functions, of polynomials, of numbers, of measures, etc.). With this first acceptation, there is a considerable overlap between the subject of "extremal problems" and that of "inequalities" (or rather "*optimal* inequalities").

2. Another one is a problem *pertaining to properties* of such suprema and infima (whether local or global, etc.), for example that of their distribution. This will not be discussed in this paper, but hopefully will be in [39].

The Bernstein-Markov inequalities are typical examples of polynomial extremal problems which find their roots in the nineteenth century theory of approximation and interpolation. But the majority of twentieth century polynomial extremal problems find their roots in the nineteenth century theory of trigonometric and Fourier series. We will (briefly) mention a few examples in this paper. The reader should, however, first read the next section on "peculiarities" of this paper.

1.2. *Peculiarities and aims of this paper*

I first explain in a page or two why this paper is a quite peculiar one. Its aims will then follow logically.

In May 2000, a few weeks before this July 2000 ASI, I realized that my original title *"Extremal problems on polynomials and trigonometric polynomials—a century of progress"* was too ambitious and unrealistic for a $2\frac{1}{2}$-hour tutorial lecture. So I toned it down to the present title. As for my oral lecture, I personally found it unsatisfactory, but in view of the audience's enthusiasm it seemed successful. I do not consider this to be my personal success but rather the merit of this delightful subject itself, where the problems and theorems are most often simple to state but the proofs are *sometimes* very hard and profound, sometimes easy and straightforward. I hope to return to this "instability phenomenon" of mathematical statements (with its sometimes very sad consequences) either at the end of this paper, or more likely in [39].

I intended to write up my lecture for the ASI proceedings but a major illness and accident in my family left me no free time in August and September of 2000. Thus I missed the submission deadline and decided to write instead a much longer version, perhaps a whole monograph. But J.-P. Kahane (who was my teacher and whose judgment I respect) pointed out that a monograph on such a huge subject might take years for a single person, and in addition the author might be tempted to stop the write-up and work on the fascinating open problems of this subject. So Kahane advised me to request a deadline extension from the editor and try to write the ASI paper anyway.

As the editor granted such extension, I could use a week or so of peace, in the house of a friend in Newport, Rhode Island, to produce the present paper at the beginning of November 2000. So this text was written within a very short period, when I had no access to a library nor to my personal notes and documents either (as I had left nearly all of them in Europe), and also the editor had imposed a 30-page bound on the length of the paper. Thus I had no choice but to write a "highly imperfect" paper, with the following features (some of which the reader might interpret as the "aims" of this paper):

1. A number of topics are *touched upon* but all of them very briefly, and most of them quite superficially. But I intend to write soon an extended version [39] of this paper.
2. All of the topics are *among* those I had wished to mention anyway, but the "choice" of them (in this paper) is nowhere near any optimal order of priority. The choice of topics here is by no means the result of thoughtful considerations, but rather the result

of time limitations only: To write this paper in the short alloted time, I had to write down whatever would come to my mind after some very quick thinking, in a nearly random way. Had I written this paper one week earlier or one week later, the choice of topics would have been quite different. However, in the extended version [39], not only the same topics will be treated in more detail and depth, but also other topics will be added.

3. Although many nice topics are not even mentioned here, all those touched upon in this paper are subjects that, I for one, do consider extremely interesting. All of them can be described as "Erdös-style harmonic analysis". Proofs are omitted (save for a couple exceptions), due to time and space limitations. Many of the topics mentioned concern my own results and conjectures, usually unpublished (and even unwritten.) This is again simply due to time limitations: After all, my own results (whether written or not) where the most accessible ones under such "space-time" pressure!

4. No systematic history of any of the topics mentioned will be presented in this paper, but hopefully this will be done (to some extent) in the extended version [39], at least for some of the topics. However, a number of historical points or anecdotes will be encountered here. One of my personal obsessions is *historical accuracy and honesty*, although I am not always successful at that. It is a very unfortunate fact that historical accuracy and honesty is *not* widespread among mathematicians, whether due to ignorance, negligence or malevolence. Authors *very frequently* reproduce, in their publications, historical errors or lies they have read elsewhere, without checking anything, thus perpetuating errors or lies. Such sad things are often unintentional, but too often intentional also. I have had my fair share of this kind of historical negligence, but it was always unintentional.

5. What was just said about historical points can also be said about *references*, and the way erroneous or inappropriate references are perpetuated and carried from publication to publication. Despite my good intentions in this respect, I cannot guarantee that all the references in *this* paper are correct, in view of the circumstances under which this paper was written. Quite often, I do not even give any reference at all to a result I am quoting only by relying on my memory, as I do not have the reference on my desk at the moment of writing. (This is the case even with *my own* papers!)

6. In conclusion, this paper is not to be viewed as a survey of extremal problems, even in a very limited sense. Its main function is to serve as a "memorandum" for myself with a view to the extended version [39]. The reader's criticisms, comments (and whatever information he/she could give me) are welcome and will be gratefully acknowledged.

2. Back to Bernstein's inequality

Two of the most useful refinements of Bernstein's inequality are Szegö's inequality [42] proved in 1928 and the Schaake-van der Corput inequality [40] proved in 1935. The latter was quickly noted to be a weaker form (and a corollary) of Szegö's inequality, but has a usefulness of its own. Both are among the very few results that I will prove in this paper. These proofs will not be those of the original authors, as they were quite complicated, but proofs of my own (as they are the simplest ones I know of.)

2.1. Statement of the Schaake-van der Corput inequality

THEOREM 2.1 (Schaake & van der Corput [40]).
For any real trigonometric polynomial

$$(2.1) \qquad f(t) = A_0 + \sum_{k=1}^{n} (A_k \cos kt + B_k \sin kt)$$

we have, for all $t \in \mathbb{R}$, the following refinement of Bernstein's inequality (1.5):

$$(2.2) \qquad (f'(t))^2 + n^2 (f(t))^2 \leqslant n^2 \|f\|_\infty^2.$$

In other words, we have the identity

$$(2.3) \qquad \|f'^2 + n^2 f^2\|_\infty = n^2 \|f\|_\infty^2.$$

Remarks on the equality case of (2.2). Although the inequality (2.2) is a refinement of Bernstein's inequality (1.5), the *equality cases* of these two inequalities are very different in nature. For Bernstein's inequality (1.5), the assumption that $|f'(t_0)| = n\|f\|_\infty$ for *at least one* point $t_0 \in \mathbb{R}$ implies that $f(t)$ is a sinusoidal function: $f(t) = A \cos nt + B \sin nt$. But for any (arbitrary) real trigonometric polynomial $f(t)$ of the form (2.1), the Schaake-van der Corput inequality (2.2) is obviously always an equality at every $t_0 \in \mathbb{R}$ where $|f(t_0)| = \|f\|_\infty$. Also (2.2) is an equality at *every* $t \in \mathbb{R}$ whenever $f(t) \equiv A \cos nt + B \sin nt$ or $f(t) \equiv$ constant, and the converse of this is also trivially true. However one can easily show that if (2.2) is an equality for at least $2n$ distinct points of the semi-open interval $[0, 2\pi[$, then either $f(t) \equiv A \cos nt + B \sin nt$ or $f(t) \equiv$ constant. In this last remark the number $2n$ is *optimal*, (*i.e.*, minimal): One can find non-sinusoidal and non-constant real trigonometric polynomials of the form (2.1) for which (2.2) is an equality at $2n - 1$ distinct points of the semi-open interval $[0, 2\pi[$.

2.2. Saffari's proof of the Schaake-van der Corput inequality

We will obtain (2.2) as a corollary to Bernstein's inequality (1.5) *together with its equality case*, that is, $\|f'\|_\infty = n\|f\|_\infty$ *if and only if $f(t)$ is a sinusoidal function of the form*

$$(2.4) \qquad f(t) = A \cos nt + B \sin nt.$$

Our proof is by contradiction. Let V_n denote the vector space (on the field $\mathbb{R}$) of all real trigonometric polynomials of the form (2.1), with degree $\leqslant 2n$. For any $f \in V_n \backslash \{0\}$ (*i.e.*,

$f \not\equiv 0$), define:

$$(2.5) \qquad H(f) := \frac{\|f'^2 + n^2 f^2\|_\infty}{n^2 \|f\|_\infty^2}$$

and let

$$(2.6) \qquad H_{\max} := \max_{f \in V_n \setminus \{0\}} H(f)$$

To prove that the maximum $H_{\max}$ defined by (2.6) indeed exists, first note that V_n is a finite-dimensional space over $\mathbb{R}$ (since $\dim_\mathbb{R} V_n = 2n + 1$). If V_n is equipped with the norm topology, then (2.5) defines a *continuous* mapping H from $V_n \setminus \{0\}$ into $\mathbb{R}$: Indeed, since $\dim_\mathbb{R} V_n$ is finite, the (linear) differentiation operator $f \to f'$ is continuous on V_n, and so the mapping $f \to f'^2 + n^2 f^2$ is a continuous function from V_n into V_{2n-1}. Also all the norms (and in particular the sup-norm function $\varphi \to \|\varphi\|_\infty$) are continuous on the finite-dimensional vector spaces V_n and V_{2n-1}, ($\dim_\mathbb{R} V_{2n-1} = 4n - 1$), so the mapping H defined by (2.5) is continuous as claimed. Also $H(\lambda f) = H(f)$ for all $\lambda \in \mathbb{R}$ with $\lambda \neq 0$, hence the maximum defined by (2.6) is the same as the maximum of $H(f)$ when f is restricted to the (compact) unit sphere of V_n for the sup-norm metric, and this last maximum is indeed attained.

Obviously $H_{\max} \geqslant 1$ (since $H(f) \geqslant 1$ for every $f \in V_n \setminus \{0\}$), and the Schaake-van der Corput inequality we wish to prove is equivalent to the equality $H_{\max} = 1$. To prove this by contradiction, we suppose henceforth that

$$(2.7) \qquad H_{\max} > 1.$$

Let $\psi \in V_n \setminus \{0\}$ be some trigonometric polynomial of the form (2.1) for which

$$(2.8) \qquad H(\psi) = H_{\max}.$$

We first note that ψ is not a constant function, since otherwise $\psi' \equiv 0$ and so, by (2.5), $H(\psi) = 1$ contrary to our assumption (2.7). Thus $\psi' \not\equiv 0$, *i.e.*, $\psi' \in V_n \setminus \{0\}$, hence we can consider

$$(2.9) \qquad H(\psi') = \frac{\|\psi''^2 + n^2 \psi'^2\|_\infty}{n^2 \|\psi'\|_\infty^2}.$$

Let the trigonometric polynomial $(\psi'(t))^2 + n^2 (\psi(t))^2$ reach its absolute maximum at $t_0 \in \mathbb{R}$:

$$(2.10) \qquad (\psi'(t_0))^2 + n^2 (\psi(t_0))^2 = \|\psi'^2 + n^2 \psi^2\|_\infty^2.$$

So its derivative vanishes at t_0:

$$(2.11) \qquad 2\psi'(t_0)\psi''(t_0) + 2n^2 \psi(t_0)\psi'(t_0) = 0.$$

By (2.9) we cannot have $\psi'(t_0) = 0$, otherwise $\|\psi'^2 + n^2\psi^2\|_\infty = n^2 (\psi(t_0))^2$ by (2.10), and so we would have

$$H(\psi) = \frac{\|\psi'^2 + n^2\psi^2\|_\infty}{n^2\|\psi\|_\infty^2} = \frac{n^2 (\psi(t_0))^2}{n^2\|\psi\|_\infty^2} \leqslant 1$$

which is impossible in view of (2.7) and (2.8). Thus $\psi'(t_0) \neq 0$, so upon dividing both sides of (2.11) by $\psi'(t_0)$ we get

$$(2.12) \qquad\qquad \psi''(t_0) + n^2\psi(t_0) = 0.$$

Note, incidentally, that our observation $\psi'(t_0) \neq 0$ is another way of seeing that $\psi' \not\equiv 0$ so as to justify (2.9). Finally, since for sinusoidal functions $f(t)$ of the form (2.4) we obviously have $H(f) = 1$, it follows from (2.7) and (2.8) that $\psi(t)$ is *not* a sinusoidal function of the form (2.4). Therefore, by the equality case of Bernstein's inequality,

$$(2.13) \qquad\qquad \|\psi'\|_\infty < n\|\psi\|_\infty.$$

We now have all we need to find a lower bound for (2.9) which will yield the desired contradiction to (2.7). By (2.9) and the *strict* inequality (2.13),

$$
\begin{aligned}
H(\psi') \;&>\; \frac{\|\psi''^2 + n^2\psi'^2\|_\infty}{n^4\|\psi\|_\infty^2} \;\geqslant\; \frac{(\psi''(t_0))^2 + n^2 (\psi'(t_0))^2}{n^4\|\psi\|_\infty^2} \\[2mm]
&=\; \frac{n^4 (\psi(t_0))^2 + n^2 (\psi'(t_0))^2}{n^4\|\psi\|_\infty^2} \qquad \text{[by using (2.12)]} \\[2mm]
&=\; \frac{n^2 (\psi(t_0))^2 + (\psi'(t_0))^2}{n^2\|\psi\|_\infty^2} \qquad \text{[upon dividing by } n^2\text{]} \\[2mm]
&=\; \frac{\|\psi'^2 + n^2\psi^2\|_\infty}{n^2\|\psi\|_\infty^2} \qquad \text{[by (2.10)]} \\[2mm]
&=\; H_{\max}. \qquad \text{[by (2.8)]}
\end{aligned}
$$

Thus $H(\psi') > H_{\max}$, which contradicts the maximality of $H_{\max}$ defined by (2.6). Therefore our assumption (2.7) leads to a contradiction, and thus we have $H_{\max} = 1$ and have proved the Schaake-van der Corput inequality. $\square$

2.3. *Other proofs or reformulations of the Schaake-van der Corput inequality*

I gave my own proof of the Schaake-van der Corput inequality as it is one of the simplest ones I know, and certainly simpler than the original one [40]. Just relying on my memory at the moment of this write-up, I think I know a good dozen different proofs of this inequality. There might be many more in the literature. I hope to give some of them in [39]. In this connection, here is an unfortunate anecdote: Just over twenty years ago, in the late 1970's, I found my above proof of the Schaake-van der Corput inequality and gave a seminar talk about it at Orsay. The following week the late S. K. Pichorides (1940-1992) talked in the same seminar on a nice application of the Schaake-van der Corput inequality. Shortly after, without Pichorides or myself knowing about it, two *physicists* happened to submit a joint mathematical note to the French "Comptes Rendus de l' Académie des Sciences" giving a beautiful number-theoretic proof of the Schaake-van der Corput inequality (which they believed to be a new result), and they had applications of this inequality to *optics*. Their note was abruptly rejected by the referees (two mathematicians at Orsay, none of whom were J.-P. Kahane or Y. Meyer), with the comment: "*This result is not new, it is a recent theorem of Saffari and Pichorides*". So much for the competence of such "referees", who never bothered to ask me or Pichorides. Only one year after, when I heard about this rejection, could I tell those "referees" that the result actually went back to 1928 [42] but such a nice new proof should not have been rejected. In [39] I will present the (unpublished) proof of those physicists, and disclose their names.

Polynomial reformulation of the Schaake-van der Corput inequality (2.2). Let the polynomial $P(x) = \sum_{k=0}^{n} a_k X^k \in \mathbb{C}[X]$ be self-inversive, that is, $a_{n-k} = \bar{a}_k$ for all $k = 0, 1, \ldots, n$. Then the Bernstein inequality $\|P'\|_\infty \leqslant n\|P\|_\infty$ can be improved to:

$$(2.14) \qquad \|P'\|_\infty = \frac{n}{2}\|P\|_\infty \qquad (yes,\ equality!)$$

Proof. The truth of (2.14) when n is *odd* follows from the truth of (2.14) when n is *even* by considering the even-degree polynomial $P(z^2)$. So, without loss, we may suppose n even, $n = 2m$. The (real-valued) trigonometric polynomial $f(t) := e^{-imt}P(e^{it})$ is then of the form (2.1), with degree m, hence upon writing $P(e^{it}) = e^{imt}f(t)$ and differentiating both sides, we get

$$ie^{it}P'(e^{it}) = e^{imt}\left(f'(t) + imf(t)\right),$$

hence

$$(2.15) \qquad \left|P'(e^{it})\right|^2 = (f(t))^2 + m^2\left(f(t)\right)^2 .$$

Thus (2.14) is indeed equivalent to (2.3). □

The above polynomial reformulation (2.14), like the original formulation (2.2), was re-discovered by many people. Some of these proofs do shed new light on the subject, see [39].

2.4. *"Self-improvement" of the Schaake-van der Corput inequality*

This is an important topic, and we will see an application of it in Section 4. Let the real trigonometric polynomial $f(t)$ be as in (2.1), and consider the trigonometric polynomial (of degree $\leqslant 2n - 1$):

$$(2.16) \qquad f_1(t) := (f'(t))^2 + n^2\left(f(t)\right)^2 .$$

By (2.3) we have $\|f_1\|_\infty = n^2\|f\|_\infty^2$. Applying the Schaake-van der Corput inequality (2.2) to this $f_1(t)$ (with n replaced here by $2n - 1$), we obtain

$$(2.17) \qquad (f_1'(t))^2 + (2n - 1)^2 (f_1(t))^2 \leqslant (2n - 1)^2\|f_1\|_\infty^2$$

which, on dividing by $(2n - 1)^2$ and using (2.16), can be rewritten:

$$(2.18) \qquad \frac{4\,(f'(t))^2}{(2n - 1)^2}\left(f''(t) + n^2 f(t)\right)^2 + \left((f'(t))^2 + n^2\,(f(t))^2\right)^2 \leqslant n^4\|f\|_\infty^4.$$

This is a refinement of (2.2). We can continue such improvements indefinitely, but they become too complicated and, mostly, this method is very *wasteful* as it ignores the fact that $f_1(t) \geqslant 0$ while for *non-negative* trigonometric polynomials there is a better form of the Schaake-van der Corput inequality than (2.2), namely inequality (2.19) below:

THEOREM 2.2 (The case of non-negative trigonometric polynomials).
If a trigonometric polynomial $g(t)$ of the form (2.1) satisfies $g(t) \geqslant 0$ for all $t \in \mathbb{R}$, then for all $t \in \mathbb{R}$

$$(2.19) \qquad (g'(t))^2 + n^2\,(g(t))^2 \leqslant n^2\|g\|_\infty \cdot g(t).$$

Proof. Apply (2.2) to the polynomial $f(t) := g(t) - \frac{1}{2}\|g\|_\infty$. $\square$

Now, applying (2.19) to the non-negative trigonometric polynomial $f_1(t)$ given by (2.16), we obtain for *any* (not necessarily non-negative) real trigonometric polynomial of the form (2.1):

$$(2.20) \qquad \frac{4\,(f'(t))^2}{(2n - 1)^2} \cdot \frac{(f''(t) + n^2 f(t))^2}{(f'(t))^2 + n^2\,(f(t))^2} + (f'(t))^2 + n^2\,(f(t))^2 \leqslant n^2\|f\|_\infty^2$$

which is a finer improvement of the Schaake-van der Corput inequality than (2.18). In turn, there is yet an improvement of (2.20) for non-negative trigonometric polynomials. Actually there is a rather nice method for refining (2.20) *indefinitely* while obtaining expressions which are not too complicated and also, in a sense, optimal. We will drop this matter here but might return to it in [39] under the (weird-sounding but appropriate) term of "*analytic bootstrapping*". Let us also note that, just as for (2.14), the other variations and refinements of the Schaake-van der Corput inequality can be reformulated in terms of algebraic polynomials, *whether self-inversive or not*. Thus, for $P(X) = \sum_{k=0}^{n} a_k X^k \in \mathbb{C}[X]$ not necessarily self-inversive,

$$(2.21) \qquad \frac{4}{n^2}\left(\frac{d}{dt}\,|P(e^{it})|\right)^2 + |P(e^{it})|^2 \leqslant \|P\|_\infty^2.$$

Proof. Divide both sides of (2.19) by $n^2 g(t)$ and take $g(t) = |P(e^{it})|^2$. $\square$

2.5. Elementary applications of the Schaake-van der Corput inequality

If $f(t)$ is a real trigonometric polynomial of the form (2.1), then one can refine the trivial inequality $\|f\|_2 \leqslant \|f\|_\infty$ into:

$$(2.22) \qquad n^{-2}\|f'\|_2^2 + \|f\|_2^2 \leqslant \|f\|_\infty^2.$$

This is obtained by integrating (2.2) on $[0, 2\pi]$. □

Although we lost information in the integration process, and (2.22) is therefore weak and wasteful, it is still useful and I hope to give some application of (2.22) in [39]. An even more useful inequality is the following:

THEOREM 2.3. *If $g(t)$ is a non-negative trigonometric polynomial of the form (2.1), then one can refine the trivial inequality $\|g\|_2^2 \leqslant \|g\|_1 \cdot \|g\|_\infty$ into:*

$$(2.23) \qquad n^{-2}\|g'\|_2^2 + \|g\|_2^2 \leqslant \|g\|_1 \cdot \|g\|_\infty.$$

Equivalently, if $P(X) = \sum_{k=0}^n a_k X^k \in \mathbb{C}[X]$ is a (not necessarily self-inversive) polynomial with complex coefficients, then the trivial inequality $\|P\|_4^2 \leqslant \|P\|_2^2 \cdot \|P\|_\infty^2$ can be refined into:

$$(2.24) \qquad n^{-2}\left\|\frac{d}{dt}\left(|P(e^{it})|^2\right)\right\|_2^2 + \|P\|_4^4 \leqslant \|P\|_2^2 \cdot \|P\|_\infty^2.$$

Proof.   To obtain (2.23) integrate (2.19) on $[0, 2\pi]$. Then take $g(t) = |P(e^{it})|^2$ to obtain (2.24). □

2.6. Szegö's inequality

One year after the appearance of their 1935 paper [40] in which (2.2) was proved, Schaake and van der Corput pointed out in a short note [41] that their result (2.2) was already contained in (*i.e.*, was a weaker form of) a more precise inequality published by Szegö [42] in 1928. The statement of Szegö's inequality is as follows:

THEOREM 2.4. *Let $f(t)$ be a real trigonometric polynomial of the form (2.1), and let $\tilde{f}(t)$ denote its conjugate trigonometric polynomial. Then, for all $t \in \mathbb{R}$,*

$$(2.25) \qquad |nf(t) - \tilde{f}'(t)| + \sqrt{(f'(t))^2 + \left(\tilde{f}'(t)\right)^2} \leqslant n\|f\|_\infty.$$

In this paper I will not give Szegö's original proof of (2.25) but, in section 2.7 below, I will give the (possibly new) simple proof *by myself and the late S. K. Pichorides* [36], via an interpolation method. Yet it is conceivable that the Saffari-Pichorides proof of (2.25) (or a similar one) is already in the literature.

A few remarks before quitting this section 2.6. First, in (2.25) $\tilde{f}'(t)$ is the conjugate function of the derivative $f'(t)$ and also the derivative of the conjugate function $\tilde{f}(t)$, as these are equal. So, no ambiguity here.

Another remark is that Szegö's inequality (2.25) is finer than the Schaake-van der Corput inequality (2.2) simply because, for all $t \in \mathbb{R}$,

$$(2.26) \qquad \sqrt{(f'(t))^2 + n^2 (f(t))^2} \leqslant \left| nf(t) - \tilde{f}'(t) \right| + \sqrt{(f'(t))^2 + \left(\tilde{f}'(t)\right)^2}.$$

To get (2.26), take $x = nf(t)$, $y = f'(t)$, $z = \tilde{f}'(t)$ in the inequality

$$(2.27) \qquad \sqrt{x^2 + y^2} \leqslant |x - z| + \sqrt{y^2 + z^2}$$

which holds for all $x, y, z \in \mathbb{R}$ and is just the ordinary triangle inequality for Euclidean norm.

Equality cases of Szegö's inequality (2.25): One can see in several ways (*e.g.*, from our proof in the next section 2.7) that equality holds *for all* $t \in \mathbb{R}$ if and only if $f(t)$ is of the form

$$(2.28) \qquad f(t) = A_0 + A \cos nt + B \sin nt,$$

i.e., for a wider class than the ones for the Schaake-van der Corput and Bernstein inequalities. Also, for a given $f(t)$, there are points $t_0 \in \mathbb{R}$ (depending on f) where equality holds in (2.25). This discussion is interesting but we drop it in this paper.

A final remark: By throwing out $\left| nf(t) - \tilde{f}'(t) \right|$ in (2.25), one obtains:

$$(2.29) \qquad (f'(t))^2 + \left(\tilde{f}'(t)\right)^2 \leqslant n^2 \|f\|_\infty^2$$

or, *equivalently*,

$$(2.30) \qquad \|P'\|_\infty \leqslant n \|\Re(P)\|_\infty$$

for all polynomials $P(X) = \sum_{k=0}^n a_k X^k \in \mathbb{C}[X]$. Both (2.29) and (2.30) had, of course, been noted by Szegö (in [42] and elsewhere). They are also in Zygmund's classical treatise [44], together with their L^p generalizations, $(p \geqslant 1)$, and therefore well known to a wide public of present-day analysts. However, (2.2) (which, incidentally, does not imply (2.29) nor is implied by it either) and (2.25) are *not* in Zygmund's treatise [44] and thus are not as widely known as they deserve. In [44], Zygmund does attribute (2.29) to Szegö and refers to Szegö's 1928 paper [42], yet the proof of (2.29) presented by Zygmund in [44] is *not* Szegö's original proof but another proof akin to one of the earliest proofs of Bernstein's inequality (1.5) by Marcel Riesz [35] (or to the interpolation method of section 2.7 below). The last time I saw the late Professor Zygmund, in 1984, I asked him why he chose to include (2.29) in his treatise [44] *but not* the similar (and even more useful) Schaake-van der Corput inequality (2.2) nor the strong Szegö inequality (2.25) either. Zygmund replied that, until my conversation with him about this matter, he had never heard of the inequalities (2.2) and (2.25), and although he had quoted Szegö's paper [42] in his treatise [44], he actually had never taken a look at Szegö's paper [42] (where (2.25) was originally proved). He had also not read some other papers quoted in his treatise [44], and in which (2.2) and (2.25) were mentioned together with some extensions. At least Zygmund was perfectly honest and candid in this respect. He was a good friend of Szegö, and the two did great joint research (some of which was on polynomials, see [43]).

2.7. The Pichorides-Saffari proof of Szegö's inequality

(Again, I note that our proof below might already be in the literature.)

The proof is based on the identity

$$(2.31) \qquad \sum_{h=1}^{n} F_{n-1}\left(\frac{2h\pi - \alpha}{n}\right) e^{ik(2h\pi - \alpha)/n} = n - k + ke^{-i\alpha}$$

for all $\alpha \in \mathbb{R}$ and all integers k, n such that $0 \leqslant k \leqslant n$, where $F_{n-1}(x)$ denotes the Fejér kernel of degree $n - 1$:

$$(2.32) \qquad F_{n-1}(x) := \sum_{r=-n}^{n}\left(1 - \frac{|r|}{n}\right) e^{irx} = \frac{1}{n} \cdot \left(\frac{\sin \frac{nx}{2}}{\sin \frac{x}{2}}\right)^2.$$

To check (2.31), use the middle sum in (2.32) with $x = (2h\pi - \alpha)/n$, plug this expression into the left hand side of (2.31) and change the order of summation, then note that the new inner sum is always zero except for $r = -k$ and for $r = n - k$.

Another form of (2.31) is

$$(2.33) \qquad \sum_{h=1}^{n} F_{n-1}\left(\frac{2h\pi - \alpha}{n}\right) e^{ik(2h\pi - \alpha)/n} = n - |k| + |k| \cos\alpha - ik \sin\alpha$$

for all $\alpha \in \mathbb{R}$ and all integers k, n such that $-n \leqslant k \leqslant n$.

[For $k \geqslant 0$, (2.33) is the same as (2.31); for $k < 0$ take the complex conjugates on both sides of (2.31)].

On taking $k = 0$, we have the following useful identity:

$$(2.34) \qquad \sum_{h=1}^{n} F_{n-1}\left(\frac{2h\pi - \alpha}{n}\right) = n \qquad \text{(for all } \alpha \in \mathbb{R}\text{)}.$$

Now, for any integer k with $1 \leqslant k \leqslant n$, multiply both sides of (2.31) by $e^{i(kt+\varphi_k)}$ (with $t, \varphi_k \in \mathbb{R}$) and then take the real parts, to get

$$\sum_{h=1}^{n} F_{n-1}\left(\frac{2h\pi - \alpha}{n}\right) \cos\left(kt + \varphi_k + k \cdot \frac{2h\pi - \alpha}{n}\right) =$$

$$(2.35) \qquad = (n - k)\cos(kt + \varphi_k) + k\cos(kt + \varphi_k - \alpha).$$

We now have all we need to prove the *interpolation formula*:

$$nf(t) - \tilde{f}'(t) + \tilde{f}'(t)\cos\alpha - f'(t)\sin\alpha =$$

$$(2.36) \qquad = \sum_{h=1}^{n} F_{n-1}\left(\frac{2h\pi - \alpha}{n}\right) f\left(t + \frac{2h\pi - \alpha}{n}\right)$$

for any real trigonometric polynomial $f(t)$ of the form (2.1). Indeed, write $f(t)$ in the more convenient form

$$(2.37) \qquad f(t) = A_0 + \sum_{k=1}^{n} r_k \cos(kt + \varphi_k), \qquad (r_k = \sqrt{A_k^2 + B_k^2})$$

and note that, by linearity of the differentiation $f \to f'$ and of the Hilbert transform $f \to \tilde{f}$, it suffices to check (2.36) for $f(t) \equiv 1$ and for $f(t) \equiv \cos(kt + \varphi_k)$, $(1 \leqslant k \leqslant n)$. For $f(t) \equiv 1$, (2.36) reduces to (2.34). For $f(t) \equiv \cos(kt + \varphi_k)$, we have $f'(t) = -k\sin(kt + \varphi_k)$ and $\tilde{f}'(t) = k\cos(kt + \varphi_k)$, so that (2.36) reduces to (2.35). Thus (2.36) is proved.

The desired inequality (2.25) follows from (2.36). Indeed by (2.32) (which implies $F_{n-1}(x) \geqslant 0$) and (2.34), for any fixed α and $t \in \mathbb{R}$, the modulus of the right-hand side of (2.36) is majorized by

$$\sum_{h=1}^{n} F_{n-1}\left(\frac{2h\pi - \alpha}{n}\right) \cdot \|f\|_\infty = n\|f\|_\infty.$$

Now, for any fixed $t \in \mathbb{R}$, the maximum modulus of the left-hand side of (2.36) (as α varies) is

$$|nf(t) - \tilde{f}'(t)| + \sqrt{\left(\tilde{f}'(t)\right)^2 + (f'(t))^2}$$

and the proof of (2.25) is complete. $\square$

2.8. *Some sources*

The very important subject of Markov-Bernstein type inequalities deserves a reasonably complete monograph. Pending the writing of such a book, which is no easy task, here are just a few good sources of information (*among many others*):

1. The 1983 book "Les inégalités de Markoff et de Bernstein" by Q. I. Rahman and G. Schmeisser [31] and its good *list of references*. This (rather short) mimeographed book is written in a delightfully reader-friendly style reminiscent of those of S. N. Bernstein [4] and C. de La Vallée Poussin [12]. A modest knowledge of French should be enough for reading this book.

2. P. Borwein and T. Erdelyi (either separately, or together, or with other co-authors) have many publications on this subject (and, by the way, on several other topics pertaining to the themes of this paper and to those of [39]). Most preprints of their publications can

be downloaded from the authors' well organized home pages at the Web sites of Simon Fraser University [9] and Texas A&M University [14], respectively. Although these reprints are not always the final forms of the publications and do sometimes contain typos, the papers are of a high level and also good sources of references (modulo a few errors).

3. Many Ph. D theses and other publications on this subject produced in the province of Québec, Canada, where there is a tradition of this type of studies. I have lost track of what is being done there, but the reader could look at the university Web sites and at the publication lists of "Presses de l'Université de Montréal.". One example (among others) is the 1983 Ph. D. thesis of C. Frappier [17].

3. Some types of flat polynomials

If a polynomial $P(X) = \sum_{k=0}^{n} a_k X^k \in \mathbb{C}[X]$ has at least two non-zero coefficients, then it cannot be "perfectly flat" (*i.e.*, have constant modulus) *on the whole unit circle*: The relation

$$(3.1) \qquad\qquad |P(e^{it})| = \text{constant} \qquad (\text{for all } t \in \mathbb{R})$$

is impossible. To see this, call $a_r X^r$ (resp. $a_s X^s$) the non-zero term of $P(X)$ of lowest (resp. highest) degree, $(0 \leqslant r < s \leqslant n)$, and note that $|P(e^{it})|^2$ is a non-zero trigonometric polynomial with leading term $2|a_r a_s| \cos(mt + \beta - \alpha)$ where $m = s - r > 0$, $a_r = |a_r|e^{i\alpha}$, $a_s = |a_s|e^{i\beta}$.

However, for a variety of classes of non-monomial polynomials $P(X)$, the ideal (and impossible!) "perfect flatness" situation (3.1) can still be approximated in various ways. The diverse notions of "flatness" of a polynomial (or, more frequently, of *a sequence or a class of polynomials*) often refer to the various ways of approximating the ideal situation (3.1). For example, given a subset Γ (usually a subgroup) of the unit circle, if

$$(3.2) \qquad\qquad |P(g)| = \text{constant} \qquad (\text{for all } g \in \Gamma)$$

then we say that the polynomial $P(X)$ is *perfectly flat* on the set Γ. There are some non-trivial open problems on that notion of perfect flatness, and we will briefly see a glimpse of them in section 3.1 below. Another example of a flatness requirement is to look for polynomials $P(X) = \sum_{k=0}^{n} a_k X^k \in \mathbb{C}[X]$ for which the moduli $|a_0|, |a_1|, \ldots, |a_n|$ are given non-negative numbers and for which the sup-norm $\|P\|_{\infty} := \max_{t \in \mathbb{R}} |P(e^{it})|$ is either as small as possible (*extremal problem*) or satisfies some smallness requirement. Thus, if we impose $|a_k| = 1$ for all $k = 0, 1, \ldots, n$ (or even without this restriction), the previous problem is equivalent to some smallness requirement on the "*crest factor*" (or "*peak factor*") $\|P\|_{\infty}/\|P\|_2$ where $\|P\|_2$ denotes the L^2-norm:

$$(3.3) \qquad \|P\|_2 := \left(\frac{1}{2\pi} \int_0^{2\pi} |P(e^{it})|^2 dt \right)^{1/2} = \left(\sum_{k=0}^{n} |a_k|^2 \right)^{1/2}.$$

Actually, many flatness problems for polynomials can be expressed in terms of *comparison of two norms* in the vector space of polynomials of degree $\leqslant n$. We shall take brief looks at some examples in the sequel.

3.1. *Perfect flatness*

As we said above, a polynomial $P(X) = \sum_{k=0}^{n} a_k X^k \in \mathbb{C}[X]$ is *"perfectly flat"* on a subset Γ of the unit circle if (3.2) holds. It is easy to see that if Γ is a (finite or infinite) *subgroup* of the unit circle with card $\Gamma > \deg P$ (where card Γ denotes the number of elements of Γ), then the constant in (3.2) equals $\|P\|_2$ defined by (3.3), but this is not always true for a subgroup Γ such that card $\Gamma \leqslant \deg P$.

So the perfect flatness (3.2) cannot hold if Γ is the whole unit circle. Assuming $P(0) \neq 0$ and $\deg P = n$, it is easy to see that (3.2) cannot hold either if card $\Gamma > 2n$, but that whenever card $\Gamma \leqslant 2n$ there are some such $P(X)$ for which (3.2) holds.

An interesting *open* problem is the following: Given an integer $n \geqslant 1$, find the largest value $\Phi(n)$ of those integers $d \geqslant 1$ such that there exists a polynomial $P(X) = \sum_{k=0}^{n} a_k X^k \in \mathbb{C}[X]$ which is *"unimodular"* (*i.e.*, $|a_k| = 1$ for all $k = 0, 1, \ldots, n$) and for which the perfect flatness (3.2) holds on the group Γ_d of d-th roots of unity in $\mathbb{C}$. The computation of $\Phi(n)$ is easy for small n, for example: $\Phi(1) = 2$, $\Phi(2) = 4$, $\Phi(3) = 5$, $\Phi(4) = \Phi(5) = 6$. A list of values of $\Phi(n)$ has been computed (Björck & Saffari, 1998, and on-going search in 2000–2001). No values of n with $\Phi(n) \geqslant n + 3$ have been found so far, but we do not have enough evidence to conjecture that one always has either $\Phi(n) = n + 1$ or $\Phi(n) = n + 2$. The following partial results are known:

$$(3.4) \qquad \Phi(n) \geqslant n + 1 \qquad \text{(for all } n \geqslant 1)$$

$$(3.5) \qquad \Phi(n) \geqslant n + 2 \qquad \text{(for infinitely many } n)$$

$$(3.6) \qquad \Phi(n) \leqslant 2n - 1 \qquad \text{(for all } n \geqslant 1).$$

(3.4) follows from elementary results (Gauss sequences), see section 3.2 on bi-unimodular sequences. (3.5) is a non-trivial result of Björck and Saffari (unpublished). (3.6) is not hard to prove for *even* n, however for *odd* n it can be shown (Saffari, unpublished) to be equivalent to a profound theorem of Dresel, White and Hunt (see also the section in [39] on "Huffman sequences").

If, instead of polynomials with *complex* unimodular coefficients, we consider polynomials with *real* unimodular coefficients (*i.e.*, $a_k = \pm 1$ for all $k = 0, 1, \ldots, n$), then (3.2) becomes a different type of problem. It is very easy to check that, in that case, (3.2) never holds if card $\Gamma \geqslant n + 3$. It can be proved (Saffari [38], unpublished) that (3.2) never holds either if card $\Gamma = n + 2$. As for the impossibility of (3.2) for card $\Gamma = n + 1$ (except for $n = 3$), it is *equivalent* to the famous "Hadamard Circulant Conjecture" due to Ryser [33], an open problem going back to the 1950's and stating that a circulant matrix of order L (with ± 1 entries) cannot be a Hadamard matrix *unless* $L = 4$.

3.2. Bi-unimodular sequences

This is a notion which we will define a little later and which will turn out to be *equivalent* to that of unimodular polynomials of degree n which are perfectly flat on the group Γ_{n+1} of $(n+1)$th roots of unity in $\mathbb{C}$ (see section 3.1 above). I mention this notion at this point because, historically, it can be considered as the oldest idea of "flat polynomials" since it goes back to Gauss. Indeed, let L be any *odd* integer $\geqslant 3$, and consider the "*Gaussian sequence*" of length L:

$$(3.7) \qquad a_k := \omega^{\lambda k^2}, \quad (k = 0, 1, \ldots, L-1)$$

where $\omega = \exp\left(2i\pi/L\right)$ is the first primitive L-th root of 1 in $\mathbb{C}$, and λ is any integer relatively prime to L. Equivalently, we may consider the infinite sequence (a_k), $(k \in \mathbb{Z})$, as periodic of period L. Defining the normalized discrete Fourier transform (DFT) of a_k as the sequence

$$(3.8) \qquad \hat{a}_r := \frac{1}{\sqrt{L}} \sum_{k=0}^{L-1} a_k \cdot \omega^{rk}, \quad (r = 0, 1, \ldots, L-1)$$

we can easily check that $(\hat{a}_r)$ is unimodular as well: $|\hat{a}_r| = 1$ for all $r = 0, 1, \ldots, L-1$. This fact was known to Gauss (to whom, by the way, the earliest ideas of Fourier Analysis can be tracked down, and not to Fourier or Clairaut). More generally, given any integer $L \geqslant 2$, whether odd or even, any finite sequence (a_k), $(k = 0, 1, \ldots, L-1)$, of L complex numbers will be called "*bi-unimodular*" if it has modulus one $(|a_k| = 1$ for all $k = 0, 1, \ldots, L-1)$ and if its normalized DFT

$$\hat{a}_r := \frac{1}{\sqrt{L}} \sum_{k=0}^{L-1} a_k \cdot \omega^{rk}, \quad (\omega = e^{2i\pi/L};\ r = 0, 1, \ldots, L-1)$$

has modulus one, too: $|\hat{a}_r| = 1$ for all $r = 0, 1, \ldots, L-1$. (The term "bi-unimodular" was coined by G. Björck and myself in our 1995 joint paper [7]). Thus, for *odd* L, the Gaussian sequence (3.7) is an example of a bi-unimodular sequence of length L. The sequence (3.7) is *not* bi-unimodular when L is *even*, but in this case we have another type of Gaussian sequence which *is* bi-unimodular:

$$(3.9) \qquad b_k := \xi^{\lambda k^2}, \quad (k = 0, 1, \ldots, L-1)$$

where $\xi = \exp\left(i\pi/L\right)$ is the first primitive root of unity of order $2L$ in $\mathbb{C}$, and λ again any integer relatively prime to L. Note that, while (3.7) is bi-unimodular for odd L but not for even L, similarly (3.9) is bi-unimodular for even L but not for odd L.

Obviously a unimodular sequence (a_k), $(k = 0, 1, \ldots, L-1)$, of length L is bi-unimodular if and only if its "associated polynomial" $P(X) = \sum_{k=0}^n a_k X^k$ (of degree $n = L-1$) is perfectly flat on the group Γ_L of L-th roots of 1 in $\mathbb{C}$.

The starting point of the theory of bi-unimodular sequences was an oral question asked by Per Enflo in 1983 at Stockholm University [13]: *If p is a given odd prime number, is it true* *that the Gaussian sequences* $a_k = \omega^{\lambda k^2 + \mu k}$, ($\omega = e^{2i\pi/p}$, λ and μ integers with p not dividing λ), ($k = 0, 1, \ldots, p - 1$), are the only unimodular sequences of length p, with $a_0 = 1$, whose normalized DFT has modulus one?

In our present vocabulary, Per Enflo was asking whether such Gaussian sequences were the only normalized ($a_0 = 1$) bi-unimodular sequences of odd prime length. If the answer had been "yes", it would have helped him with estimations of some exponential sums. Later on he found another method (not requiring an answer to the above question) to carry out the estimations of his exponential sums anyway.

For $p = 3$ the answer is trivially "yes", and for $p = 5$ Lovász [24] checked that the answer is "yes" as well.

In 1984 G. Björck (Stockholm University) was trying, by computer search, to check that for $p = 7$ the answer to Per Enflo's question was "yes" as well, when suddenly the counter-example

(3.10) $\qquad (1, 1, 1, e^{i\theta}, 1, e^{i\theta}, e^{i\theta})$ \qquad (with $\theta = \arccos(-3/4)$)

"popped out" (as Björck put it!). Later on in 1984 Björck found, again by computer search, other counter-examples to Per Enflo's question, including

(3.11) $\qquad (1, 1, e^{i\sigma}, 1, 1, 1, e^{i\sigma}, e^{i\sigma}, e^{i\sigma}, 1, e^{i\sigma})$

(with $\sigma = \arccos(-5/6)$). When early in 1985 Björck presented these counter-examples at the A. Haar memorial Conference [5], he still had no idea of the structure of the sequences (3.10) and (3.11). Actually, if in the sequence (3.10) [resp. (3.11)] we replace the first term by zero and the terms $e^{i\theta}$ [resp. $e^{i\sigma}$] by -1, we get the *"Legendre symbol"* sequence modulo 7 [resp. mod 11]: The terms $e^{i\theta}$ [resp. $e^{i\sigma}$] are located at the quadratic non-residues (modulo 7, resp. modulo 11). This is a fact that Björck observed a little later, in the fall of 1985, and that he subsequently generalized ([6], 1990) to every prime $\equiv -1 \pmod 4$:

If in the p-term "Legendre symbol" sequence $(0, 1, \ldots, -1)$, (p any prime $\equiv -1$ modulo 4) we replace the first term zero by 1 and every -1 by

(3.12) $\qquad \exp\left(i \arccos \dfrac{1-p}{1+p}\right) = \dfrac{1-p}{1+p} + i\dfrac{2\sqrt{p}}{1+p}$

we obtain a bi-unimodular sequence of length p, with only two values, namely 1 and the number given by (3.12).

In sections 3.5 and 3.6 below I shall determine *all* bi-unimodular sequences with only two values: it will turn out, in particular, that Björck's above theorem is not specific to prime numbers but can be extended to all integers v (necessarily $\equiv -1$ modulo 4) for which there exists a so-called "Hadamard-Paley cyclic difference set." Thus it will be seen to work for $v = 15$, which is the smallest non-prime v with this property. From the same discussion of sections 3.5 and 3.6 it will follow that such a bi-unimodular sequence (with *only two values*) cannot exist for any length $\equiv 1 \pmod 4$. However, for *prime* lengths $\equiv 1 \pmod 4$, Björck proved (in the same paper [6]) the next best thing:

If in the p-term "Legendre symbol" sequence $(0, 1, \ldots, -1, \ldots, 1)$, *(p any prime* $\equiv 1$ *modulo 4) we replace the first term 0 by 1, every term 1 by*

$$(3.13) \qquad \eta := \exp\left(i \arccos \frac{\delta\sqrt{p} - 1}{p - 1}\right) = \frac{\delta\sqrt{p} - 1}{p - 1} + i\frac{\sqrt{p^2 - 3p + 2\delta\sqrt{p}}}{p - 1}$$

(with any choice of $\delta = \pm 1$*) and every* -1 *by the complex conjugate* $\bar{\eta}$, *with the same choice of* $\delta = \pm 1$, *then we obtain a bi-unimodular sequence of length p (which thus has first term 1 and only two other values, namely* η *and* $\bar{\eta}$*).*

In the 1990's more research was done on bi-unimodular sequences (Saffari [37], Björck & Saffari [7], Haagerup [19], ...) and there are some very nice results and open problems on this subject, that I hope to discuss in [39].

3.3. *Digression on* (v, k, λ) *difference sets*

This brief digression into Combinatorics (in this section and in next section 3.4) is just intended to recall, for the benefit of the harmonic analyst reader, some definitions and elementary facts which will be useful in the discussion (in sections 3.5 and 3.6) of polynomials whose coefficients only take *two* values (whether unimodular or not) and which are perfectly flat on the group Γ_L of L-th roots of 1 in $\mathbb{C}$, $(L = 1 + \deg P)$.

There are currently two *entirely different and unrelated* types of mathematical objects, both carrying (unfortunately) the same name of "difference sets". We will be concerned only with the second notion, yet I will define both just to avoid any possible confusion (as I have often witnessed instances of such confusion between the two notions).

The set $\Delta(S) := \{x - y : x, y \in S\}$, where S is any subset of an (additive) abelian group G, is often called the "difference set" of S. This notion is well known to all analysts because of the result saying that if $G = \mathbb{R}^n$ and if S has positive Lebesgue measure, then the origin belongs to the interior of $\Delta(S)$ and therefore $\Delta(S)$ has non-empty interior. There is an enormous literature on this notion and its extensions, in such areas as analysis, algebra, combinatorics and number theory. I will not say anything else on this, as it is of no concern to us here.

The second notion of "difference sets" (the one which *does* interest us here) pertains to *finite* groups only. Let G be any finite group, abelian or not, with neutral element denoted by e. A subset D of G is called a left (resp. right) (v, k, λ) *difference set* if card $G = v$, card $D = k$ and the intersection $(uD) \cap D$ (resp. $(Du) \cap D$) has cardinality λ whenever $u \in G$, $u \neq e$. The term "difference set" is due to the fact that such sets were first considered in additive groups $\mathbb{Z}/n\mathbb{Z}$. Note that $D \subset G$ is a left (v, k, λ) difference set if and only if $D^{-1} := \{x^{-1} : x \in D\}$ is a right (v, k, λ) difference set. Also if D is a left (resp. right) (v, k, λ) difference set, then its complement $D' := G \backslash D$ is also a left (resp. right) (v', k', λ') difference set, with

$$(3.14) \qquad\qquad v' = v, \quad k' = v - k, \quad \lambda' = v - 2k + \lambda.$$

The most obvious examples of difference sets are the following four types (called the "*trivial difference sets*"):

1. The empty set $\emptyset$, which is a $(v, 0, 0)$ difference set.
2. Singletons, which are $(v, 1, 0)$ difference sets.
3. $D = G$, which is a (v, v, v) difference set.
4. Complements of singletons, which are $(v, v - 1, v - 2)$ difference sets (if $v \geqslant 2$).

The simplest (and, historically, the first) example of a *non-trivial* difference set is due to Paley [29]: *If p is a prime $\equiv -1 \ (mod\,4)$, then the set of all (non-zero) quadratic residues modulo p, and also the set of all quadratic non-residues modulo p, are (v, k, λ) difference sets with $v = p$, $k = (p - 1)/2$, $\lambda = (p - 3)/4$.* (These sets are non-trivial difference sets if $p \geqslant 7$, and trivial $(3, 1, 0)$ difference sets if $p = 3$).

In the general case, a simple counting argument shows that

$$(3.15) \qquad\qquad k(k - 1) = (v - 1)\lambda.$$

Let $F : G \to \mathbb{C}$ be any complex-valued function defined on G. Its right (resp. left) *autocorrelation function* is the function $\gamma_F : G \to \mathbb{C}$ (resp. $_F\gamma : G \to \mathbb{C}$) defined by

$$(3.16) \qquad \gamma_F(u) := \sum_{g \in G} \overline{F(g)}F(gu) \qquad _F\gamma(u) := \sum_{g \in G} \overline{F(g)}F(ug) \ .$$

If $\chi = \chi_D$ is the characteristic function of any (a priori arbitrary) subset D of G, i.e., $\chi(g) = 1$ if $g \in D$ and $\chi(g) = 0$ if $g \notin D$, then D is obviously a right (resp. left) (v, k, λ) difference set if and only if

$$(3.17) \qquad\qquad \gamma_\chi(u) = k \quad \text{if } u = e, \qquad \gamma_\chi(u) = \lambda \quad \text{if } u \neq e,$$
$$(3.18) \qquad \text{resp.} \quad _\chi\gamma(u) = k \quad \text{if } u = e, \qquad _\chi\gamma(u) = \lambda \quad \text{if } u \neq e.$$

This yields another proof of (3.15). Indeed, for example by (3.17),

$$k^2 = \left(\sum_{g \in G} \chi(g)\right)^2 = \sum_{g \in G}\chi(g)\sum_{h \in G}\chi(h) = \sum_{g \in G}\chi(g)\sum_{u \in G}\chi(gu)$$
$$= \sum_{u \in G}\gamma_\chi(u) = \gamma_\chi(e) + \sum_{u \neq e}\gamma_\chi(u) = k + (v - 1)\lambda.$$

(This was, of course, just the classical convolution product argument).

3.4. Binary functions on finite groups with autocorrelation functions constant outside the neutral element

By a "*binary function*" we mean complex-valued functions only taking two values, which can be arbitrary complex numbers, possibly of modulus > 1. If $\alpha \in \mathbb{C}$ and $\beta \in \mathbb{C}$ are these values, we call such a function an $\{\alpha, \beta\}$-function. Examples of binary functions are $\{0, 1\}$-functions, $\{1, -1\}$-functions (also called ± 1 functions), etc.

Although the following result might conceivably not have appeared elsewhere in the explicit form given below, it is nevertheless probably well known (in essence) to many people in combinatorics and signal processing, at least in the case of abelian or cyclic groups.

THEOREM 3.1. *Let G be a (not necessarily abelian) finite group of order v with neutral element e, and let $F : G \rightarrow \mathbb{C}$ be any binary function with values $\{\alpha, \beta\}$, $(\alpha \neq \beta)$. Then the right (resp. left) autocorrelation function of F is constant on $G\backslash\{e\}$ if and only if $F^{-1}(\alpha) := \{x \in G : F(x) = \alpha\}$ is a right (resp. left) (v, k, λ) difference set. In that case the right (resp. left) autocorrelation function has value $\gamma_0 > 0$ at $x = e$, and the same (real) value $\gamma < \gamma_0$ at all $x \in G\backslash\{e\}$, where*

$$(3.19) \qquad \gamma_0 = k|\alpha|^2 + (v - k)|\beta|^2$$

and γ is defined by any of the three (equivalent) relations

$$(3.20) \qquad \gamma = \gamma_0 - (k - \lambda)|\alpha - \beta|^2$$

$$(3.21) \qquad \gamma = \lambda|\alpha|^2 + (v - 2k + \lambda)|\beta|^2$$

$$(3.22) \qquad |k\alpha + (v - k)\beta|^2 = \gamma_0 + (v - 1)\gamma,$$

so that γ satisfies the inequalities

$$(3.23) \qquad -\frac{\gamma_0}{v - 1} \leqslant \gamma < \gamma_0.$$

The proof is based on the following lemma, which is also useful elsewhere.

LEMMA 3.1. *Let $H : G \rightarrow \mathbb{C}$ be any complex-valued function on a (not necessarily abelian) group of order v. Put $J(x) = aH(x) + b$ with $a, b \in \mathbb{C}$. Then the right autocorrelation functions $\gamma_H(u)$ and $\gamma_J(u)$ satisfy*

$$(3.24) \qquad \gamma_J(u) = |a|^2\gamma_H(u) + |b|^2 \cdot v + 2\Re\left(a\bar{b}\sum_{x \in G} H(x)\right),$$

and a similar identity holds for the left autocorrelation functions.

Proof. The proof of the lemma is straightforward. To prove the above theorem, we may just consider the right autocorrelation function, as the change of variable $x \rightarrow x^{-1}$ reduces the left case to the right one. Let $\chi := \chi_D$ where $D := F^{-1}(\alpha)$. Then the theorem follows from the above lemma by straightforward calculations, upon noting that each of $\chi(x)$ and $F(x)$ can be expressed in terms of the other one from the identity $F(x) = (\alpha - \beta)\chi(x) + \beta$, since $\alpha - \beta \neq 0$. $\square$

3.5. Binary functions on finite groups with autocorrelation functions vanishing outside the neutral element

Let G be any finite (not necessarily abelian) group of order $v \geqslant 2$, with neutral element e. Our purpose in this section is to find all *binary* functions (and, in particular, all *unimodular binary* functions) $F : G \to \mathbb{C}$ for which the, say, right autocorrelation function γ_F satisfies $\gamma_F(u) = 0$ for all $u \in G \backslash \{e\}$. Without loss we may assume that the two values of F are 1 and some $\beta \in \mathbb{C}$, $(\beta \neq 1)$. As a special case of the result of section 3.4, with $\gamma = 0$, G contains a (right) (v, k, λ) difference set D such that $F(x) = 1$ if $x \in D$ and $F(x) = \beta$ if $x \in G \backslash D$. Since the case $|\beta| = 1$ (of binary *unimodular* functions) is an important special case, our purpose consists of addressing the following two problems:

A) *If $D \subset G$, $(D \neq \emptyset)$, is some (right) (v, k, λ) difference set, find all those $\beta \in \mathbb{C}$ such that the binary function $F : G \to \mathbb{C}$ defined by $F(x) = 1$ if $x \in D$ and $F(x) = \beta$ if $x \in G \backslash D$ satisfies $\gamma_F(u) = 0$ for all $u \in G \backslash \{e\}$.*

B) *Find all those (right) (v, k, λ) difference sets $D \subset G$ for which there exists at least one complex number β, with $|\beta| = 1$, so that the above (unimodular) function F satisfies $\gamma_F(u) = 0$ for all $u \notin G \backslash \{e\}$.*

Solving Problem A): Put $\alpha = 1$ and $\gamma = 0$ in (3.21), to get

$$(3.25) \qquad \lambda + (v - 2k + \lambda)|\beta|^2 + 2(k - \lambda)\Re\beta = 0.$$

Recall that $G \backslash D$ is a (right) (v', k', λ') difference set with $v' = v$, $k' = v - k$, $\lambda' = v - 2k + \lambda$. So if the coefficient $\lambda' = v - 2k + \lambda$ of $|\beta|^2$ in (3.25) is zero, then D is a "trivial" difference set (see section 3.3) for which either $k = \lambda = v$ or $k = v - 1$, $\lambda = v - 2$. In the former case the left side of (3.25) reduces to $\lambda = v$, which is impossible since $v \neq 0$. In the latter case (3.25) reduces to

$$(3.26) \qquad \Re\beta = 1 - \frac{v}{2} \qquad \text{(with } v \geqslant 2\text{)}$$

and those $\beta \in \mathbb{C}$ which satisfy (3.26) constitute a vertical line. The case $k = \lambda = 0$ is excluded since we supposed $D \neq \emptyset$. The last possibility for D to be a "trivial" difference set (*i.e.*, now a singleton) is $k = 1$ and $\lambda = 0$, in which case (3.25) reduces to

$$(3.27) \qquad (v - 2)|\beta|^2 + 2\Re\beta = 0.$$

If $v = 2$, then (3.27) is the same as (3.26) and the acceptable $\beta \in \mathbb{C}$ are all the $\beta = it$, $(t \in \mathbb{R})$. If $v \geqslant 3$, then the set of those $\beta \in \mathbb{C}$ which satisfy (3.27) is the circle of radius $1/(v - 2)$ with center of abscissa $-1/(v - 2)$ on the real axis.

Now if D is a *non-trivial* difference set (which easily implies $v \geqslant 7$), then in (3.25) we have $\lambda \geqslant 1$, $\lambda' \geqslant 1$ and $k - \lambda \geqslant 1$. The set of these $\beta \in \mathbb{C}$ satisfying (3.25) is then the circle of radius $\sqrt{k - \lambda}/\lambda'$ with center of abscissa $-(k - \lambda)/\lambda'$ on the real axis, $(\lambda' = v - 2k + \lambda)$. Let us recapitulate the result obtained regarding problem A):

THEOREM 3.2. *Let G be a (not necessarily abelian) finite group of order $v \geqslant 2$, with neutral element e. Let $D \subset G$, $(D \neq \emptyset)$, be a (right) (v, k, λ) difference set. Define a binary function $F : G \rightarrow \mathbb{C}$ by $F(x) = 1$ if $x \in D$ and $F(x) = \beta$ if $x \in G\backslash\{e\}$, where $\beta \in \mathbb{C}$ is arbitrary. Then the right autocorrelation function γ_F identically vanishes on $D\backslash\{e\}$ if and only if β is in the set S_D of those $\beta \in \mathbb{C}$ such that*

$$(3.28) \qquad \lambda + \lambda'|\beta|^2 + 2(k - \lambda)\Re\beta = 0$$

with $\lambda' = v - 2k + \lambda$. Also S_D is non-empty if and only if:

a) *Either D is the complement of a singleton, in which case S_D is the vertical line $\Re\beta = 1 - \frac{v}{2}$.*
b) *Or $v \geqslant 3$ and D is a singleton, in which case S_D is the circle of radius $1/(v - 2)$ with center of abscissa $-1/(v - 2)$ on the real axis.*
c) *Or D is a non-trivial difference set, in which case S_D is the circle of radius $\sqrt{k - \lambda}/\lambda'$ with center of abscissa $-(k - \lambda)/\lambda'$ on the real axis.*

Before proceeding to address Problem B, let us make an interesting remark: Putting $D' := G\backslash D$, then in the above theorem, whether D is a trivial or a non-trivial difference set, the set $S_{D'} \subset \mathbb{C}$ is always the transform of S_D by the inversion whose inversion-circle is the unit circle.

Solving Problem B). This boils down to deciding whether the set S_D of the above theorem intersects the unit circle. So let us examine the various possibilities. If D is the complement of a singleton (with $v \geqslant 2$), the vertical line $\Re\beta = 1 - \frac{v}{2}$ intersects the unit circle if and only if $1 - \frac{v}{2} \geqslant -1$, that is, either $v = 2$ or $v = 3$ or $v = 4$. For $v = 2$ the intersection points are $\beta = \pm i$, yielding the Gauss sequences $(1, \pm i)$ of length 2 (see section 3.2). For $v = 3$ the intersection points are $\beta = \exp(\pm 2i\pi/3)$, yielding the Gauss sequences $(1, 1, j)$ and $(1, 1, j^2)$, $(j = e^{2i\pi/3})$, and their obvious modifications. For $v = 4$ the intersection point is at $\beta = -1$, yielding the ± 1 functions whose values a, b, c, d are ± 1 and satisfy $abcd = -1$. For $v = 4$ the group G can be non-cyclic. If G *is* cyclic, then we get ordinary ± 1 *sequences* of length 4 which are special cases of "Barker sequences", "Golay sequences", "Shapiro sequences", "PONS sequences", etc. If $v \geqslant 3$ and D is a singleton, again $S_D \neq \emptyset$ if and only if $v = 3$ or $v = 4$, and we once again obtain the same (Gauss) sequences of length 3 and the same ± 1 functions as above.

We now come to the case when D is a non-trivial difference set, and this is the only non-trivial part (!) of this section 3.5. From assertion c) in the above theorem it easily follows by elementary calculations that S_D intersects the unit circle if and only if

$$(3.29) \qquad v - 4(k - \lambda) \leqslant 0.$$

On the other hand, upon choosing $\alpha = 1$ and $\beta = -1$ in Theorem 3.1, we infer from (3.19), (3.20) and (3.23) that, for *any* (v, k, λ) difference set with $v \geqslant 3$,

$$(3.30) \qquad v - 4(k - \lambda) \geqslant -\frac{v}{v-1}$$

and therefore that $v - 4(k - \lambda) \geqslant -1$, since the left side of (3.30) is an integer. Thus (3.29) can be satisfied in only two cases:

$$(3.31) \qquad \text{Case 1.} \qquad v - 4(k - \lambda) = -1$$

$$(3.32) \qquad \text{Case 2.} \qquad v - 4(k - \lambda) = 0.$$

Before studying these two (*very important*) cases, let us first clarify some terminology. Many authoritative experts (such as [2] and [18], for example) call (v, k, λ) difference sets satisfying (3.31) "Hadamard difference sets" and those satisfying (3.32) "Menon difference sets". Other authoritative experts call those satisfying (3.32) "Hadamard difference sets" and those satisfying (3.31) "Hadamard-Paley difference sets". All these choices of names are historically justified, yet this discrepancy is unfortunate and confusing. So I humbly propose to adopt the following choices (and make everyone happy):

Definitions. *Difference sets satisfying (3.31) will be called "Hadamard-Paley difference sets". Those satisfying (3.32) will be called "Hadamard-Menon difference sets".*

Let us now study these two types of difference sets and their incidences on our Problem B). Elementary arithmetical calculations show that (assuming $k \leqslant v/2$ *without loss*) the diophantine identities (3.15) and (3.31) hold simultaneously if and only if the parameters (v, k, λ) of the (Hadamard-Paley) difference set have the form

$$(3.33) \qquad v = 4n - 1, \qquad k = 2n - 1, \qquad \lambda = n - 1$$

and that the diophantine identities (3.15) and (3.32) hold simultaneously if and only if the parameters (v, k, λ) of the (Hadamard-Menon) difference set have the form

$$(3.34) \qquad v = 4N^2, \qquad k = 2N^2 - N, \qquad \lambda = N^2 - N.$$

Now some elementary calculations show that if D is a (right, say) Hadamard-Paley difference set as defined by (3.33), then the only *unimodular* solutions $\beta \in \mathbb{C}$ of our above Problem B) are

$$(3.35) \qquad \beta = \exp\left(i \arccos \frac{1-v}{1+v}\right) = \exp\left(i \arccos \frac{1-2n}{1+2n}\right)$$

and, of course, the complex conjugate of this number. Also elementary calculations show that if D is a (right, say) Hadamard-Menon difference set as defined by (3.34), then the only unimodular solution $\beta \in \mathbb{C}$ of our above Problem B) is $\beta = -1$.

Also observe that all the solutions of Problem B) in the case of *trivial* difference sets are special cases of the above discussion (for Hadamard-Paley and Hadamard-Menon difference sets), *except* for the Gaussian sequences $(1, \pm i)$ of length 2. Let us now recapitulate:

THEOREM 3.3. *Let G be any finite (not necessarily abelian) finite group of order $v \geqslant 2$, with neutral element e. Let $D \subset G$, $(D \neq \emptyset)$, be a (right) (v, k, λ) difference set (satisfying $k \leqslant v/2$ without loss). Define a binary unimodular function $F : G \to \mathbb{C}$ by $F(x) = 1$ if $x \in D$ and $F(x) = \beta$ if $x \in G \backslash D$ where $\beta \in \mathbb{C}$ and $|\beta| = 1$. Then the (right) autocorrelation function γ_F satisfies $\gamma_F(u) = 0$ for all $u \in G \backslash \{e\}$ if and only if we are in one of these three situations:*

a) *$v = 2$ and F is one of the (Gauss) sequences $(1, i)$ and $(1, -i)$.*

b) *D is a Hadamard-Paley difference set given by (3.33) and β is defined by (3.35) or is the complex conjugate of that number.*

c) *D is a Hadamard-Menon difference set given by (3.34), and $\beta = -1$.*

3.6. Binary bi-unimodular sequences

The case of binary bi-unimodular *sequences* is just the special case of the discussions and results of section 3.5 when the group G is *cyclic*. The extension of the result (3.12) of Björck [6], promised in section 3.2, is the special case of Theorem 3.3 when G is cyclic.

This now leads us to the crucial problem of the *existence and determination* of such Hadamard-Paley and Hadamard-Menon difference sets. Even in the case of cyclic groups these existence problems are far from being entirely solved. Several classes of *cyclic* Hadamard-Paley difference sets are known. See *for example* [2] and [18]. As for *cyclic* Hadamard-Menon difference sets, no example is known (except, of course, for $v = 4$), and a long-standing conjecture (due to Ryser [33]) is that none exists if $v > 4$. This is equivalent to the famous "*Hadamard circulant conjecture*" which states that no circulant matrix of order v with entries ± 1 can be a Hadamard matrix *unless* $v = 4$.

The interpretation of binary bi-unimodular sequences in terms of polynomials of degree $v - 1$ (with binary unimodular coefficient sequences), which are perfectly flat on the group Γ_v of v-th roots of 1, is by now obvious. We will come back to these in some depth in [39].

3.7. Sup-norms of bi-unimodular polynomials on the unit circle

A "*bi-unimodular polynomial*" is, of course, the associated polynomial

$$(3.36) \qquad P(X) = \sum_{k=0}^{L-1} a_k X^k$$

of a bi-unimodular sequence $(a_0, a_1, \ldots, a_{L-1})$.

A useful theorem of Landau [21] says that if $Q(X) = \sum_{k=0}^{L-1} c_k X^k \in \mathbb{C}[X]$ is *any* polynomial with complex coefficients, then its maximum modulus $\|Q\|_\infty$ on the whole unit circle is majorized in terms of its maximum modulus on the group Γ_L of L-the roots of 1 as follows:

(3.37)
$$\|Q\|_\infty \leqslant C \cdot \left(\max_{g\in\Gamma_L} |Q(g)|\right) \cdot \log L$$

where C is some absolute constant, and the $\log L$ cannot be replaced (in the general case) by a smaller factor. This can be proved by Lagrange interpolation, and also otherwise.

Since our bi-unimodular polynomial (3.36) satisfies $|P(g)| = \sqrt{L}$ for all $g \in \Gamma_L$, (3.37) implies that

(3.38)
$$\|P\|_\infty \leqslant C\sqrt{L} \log L.$$

After Björck [6] gave, at the 1989 ASI on *"Recent Advances in Fourier Analysis"*, his lecture on (what we subsequently called the) bi-unimodular sequences, one of the participants (*perhaps* H. S. Shapiro or D. J. Newman, if my memory is correct) conjectured that, for *bi-unimodular* polynomials $P(X)$, (3.38) could be improved to:

(3.39)
$$\|P\|_\infty \leqslant C_1\sqrt{L} \qquad (C_1 =\text{some absolute constant}),$$

i.e., $P(X)$ has *bounded crest factor* (see beginning of part 3). His "evidence" for the conjecture (3.39) was twofold:

A) On one hand the conjecture (3.39), in the special case of the Gauss sequences

(3.40)
$$a_k = \omega^{ak^2+bk}$$

   (L odd, $\omega = e^{2i\pi/L}$, a relatively prime to L), had been around for many decades, perhaps even before the 1920's and, incidentally, is *still open* in 2000–2001.
B) On the other hand (3.39) was known to be true for two particular classes of bi-unimodular sequences/polynomials, namely the Gauss sequences (3.40) with the choices $a = (L \pm 1)/2$ (see Littlewood [22]) and the bi-unimodular polynomials (of length $L = M^2$) studied in 1977 by Byrnes [10]:

(3.41)
$$P(X) = \sum_{h=0}^{M-1} \sum_{r=0}^{M-1} e^{2hri\pi/M} X^{Mh+r}.$$

I did some thinking on the conjecture (3.39) from time to time, until I saw how to *disprove* it in June 2000:

THEOREM 3.4. *If p is a prime $\equiv -1(mod\,4)$ and a_k is the binary bi-unimodular sequence of length p (originally introduced by Björck) defined by*

$$a_k = \exp\left(i \arccos \frac{1-p}{1+p}\right)$$

if k is a quadratic non-residue (mod p) and $a_k = 1$ *otherwise,* $(0 \leqslant k \leqslant p - 1)$, *then the associated (bi-unimodular) polynomial satisfies, for all sufficiently large p,*

$$(3.42) \qquad \|P\|_\infty > \frac{2}{\pi}\sqrt{p}\log\log p - \frac{1}{2}$$

and, if $\epsilon > 0$ *is fixed, then for infinitely many such primes p*

$$(3.43) \qquad \|P\|_\infty > \left(\frac{2}{\pi} \cdot e^\gamma - \epsilon\right)\sqrt{p}\log\log p$$

where γ *is Euler's constant.*

These results are immediate consequences of Montgomery's 1980 work [28] on the Fekete polynomials $\sum_{k=0}^{p-1} \chi\left(\frac{k}{p}\right) X^k$, where $\chi\left(\frac{k}{p}\right)$ is Legendre's symbol. It is unforgivable that I have not seen earlier this straightforward disproof of conjecture (3.39), since Montgomery had given me a copy of his paper [28] as early as 1988.

This disproof, in turn, gives rise to interesting results and problems on binary bi-unimodular sequences, which I hope to discuss in [39].

4. The drunkard, the bar owner and the police

Sorry, dear reader, I must abruptly stop my write-up at this point. I have a deadline to respect and I had underestimated the time and space needed to just *touch upon* some most interesting topics I had in mind when writing the introduction. Below is a list of *some* of these topics, and I am also omitting some of them, with the intention of including most of them in the extended version [39] of this paper (which, in a way, I have barely started to write).

4.1. *Some topics to be found in the extended version [39]*

- The genesis of the *analytic* theory of flat polynomials from nineteenth century Fourier analysis. The origins from the absolute convergence of trigonometric and Fourier series: Dirichlet, Fejér, Bernstein and many others up to the present days.
- The various Hardy-Littlewood unimodular polynomials with bounded crest factors. Their connections with number theory and Fourier analysis.
- Bounded crest factor properties of Gauss and Byrnes polynomials and open problems. The van der Corput methods and their discrete analogs (Kuzmin, Landau). D.J. Newman's method.
- D. J. Newman and H. S. Shapiro in New York in the late 1940's. The making of the *Shapiro polynomials* at M.I.T. (1950–1951). Raphaël Salem and the Fekete polynomials. Infra-red spectrometry and the Golay sequences. History of Shapiro polynomials and Golay sequences: Rediscoveries, truths, lies, counter-truths and folklore galore.
- Barker sequences and their generalizations. The Hadamard Circulant Conjecture, from Reiser to B. Schmidt via R. Turyn.

- D.J. Newman's conjecture and Barker sequences. Recent (published and unpublished) research.
- Norms of unimodular polynomials. History of Erdös-Newman conjectures. Little-wood's work and his various conjectures and *"counter-conjectures"*. The strange story of the Byrnes pseudo-polynomials, Körner's "true-false" theorem, Kahane's ultra-flat polynomials and further on-going research.
- Littlewood's conjecture on L^p-norms of exponential sums: Exciting research from Paul Cohen (late 1950's) to Ivo Klemes (early 2000's).

4.2. *Time to stop*

The above list is incomplete. It is pointless to try to remember all the topics that I had *intended* to put in this paper, as I have to stop anyway. I deliberately *refrained from giving any references* in the above section 4.1, as there is just too much exciting work (whether old, recent, unpublished or on-going) to quote. It is better to do a rather thorough job in [39] than a sloppy job here.

I find myself in the same situation as a drunkard sitting late at night in a bar with a (nearly) full bottle of whiskey he intends to drink, who is suddenly told by the bar owner that it's now closing time and that the police are about to arrive any second and enforce the regulations. The drunkard has to leave the bar, and I have to stop here. By the time this (quasi-aborted) paper appears, I hope the extended version [39] will be available for whoever is interested.

References

[1]   J. Arsac. *Fourier Transforms and the Theory of Distributions.* Prentice Hall, 1966.

[2]   L. D. Baumert. *Cyclic Difference Sets.* Lecture Notes in Mathematics 182, Springer-Verlag, 1971.

[3]   S. N. Bernstein. *On the order of the best approximation of continuous functions by polynomials of given degree.* Memoirs of the Royal Academy of Belgium (2), 4, pages 1–103, 1912. (French)

[4]   S. N. Bernstein. *Lectures on extremal properties and the best approximation of analytic functions of a real variable* Gauthiers-Villars, 1926. (Reprinted as the first part of *"Approximation"* by S. N. Bernstein and C. de La Vallée Poussin, Chelsea, New York, 1970). (French).

[5]   G. Björck. Functions of modulus one on $\mathbb{Z}_p$ whose Fourier transforms have constant modulus. *Proceedings of A. Haar Memorial Conference,* 1985, Colloq. Math. Soc. János Bolyai 49, pages 193–197, 1985.

[6]   G. Björck. Functions of modulus 1 on $\mathbb{Z}_n$ whose Fourier transforms have constant modulus, and "cyclic n-roots". *Proceedings of the 1989 NATO Advanced Study Institute on "Recent Advances in Fourier Analysis and Its Applications",* J. S. Byrnes & J. L. Byrnes, ed., pages 131–140, 1990.

[7]   G. Björck & B. Saffari. *New classes of finite unimodular sequences with unimodular Fourier transforms. Circulant Hadamard matrices with complex entries.* C. R. Acad. Sci. Paris, Ser. I Math. 320, No. 3, pages 319–324, 1995.

[8]   R. P. Boas, Jr. Inequalities for the derivatives of polynomials, *Math. Mag.* 42, pages 165–174, 1969.

[9]   P. Borwein. Home page, with (fairly) complete list of publications, on Web site of Math. Dept. , Simon Fraser University. Updated 2000-2001. (*c.f.* Ref. [14]).

[10]   J. S. Byrnes. On polynomials with coefficients of modulus one. Bull. London Math. Soc. 9, pages 171–176, 1977.

[11]   C. De La Vallée Poussin. Sur la convergence des formules d'interpolation entre ordonnées équidistantes, Acad. Roy. Belg. Bull. Cl. Sci (5), pages 319–410, 1908. (French)

[12]   C. De La Vallée Poussin. *Lectures on approximation of functions of a real variable*, 1919. (Reprinted as the second part of "Approximation" by S. N. Bernstein and C. de La Vallée Poussin, Chelsea, New York, 1970). (French)

[13]   P. Enflo. Oral question about a uniqueness property of Gauss sequences of prime length, Stockholm University, 1983.

[14]   T. Erdelyi. Home, page with (fairly) complete list of publications, on Web site of Math. Dept., Texas A&M University. Updated 2000–2001. (*c.f.* Ref. [9]).

[15]   L. Fejér. *Über konjugierte trigonometrische Reihen*, J. Reine Angew. Math. 144, pages 48–56, 1914.

[16]   M. Fekete. *Über einen Satz des Herrn Serge Bernstein*, J. Reine Angew. Math. 146, pages 88–94, 1916.

[17]   C. Frappier. Some extremal problems for polynomials and entire functions of exponential type. Ph. D. thesis, University of Montreal, 1983. (French).

[18]   S. W. Golomb. Cyclic Hadamard difference sets. *SETA'98* (*Proceedings of December 1998 Conference on "Sequences and their applications", Singapore*). World Scientific, 2000.

[19]   U. Haagerup. Personal letter (in Danish) to G. Björck, with copy to B. Saffari, 1995.

[20]   E. Landau. Personal letter to S. N. Bernstein, 1912.

[21]   E. Landau. Bemerkungen zu einer Arbeit von Herrn Carlemann. Math. Zeit. Bd 5, 1919.

[22]   J. E. Littlewood. *Some Problems in Real and Complex Analysis*, Heath Mathematical Monographs, Heath & Co., 1968.

[23]   G. G. Lorentz. *Approximation of Functions*. Holt, Rinehart & Winston, 1966.

[24]   L. Lovász. Private communication to G. Björck, 1983.

[25]   A. A. Markov. On a question posed by D. I. Mendeleiev. Bulletin of the Academy of Sciences of St. Petersburg 62, pages 1–24, 1889. (Reprinted in "Selected Works", Izdat. Akad. SSSR, Moscow, pages 51–75, 1948. (Russian)

[26]   Z. A. Melzak. *Companion to Concrete Mathematics*, vol II, Wiley, 1976.

[27]   D. I. Mendeleiev. Study of aqueous dissolutions, based on changes of their specific weights. St. Petersburg, 1887. (Russian).

[28]   H. L. Montgomery. An exponential polynomial formed with the Legendre symbol. Acta Arithmetica, vol 38, pages 375–380, 1980.

[29]   R. E. A. C. Paley. On Orthogonal Matrices, J. Math. and Phys. 12, pages 311-320, 1933.

[30]   S. K. Pichorides & B. Saffari. A proof by interpolation of a theorem of Szegö. Private manuscript, 1980. (presented in Ref. [36], 1980–1981).

[31]   Q. I. Rahman. Applications of Functional Analysis to Extremal Problems for Polynomials. Presses. Univ. Montréal, 1968.

[32]   Q. I. Rahman & G. Schmeisser. Les inégalités de Markov et de Bernstein. Presses Univ. Montréal, 1983.

[33]   H. J. Ryser. Combinatorial Mathematics. Carus Math. Monographs 14, Wiley, 1963.

[34]   F. Riesz. On trigonometric polynomials. C. R. Acad. Sci. Paris, vol 158, pages 1657–1661, 1914. (French)

[35]   M. Riesz. Interpolation formula for the derivative of a trigonometric polynomial. C. R. Acad. Sci. Paris vol. 158, pp. 1152–1154, 1914. (French)

[36]   B. Saffari. An interpolation proof of Szegö's inequality by S. K. Pichorides and myself. In *"Extremal problems on trigonometric polynomials"*, Postgraduate Course at Univ. of Geneva, Switzerland, 1980–1981.

[37]   B. Saffari. New unimodular polynomials with vanishing periodic autocorrelations. Research report, Prometheus Inc., Newport, RI, 1990.

[38]   B. Saffari. Perfect flatness of polynomials with coefficients ± 1 on groups of roots of unity. Private manuscript, 1998.

[39]   B. Saffari, Twentieth century extremal problems on polynomials and exponential sums, extended version of this paper, (in preparation).

[40]   G. Schaake & J. G. van der Corput. Ungleichungen für Polynome und trigonometrische Polynome. *Compositio Math.* vol 2, pages 321–361, 1935.

[41]   G. Schaake & J. G. van der Corput. Berichtigung zu: Ungleichungen für Polynome und trigonometrische Polynome. Compositio Math. vol 3, page 128, 1936.

[42]   G. Szegö. Über einen Satz des Herrn Serge Bernstein. Schriften der Königsberger Gelehrten Gesellschaft, pages 59–70, 1928.

[43]   G. Szegö & A. Zygmund. On certain mean values of Polynomials, J. d'Analyse Math. vol 3, pages 225–244, 1953–1954.

[44]   A. Zygmund. Trigonometric Series. Cambridge at the University Press, 1959.

The Problem of Efficient Inversions and Bezout Equations

Nikolai Nikolski

Steklov Institute of Mathematics
27 Fontanka
191011 St. Petersburg, Russia

Laboratoire de Maths Pures
Université Bordeaux I
351 cours de la Libération
33405 Talence France
`nikolski@math.u-bordeaux.fr`

ABSTRACT. This is a survey of some recent results on the phenomenon of the "invisible spectrum" for Banach algebras. Function algebras, formal power series and operator algebras are considered. This includes a quantitative treatment of the famous Wiener-Pitt-Sreider phenomenon for convolution measure algebras on locally compact abelian (LCA) groups. Efficient sharp estimates for resolvents and solutions of higher Bezout equations in terms of their spectral bounds are considered. The smallest spectral "efficiency hull" of a given closed set is introduced and studied. Using the spectral hulls we define uniformly bounded functional calculi for elements of the algebras in question. This program is realized for the measure algebras of LCA groups and for the measure algebras of a large class of topological abelian semigroups; for their subalgebras—the (semi)group algebra of LCA (semi)groups, the algebra of almost periodic functions, the algebra of absolutely convergent Dirichlet series, as well as for the weighted Beurling-Sobolev algebras, for H^∞ quotient algebras, and for some finite dimensional algebras.

235

J.S. Byrnes (ed.), Twentieth Century Harmonic Analysis - A Celebration, 235–269.
© 2001 *Kluwer Academic Publishers. Printed in the Netherlands.*

Part I. Quantitative version of the Gelfand theory

1. Introduction: Efficient Inversion

1.1. *Basic motivations*

There are three classical problems of harmonic analysis and function theory related to a phenomenon we call *invisible spectrum*. Formal definitions are contained in Subsection 1.2 below.

The *first problem* comes from convolution equations and is related to what is usually called *"the Wiener-Pitt phenomenon"*. Namely, let G be a locally compact abelian group (*LCA* group) written additively, and $\mathcal{M}(G)$ the convolution algebra of all complex measures on G. The fundamental problem is to find an invertibility criterion for measures $\mu \in \mathcal{M}(G)$, that is, a criterion for the existence of $\nu \in \mathcal{M}(G)$ such that $\mu * \nu = \delta_0$, δ_0 being the unit of $\mathcal{M}(G)$ (the Dirac δ-measure at 0). An obvious obstruction for invertibility is the vanishing of the Fourier transform $\hat{\mu}(\gamma) = 0$ at a point γ of the dual group $\hat{G}$, since the equality $\hat{\mu}\hat{\nu} \equiv 1$ is equivalent to the initial convolution equation. Generally, the boundedness away from zero

$$(1.1) \qquad \delta = \inf_{\gamma \in \hat{G}} |\hat{\mu}(\gamma)| > 0$$

is necessary for μ to be invertible. N. Wiener and R. Pitt (1938), and Yu. Sreider (1950, a corrected version of Wiener and Pitt's result) discovered that, in general, this is not sufficient. Namely, there exists a measure μ on the line $\mathbb{R}$ whose Fourier transform

$$(1.2) \qquad \hat{\mu}(y) = \int_{\mathbb{R}} e^{-ixy} \, d\mu(x), \quad y \in \mathbb{R}$$

is bounded away from zero but there exists no measure $\nu \in \mathcal{M}(\mathbb{R})$ such that $\hat{\nu}(y) = 1/\hat{\mu}(y)$ for $y \in \mathbb{R}$. This result still holds true for an arbitrary *LCA* group G which is not discrete, see [Ru1], [GRS], [HR] for references and historical remarks. Using the Banach algebra language, one can say that the dual group $\hat{G}$, being the "visible part" of the maximal ideal space $\mathfrak{M} = \mathfrak{M}(A)$ of the algebra $A = \mathcal{M}(\mathbb{T})$, is far from being a dense subset of $\mathfrak{M}$. Nonetheless, later on we will see that some quantitative precisions of (1.1), namely a closeness of the norm $\|\mu\|$ and δ of (1.1), lead to the desired invertibility, and even to a norm control of the inverse.

On the contrary, for a discrete group G, the classical Wiener theorem on absolutely convergent Fourier series tells that condition (1.1) implies the invertibility of μ, and, moreover, $\mathfrak{M} = \widetilde{G}$. However, in this setting too, the problem of the norm control for inverses μ^{-1} is still meaningful and interesting, in spite of the Wiener theorem, since the latter does not yield any estimate. In fact, from the quantitative point of view, there is no big difference between these qualitatively polar cases (we mean the cases of nondiscrete and discrete groups). It turns out that, in both cases, one can control the norms $\|\mu^{-1}\|$ for $\delta > 0$ close enough to the norm $\|\mu\|$, but this is not the case for small $\delta > 0$.

The *second problem* we are interested in is to distinguish, among all unital Banach algebras A, those permitting an estimate of the resolvents only in terms of the distance to the spectrum. More precisely, we want to know for which algebras A there exists a function φ such that

$$(1.3) \qquad \left\| (\lambda e - f)^{-1} \right\| \leqslant \varphi(\text{dist}\,(\lambda, \sigma(f)))$$

for all $\lambda \in \mathbb{C} \backslash \sigma(f)$ and all $f \in A$, $\|f\| \leqslant 1$. Here e stands for the unit of A, and $\sigma(f)$ for the spectrum of f in the algebra A. We treat this problem in a more general context of norm-controlled functional calculi, that is, as a partial case of the norm estimates problem for functions operating on a Banach algebra. See Chapter 2 for more details.

The *third problem* is a multi-element version of the previous two. Postponing precise definitions and discussions to Subsection 1.2, we mention here the classical corona problem for the algebra $H^\infty(\Omega)$ of all bounded holomorphic functions on Ω, an open subset of $\mathbb{C}^n$ or of a complex manifold. Recall that the problem is to solve the Bezout equations

$$(1.4) \qquad \sum_{k=1}^{n} g_k f_k = 1$$

in the algebra $H^\infty(\Omega)$, where the data $f_k \in H^\infty(\Omega)$ satisfy an analogue of condition (1.1),

$$(1.5) \qquad \delta^2 = \inf_{z \in \Omega} \sum_{k=1}^{n} |f_k(z)|^2 > 0,$$

and to estimate solutions $g_k \in H^\infty(\Omega)$. The Banach algebra meaning of the corona problem is well known; namely, the existence of $H^\infty(\Omega)$ solutions for any data satisfying (1.5) is equivalent to the density of Ω, the "visible part" of the spectrum $\mathfrak{M} = \mathfrak{M}(H^\infty(\Omega))$, in $\mathfrak{M}$. In what follows, we consider a norm refinement of this problem for several algebras different from $H^\infty(\Omega)$.

1.2. (In)Visibility levels. Main problems

Let A be a commutative unital Banach algebra and X be a subset of the maximal ideal space of A (the spectrum of A), $\mathfrak{M} = \mathfrak{M}(A)$, $X \subset \mathfrak{M}(A)$. We write $f \longmapsto \hat{f}(\mathrm{m})$ for $\mathrm{m} \in M$, or simply $f \longmapsto f(\mathrm{m})$, for the Gelfand transform of an element $f \in A$, and hence, staying on X, we can embed A into $C(X)$, $f \longmapsto f(x) = \delta_x(f)$ for $x \in X$, where $\delta_x \in \mathfrak{M}$ stands for the point evaluation $\delta_x(f) = f(x)$, $f \in A$. In what follows, we regard $\mathrm{clos}\, X$ as the *visible part* of $\mathfrak{M}$.

Recall that the spectrum $\sigma(f)$ of an element $f \in A$ coincides with the range $f(\mathfrak{M})$ of the Gelfand transform. The following definition formalizes different levels of "visibility" of the spectrum.

DEFINITION 1.2.1. *The spectrum of A is called n-visible* (or, *n-visible from X*), $n = 1, 2, \ldots$, *if $f(\mathfrak{M}) = \mathrm{clos}\,(f(X))$ for all n-tuples $f = (f_1, \ldots, f_n) \in A^n = A \times \ldots \times A$; and it is called* completely visible *if $\mathfrak{M} = \mathrm{clos}\, X$.*

It is clear that $(n + 1)$-visibility implies n-visibilty for any $n \geqslant 1$, and the complete visibility is equivalent to n-visibility for all $n \geqslant 1$. Moreover, the Gelfand theory of maximal ideals makes evident the following lemma.

LEMMA 1.2.2. *For a commutative unital Banach algebra A, the following properties are equivalent.*

(i) The spectrum of A is n-visible.

(ii) For every $f = (f_1, \ldots, f_n) \in A^n$ satisfying

$$(1.6) \qquad \delta^2 =: \inf_{x \in X} \sum_{k=1}^{n} |f_k(x)|^2 > 0,$$

there exists an n-tuple $g \in A^n$ solving the Bezout equation

$$(1.7) \qquad \sum_{k=1}^{n} g_k f_k = e.$$

In this language, the Wiener-Pitt phenomenon is exactly the 1-invisibility of the spectrum for the measure algebra $\mathcal{M}(G)$, if we stay on the dual group $X = \hat{G}$. Rigorously speaking, we mean the algebra of Fourier transforms $\mathcal{F}M(G) = \{\hat{\mu} : \mu \in \mathcal{M}(G)\}$ endowed with the norm $\|\hat{\mu}\| = \|\mu\|$ and embedded into $C(\hat{G})$. In what follows, we systematically identify these algebras.

The next definition specifies the previous one in the case of norm controlled invertibility instead of simple invertibility.

DEFINITION 1.2.3. Let A and X be as above, and let $0 < \delta \leqslant 1$. The spectrum of A is called $(\delta - n)$-*visible (from X)* if there exists a constant c_n such that any Bezout equation (1.7) with data $f = (f_1, \ldots, f_n) \in A^n$ satisfying (1.6) and the normalizing condition

$$(1.8) \qquad \|f\|^2 =: \sum_{k=1}^{n} \|f_k\|^2 \leqslant 1$$

has a solution $g \in A^n$ with $\|g\| \leqslant c_n$. The spectrum is called *completely δ-visible* if it is $(\delta - n)$-*visible for all $n \geqslant 1$ and the constants c_n can be chosen in such a way that* $\sup_{n \geqslant 1} c_n < \infty$.

Clearly, there exist the *best possible constants*, in the following sense. Setting

$$(1.9) \qquad c_n(\delta) =: c_n(\delta, A) = c_n(\delta, A, X) = \sup_f \{\inf(\|g\| : \sum_{k=1}^{n} g_k f_k = e, \ g \in A^n)\},$$

where the *supremum* is taken over all $f \in A^n$ satisfying (1.8) and (1.6), we get the smallest number for which $c_n = c_n(\delta) + \epsilon$ meets the requirements of Definition 1.2.3 for every $\epsilon > 0$. In particular,

$$(1.10) \qquad c_1(\delta) = \sup\{\|f^{-1}\| : f \in A; \ \delta \leqslant |f(x)| \leqslant \|f\| \leqslant 1, \ x \in X\},$$

and, in this case, we can take $c_1 = c_1(\delta)$ in Definition 1.2.3; here and in what follows we formally set $\|f^{-1}\| = \infty$ for noninvertible elements of A.

We define the n-th critical constant $\delta_n(A, X)$ by the relation

$$\delta_n(A, X) = \inf\{\delta : c_n(\delta, A, X) < \infty\}$$

1.2.4. Main problems. Our main objective is to estimate from above and from below, and (if possible) to compute the critical constants $\delta_n(A, X)$ and the majorants $c_n(\delta, A, X)$ for basic Banach algebras A and, thus, to study norm controlled visibility properties for these algebras.

For $n = 1$ we deal with more general objects. Namely, we are interested in describing functions boundedly acting on a given algebra. Supposing that a "visible part" of the spectrum is fixed, $X \subset \mathfrak{M}(A)$, we can say that a function (say, φ) defined on $\sigma \subset \mathbb{C}$ acts on an algebra A if for every $a \in A$ with $\hat{a}(X) \subset \sigma$ there exists $b \in A$ such that $\varphi \circ \hat{a} = \hat{b}$ on X. A "bounded action" is defined in the same way but adding an estimate of the form $\|b\| \leqslant k\|\varphi\|_*$, where $k = k(\sigma, A, X)$, and $\|\cdot\|_*$ is an appropriate norm majorizing $\|\varphi\|_\sigma = \sup_\sigma |\varphi|$. We mention, however, that such a definition is too broad to be useful: *the spectral inclusion $\hat{a}(X) \subset \sigma$ alone, without any norm restrictions, cannot imply the boundedness of compositions.* We formalize this statement as follows.

LEMMA 1.2.5. *Assume that $|\hat{a}(x)| \geqslant \delta(x \in X)$ always implies $\|a^{-1}\| \leqslant C$, for some positive δ and C. Then A is a uniform algebra whose norm is equivalent to the sup norm on X: $\|\hat{a}\|_X \leqslant \|a\| \leqslant (2C + \delta)\|\hat{a}\|_X$ for all $a \in A$.*

The lemma shows that certain norm requirements are necessary. The classical results on functions operating on Fourier transforms show many specific examples of this kind, see [HKKR], [Ru1] and further references therein.

In paper [N4], adding the normalizing condition $\|a\| \leqslant 1$ to the spectral inclusion $\hat{a}(X) \subset \sigma$ we look for a "minimal spectral hull" $h(\sigma) = h(\sigma, A, X)$ such that functions holomorphic on $h(\sigma)$ boundedly act on a given algebra. Our approach to this problem is explained in Section 2.

1.3. Outline of the theory

The main goal of the theory presented below is to estimate, from above and from below, and (if possible) to compute, the critical constants $\delta_n(A, X)$ and the majorants $c_n(\delta, A, X)$ for some commutative Banach algebras frequently used in harmonic analysis and for the corresponding (customary) visible parts X of their spectra. The basic algebras are the following ones:

(i) the measure algebra $\mathcal{M}(G)$ on an infinite LCA group G with $X = \hat{G}$, and in particular, the Wiener algebra $W = \mathcal{F}l^1(\mathbb{Z})$ of absolutely convergent Fourier series with $X = \mathbb{T}$ (Section 5);

(ii) the analytic Wiener algebra $W_+ = \mathcal{F}l^1(\mathbb{Z}_+)$ with $X = \overline{\mathbb{D}} = \{z \in \mathbb{C} : |z| \leqslant 1\}$ (Section 5);

(iii) the weighted Beurling-Sobolev algebras of positive spectral radius, both analytic $\mathcal{F}l^p(\mathbb{Z}_+, w)$ (with $X = \overline{\mathbb{D}}$), and "symmetric" $\mathcal{F}l^p(\mathbb{Z}, w)$ (with $X = \mathbb{T}$), for some class of regularly growing weights $w(n)$ tending to ∞ as $|n| \longrightarrow \infty$ (Section 7).

In Section 2, following [N4], we introduce two general methods: a method for upper estimates of the visibility constants $c_n(\delta, A, X)$, and a method for their lower estimates. The method for lower estimates for $c_n(\delta, A, X)$ described in Subsection 2.3 refers to elements of A with "almost independent" powers. In Section 5 we specify this method for the algebras $\mathcal{M}(G)$, $\mathcal{M}(S)$ using the Sreider measures, which are defined as measures with real Fourier transforms and the spectrum filling in a disc. In particular, a short proof is given, following [N4], to the fact that $\delta_1(W_+, \overline{\mathbb{D}}) = 1/2$ and $c_1(\delta) = (2\delta - 1)^{-1}$ for $1/2 < \delta \leqslant 1$ and $c_1(\delta) = \infty$ for $\delta \leqslant 1/2$. A different, longer but more elementary proof of this fact is contained in [ENZ]. Independently, another elementary proof of the equality $\delta_1(W_+, \overline{\mathbb{D}}) = 1/2$ was recently presented by H.S. Shapiro at the 6th St. Petersburg Summer Analysis Conference, see [Sh2].

In the same Section 5, we show that $c_1(\delta, \mathcal{F}M(G), \hat{G}) = \infty$ for $0 < \delta \leqslant 1/2$, and thus $\delta_1(\mathcal{F}M(G), \hat{G}) \geqslant 1/2$, for every infinite LCA group G. Moreover, $(2\delta - 1)^{-1} \leqslant c_1(\delta, \mathcal{F}M(G), \hat{G}) \leqslant c_n(\delta, \mathcal{F}M(G), \hat{G})$ for $1/2 < \delta \leqslant 1$, and $c_n(\delta, \mathcal{F}M(G), \hat{G}) \leqslant (2\delta^2 - 1)^{-1}$ for $1/\sqrt{2} < \delta \leqslant 1$. Again, these results are contained both in [N4] and [ENZ] but the proofs are different.

It should be noted that some of these results were mentioned even in Shapiro's paper [Sh1] (Remarks 2 and 3, and a footnote on page 235 of [Sh1]), where they were attributed to Y. Katznelson (for $c_1(\delta, W, \mathbb{T}) = \infty$ for $\delta < 1/2$), to Y. Katznelson and D. J. Newman (for $c_1(\delta, W, \mathbb{T}) < \infty$ for $\delta > 1/\sqrt{2}$), and to Bell (for $c_1(\delta, W_+, \overline{\mathbb{D}}) < \infty$ for $\delta > 1/2$), but at present we cannot specify references.

Section 6 contains some results on the efficient inversion on some finite groups and semi-groups (again, following [N4]).

In Section 4, we deal with a more general framework, namely with the norm-controlled functional calculi and the so-called spectral efficiency hulls $h(\sigma, A, X)$ introduced (following [N4]) in the same section. The links of efficiency hulls with the norm-controlled calculi are established and a description of these hulls for the measure algebra $\mathcal{M}(S)$ on a semigroup S is given using the horodisc expansions.

The results and the methods employed for *weighted* convolution algebras are completely different. In Section 3, following [ENZ], a general method is developed that allows us to control the inverses in terms of $\delta = \inf_{\mathfrak{M}(A)} |\hat{x}|$ for rotation invariant topological Banach algebras A. More precisely, our goal in Section 3 is to give a method for proving the equality $\delta_1(A, \mathfrak{M}) = 0$. This method relies on two ideas. First, to estimate $\|x^{-1}\|_A$, we use the multipliers $\text{mult}(\mathcal{D}A)$ of the space $\mathcal{D}A$ of derivatives of our algebra A, where $\mathcal{D} = z\frac{d}{dz}$. The second idea is to deduce estimates from the compactness of the embedding $A \subset \mathrm{mult}\,(\mathcal{D}A)$. In applications, the main point is precisely in the proof of this compactness and in estimating the rate of decay of relevant best polynomial approximations in the $\mathrm{mult}\,(\mathcal{D}A)$ norm. Following [ENZ], we realize this approach in Section 7 for the Beurling-Sobolev algebras $A = l^p(w)$ of positive spectral radius, $r(A) = \lim_n w(n)^{1/n} > 0$, both on $\mathbb{Z}$ and $\mathbb{Z}_+$.

Historical remarks. As is already mentioned, the prehistory of the ideas presented in this paper was started with the classical theorems of Wiener-Lévy and Wiener-Pitt-Sreider quoted above.

The second wave of results, sharpening the Gelfand and Riesz-Dunford functional calculi, was devoted to functions operating on Fourier transforms, and was mostly due to H. Helson, J.-P. Kahane, Y. Katznelson, and W. Rudin. The main problem considered and resolved was to describe functions φ defined on an interval $\sigma = [a, b] \subset \mathbb{R}$ and such that $\varphi(\mathcal{F}f) \in \mathcal{F}A$ for every $f \in A$ with $\mathcal{F}f(\hat{G}) \subset \sigma$, where $A = \mathcal{M}(G)$ or $A = L^1(G)$. See [HKKR], [Ru1], [GMG], [K1], [HR] for exhaustive presentations and further references. Nonanalytic functions operating on certain weighted algebras of Fourier transforms $\mathcal{F}l^1(\mathbb{Z}, w)$ occurred in the papers of J.-P. Kahane, [K2], and N. Leblanc [L]. However, no quantitative aspects similar to those of Sections 1–2 were explicitly presented.

The third wave of results related to norm-controlled calculi can be linked with constructive proofs of the Wiener-Lévy theorem on inverses. We mention the proofs by A. Calderon, presented in [Z], by P. Cohen [C], and by D. Newman [New]. The Calderon approach was developed by E. Dyn'kin [D].

The problem of norm-controlled inversion (for the Wiener algebra $A = l^1(\mathbb{Z})$) was first mentioned by J. Stafney in [St], where the existence of $a, b, K > 0$ was proved such that $\sup\{\|f^{-1}\|_A : \|f\| \leqslant K, \hat{f}(\mathbb{T}) \subset [a, b]\} = \infty$. This implies that $c_1(\delta, A, \mathbb{T}) = \infty$ for *some* $\delta > 0$. The proof, based on Y. Katznelson's results, does not permit to specify the value of δ. Independently, and in a more constructive way, the result was obtained by H. Shapiro [Sh1] (in response to a question by a physicist G. Ehrling), but also without any concrete value of δ. Several remarks were made in Shapiro's note attributing to various authors certain estimates for quantities we call the critical constants $\delta_1(l^1(\mathbb{Z}_+), \overline{\mathbb{D}})$ and $\delta_1(l^1(\mathbb{Z}), \mathbb{T})$; no precise references were given. In fact, the paper [N4] was inspired by Shapiro's construction. In the paper of J.-E. Björk [B], a problem related to uniform functional calculi was considered. In our language, it is equivalent to an estimate for $\delta_1^0(A, \mathfrak{M}(A))$, a microlocal version of $\delta_1(A, \mathfrak{M}(A))$ studied in Section 4 below. In [B], a criterion was given in terms of another quantity which can be regarded as a "uniform spectral radius" of the algebra. O. El-Fallah [E] recently gave an application of Bkörk's result to the algebras $l^p(w)$ with slowly growing w.

In the paper of S. Vinogradov and A. Petrov [VP], a description was given of Banach spaces A of functions on $\mathbb{T}$ satisfying the following property: $f \in A$ and $|f| \geqslant \delta$ on $\mathbb{T}$ imply $1/f \in A$.

We finish these remarks mentioning that the case of higher Bezout equations (the "corona problems") related to the constants $c_n(\delta, A, X)$ and $\delta_n(A, X)$ for $n > 1$ is considered in this paper very briefly. For their history see [Gar], [N1], as well as [To1] and [To2].

2. How and Why One Can(not) Control the Inverses

Let A be a commutative Banach algebra with unit e, and let X be a Hausdorff topological space such that A is continuously embedded into $C(X)$, so that $X \subset \mathfrak{M}(A)$. We also use other notation introduced in Section 1. The set of invertible elements of A is denoted by $\mathcal{G}(A)$. We start with simple observations on the critical constant δ_1.

2.1. *First observations*

We say that A is an algebra of *distance controlled resolvent growth* if there exists a monotone decreasing function $\varphi : \mathbb{R}^*_+ \longrightarrow \mathbb{R}^*_+ = (0, \infty)$ such that

(2.1) $$\left\| (\lambda e - f)^{-1} \right\| \leqslant \varphi(\text{dist} \, (\lambda, \sigma(f)), \quad \lambda \in \mathbb{C} \backslash \sigma(f)$$

for all $f \in A, \|f\| \leqslant 1$. It is easy to see that this estimate implies $\left\| (\lambda e - f)^{-1} \right\| \leqslant \frac{1}{\|f\|} \varphi(\frac{1}{\|f\|}\text{dist} \, (\lambda, \sigma(f)))$ for all $f \in A$ and $\lambda \in \mathbb{C} \backslash \sigma(f)$. It is shown in [N4] that a commutative Banach algebra A obeys the distance controlled resolvent growth if and only if $\delta_1(A, \mathfrak{M}(A)) = 0$ (the critical constant introduced in Section 1). In fact, $c_1(\delta, A, \mathfrak{M}) \leqslant \varphi(\delta) \leqslant c_1(\delta(2 + \delta)^{-1}, A, \mathfrak{M})$. Similar equivalences hold for the constant $\delta_1(A, X)$ related to a subset $X \subset \mathfrak{M}(A)$, and for $(\delta - n)$-visibility and the "n-resolvent" $(\lambda e_n - f)^{-1}$, where $\lambda e_n = (\lambda_1 e, \ldots, \lambda_n e) \in A^n, \lambda_k \in \mathbb{C}$.

For many examples of function Banach algebras with various behaviour of constants δ_n and $c_n(\delta)$, including the classical ones (like $A = H^\infty(\Omega)$), we refer to [N4], [ENZ].

2.2. *Splitting X-symmetric algebras for upper estimates*

As before, we consider a commutative unital Banach algebra A continuously embedded into the space $C(X)$, where $X \subset \mathfrak{M}(A)$. Now, we need the following definition, see [N4].

DEFINITION 2.2.1. We say that an algebra A *splits at the unit* if there exists a subspace $A_0 \subset A$ such that $A = e \cdot \mathbb{C} + A_0$ (a direct sum) and for $f = \lambda e + f_0, f_0 \in A_0$ we have $\|f\| = |\lambda| + \|f_0\|$.

An algebra A is said to be *X-symmetric* if for every $f \in A$ there exists an element $g \in A$ such that $\|g\| \leqslant \|f\|$ and $g(x) = \overline{f(x)}$ for all $x \in X$.

Obviously, the splitting property is satisfied by the algebras A obtained by the standard adjoining of unity to a Banach algebra A_0 without unit. For instance, this is the case for the group algebra $A_0 = L^1(G)$ of a nondiscrete LCA group G. Also, it should be mentioned that the classical symmetry property of Banach algebras corresponds, in our language, to the $\mathfrak{M}(A)$-symmetry. For $X \neq \mathfrak{M}(A)$, X-symmetry may happen to be a considerably weaker property. For instance, the algebra $A = \mathcal{F}M(G)$ is obviously $\hat{G}$-symmetric: for $f = \hat{\mu} \in A$

take $g = \hat{\mu}_*$, where $\mu_*(\sigma) = \overline{\mu(-\sigma)}, \sigma \subset G$. But it is not $\mathfrak{M}$-symmetric for a nondiscrete LCA group G. This latter property is essentially equivalent to the Wiener-Pitt phenomenon.

For X-symmetric algebras, the $(\delta - n)$-visibility properties are related to each other as follows.

LEMMA 2.2.2. *[N4] For an X-symmetric algebra A, the $(\delta - 1)$-visibility of the spectrum implies the $(\sqrt{\delta} - n)$-visibility for all $n \geqslant 1$, and even the complete $\sqrt{\delta}$-visibility with* $\sup_{n \geqslant 1} c_n(\sqrt{\delta}, A, X) \leqslant c_1(\delta, A, X).$

Another feature of X-symmetric algebras is that one can always control the inverses of elements whose lower bound $\delta = \inf_X |f|$ is sufficiently close to the norm $\|f\|$. To show this we use the following obvious observation.

LEMMA 2.2.3. *Let A be an algebra splitting at the unit, and let $f = \lambda e + f_0, 1/2 < \delta \leqslant |\lambda| \leqslant \|f\| \leqslant 1$. Then f is invertible in A, and $\|f^{-1}\| \leqslant (2\delta - 1)^{-1}$.* ∎

2.2.4. **X-domination for the unit evaluation functional.** Let φ_e be the following functional evaluating the coefficient of the unit in the standard expansion of an element of a splitting algebra:

$$(2.2) \qquad \varphi_e(\lambda e + f_0) = \lambda$$

for $\lambda e + f_0 \in A = \mathbb{C} \cdot e + A_0$. Note that φ_e is a norm 1 linear functional on A, not necessarily multiplicative. The following definition will be useful.

DEFINITION 2.2.5. *Let $A = e \cdot \mathbb{C} + A_0$ be a direct sum decomposition of a Banach algebra A, and let $X \subset \mathfrak{M}(A)$. We say that the unit evaluation functional (2.2) is X-dominated if $|\varphi_e(f)| \leqslant \|f\|_X$ for every $f \in A$, where $\|f\|_X = \sup_X |f|$.*

Standard Hahn-Banach arguments show the following lemma, where, as above, δ_x means the evaluation functional at a point $x \in X$: $\delta_x(f) = f(x)$ for $f \in A$.

LEMMA 2.2.6. *The following assertions are equivalent.*
(i) The functional φ_e of (2.2) is X-dominated.
(ii) $\varphi_e \in \mathrm{conv}\,(\delta_x : x \in X)$, where $\mathrm{conv}\,(\cdot)$ stands for the weak-$*$ closed convex hull of $(\cdot)$. ∎

THEOREM 2.2.7. *Let A be a commutative unital Banach algebra and $X \subset \mathfrak{M}(A)$ such that (i) A is X-symmetric; (ii) A splits at the unit; and (iii) φ_e is X-dominated.*
Then, the spectrum of A is $(\delta - n)$-visible for all $n \geqslant 1$ and all δ satisfying $1/\sqrt{2} < \delta \leqslant 1$, and even completely δ-visible with $c_n(\delta, A, X) \leqslant (2\delta^2 - 1)^{-1}$.

Proof. Let $f = (f_1, \ldots, f_n) \in A^n$ be such that $\delta \leqslant |f(x)| \leqslant \|f\| \leqslant 1$ for all $x \in X$, and let g_k be elements of A corresponding to the f_k as in the definition of X-symmetric algebras. Then $h \in A$, where $h = \sum_{k=1}^{n} f_k g_k$, and $1/2 < \delta^2 \leqslant h(x) \leqslant \|h\| \leqslant 1$ for all $x \in X$. Using condition *(iii)*, Lemma 2.2.6, and an obvious fact that the interval $[\delta^2, 1]$ is a convex set, we obtain $\delta^2 \leqslant \varphi_e(h) \leqslant 1$. Condition *(ii)* and Lemma 2.2.3 imply that h is invertible and $\|h^{-1}\| \leqslant (2\delta^2 - 1)^{-1}$. Hence, $g = (g_1 h^{-1}, \ldots, g_n h^{-1}) \in A^n$, $\sum_{k=1}^{n} f_k(g_k h^{-1}) = e$, and

$$\|g\| = \Big(\sum_{k=1}^{n} \|g_k h^{-1}\|^2\Big)^{1/2} \leqslant \|h^{-1}\| \cdot \|f\| \leqslant (2\delta^2 - 1)^{-1},$$

as desired. ∎

2.3. *A method for lower estimates*

The method to get a lower estimate for $c_1(\delta, A, X)$ stated in theorem 2.3.1 below is inspired by Shapiro's example [Sh1] mentioned above. Essentially, it reduces to the existence of elements $a \in A$ whose normalized powers $a^k/\|a\|^k$, $0 \leqslant k \leqslant p$, for a given p are ϵ-equivalent to the standard basis of an l^1−space, whereas asymptotically they tend to zero faster than a given exponential. We use this method in Section 5; see also [ENZ].

THEOREM 2.3.1. *[N4] Let A be a unital commutative Banach algebra, $X \subset \mathfrak{M}(A)$, and given $\epsilon > 0$ let A^ϵ be the set of all elements $a \in A$ such that $\|a\|_{C(X)} < \epsilon$ and $\|a\| = 1$. Suppose that*

$$(2.3) \qquad\qquad \sup_{a \in A^\epsilon} \left\| \sum_{k=0}^{p} b_k a^k \right\| \geqslant \sum_{k=0}^{p} b_k$$

for all $p \geqslant 0$, $\epsilon > 0$, and $b_k \geqslant 0$. Then $c_1(\delta, A, X) \geqslant (2\delta - 1)^{-1}$ for all δ, $1/2 < \delta < 1$. In particular, $\delta_1(A, X) \geqslant 1/2$.

The proof consists of setting $f_t = (1 + t)^{-1}(e - ta)$ for $t > 0$ such that $(1 + t)^{-1} > \delta$ and $1/2 < \delta < 1$, then estimating $\|f_t^{-1}\|$ from below and maximizing in t. See [N4], theorem 1.5.1, for details.

3. Estimates of Inverses for Rotation Invariant Algebras

In this Section we describe, following [ENZ], a general method permitting to control the inverses in terms of $\delta = \inf_{\mathfrak{M}(A)} |\hat{x}|$ for rotation invariant topological Banach algebras A on the circle $\mathbb{T}$. In other words, our goal is to give a method for proving the equality $\delta_1(A, \mathfrak{M}) = 0$. As mentioned above, the latter property is equivalent to a distance controlled estimate for resolvents, $\|(\lambda e - x)^{-1}\| \leqslant \varphi(\text{dist}\,(\lambda, \sigma(x))$.

In short, the main idea how to estimate $\|x^{-1}\|$, $x \in A$, is to "reduce the smoothness" of x^{-1} by applying a first order differential operator $\mathcal{D}$, and then to use the formula $\mathcal{D}x^{-1} = -x^{-2}\mathcal{D}x$. To estimate the norm of the product $x^{-2}\mathcal{D}x$ we use the range norm $\|\cdot\|_{\mathcal{D}A}$ on $\mathcal{D}A$ and the multiplier norm related to $\mathcal{D}A$; namely $\|\mathcal{D}x^{-1}\|_{\mathcal{D}A} \leqslant \|\mathcal{D}x\|_{\mathcal{D}A}\|x^{-2}\|_{\text{mult}(\mathcal{D}A)}$.

The second idea is to rely on the compactness of the embedding $A \subset \text{mult}\,(\mathcal{D}A)$ to ensure a uniform estimate for $\|x^{-2}\|_{\text{mult}(\mathcal{D}A)}$. We obtain an estimate for $c_1(\delta, A)$ measuring the size of a compact subset of mult $(\mathcal{D}A)$ in terms of the decreasing rate of the best polynomial approximations, see Subsection 3.2 below.

In Section 7 we apply this method to Beurling-Sobolev algebras

$$A = l^p(\mathbb{Z}, w), l^p(\mathbb{Z}_+, w).$$

3.1. *How to use multipliers*

Having in mind applications to the case $A = l^p(w)$ (Section 7), from now on we distinguish between the Banach algebras and the algebras that become a Banach algebra after equivalent norming. More precisely, *a unital Banach algebra* is a Banach space A endowed with a multiplication such that $\|xy\| \leqslant \|x\| \cdot \|y\|$ for all $x, y \in A$, and $\|e\| = 1$, e stands for the unit of A. A *unital topological Banach algebra* A is a Banach space A endowed with a continuous multiplication so that $\|xy\| \leqslant C\|x\| \cdot \|y\|$ for all $x, y \in A$ and for some constant C. Clearly, every commutative topological Banach algebra is a Banach algebra with respect to the operator norm $\|\cdot\|^*$,

$$\|x\|^* = \sup\{\|xy\| : y \in A, \|y\| \leqslant 1\}.$$

It is clear that some properties of A, such as, e.g., the property $\delta_1(A, X) = 0$, are renorming stable. Some others may be greatly affected by such a renorming. For instance, so is the sharp value of the critical constant $\delta_1(A, X)$ if it is positive, or, in the case where $\delta_1(A, X) = 0$, so is the growth rate of the constants $c_1(\delta, A, X)$ as $\delta \longrightarrow 0$.

Now, let A be a unital topological Banach algebra of sequences $x = (x_n)$ on $\mathbb{Z}$ or $\mathbb{Z}_+$, which means that $x \longmapsto x_n$ is a continuous functional for every n, with the convolution $*$ as an algebra operation, and such that

(i) the set S_0 of finitely supported sequences (on $\mathbb{Z}$ or $\mathbb{Z}_+$, respectively) is a dense subset of A;

(ii) A is a rotation invariant (homogeneous) space of sequences, that is, if $x \in A$ then $x_t = (x_n \bar{t}^n)_n \in A$ and $\|x_t\| = \|x\|$ for every $t \in \mathbb{T}$.

Conditions *(i)* and *(ii)* guarantee that the rotation $t \longmapsto x_t$ is a norm continuous mapping from $\mathbb{T}$ to A, for every $x \in A$. Consequently, the Césaro (Fejér) or Abel-Poisson averages of the series $x \sim \sum_k x_k e_k$ converge to x for every $x \in A$; here $e_k = (\delta_{kn})_n$ is the standard $0-1$ algebraic basis of S_0.

A complex homomorphism φ of A is uniquely determined by its value $\lambda = \varphi(e_1)$ on the generator e_1 of the algebra A, and the Gelfand (Fourier) transformation is given by the formula,

$$x \longmapsto \hat{x}(\lambda) = \mathcal{F}x(\lambda) = \sum_k x_k \lambda^k,$$

at least for $x \in S_0$. Hence, $\varphi \longmapsto \lambda = \varphi(e_1)$ is a bijection of $\mathfrak{M}(A)$ on a compact subset of $\mathbb{C}$. We identify this subset with $\mathfrak{M}(A)$. In can be easily seen that $\mathfrak{M}(A) = A(r_-, r_+) = \{z \in \mathbb{C} : r_- \leqslant |z| \leqslant r_+\}$, where $r_+ = \lim_n \|e_n\|^{1/n} < \infty$ and $r_- = \lim_n \|e_{-n}\|^{-1/n} > 0$. We refer to $r_{\pm} = r_{\pm}(A)$ as to the lower and the upper spectral radius of A. The normalized case, where $r_+ = 1$, or $r_+ = r_- = 1$ in the case of $\mathbb{Z}$, will be the main one for us.

3.1.1. A Green type norm. The *operator* $\mathcal{D}$ is defined by $\mathcal{D}x = (n x_n)_n$, $x \in A$. It maps A to S, the space of all sequences on $\mathbb{Z}$ (respectively, on $\mathbb{Z}_+$). On the Gelfand transforms $\hat{x} = \sum_k x_k z^k$, the operator $\mathcal{D}$ acts as $\hat{\mathcal{D}}\hat{x} = (\mathcal{D}x)^{\hat{}}$. This is a formal first order differential operator, $\hat{\mathcal{D}} = z\frac{d}{dz}$. In particular, it obeys the Leibnitz rule for products $\hat{\mathcal{D}}(fg) = f\hat{\mathcal{D}}g + g\hat{\mathcal{D}}f$, at least for "trigonometric polynomials" $f, g \in \hat{S}_0$. Hence, $\mathcal{D}(x * y) = x * \mathcal{D}y + y * \mathcal{D}x$ for every $x, y \in S_0$. In fact, the same formula works for $x \in S_0$, $y \in A$, and in particular, $\mathcal{D}x^{-1} = -x^{-2} * \mathcal{D}x$ for all $x \in \mathcal{G}(A) \cap S_0$.

The range $\mathcal{D}A$ is a Banach space with respect to the range norm $\|\mathcal{D}x\|_{\mathcal{D}A} = \|x\|_A$, where $x \in A$, $x_0 = 0$. For convenience reasons, we add the unit $e = e_0$ to $\mathcal{D}A$ and will consider the sum $A' = \mathcal{D}A + \mathbb{C} \cdot e$ as the *range space of* $\mathcal{D}$ endowed with the following range norm $\|\lambda e + \mathcal{D}x\|_{A'} = \|\lambda e + x\|_A$, where $x \in A$, $x_0 = 0$ and $\lambda \in \mathbb{C}$. Clearly, the space A' contains S_0 as a dense subset, and is rotation invariant.

It is shown, see [ENZ], that $A \subset A'$ and every element $x \in S_0$ defines a continuous convolution operator $y \longmapsto x * y$ on A'. Hence, one can introduce the *convolution multiplier norm* on finitely supported elements $x \in S_0$ as the operator norm,

$$\|x\|_{\text{mult}(A')} = \sup\{\|x * y\|_{A'} : y \in S_0, \|y\|_{A'} \leqslant 1\}.$$

By definition, the *space* mult (A') *of (little) convolution multipliers* of A' is the completion of S_0 with respect to this norm. Clearly, mult (A') is a unital rotation invariant Banach algebra.

LEMMA 3.1.2. *Let A be a topological Banach algebra satisfying the above conditions (i)–(ii), and let $x \in \mathcal{G}(A) \cap \mathcal{G}(\text{mult}(A'))$ such that $\delta = \min_{\lambda \in \mathfrak{M}(A)} |\hat{x}(\lambda)| > 0$. Then*

$$\|x^{-1}\|_A \leqslant \|e\|_A \delta^{-1} + 2\|x\|_A \cdot \|x^{-2}\|_{\text{mult}(A')}.$$

3.2. How to use compactness

3.2.1. A multiplier estimate. Lemma 3.1.2 makes evident the following sufficient condition for $(\delta - 1)$-*visibility of the spectrum*: given a topological Banach algebra A of sequences on $\mathbb{Z}$, or $\mathbb{Z}_+$, satisfying conditions *(i)–(ii)* of Subsection 3.1, and *compactly* embedded into mult (A'): $A \subset_c \text{mult}(A')$, then $\delta_1(A, \mathfrak{M}(A)) = 0$, and, moreover,

$$c_1(\delta, A, \mathfrak{M}) \leqslant \delta^{-1}\|e\|_A + 2C(K_\delta)$$

for all $\delta > 0$, where $K_\delta = A_\delta^2$, $A_\delta = \{x \in A : \|x\|_A \leqslant 1, |\hat{x}(\lambda)| \geqslant \delta \text{ for all } \lambda \in \mathfrak{M}(A)\}$ and $C(K_\delta)$ is defined by the formula

(3.1) $$C(K_\delta) = \sup\{\|x^{-1}\|_{\text{mult}(A')} : x \in K_\delta\} < \infty.$$

It remains to estimate the constant $C(K_\delta)$ of formula (3.1).

3.2.2. **Approximate characteristics of compact sets.** We use the standard classification of compact sets in terms of the best polynomial approximations. Let B be a Banach space and let $L_n \subset B$ be subspaces of B such that $L_n \subset L_{n+1}$, $\dim L_n < \infty$ for $n \geq 1$, and $\mathrm{clos}\,(\bigcup_n L_n) = B$. Further, let $K \subset B$ and

$$(3.2) \qquad \epsilon_n(K) = \epsilon_n(K, B) = \sup\{\mathrm{dist}\,(x, L_n) : x \in K\}$$

be the best approximations of K by elements of L_n. It is well known that *a bounded subset $K \subset B$ is relatively compact if and only if* $\lim_n \epsilon_n(K) = 0$. We apply this criterion to the space $B = \mathrm{mult}\,(A')$ and to subspaces $L_n = \mathcal{P}_n$ of all trigonometric polynomials of degree less than or equal to n. Then $\epsilon_n(K)$ of (3.2) are the best polynomial approximations of elements of K. The axioms *(i)* and *(ii)* of Subsection 3.1 and the definition of $\mathrm{mult}\,(A')$ imply that $\lim_n \|\Phi_n x - x\|_B = 0$ for all $x \in B$, where $\Phi_n x$ stands for the Fejer mean of a sequence $x \in B$. Therefore, we can use the above compactness criterion for $B = \mathrm{mult}\,(A')$.

Now, our aim is to specify constants $C(K_\delta)$ from formula (3.1) in terms of $\epsilon_n(K_\delta)$, $n \geq 1$.

3.2.3. **Calderon-Cohen-Dyn'kin "constructive inversions".** As mentioned in the beginning of Section 3, the use of compactness for estimates of inverses was started with various "constructive proofs" of the classical Wiener-Lévy inversion theorem for absolutely convergent Fourier series. Here "constructive" means "without using the Gelfand theory of maximal ideals". The basic A. Calderon proof, [Z] Chapter VI, theorem (5.2), exploits the Cauchy formula, and, thus, is applicable not only to inverses x^{-1}, but also to compositions $\varphi \circ x$. P. Cohen [C] used the same techniques for inverses and for higher Bezout equations. In [C], a possibility to estimate the norms $\|x^{-1}\|_W$ in terms of $\delta = \inf_{\mathbb{T}} |\mathcal{F}x|$ and certain characteristics similar to the quantities ϵ_n of (3.2) was mentioned explicitly; here and below the symbol $\mathcal{F}x$ is used for the Gelfand (discrete Fourier) transform, keeping the notation $\hat{x}$ for the classical Fourier coefficients. In [D], E. Dyn'kin applied Calderon's method to the Beurling algebras $A = l^1(\mathbb{Z}, w)$ to get estimates for the inverses $\|x^{-1}\|_A$ in terms of the same characteristics $\epsilon_n(\{x\})$. D. Newman [New] gave a completely elementary proof of the Wiener-Lévy theorem also based on polynomial approximations.

All authors mentioned above obtained some estimates for the inverses $\|x^{-1}\|_A$ assuming, or implicitly assuming, that x runs over some compact subset $K \subset A$. The approach of [ENZ] is different: we *prove* the compactness of the set $A_\delta = \{x \in A : \|x\|_A \leq 1, |\mathcal{F}x| \geq \delta\}$ in the algebra $\mathrm{mult}\,(A')$ and use this compactness for obtaining a uniform estimate of $\|x^{-1}\|_A$ for $x \in A_\delta$. The next theorem is proved in [ENZ] using the Dyn'kin method from [D]. The latter paper contains a similar result for the special case of the algebra $B = l^1(\mathbb{Z}, w)$ with $w(n) = w(-n)$ and $r_- = r_+ = \lim_n w(n)^{1/n} = 1$.

THEOREM 3.2.4. *Let B be a convolution Banach algebra on $\mathbb{Z}_+$, or on $\mathbb{Z}$, satisfying conditions (i) and (ii) of Subsection 3.1, with the spectral radius r_+(respectively spectral radii $r_- \leq r_+$). Let K be a relatively compact subset of $B_\delta = \{x \in B : \|x\|_B \leq 1$, and $\delta \leq |\mathcal{F}x(\zeta)|$ for $\zeta \in \mathfrak{M}(B)\}$ and let $\lambda = \lambda_K$ be the distribution function of the sequence*

N. Nikolski / The Problem of Efficient Inversions and Bezout Equations

$(\epsilon_n(K))_{n\geq 0}$:

$$\lambda_K(t) = \text{card}\,\{n : n \geq 1,\ \epsilon_{n-1}(K) > t\}, \quad t > 0.$$

Then

$$\|x^{-1}\|_B \leq M(\delta, \lambda) = \frac{16}{\delta} \sum_{j \geq 0} (r_+^{-j}\|e_j\|_B + r_-^{j}\|e_{-j}\|_B)e^{-\delta j/17\lambda(\delta/4)}$$

for every $x \in K$.

For the case of an algebra on $\mathbb{Z}$ satisfying $r_- = r_+ = 1$, the proof starts by using the Calderon approach: we choose a polynomial $\mathcal{F}y \in \mathcal{P}_n$, $n = \lambda(\delta/4)$ with $\|x - y\|_B \leq \delta/4$ and write the Cauchy formula

$$x^{-1} = (2\pi i)^{-1} \int_{|z|=\delta/2} (y + ze)^{-1}(ze + (y - x))^{-1}\,dz,$$

where $(y + ze) \in \mathcal{G}(B)$ since $|\mathcal{F}y + z| \geq \delta/4$ on the space $\mathfrak{M}(B)$. To estimate $\|(y + ze)^{-1}\|_B$, we use Dyn'kin's method involving the S. Bernstein inequality, see [ENZ] for details.

Now, having in mind applications to the Beurling-Sobolev algebras in Section 7, we combine 3.2.1 and theorem 3.2.4.

THEOREM 3.2.5. *Let A be a convolution topological Banach algebra on $\mathbb{Z}$ or on $\mathbb{Z}_+$, satisfying conditions (i)–(ii) of Subsection 3.1 and compactly embedded into $B = \text{mult}\,(A')$, that is $A \subset_c B = \text{mult}\,(A')$. Let constants C and $\mathcal{E}$ be defined by the inequalities $\|x * y\|_A \leq C\|x\|_A\|y\|_A$ and $\|x\|_B \leq \mathcal{E}\|x\|_A$ for all $x, y \in A$, and let $\epsilon_n(A_0, B)$ be the best polynomial approximations of the unit ball $A_0 \subset A$, and $\lambda_0 = \lambda_{A_0}$ be their distribution function. Then $\delta_1(A, \mathfrak{M}(A)) = 0$, and $c_1(\delta, A, \mathfrak{M}(A)) \leq \frac{\|e\|_A}{\delta} + M(\delta)$ for all $\delta > 0$, where*

$$M(\delta) = \frac{2^5\mathcal{E}^2}{\delta^2} \sum_{j \geq 0} (r_+^{-j}\|e_j\|_B + r_-^{j}\|e_{-j}\|_B)e^{-\delta^2 j/17\mathcal{E}^2\lambda_0(\delta^2/4C)}.$$

4. Spectral Hulls and Norm-Controlled Functional Calculi

In the three preceding sections, we considered the problem of uniform upper bounds for inverses when staying on a given subset $X \subset \mathfrak{M}$ of the maximal ideal space $\mathfrak{M}$. Following [N4], now we treat a more general form of the same problem defining and studying the so-called X-spectral efficiency hull $h(\sigma, X)$ of a given set $\sigma \subset \mathbb{C}$. In this language, the uniform boundedness of inverses is equivalent to the property $0 \notin h(\sigma_\delta, X)$, where σ_δ stands for the annulus $\{z \in \mathbb{C} : \delta \leq |z| \leq 1\}$. Yet another reason to study the hulls $h(\sigma, X)$ is that $h(\sigma, X)$ is the minimal set satisfying the "uniform calculus property". The latter means that for every open set $\Omega \subset \mathbb{C}$ containing $h(\sigma, X)$ there exists a constant $k = k(\Omega, X)$ such that $\|f(a)\| \leq k\|f\|_\Omega$ for every $f \in Hol(\Omega)$ and for every $a \in A$ with $\sigma(a) \subset \sigma$ and $\|a\| \leq 1$. Similar uniformly bounded calculi were implicitly involved in classical studies of functions operating on Fourier algebras, see [HKKR], [Ru1].

The main results of this section are theorems 4.1.5 and 4.2.2. Examples of spectral hulls are gathered in subsection 4.3.

4.1. *Spectral hulls and resolvent majorants*

Let A be a unital Banach algebra, and let $X \subset \mathfrak{M}(A)$.

DEFINITION 4.1.1. Let $\sigma \subset \overline{\mathbb{D}}$. We set $A(\sigma; X) = \{a \in A : \|a\| \leqslant 1, \, \hat{a}(X) \subset \sigma\}$, $C(\lambda, \sigma; X) = C(\lambda, \sigma; A, X) = \sup\{\|(\lambda e - a)^{-1}\| : a \in A(\sigma; X)\}$ for $\lambda \in \mathbb{C}$, where we take $\|(\lambda e - a)^{-1}\| = \infty$ for $\lambda \in \sigma(a)$. Let also $h(\sigma; X) = h(\sigma; A, X) = \{\lambda \in \mathbb{C} : C(\lambda, \sigma; X) = \infty\}$.

The set $h(\sigma; X)$ is called the *X-spectral hull of σ*; the *full spectral hull of σ* is $h(\sigma) = h(\sigma; A, \mathfrak{M}(A))$. The complement $\rho(\sigma; X) = \mathbb{C} \backslash h(\sigma; X)$ is called the *norm-controlled (or efficient) resolvent complement of σ*. For a positive constant $k > 0$, we also consider the sets $h(\sigma, k; X) = \{\lambda \in \mathbb{C} : C(\lambda, \sigma) > k\}$ and $\rho(\sigma, k; X) = \mathbb{C} \backslash h(\sigma, k; X)$. For $X = \mathfrak{M}(A)$ we simplify the notation in a natural way: $A(\sigma) = A(\sigma; \mathfrak{M}(A))$, $h(\sigma, k) = h(\sigma, k; \mathfrak{M}(A))$, $\rho(\sigma, k) = \rho(\sigma, k; \mathfrak{M}(A))$.

4.1.2. **First properties.** It is clear that the definition of the $(\delta - 1)$-*visibility* (Section 1) is a special case of those of the efficient resolvent complement. Indeed, let σ_δ be an annulus, $\sigma_\delta = \{z \in \mathbb{C} : \delta \leqslant |z| \leqslant 1\}$. Then the spectrum of A is $(\delta - 1)$-visible if and only if $0 \in \rho(\sigma_\delta; X)$. Moreover, $c_1(\delta, A, X) = C(0, \sigma_\delta; A, X), 0 < \delta \leqslant 1$.

Note that n-variables counterparts of $A(\sigma; X), C(\lambda, \sigma; X)$, etc., could be considered as well, but we restrict ourselves to the case $n = 1$.

The following property of $C(\lambda, \sigma; A, X)$ and $h(\sigma, k; A, X)$ is a more or less straightforward consequence of the definitions, see [N4] for the proofs: $h(\sigma; X) = \sigma$ for every closed $\sigma \subset \overline{\mathbb{D}}$ if and only if $\delta_1(A, X) = 0$, where $\delta_1(A, X)$ is the critical constant for the pair (A, X).

4.1.3. **Microlocalization.** It is clear that lower estimates for spectral hulls and resolvent majorants must depend on the geometry of the subset $\sigma \subset \overline{\mathbb{D}}$ under consideration. For instance, $A(\sigma; X) = \{\text{const}\}$ for every subset $X \subset \mathfrak{M}(A)$ equipped with an analytic structure and for every σ with int $(\sigma) = \emptyset$, see 4.3.1 for examples. Having in mind this last constraint, we restrict ourselves to the case, where $\sigma = \text{clos}\,(\text{int}\,(\sigma))$.

In this case, the behaviour of $C(\lambda, \sigma; X)$ depends on the following microlocal version of the critical constants and the inversion majorants.

Let A be a unital Banach algebra, $X \subset \mathfrak{M}(A)$, and let $0 < \delta \leqslant 1$. A *microlocal upper bound (majorant)* for inverses is defined by

$$c_1^0(\delta, A, X) = \inf_{\epsilon > 0}(\sup\{\|f^{-1}\| : \|f\| \leqslant 1, \, \hat{f}(X) \subset \sigma_\delta, \, \text{diam}\, \hat{f}(X) < \epsilon\}),$$

and the *microlocal critical constant* is defined by $\delta_1^0(A, X) = \inf\{\delta : c_1^0(\delta, A, X) < \infty\}$; here, as before, $\sigma_\delta = \{z \in \mathbb{C} : \delta \leqslant |z| \leqslant 1\}$.

Properties of these microlocal majorants are similar to those of the global ones, that is, to the properties of $\delta_1(A, X)$ and $c_1(\delta, A, X)$ defined in Sections 1 and 2; see [N4] for details.

4.1.4. Horodisc expansions. Denote $D(z,r) = \{\zeta \in \mathbb{C} : |\zeta - z| < r\}, \overline{D}(z,r) = \{\zeta \in \mathbb{C} : |\zeta - z| \leqslant r\}$. In hyperbolic geometry of the unit disc $\mathbb{D}$, the discs $\overline{D}(z, 1 - |z|)$, $z \in \mathbb{D}$ are called *horodiscs*. Given a closed set $\sigma \subset \overline{\mathbb{D}}$, we call the set

$$\text{hor}(\sigma) = \bigcup_{z \in \sigma} (\overline{D}(z, 1 - |z|))$$

the *horodisc expansion* of σ. In order to apply the microlocal version of the critical constants we need one more notation, namely,

$$\text{hor}(\sigma, \delta) = \bigcup_{z \in \sigma} (\overline{D}(z, \frac{\delta}{1-\delta}(1 - |z|))).$$

THEOREM 4.1.5. *Let A be a unital splitting Banach algebra, X be a subset of $\mathfrak{M}(A)$ dominating φ_e, and let $\sigma = \text{clos}(\text{int}(\sigma)) \subset \overline{\mathbb{D}}$. Then*

(i) $\text{hor}(\sigma, \delta_1^0) \subset h(\sigma; A, X) \subset \text{hor}(\sigma, 1/2) = \text{hor}(\sigma)$;

(ii) if $\delta_1^0(A, X) = 1/2$, then $h(\sigma; A, X) = \text{hor}(\sigma)$;

(iii) if $\delta_1^0(A, X) = 1/2$ and $c_1^0(\delta, A, X) = (2\delta - 1)^{-1}$ (see Section 5 for examples), then $h(\sigma; A, X) = \text{hor}(\sigma)$ and $C(\lambda, \sigma; X) = 1/\text{dist}(\lambda, \text{hor}(\sigma))$.

Observe that under the splitting condition we always have $\delta_1^0(A, X) \leqslant \delta_1(A, X) \leqslant 1/2$ and $\text{hor}(\sigma, \delta_1^0) \subset \text{hor}(\sigma, 1/2) = \text{hor}(\sigma)$. The equality $\text{hor}(\sigma, \delta_1^0) = \text{hor}(\sigma)$ holds for all $\sigma \subset \overline{\mathbb{D}}$ if and only if $\delta_1^0(A, X) = 1/2$. For examples of computations and for pictures of the horodisc expansions of various sets see [N4].

4.2. *Spectral hulls and norm-controlled calculi*

The Gelfand theory guarantees the existence of a holomorphic calculus on the spectrum of every element of a Banach algebra A. Namely, if $a \in A$, the function of a

$$f(a) = \frac{1}{2\pi i} \int_{\partial \Omega} f(\lambda)(\lambda e - a)^{-1} \, d\lambda$$

is well defined for every $f \in Hol(\overline{\Omega})$ and every open neighbourhood of the spectrum $\Omega \supset \sigma(a)$ (the Riesz-Dunford calculus). As is shown in Sections 1, 2, and 5, this does not guarantee any estimate of the norm $\|f(a)\|$, even for the simplest functions like $f(z) = 1/z$, and even if we add the normalizing condition $\|a\| \leqslant 1$ to the spectral inclusion $\sigma(a) \subset \Omega$.

The true question is the following: what are the relations between the spectrum $\sigma(a)$ (or a visible part of the spectrum $\hat{a}(X)$) and a domain Ω that guarantee the uniform continuity of the Ω-calculus? In other words, does there exist a compact set $K \subset \Omega$ and a constant $c > 0$ such that $\|f(a)\| \leqslant c \cdot \sup_K |f|$ for every $f \in Hol(\Omega)$ and every $a \in A(\sigma; X) = \{a \in A : \hat{a}(X) \subset \sigma, \|a\| \leqslant 1\}$? We formalize this setting in the following definition, see [N4] for more details.

DEFINITION 4.2.1. Let A and X be as above, and let $\sigma = \overline{\sigma} \subset \overline{\mathbb{D}}$. We say that an open set $\Omega \subset \mathbb{C}$ *X-dominates the set σ*, or *is a norm-controlled calculus domain for σ*, if $\sigma \subset \Omega$ and the calculi $f \longmapsto f(a)$, $f \in Hol(\Omega)$ are well defined and uniformly continuous for all $a \in A(\sigma; X) = \{a \in A : \hat{a}(X) \subset \sigma, \|a\| \leqslant 1\}$.

In fact, this property is equivalent to the existence of a compact set $K \subset \Omega$ and a constant $c(K) = c(K, X) > 0$ such that

$$\|f(a)\| \leqslant c(K)\|f\|_K$$

for every function $f \in Hol(\Omega)$ and every $a \in A(\sigma; X)$. Here $\|f\|_K = \max\{|f(z)| : z \in K\}$.

Obviously, the definition of $(\delta - 1)$-visibility, as well as the definition of the distance controlled resolvent growth, see Subsection 1.1, are special cases of the latter concept. The following theorem shows that X-dominating domains for a given set σ can be described in terms of the spectral hulls $h(\sigma; X)$.

THEOREM 4.2.2. *Let A be a unital Banach algebra, $X \subset \mathfrak{M}(A)$, and let $\sigma \subset \overline{\mathbb{D}}$ and $\Omega \subset \mathbb{C}$ be a closed and an open set, respectively. The following assertions are equivalent.*
(i) Ω is an X-dominating domain for σ.
(ii) $\Omega \supset h(\sigma; X)$.
(iii) There exists $k > 0$ such that $\Omega \supset h(\sigma, k; X)$.

4.3. *Examples of spectral hulls*

Here we describe examples of three different types. As above, we refer to [N4] for details.

The first type pertains to algebras A whose hull operation is trivial in the sense that $h(\sigma; \mathfrak{M}(A)) = \sigma$ for every $\sigma = \overline{\sigma} \subset \overline{\mathbb{D}}$; see 4.3.5 below.

For algebras A of the second type, the same operation $\sigma \longmapsto h(\sigma; \mathfrak{M}(A))$ is also trivial, but in a different way, namely, $h(\sigma; \mathfrak{M}(A)) = \overline{\mathbb{D}}$ for every nonempty $\sigma = \overline{\sigma} \subset \overline{\mathbb{D}}$, even for singletons; see 4.3.4 below.

For the middle type algebras the full spectral hull $h(\sigma; \mathfrak{M}(A))$ essentially depends on σ but is different from it. For instance, this is the case for the measure algebras $A = \mathcal{M}(S)$ on semigroups considered in Section 5 below. In this case we can compute completely the full spectral hulls $h(\sigma; \hat{S}_b)$ and the resolvent majorants $C(\cdot, \sigma; \hat{S}_b)$, see theorem 4.3.2 below.

4.3.1. **Spectral hulls for measure algebras on semigroups.** Anticipating the systematic study of measure algebras (see Section 5 below), here we apply the above theory to compute relevant spectral hulls. Let $\mathcal{M}(S)$ be the convolution algebra of measures on a sub-semigroup of a locally compact abelian group G, and $\hat{S}_b$ be the set of all bounded semi-characters on S which we consider as the visible part of the spectrum of $\mathcal{M}(S)$ (see Section 5 for details). The corresponding Gelfand (Fourier-Laplace) transformation is denoted by $\mathcal{F}$. For example, the analytic Wiener algebra $A = \mathcal{F}M(\mathbb{Z}_+) = \mathcal{F}l_1(\mathbb{Z}_+) = W_+$ corresponds to $S = \mathbb{Z}_+$ and $\hat{S}_b = \mathfrak{M}(A) = \overline{\mathbb{D}}$, with $f \longmapsto f(z)$, $z \in \overline{\mathbb{D}}$, as the Gelfand transformation. Since a

nonconstant holomorphic function is an open mapping, the sets $A(\sigma) = A(\sigma, \overline{\mathbb{D}}) = \{f \in W_+ : \|f\| \leqslant 1, \, f(\overline{\mathbb{D}}) \subset \sigma\}$ and $A(\sigma')$, where $\sigma' = \text{clos}\,(\text{int}\,(\sigma))$, differ from each other by constant functions taking values in $\sigma \backslash \sigma'$. Hence, we can restrict ourselves to the case, where $\sigma = \sigma'$. Supposing $\sigma = \sigma'$, we can easily deduce from theorem 4.1.5 a description of $h(\sigma; \hat{S}_b)$ for $\mathcal{M}(S)$, as well as for all subalgebras of $\mathcal{M}(S)$ considered in Section 5.

THEOREM 4.3.2. *[N4] Let $A = \mathcal{F}M(S)$, or let A be any algebra $A \subset \mathcal{M}(S)$ satisfying the conditions of theorem 4.3.4 below. Let σ be a closed subset of $\mathbb{D}$ such that $\sigma = \text{clos}\,(\text{int}\,(\sigma))$. Then*

(i) $h(\sigma; \hat{S}_b) = \text{hor}\,(\sigma)$;
(ii) $C(\lambda, \sigma; \hat{S}_b) = 1/\text{dist}\,(\lambda, \text{hor}\,(\sigma))$ for $\lambda \in \mathbb{C}$;
(iii) An open set Ω is a norm-controlled calculus domain for σ if and only if $\Omega \supset \text{hor}\,(\sigma)$.

Indeed, assertion *(iii)* is a straightforward consequence of *(i)* and theorem 4.2.2. Assertions *(i)* and *(ii)* are special cases of theorem 4.1.5, because $\delta_1^0(A, \hat{S}_b) = 1/2$ and $\varphi_e(f) = f(0) \in \hat{f}(\hat{S}_b)$ for every $f \in \mathcal{M}(S)$. ∎

4.3.3. **Full spectral hulls equal to $\overline{\mathbb{D}}$.** Here we describe, following [N4], a class of "bad" Banach algebras for which the full spectral hull of *every* spectrum is equal to $\overline{\mathbb{D}}$. To this end, we need a bit of model operators. All properties claimed below can be found in [N1].

Let Θ be a singular inner function in $\mathbb{D}$, and let

$$\Theta = \Theta_\mu = exp(\int_{\mathbb{T}} \frac{z+\zeta}{z-\zeta}\, d\mu(\zeta)), \quad z \in \mathbb{D},$$

be its canonical integral representation, where μ is a positive measure on $\mathbb{T}$ singular with respect to the Lebesgue measure m. Further, let $K_\Theta = H^2 \ominus \Theta H^2$ be the orthogonal complement of the corresponding z-invariant subspace ΘH^2 of the Hardy space H^2. The compression

$$f \longmapsto M_\Theta f = P_\Theta z f, \quad f \in K_\Theta,$$

of the multiplication operator is called a model operator; P_Θ stands for the orthogonal projection to K_Θ. The Sz.-Nagy-Foias model theory tells that every completely non unitary contraction T with $rank(1 - T^*T) = 1$ and $\sigma(T) \neq \overline{\mathbb{D}}$ is unitarily equivalent to an operator M_Θ, see [SzNF].

Now, we define the algebra $A = A_\Theta$ as *the norm closure of polynomials in M_Θ.* It is an exercise to show that always $\mathfrak{M}(A_\Theta) = \sigma(M_\Theta) = \text{supp}\,(\mu) \subset \mathbb{T}$ and $\|M_\Theta\| = 1$. It is worth mentioning that, for $\Theta = \Theta_\mu$, where $\mu = \delta_1$, there exists a unitary operator $U : K_\Theta \longrightarrow L_2(0, 1)$ such that the algebra $U A_\Theta U^{-1}$ is the norm closure of polynomials in the integration operator $Jf(x) = \int_0^x f(t)\, dt$, $x \in (0, 1)$, on the space $L^2(0, 1)$.

THEOREM 4.3.4. *[N4] Let $A = A_\Theta$ be the above Banach algebra corresponding to a singular inner function $\Theta = \Theta_\mu$ with the representing measure μ whose closed support is a* set of zero Lebesgue measure, $m(\text{supp}\,(\mu)) = 0$. Then for every closed subset $\sigma \subset \overline{\mathbb{D}}, \sigma \neq \emptyset$, we have $h(\sigma; \mathfrak{M}(A)) = \overline{\mathbb{D}}$.

4.3.5. Spectrally closed sets. It is natural to call a subset $\sigma \subset \mathbb{C}$ (A, X)-*spectrally closed* (or A-*spectrally closed* in the case where $X = \mathfrak{M}(A)$) if $h(\sigma; A, X) = \sigma$. Property 4.1.2 tells us that every closed subset $\sigma \subset \overline{\mathbb{D}}$ is (A, X)-closed if and only if $\delta_1(A, X) = 0$. In Section 7 we give an account of known results about weighted Beurling-Sobolev algebras satisfying this property.

Part II. Convolution Algebras

5. Unweighted Convolution Algebras on Groups and Semigroups

Let G be an LCA group and $\mathcal{M}(G)$ be the convolution algebra of all complex Borel measures on G endowed with the standard variation norm $\|\mu\| = \text{Var}\,(\mu)$. The Fourier transforms $\mathcal{F}\mu = \hat{\mu}$,

$$\mathcal{F}\mu(\gamma) = \hat{\mu}(\gamma) = \int_G (-x, \gamma)\, d\mu(x), \quad \gamma \in \hat{G},$$

form an algebra of functions on the dual group $\hat{G}$. We denote it by $\mathcal{F}M(G)$ and endow it with the range norm $\|\mathcal{F}\mu\| = \|\mu\|$. Clearly, $\mathcal{F}(\mu * \nu) = (\mathcal{F}\mu) \cdot (\mathcal{F}\nu)$ for all $\mu, \nu \in \mathcal{M}(G)$, and so the Banach algebras $\mathcal{M}(G)$ and $\mathcal{F}M(G)$ are isometrically isomorphic. The term "the measure algebra of G" will be referred to the both.

Since the mapping $\mu \longmapsto \hat{\mu}(\gamma)$ is a complex homeomorphism of $\mathcal{M}(G)$, we can injectively embed $\hat{G}$ into $\mathfrak{M}(\mathcal{M}(G))$, and regard $\hat{G}$ as the *visible spectrum* of $\mathcal{M}(G)$ in the sense of the Introduction. In particular, $\text{clos}\,\hat{\mu}(\hat{G}) \subset \sigma(\mu)$ for any $\mu \in \mathcal{M}(G)$. The Wiener-Pitt-Sreider theorem implies (see the Introduction) that $\mathfrak{M}(\mathcal{M}(G)) = \text{clos}\,\hat{G}$ if and only if G is a discrete group; in fact, in this case we simply have $\mathfrak{M}(\mathcal{M}(G)) = \hat{G}$.

However, even in the latter case, the problem of the norm-controlled inversion, as described in previous sections, is still of interest. Moreover, from the *quantitative point of view*, we cannot distinguish any advantage of discrete groups, for which the spectrum $\hat{G} = \mathfrak{M}(\mathcal{M}(G))$ is completely visible (in the sense of Definition 1.2.1), as compared with the general LCA groups, for which $X = \hat{G} \neq \mathfrak{M}(\mathcal{M}(G))$, and the spectrum is even 1-invisible. This is an essential distinction between the concepts of n-visibility (without any norm control) and $(\delta - n)$-visibility. Speaking informally, the 1-visibility of $\mathfrak{M}(\mathcal{M}(G))$ for discrete groups, guaranteed by the classical Wiener-Lévy theorem, is illusory because it does not endure quantitative specifications by $(\delta - 1)$-estimates of inverses.

5.1. *An upper estimate for the measure algebra on a group*

The constants $c_n(\delta, \mathcal{M}(G), \hat{G})$, defined in 1.2.3, can be estimated by using theorem 2.2.7, because the algebra $\mathcal{M}(G)$ is $\hat{G}$-symmetric in the sense of 2.2.1. The main condition *(iii)* of theorem 2.2.7 can be checked with the help of lemma 2.2.6. Namely, one can prove that $\varphi_e \in \text{conv}\,(\delta_\gamma : \gamma \in \hat{G})$ by means of "triangular" positive semidefinite kernels k_α on G, such that $\hat{k}_\alpha(\gamma) \geqslant 0$ for $\gamma \in \hat{G}$, $0 \leqslant k_\alpha(x) \leqslant 1 = k_\alpha(0)$ for $x \in G$, and $k_\alpha(x) = 0$ outside of a neighborhood V_α of the origin such that $\bigcap_\alpha V_\alpha = \{0\}$. Indeed, setting

$$\varphi_\alpha(\mu) = \int_{\hat{G}} \hat{k}_\alpha(\gamma) \hat{\mu}(\gamma)\, dm_{\hat{G}}(\gamma),$$

we have $\varphi_\alpha \in \text{conv}\,(\delta_\gamma : \gamma \in \hat{G})$ and $\lim_\alpha \varphi_\alpha(\mu) = \lim_\alpha \int_G k_\alpha\, d\mu = \mu(\{0\}) = \varphi_e(\mu)$ for all $\mu \in \mathcal{M}(G)$. This proves the following theorem.

THEOREM 5.1.1. *For every LCA group G, the spectrum of $\mathcal{M}(G)$ is completely δ-visible for every δ satisfying $\frac{1}{\sqrt{2}} < \delta \leqslant 1$, and, consequently, $\delta_n(\mathcal{M}(G), \hat{G}) \leqslant \frac{1}{\sqrt{2}}$ for all $n \geqslant 1$. Moreover,*

$$(5.1) \qquad\qquad c_n(\delta, \mathcal{M}(G), \hat{G}) \leqslant \frac{1}{2\delta^2 - 1}$$

for $1/\sqrt{2} < \delta \leqslant 1$ and for all $n \geqslant 1$.

5.2. *The measure algebra on a semigroup*

Here we consider the convolution measure algebras $\mathcal{M}(S)$ on semigroups S. Since the language of the semigroup theory is not canonically fixed, we define exactly which objects we are dealing with. For general facts of harmonic analysis on semigroups we refer to [T].

5.2.1. **Definitions.** By a *semigroup* S we mean the following.
(i) *S is a Borel subset of a LCA group G such that $x, y \in S \Rightarrow x + y \in S$,*
(ii) *$0 \in S$.*
A *bounded character* (also called semicharacter) on S is a bounded continuous function $\gamma : S \longrightarrow \mathbb{C}$ such that $\gamma(0) = 1$ and $\gamma(x + y) = \gamma(x)\gamma(y)$ for all $x, y \in S$. It is clear that every such function is bounded by 1: $|\gamma(x)| \leqslant 1$, $x \in S$. The *set of all bounded characters* of S is denoted by $\hat{S}_b$. Obviously, $\hat{G} \subset \hat{S}_b$, in the sense that the restriction $\gamma|S$ of a character $\gamma \in \hat{G}$ is a bounded character of S. In what follows, we assume that the following *separation property* holds.
(iii) *For every $x \in S$, $x \neq 0$, there exists a bounded character $\gamma \in \hat{S}_b$ such that $|\gamma(x)| < 1$.*

In particular, $S \cap (-S) = \{0\}$ if a semigroup S satisfies condition *(iii)*. Let $\mathcal{M}(S) = \{\mu \in \mathcal{M}(G) : \mu|(G \backslash S) \equiv 0\}$ be the subspace of $\mathcal{M}(G)$ consisting of all measures supported by S. An immediate verification shows that $\mathcal{M}(S)$ is a (closed) subalgebra of $\mathcal{M}(G)$. Now, we define the *Fourier-Laplace transformation* on $\hat{S}_b$ setting $\mathcal{L}\mu(\gamma) = \hat{\mu}(\gamma) = \int_S \gamma(x)\, d\mu(x)$ for $\mu \in \mathcal{M}(S)$ and $\gamma \in \hat{S}_b$. Clearly, the functional

$$(5.2) \qquad \mu \longmapsto \mathcal{L}\mu(\gamma), \quad \mu \in \mathcal{M}(S),$$

is a norm continuous homomorphism of the algebra $\mathcal{M}(S)$ for every $\gamma \in \hat{S}_b$. Moreover, $\mathcal{L}\mu(\gamma) = \mathcal{F}\mu(-\gamma)$ for all $\gamma \in \hat{G}$. Hence, it is natural to consider the space of bounded characters $\hat{S}_b$ as *the visible part* of $\mathfrak{M}(\mathcal{M}(S))$, and write $\hat{S}_b \subset \mathfrak{M}(\mathcal{M}(S))$.

The following theorem can be proved in a similar way to 5.1.1. However, the result is two times better as compared with the algebras $\mathcal{M}(G)$.

THEOREM 5.2.2. *Let S be a semigroup satisfying conditions (i)–(iii). Then,*

$$(5.3) \qquad \delta_1(\mathcal{M}(S), \hat{S}_b) \leqslant \frac{1}{2},$$

and

$$(5.4) \qquad c_1(\delta, \mathcal{M}(S), \hat{S}_b) \leqslant \frac{1}{2\delta - 1}$$

for all δ, $\frac{1}{2} < \delta \leqslant 1$.

5.3. *Examples and comments*

5.3.1. **Symmetric and analytic Wiener algebras.** Let $G = \mathbb{Z}$ be the additive group of integers, and let $S = \mathbb{Z}_+ = \{n \in \mathbb{Z} : n \geqslant 0\}$. Then $\mathcal{M}(\mathbb{Z}) = l^1(\mathbb{Z})$, $\mathcal{M}(\mathbb{Z}_+) = l^1(\mathbb{Z}_+)$, and $W = FM(Z) = \{\hat{\mu} = \sum_{n \in Z} \mu(n)\zeta^n : \mu \in l^1(Z)\}$ is the Wiener algebra of absolutely converging Fourier series on the circle group $\hat{Z} = \mathbb{T} = \{\zeta \in \mathbb{C} : |\zeta| = 1\}$. The bounded characters of $\mathbb{Z}_+$ fill in the closed unit disc $\hat{S}_b = \overline{\mathbb{D}} = \{z \in \mathbb{C} : |z| \leqslant 1\}$. The corresponding Fourier-Laplace transformation is $\mu \longmapsto \hat{\mu}(z) = \sum_{n \geqslant 0} \mu(n)z^n$, and $W_+ = LM(Z_+) = \{\hat{\mu} = \sum_{n \geqslant 0} \mu(n)z^n : \mu \in l^1(Z_+)\}$ is the analytic Wiener algebra on $\overline{\mathbb{D}}$. Preceding theorems tell that $\|f^{-1}\|_W \leqslant (2\delta^2 - 1)^{-1}$ if $f \in W$ and $1/\sqrt{2} < \delta \leqslant |f(\zeta)| \leqslant \|f\|_W \leqslant 1$ for $|\zeta| = 1$, and $\|f^{-1}\| \leqslant (2\delta - 1)^{-1}$ if $f \in W_+$ and $1/\sqrt{2} < \delta \leqslant |f(z)| \leqslant \|f\|_{W_+} \leqslant 1$ for $|z| \leqslant 1$.

5.3.2. **Several variables, continuous versions, and cones in $\mathbb{Z}^N$ and $\mathbb{R}^N$.** The same estimates of inverses as in 5.3.1 hold for the multivariate Wiener algebra on the torus $\mathbb{T}^N$,

$$W = \mathcal{F}M(\mathbb{Z}^N) = \{\hat{\mu} = \sum_{n \in \mathbb{Z}^N} \mu(n)\zeta^n : \mu \in l^1(\mathbb{Z}^N)\},$$

for the analytic Wiener algebra on the polydisc $\overline{\mathbb{D}}^N$,

$$W_+ = \mathcal{L}M(\mathbb{Z}_+^N) = \{\hat{\mu} = \sum_{n \in \mathbb{Z}_+^N} \mu(n)z^n : \mu \in l^1(\mathbb{Z}_+^N)\},$$

and for continuous versions of the Wiener algebras. The latter correspond to $G = \mathbb{R}$, $S = \mathbb{R}_+ = \{x \in \mathbb{R} : x \geqslant 0\}$, and—in several variables—to $G = \mathbb{R}^N$, $S = \mathbb{R}_+^N$. Here $\hat{G} = \mathbb{R}$, $\hat{S}_b = \overline{\mathbb{C}}_+ = \{z \in \mathbb{C} : \text{Im}(z) \geqslant 0\}$ - the closed upper half-plane, and, for several variables, $\hat{G} = \mathbb{R}^N$, $\hat{S}_b = \overline{\mathbb{C}}_+^N$.

Instead of the cones $\mathbb{R}_+^N \subset \mathbb{R}^N$ and $\mathbb{Z}_+^N = \mathbb{R}_+^N \cap \mathbb{Z}^N$, we may consider an arbitrary subsemigroup S of $\mathbb{R}^N$ containing 0 and such that $S\backslash\{0\}$ is contained in an *open* halfspace (this requirement guarantees the separation property *(iii)* of Subsection 5.2.1). Usually, S is a convex cone satisfying the latter property, or, in its discrete version, the intersection of such a cone with $\mathbb{Z}^N$. For the continuous version, the semicharacters are $x \longmapsto e^{i(x \cdot z)}$ with $z \in \hat{S}_b$, where $\hat{S}_b = \{z \in \mathbb{C}^N : \text{Im}(x \cdot z) \geqslant 0 \text{ for all } x \in S\}$ and $x \cdot z = \sum_{k=1}^N x_k z_k$. The Fourier-Laplace transformation is again the classical one. For instance, taking $S = \{(x_1, x_2) \in \mathbb{R}^2 : x_1 \geqslant 0, -kx_1 \leqslant x_2 \leqslant kx_1\}$, where $k > 0$, we have $\hat{S}_b = \{(z_1, z_2) \in \mathbb{C}^2 : k|\text{Im}(z_2)| \leqslant \text{Im}(z_1)\}$.

Another example is the halfspace $S = \{(x_1, x_2) \in \mathbb{R}^2 : x_1 > 0\} \cup \{0\}$, with the corresponding dual set of characters $\hat{S}_b = \{(z_1, z_2) \in \mathbb{C}^2 : \Im(z_2) = 0, \text{Im}(z_1) \geqslant 0\} = \mathbb{C}_+ \times \mathbb{R}$.

5.4. *Some subalgebras of $\mathcal{M}(G)$ and $\mathcal{M}(S)$*

Here we briefly consider two classical subalgebras of $\mathcal{M}(G)$ (or $\mathcal{M}(S)$), namely, the algebras of absolutely continuous, respectively, discrete measures on G (or on S).

5.4.1. The group algebra $L^1(G)$. Let G be an LCA group, and $L^1(G)$ the convolution group algebra on G, which becomes a unital algebra if we adjoin the Dirac point mass at the origin $e = \delta_0$ (if G is not discrete). Since the Riemann-Lebesgue lemma implies that $\varphi_e(\mu) = \lambda = \lim_{\gamma \to \infty} \hat{\mu}(\gamma)$ for $\mu = \lambda e + f dm \in (L^1(G) + \mathbb{C} \cdot e)$, we can directly apply lemma 2.2.3 and obtain the following improvement of theorem 5.1.1: for any nondiscrete LCA group G, one has $\delta_1(L^1(G) + \mathbb{C} \cdot e, \hat{G}) \leqslant 1/2$; moreover, $c_1(\delta, L^1(G) + \mathbb{C} \cdot e, \hat{G}) \leqslant (2\delta - 1)^{-1}$ for all $1/2 < \delta \leqslant 1$. (For $n \geqslant 2$ we still have estimates (5.1)). In Subsection 5.5 we show that this new estimate for $L^1(G) + \mathbb{C} \cdot e$ is sharp.

5.4.2. The algebra of almost periodic functions $\mathcal{F}M_d(G)$, and the algebra of Dirichlet series $\mathcal{L}M_d(S)$. Let $M_d(G)$ be the algebra of discrete measures on an LCA group G, and $M_d(S)$ be its subalgebra of measures supported by a subsemigroup S satisfying hypotheses *(i)–(iii)* of Subsection 5.2.1.

The algebra $\mathcal{F}M_d(G)$ of Fourier transforms of discrete measures is the algebra of almost periodic functions with absolutely convergent Fourier series. Theorem 5.1.1 works for this algebra too (the visible spectrum is, of course, $X = \hat{G}$). As for the entire algebra $\mathcal{M}(G)$, we will see in Subsection 5.5 below that $\delta_1(\mathcal{F}M_d(G), \hat{G}) \geqslant 1/2$. It should be mentioned that, as before, impossibility of the norm control of inverses for small δ, i.e., the fact that $c_1(\delta, M_d(G), \hat{G}) = \infty$ for $0 < \delta \leqslant 1/2$, is not related to the evident fact that the spectrum

$\mathfrak{M}(\mathcal{M}_d(G)) = (\hat{G})^-$ (the Bohr compactum) is much larger than $X = \hat{G}$. Indeed, we show that even staying on the Bohr compactum we still have $\delta_1(\mathcal{M}_d(G), (\hat{G})^-) \geqslant 1/2$, see below.

Similarly, $\mathcal{F}M_d(S)$ is the algebra of absolutely convergent Dirichlet series

$$f(\gamma) = \sum_{x \in S} \mu(\{x\})(x, \gamma), \gamma \in \hat{S}_b$$

with $\sum_{x \in S} |\mu(\{x\})| < \infty$. The case of the classical Dirichlet series corresponds to the case $S = \mathbb{R}_+$, $\hat{S}_b = \{z \in \mathbb{C} : Re(z) \geqslant 0\}$ with the pairing $(x, z) \longmapsto e^{-zx}$. Like $\mathcal{M}_d(G)$, the algebra $\mathcal{M}_d(S)$ is an inversion stable subalgebra of $\mathcal{M}(S)$. Hence, Theorem 5.2.2 is still valid for this algebra as well.

5.5. A lower estimate for the measure algebras

In this section, we show how to get a common lower estimate of $c_1(\delta, A, X)$ for the measure algebras of groups and semigroups, and for their subalgebras considered above. In particular, we show that the critical constant δ_1 is greater than or equal to $\frac{1}{2}$ for every infinite LCA group and for the most part of semigroups.

In fact, we dispose two approaches to lower estimates. The first one is based directly on the existence and properties of measures carried by independent Cantor sets; in what follows a special kind of such measures (Sreider measures) is used. The corresponding theory is surely one of the most subtle chapters of measure algebra theory, see [GRS], [Ru1], [GMG]. The second method is inspired by H. Shapiro's example [Sh1]. In our setting, it depends on some improvements of the technique of exponentials norm behaviour $\|e^{it\mu}\|_{\mathcal{M}(G)}$ for $t > 0$. This technique is well known to be the ground of the theory of functions operating on an algebra, see [Ru1], [GMG]. The first way is faster, and, following [N4], we use it in this section. For the second one, more explicit, we refer to [ENZ] and [Sh2].

We start by recalling some classical facts on measure algebras.

5.5.1. Sreider measures. Let G be a nondiscrete LCA group. It is known that there exists a positive measure $\mu \in \mathcal{M}(G)$ such that $\hat{\mu}(\hat{G}) \subset [-1, 1]$ and $\hat{\mu}(\mathfrak{M}) = \overline{\mathbb{D}}$, $\|\mu\| = 1$; see Sreider [Sr] for $G = \mathbb{R}$, and [Ru1], Theorem 5.3.4, for the general case and history. We call such μ's Sreider measures. Moreover, it is known that there exist continuous Sreider measures μ such that the convolution powers and their translates, μ^k, $\delta_x * \mu^k$, are all mutually singular, see [Ru1], [GMG]. It is worth mentioning that $0 \in \hat{\mu}(\hat{G})$ for all known examples of such measures.

5.5.2. Bohr compactification. For an LCA group G, we denote by $\hat{G}_d$ the dual group $\hat{G}$ endowed with the discrete topology, and set $\overline{G} = (\hat{G}_d)^{\hat{}}$. The compact group $\overline{G}$ is called the Bohr compactification of G; in fact, G is a dense subset of $\overline{G}$, and $\overline{G} = \mathfrak{M}(\mathcal{M}(\hat{G}_d)) = \mathfrak{M}(\mathcal{M}_d(\hat{G}))$.

5.5.3. **How to get lower estimates on groups.** Here we explain a method to prove lower estimates for critical constants and norm-controlling constants, making use of Sreider measures and Bohr compactification. One can call the method "a walk to the Bohr compactum". In particular, we get the needed estimates for all three algebras on groups we considered above, namely, for $\mathcal{M}(G)$, $L^1(G) + \mathbb{C} \cdot \delta_0$, and $M_d(G)$.

The method consists of the following steps (see [N4] for more details):

1) staying on an infinite group G, we lift ourselves up to the Bohr group $\overline{G} \supset G$;

2) being nondiscrete, $\overline{G}$ carries a Sreider measure, say ν, whose polynomials give the required lower estimate, see Lemma 5.5.4 below;

3) using almost periodic approximations of $\mathcal{F}\nu$, we obtain the same lower estimate on G (via the classical Bochner criterion for the membership in $\mathcal{F}M(G)$, see [Ru1], Theorem 1.9.1).

Of course, for a nondiscrete group G, steps 1) and 3) are not necessary. To accomplish step 3) for general G, we need a kind of a strengthened weak topology, precisely, a version of Beurling's narrow topology. Namely, we say that a net $(\mu_i)_{i \in I} \subset \mathcal{M}(G)$ is $\hat{G}$-*convergent* to a measure $\mu \in \mathcal{M}(G)$ if $\underline{\lim}_{i \in I} \|\mu_i\| \leqslant \|\mu\|$ and $\lim_{i \in I} \hat{\mu}_i(\gamma) = \hat{\mu}(\gamma)$ for $\gamma \in \hat{G}$. A set $A \subset \mathcal{M}(G)$ is said to be $\hat{G}$-*dense* in a set $B \subset \mathcal{M}(G)$ if every $\mu \in B$ is a $\hat{G}$-limit of a net of measures belonging to A.

The following lemma is a straightforward consequence of the spectral mapping theorem.

LEMMA 5.5.4. *Let ν be a Sreider measure on a nondiscrete LCA group G, and let $\mu = \delta e + (1 - \delta)\nu^2$, where $1/2 < \delta \leqslant 1$, and $e = \delta_0$ stands for the unit of $\mathcal{M}(G)$. Then $\|\mu\| = 1$, $\hat{\mu}(\hat{G}) \subset [\delta, 1]$ and $\|\mu^{-1}\| = (2\delta - 1)^{-1}$.*

THEOREM 5.5.5. *Let G be an infinite LCA group, and let A be a unital subalgebra of $\mathcal{M}(G)$ (not necessarily closed) satisfying the following conditions: (i) A is $\hat{G}$-symmetric; (ii) A is $\hat{G}$-dense in $\mathcal{M}(G)$. Further, given a number δ, $1/2 < \delta \leqslant 1$, let $A([\delta, 1]) = \{\alpha \in A : \|\alpha\| \leqslant 1, \hat{\alpha}(\hat{G}) \subset [\delta, 1]\}$. Then*

$$\sup_{\alpha \in A([\delta,1])} \|\alpha^{-1}\| \geqslant (2\delta - 1)^{-1}.$$

In particular, $\delta_1(A, \hat{G}) \geqslant 1/2$, and $c_1(\delta, A, \hat{G}) \geqslant (2\delta - 1)^{-1}$ for $\frac{1}{2} < \delta \leqslant 1$.

The algebras $A = \mathcal{M}(G)$, $A = L^1(G) + \mathbb{C} \cdot e$, and $A = M_d(G)$ satisfy conditions (i) and (ii), and hence the conclusions hold for these algebras.

5.5.6. **How to get a lower estimate on semigroups.** Here we describe a semigroup counterpart of theorem 5.5.5 by using the same method of moving to the Bohr compactum. However, the construction of measures giving the maximum to $\|\mu^{-1}\|$ when $\inf |\hat{\mu}|$ is fixed is necessarily different. Indeed, the previous construction is based on the $\hat{G}-$ symmetry of $\mathcal{M}(G)$ and on positive semidefiniteness of the corresponding Fourier transforms. Both reasons fail for semigroups. Instead, we are using a more complicated, but general construction of theorem 2.3.1. By the way, this gives another proof to theorem 5.5.5; namely, theorem 2.3.1 and lemma 5.5.7 below imply 5.5.5.

In order to realize the technique of narrow approximations on a semigroup instead of the entire group, we restrict ourselves to a class of semigroups S described in Subsection 5.2.1. This class contains all frequently used semigroups, in particular, all examples of Subsection 5.3 above.

Until the end of this Section, we denote by $\mathcal{A}$ a subalgebra of $\mathcal{M}(G)$, and by A a subalgebra of $\mathcal{M}(S)$.

LEMMA 5.5.7. *Let G be an infinite LCA group, and let $\mathcal{A}$ be a subalgebra of $\mathcal{M}(G)$ which is $\hat{G}$−symmetric and $\hat{G}$−dense in $\mathcal{M}(G)$ (but not necessarily closed and/or unital). Then, $\mathcal{A}$ satisfies (2.3) with $X = \hat{G}$.*

Now, we describe a class of semigroups S and a class of subalgebras $A \subset \mathcal{M}(S)$ where the method based on theorem 2.3.1 works; see [N4] for more details.

5.5.8. **Absorbing semigroups and S-subalgebras.** Let S be a semigroup embedded into an LCA group G, $S \subset G$, and satisfying hypotheses *(i)–(iii)* of Subsection 5.5.1. We say that G satisfies the *absorbtion condition* if the following is true.

(iv) For every compact set $K \subset G$ there exists an element $x \in G$ such that $x + K \subset S$.

If S is absorbing, there exists an element $x \in S$ such that $x + K \subset S$ (consider $K \cup \{0\}$), and, therefore, *(iv)* implies that S generates G in the sense $G = S - S$. On the other hand, if S is generating and $S \backslash \{0\}$ is open, or S contains an open generating part, then S is absorbing. For instance, this is the case for all examples of Subsection 5.3.

Now, we describe the class of subalgebras of $\mathcal{M}(S)$ we are working with. Namely, let S be a semigroup, $S \subset G$. We say that A is *an S-subalgebra of $\mathcal{M}(S)$* if A is *a subalgebra of $\mathcal{M}(S)$* containing a "small" subalgebra of the form $P_S \mathcal{A} + \mathbb{C} \cdot e$, where $\mathcal{A} \subset M(G)$ stands for a subalgebra of $\mathcal{M}(G)$ verifying the following conditions (compare with the conditions of theorem 5.5.5):

(i) $\mathcal{A}$ is $\hat{G}$−symmetric;

(ii) $\mathcal{A}$ is $\hat{G}$−dense in $\mathcal{M}(G)$;

(iii) $\mathcal{A}$ is S-invariant, that is, $\delta_x * \mathcal{A} \subset \mathcal{A}$ for $x \in S$;

(iv) $\hat{G}$ is a boundary for $\mathfrak{M}(\mathcal{A})$, that is, $\lim_n \|\mu^n\|^{1/n} = \sup_{\hat{G}} |\hat{\mu}|$ for every $\mu \in \mathcal{A}$.

Observe that $\mathcal{A}$ is not assumed to be unital. The standard subalgebras $A = \mathcal{M}_d(S)$ and $A = L^1(S) + \mathbb{C} \cdot e$, with obvious group counterparts $\mathcal{A} = \mathcal{M}_d(G)$ and $\mathcal{A} = L^1(G)$, respectively, are S-subalgebras of $\mathcal{M}(S)$. Hence, the same is true of any bigger subalgebra. For instance, $\mathcal{M}_d(S) + L^1(S)$, and $\mathcal{M}(S)$ itself, are S-subalgebras. Another example is the algebra $\mathcal{M}_f(S)$ of finitely supported measures on S.

Now, we are ready to state the following lower estimate for semigroups.

THEOREM 5.5.9. *Let S be a semigroup satisfying conditions (i)–(iii) of Subsection 5.2.1 and the absorbtion condition (iv). Let A be an S-subalgebra of $\mathcal{M}(S)$ (not necessarily closed). Then $\delta_1(A, \hat{S}_b) \geqslant 1/2$ and $c_1(\delta, A, \hat{S}_b) \geqslant (2\delta - 1)^{-1}$ for all δ, $1/2 < \delta < 1$.*

For the proof, we check (2.3) with $X = \hat{S}_b$, see [N4], theorem 3.3.7.

6. Some Finite Groups and Semigroups

In this section we briefly consider the measure algebras $\mathcal{M}$ on finite groups and on finite semigroups. In general, exact computations of the majorants $c_n(\delta, \mathcal{M}, X)$ for these groups and semigroups are, probably, even more complicated than in the infinite case. This is why we mainly restrict ourselves to two examples: to cyclic groups $C_d = \mathbb{Z}/d\mathbb{Z}$ of order $d \geqslant 1$, to nilpotent semigroups $Z_d = \mathbb{Z}_+/(d + \mathbb{Z}_+)$ of order d. In a sense, the groups C_d "exhaust" the group $\mathbb{Z}$ and the semigroups Z_d "exhaust" $\mathbb{Z}_+$. Essentially, we consider the asymptotics of $c_n(\delta, \mathcal{M}(C_d), \hat{C}_d)$ and $c_n(\delta, \mathcal{M}(Z_d), \hat{Z}_d)$ as $d \longrightarrow \infty$.

6.1. Finite groups

6.1.1. Preliminaries. Let G be a finite group written additively, and m_G the invariant (Haar) measure normalized by $m_G(\{x\}) = 1$ for every $x \in G$. Clearly, the space $\mathcal{M}(G) = L^1(G) = l^1(G)$ is a convolution Banach algebra with the unit $e = \delta_0$. All complex homomorphisms are given by the Fourier (Gelfand) transformation, $f \longmapsto \mathcal{F}f(\gamma) = \hat{f}(\gamma) = \sum_{x \in G} f(x)(-x, \gamma)$, $\gamma \in \hat{G}$, where $\hat{G}$ is the dual group of unimodular characters written multiplicatively. Therefore, $\mathfrak{M}(\mathcal{M}(G)) = \hat{G}$. The Haar measure $m = m_{\hat{G}}$ is normalized to have total mass 1, so that the Fourier transformation $\mathcal{F}$ is a unitary operator from $L^2(G)$ to $L^2(\hat{G})$. It is easy to see that we can regard the dual group $\hat{G}$ as the visible spectrum for any convolution algebra on G. That is, in our previous notation, we set $X = \hat{G}$.

Using equivalence of every two norms on a finite dimensional vector space, one can show that for finite groups the majorants $c_1(\delta, A, \hat{G})$ always have the linear growth rate as $\delta \longrightarrow 0$. For example, $\|\mu^{-1}\| \leqslant k(A)/\delta$ for every $\mu \in A$ satisfying $|\hat{\mu}(\gamma)| \geqslant \delta$ for all $\gamma \in \hat{G}$. Here $k(A)$ is a constant depending only on a convolution algebra A on a finite group G, and $k(\mathcal{M}(G)) \leqslant (\operatorname{card}(G))^{1/2}$.

We can say more for the special case of cyclic groups C_d.

6.2. The cyclic group C_d

Let $G = C_d = \mathbb{Z}/d\mathbb{Z} = \{0, 1, \ldots, d - 1\}$ be the cyclic group endowed with the quotient composition. The dual group $\hat{C}_d$ is the group of d-th roots of unity $\hat{C}_d = \{\zeta_k = \zeta^k : 0 \leqslant k \leqslant d - 1\}$, where $\zeta = \zeta(d) = e^{2\pi i/d}$. The Fourier transform of an element $f \in \mathcal{M}(C_d)$ is $Ff(\zeta_k) = \hat{f}(\zeta_k) = \sum_{0 \leqslant s < d} f(s)\zeta_k^s$, $\zeta_k \in \hat{C}_d$, and the norm is $\|f\| = \sum_{0 \leqslant s < d} |f(s)|$. Let $e_k = \chi_{\{k\}}$ be the basic functions on C_d. The convolution on C_d, which we denote by $\circ$, follows the rule $e_r \circ e_s = e_t$, where $t \in C_d$, $t \equiv (r + s) \bmod (d)$. Since we cannot compute $c_n(\delta, \mathcal{M}(C_d), \hat{C}_d)$, we consider the behaviour of the upper bound $c_n(\delta, \{C_d\})$ defined by

$$(6.1) \qquad c_n(\delta, \{C_d\}) = \sup\{c_n(\delta, \mathcal{M}(C_d), \hat{C}_d) : d \geqslant 1\}, \quad 0 < \delta \leqslant 1.$$

The following theorem shows that, in a sense, the algebras $\mathcal{M}(C_d)$ approximate the algebra $\mathcal{M}(\mathbb{Z}) = l^1(\mathbb{Z})$ as $d \longrightarrow \infty$.

THEOREM 6.2.1. *(i) $c_n(\delta, \mathcal{M}(C_d), \hat{C}_d) \leqslant d^{1/2} \cdot \min(\delta^{-2}, \delta^{-1}n^{1/2})$ for all $0 < \delta \leqslant 1$ and $n \geqslant 1$.*

(ii) $c_n(\delta, \mathcal{M}(C_d), \hat{C}_d) \leqslant (2\delta^2 - 1)^{-1}$ for all $1/\sqrt{2} < \delta \leqslant 1$ and $n \geqslant 1$.

(iii) $c_n(\delta, \{C_d\}) \geqslant c_n(\delta, \mathcal{M}(\mathbb{Z}), \mathbb{T})$ for all $0 < \delta \leqslant 1$, where $c_n(\delta, \{C_d\})$ is defined in (6.1). In particular, $c_1(\delta, \{C_d\}) = \infty$ for $0 < \delta \leqslant 1/2$, and $c_1(\delta, \{C_d\}) \geqslant (2\delta - 1)^{-1}$ for $1/2 < \delta \leqslant 1$.

6.3. *The nilpotent semigroup Z_d*

The semigroup $Z_d = \mathbb{Z}_+/(d + \mathbb{Z}_+)$ is defined as the set $Z_d = \{0, 1, \ldots, d-1, d\}$ endowed with the operation $(s, t) \longmapsto \min(s + t, d)$, and with the measure $m(\{s\}) = 1$ for $0 \leqslant s < d$ and $m(\{d\}) = 0$. Therefore, on the basic functions e_s, $e_s(t) = \delta_{s,t}$ (the Kronecker delta), the convolution is defined by the formula

$$e_s \circ e_t = e_{s+t} \quad \text{for} \quad s + t < d, \quad \text{and} \quad e_s \circ e_t = 0 \quad \text{for} \quad s + t \geqslant d.$$

The space $\mathcal{M}(Z_d) = L^1(Z_d, m)$ of all measures (functions) on Z_d endowed with this convolution and with the usual L^1 norm is a unital d-nilpotent Banach algebra. Namely, the algebra $\mathcal{M}(Z_d)$ has a generator e_1 such that $e_1^d = 0$. Hence, the only character on Z_d is the trivial one: $0 \longmapsto 1$ and $s \longmapsto 0$ for $s > 0$. We write $\hat{Z}_d = \{0\}$, and $\mathfrak{M}(\mathcal{M}(Z_d)) = \{0\}$ with the only homomorphism on $\mathcal{M}(Z_d)$, namely $\mu = (\mu(s))_{0 \leqslant s < d} \longmapsto \mu(0)$.

Theorem 6.3.1 below gives the exact value of the majorant $c_1(\delta, \mathcal{M}(Z_d), \{0\})$ and shows that the algebra $\mathcal{M}(\mathbb{Z}_+) = l^1(\mathbb{Z}_+)$ is, in a sense, the limit of $\mathcal{M}(Z_d)$ as $d \longrightarrow \infty$.

THEOREM 6.3.1. *For all $0 < \delta \leqslant 1$, $c_1(\delta, M(Z_d), \{0\}) = \delta^{-1}\sum_{0 \leqslant k < d}(\frac{1-\delta}{\delta})^k$, and, therefore, $c_1(\delta, \mathcal{M}(Z_d), \{0\}) \sim \delta^{-d}$ as $\delta \longrightarrow 0$. Moreover,*

$$c_1(\delta, \{Z_d\}) =: \sup\{c_1(\delta, \mathcal{M}(Z_d), \{0\}) : d \geqslant 1\} = c_1(\delta, \mathcal{M}(\mathbb{Z}_+), \overline{\mathbb{D}})$$

for all $0 < \delta \leqslant 1$, that is, $c_1(\delta, \{Z_d\}) = \infty$ for $0 < \delta \leqslant 1/2$, and $c_1(\delta, \{Z_d\}) = (2\delta - 1)^{-1}$ for $1/2 < \delta \leqslant 1$.

7. The Weighted Beurling-Sobolev Algebras

In this section, we present some results on estimates of the inverses in Beurling-Sobolev algebras contained in [ENZ]. The principal conclusion is that the spectrum $\mathfrak{M}(A)$ of a "sufficiently smooth" Beurling-Sobolev algebra $A = l^p(\mathbb{Z}, w)$, as well as of its analytic counterpart $A = l^p(\mathbb{Z}_+, w)$, is $(\delta - 1)$-visible for every $\delta > 0$, that is $\delta_1(A, \mathfrak{M}(A)) = 0$. We also show explicit upper bounds for $c_1(\delta, A, \mathfrak{M}(A))$ for $\delta > 0$. The weights w are subject to some regularity conditions.

In particular, the analytic Wiener algebra $l^1(\mathbb{Z}_+)$, which admits no norm control for the inverses for $0 < \delta \leqslant 1/2$ (see Section 5), turns out to be the only exception in the scale of weighted algebras $l^1(\mathbb{Z}_+, w)$, up to the weight regularity mentioned above. The reason for this difference lies in a sort of "asymptotic compactness" of the algebra multiplication in the presence of a non-trivial weight regularly tending to infinity, and in its absence for the un-weighted case; see comments to theorem 7.2.4 for the definitions. Technically, the phenome-non mentioned is measured by the compactness of the embedding of the algebra $l^p(\mathbb{Z}_+, w)$ or $l^p(\mathbb{Z}, w)$ in the multiplier algebra mult $(l^p(w'))$ of the space of derivatives $l^p(w') = \mathcal{D}l^p(w)$, as is required by the method of Section 3.

7.1. The Beurling-Sobolev algebras $l^p(w)$

We start by fixing notation and recalling some basic facts on the weighted convolution algebras $l^p(w)$, and on the Beurling-Sobolev algebras; by our definition, the latter represent a special case of $l^p(w)$. Then we describe regularly growing weights generating a Beurling-Sobolev algebra.

7.1.1. The topological Banach algebras $l^p(\mathbb{Z}, w)$.
For a positive function (a weight) $w : \mathbb{Z} \longrightarrow (0, \infty)$ and an exponent $1 \leqslant p < \infty$, we denote

$$l^p(\mathbb{Z}, w) = \{x = (x_n)_{n \in \mathbb{Z}} : xw \in l^p(\mathbb{Z})\} = \{x : \|x\|_{p,w} < \infty\},$$

where $\|x\|_{p,w} = (\sum_{n \in \mathbb{Z}} |x_n|^p w(n)^p)^{1/p} < \infty$, with the usual modification for $p = \infty$, $\|x\|_{\infty,w} = \sup_{n \in \mathbb{Z}} |x_n w(n)| < \infty$. The convolution of two finitely supported sequences is defined in the usual way, $x * y = (\sum_{k \in \mathbb{Z}} x_k y_{n-k})_{n \in \mathbb{Z}}$. It is well known that the convolution can be extended to a continuous multiplication on $l^p(\mathbb{Z}, w)$ provided that

$$(7.1) \qquad C_{p,w} = \sup_{n \in \mathbb{Z}} \left(\sum_{k \in \mathbb{Z}} \left(\frac{w(n)}{w(k)w(n-k)} \right)^{p'} \right)^{1/p'} < \infty,$$

where p' stands for the conjugate exponent, $1/p + 1/p' = 1$. The case $p = 1$ is classical. For $p > 1$, the sufficiency of the continuous analogue of (7.1) for the weighted space $L^p(\mathbb{R}, w)$ to be a convolution algebra was proved by J. Wermer in [W]. For the case of $l^p(\mathbb{Z}, w)$, the suffi-ciency of (7.1), as well as the necessity of it in the case $p = \infty$, was proved in [N3] (without knowing about Wermer's result). In [KL], it was shown that, in general, (7.1) is not necessary, but for an even $(w(-n) = w(n))$ and log-concave weight w this condition is necessary for any p. However, a practically useful necessary and sufficient condition for regularly varying weights first appeared only in [ENZ], see below theorem 7.1.5. For information on weighted convolution L^p algebras on $\mathbb{R}^n$, see [KS].

Assuming (7.1), we get $\|x * y\|_{p,w} \leqslant C_{p,w} \|x\|_{p,w} \|y\|_{p,w}$ for every $x, y \in l^p(\mathbb{Z}, w)$. Hence, $l^p(\mathbb{Z}, w)$ becomes a unital topological Banach algebra with the unit $e = e_0 = (\delta_{0k})_{k \in \mathbb{Z}}$. For

$p = 1$, the condition $w(0) = 1$, $w(n + k) \leqslant w(n)w(k)$ obviously is necessary and sufficient for $l^1(\mathbb{Z}, w)$ to be a Banach algebra. It is an exercise to show that for $p > 1$ the norm $\| \cdot \|_{p,w}$ is *never* a Banach algebra norm; on the contrary, for $p = 1$ every topological Banach algebra weight is equivalent to a Banach algebra weight.

By a *Beurling-Sobolev algebra* we mean the space $l^p(\mathbb{Z}, w)$ satisfying condition (7.1) and endowed with the norm $\| \cdot \|_{p,w}$. We recall that the l^1 weighted Banach algebras are usually called *Beurling algebras*; in the general case of the algebras $l^p(\mathbb{Z}, w)$, there are reasons for adding the name of S. L. Sobolev, because for $p = 2$ and $w(n) = (1 + |n|)^\alpha$ we get the standard Sobolev spaces (algebras) on $\mathbb{T}$ as the spaces $\mathcal{F}l^p(\mathbb{Z}, w)$ of Fourier transforms.

7.1.2. Maximal ideals. A necessary condition.
Since $A = l^p(\mathbb{Z}, w)$ satisfies axioms *(i)* and *(ii)* of Subsection 3.1, the spectrum of $l^p(\mathbb{Z}, w)$ is the annulus $A(r_-, r_+)$, $A(r_-, r_+) = \{\lambda \in \mathbb{C} : r_-(w) \leqslant |\lambda| \leqslant r_+(w)\}$, where $r_- = r_-(w) = \lim_{k \to -\infty} w(k)^{1/k} > 0$, $r_+ = r_+(w) = \lim_{k \to +\infty} w(k)^{1/k} < \infty$.

The Gelfand transform is given by $x \longmapsto \hat{x}$,

$$\hat{x}(\lambda) = \mathcal{F}x(\lambda) = \sum_{k \in \mathbb{Z}} x_k \lambda^k, \lambda \in A(r_-, r_+).$$

Moreover, it follows from the inclusion $\mathcal{F}l^p(\mathbb{Z}, w) \subset C(A(r_-, r_+))$ that the algebra $l^p(\mathbb{Z}, w)$ is always embedded into the weighted algebra

$$l^1(\mathbb{Z}, r_\pm^n) = \{x : \sum_{k<0} |x_k| r_-^k + \sum_{k \geqslant 0} |x_k| r_+^k < \infty\}.$$

This provides a *necessary condition* for $l^p(\mathbb{Z}, w)$ to be an algebra, namely, $(r_\pm^n / w(n)) \in l^{p'}(\mathbb{Z})$, or, equivalently,

$$\sum_{k \geqslant 0} (\frac{r_+^k}{w(k)})^{p'} < \infty, \quad \sum_{k<0} (\frac{r_-^k}{w(k)})^{p'} < \infty,$$

with the usual modification for $p' = \infty$.

7.1.3. Remarks and notation.
With appropriate modifications, similar facts are true for the *analytic Beurling-Sobolev algebras* $l^p(\mathbb{Z}_+, w)$. In this Section, we consider the case of *positive spectral radius only*, $r_+(w) > 0$.

In the case of $l^p(\mathbb{Z}, w)$, we often assume a sort of weak symmetry of w, requiring that $r_- = r_+$. Without loss of generality, we can normalize the weight so as to have $r_+ = 1$. A general method for obtaining algebras $l^p(\mathbb{Z}, w)$ with arbitrary spectral radii is as follows. Let $l^p(\mathbb{Z}, w)$ be a Beurling-Sobolev algebra, and $R = (R_-, R_+)$ be two positive numbers such that $R_- \leqslant R_+$. Then, if we set $W_R(n) = R_{\text{sign}(n)}^n w(n)$ for $n \in \mathbb{Z}$, we get an algebra weight satisfying $C_{p,W_R} \leqslant C_{p,w}$ and $r_\pm(W_R) = R_\pm r_\pm$. In particular, starting with $r_- = r_+ = 1$, we can obtain an arbitrary pair of spectral radii.

Notation. We use the symbol l^p, respectively $l^p(w)$, as common notation for $l^p(\mathbb{Z})$ and $l^p(\mathbb{Z}_+)$, respectively for $l^p(\mathbb{Z}, w)$ and $l^p(\mathbb{Z}_+, w)$.

7.1.4. The Beurling-Sobolev algebras $l^p(w)$.

Here we give regularity conditions on w guaranteeing that $l^p(w)$ is a Beurling-Sobolev algebra.

THEOREM 7.1.5. *I. Let $1 \leqslant p \leqslant \infty$, and $v = \log(w)$ be a positive eventually concave function on $\mathbb{R}_+$ such that $r_+ = \lim_{x \to \infty} w(x)^{1/x} = 1$. Assume that the following condition is fulfilled: either, (a) there exists $\alpha > 0$ such that $w(x)/x^\alpha$ eventually decreases; or, (b) there exists $\alpha > 1/p'$ such that $w(x)/x^\alpha$ eventually increases and has concave logarithm.*

Then the following assertions are equivalent.

(i) The space $l^p(\mathbb{Z}_+, w)$ is an algebra.

(ii) The space $l^p(\mathbb{Z}_+, w)$ is a Beurling-Sobolev algebra.

(iii) $1/w \in L^{p'}(\mathbb{R}_+)$, or, equivalently, $(1/w(n))_{n \geqslant 0} \in l^{p'}$.

II. Let w be a weight function on $\mathbb{Z}$ such that $r_- = r_+ = 1$. Assume that $(w(n))_{n \geqslant 0}$ and $(w(-n))_{n \geqslant 0}$ are quasiincreasing sequences, that is $\sup_{n,j \geqslant 0} w(n)w(n+j) < \infty$, and similarly for $(w(-n))_{n \geqslant 0}$. Then $l^p(\mathbb{Z}, w)$ is a Beurling-Sobolev algebra if and only if so are $l^p(\mathbb{Z}_+, w)$ and $l^p(\mathbb{Z}_-, w)$.

7.1.6. Examples.

(i) $l^p(|n|_*^\alpha)$ *is a Beurling-Sobolev algebra if and only if $\alpha p' > 1$ for $p > 1$, and $\alpha \geqslant 0$ for $p = 1$. Here $x_* = \max(x, 1)$ for $x \in \mathbb{R}$.*

(ii) If $l^1(w_\alpha)$ is a Beurling algebra, $w_\alpha(n) = w(n)/|n|_*^\alpha$, and $\alpha p' > 1$, then $l^p(w)$ is a Beurling-Sobolev algebra.

7.2. When is the embedding $l^p(w) \subset \mathrm{mult}\,(\mathcal{D}l^p(w))$ compact?

As was shown in Section 3, the compactness of this embedding plays a crucial role in the estimates of the norms of inverses. The best polynomial approximations, measured by the ϵ_n characteristics of this embedding, make it possible to control the norms of the inverses. This completes the program proposed in Section 3 for estimates of $c_1(\delta, A, \mathfrak{M}(A))$ in the case of Beurling-Sobolev algebras $A = l^p(w)$. We consider the special case $p = 1$ of Beurling algebras separately, because in this case we can state an explicit necessary and sufficient condition for compactness. Similar analysis can be made for $l^\infty(w)$. For the Beurling-Sobolev algebras $l^p(w)$, $1 < p < \infty$, some broad sufficient conditions are obtained. In this Subsection, we follow the paper [ENZ].

The case of the group $\mathbb{Z}$ is slightly different from the semigroup $\mathbb{Z}_+$, because in these two cases the convolutions differ in nature. Namely, on $\mathbb{Z}$, to form $(x * y)_n$ we add products $x_k y_j$ with k and j both unbounded, and for $\mathbb{Z}_+$ this is not the case. In particular, this elementary remark explains the difference between the nature of the weight $\sigma(n) = \|e_n\|_{\mathrm{mult}}$ on $\mathbb{Z}_+$ from that on $\mathbb{Z}$, see 7.2.5–7.2.6. For instance, for any analytic Beurling-Sobolev algebra $A = l^p(\mathbb{Z}_+, w)$ we always have $A \subset \mathrm{mult}\,(A')$, but for the algebras $A = l^p(\mathbb{Z}, w)$, in general, this inclusion fails.

As before, the *weight* w' is defined by the formula $w'(n) = w(n)/|n|_*$, where $|n|_* = \max(|n|, 1)$. It is clear that $A' = \mathcal{D}A = l^p(w')$ for $A = l^p(w)$.

7.2.1. **Continuous embeddings** $l^p(w) \subset \mathrm{mult}\,(l^p(w'))$. As was just mentioned, an *analytic* Beurling-Sobolev algebra $l^p(\mathbb{Z}_+, w)$ is always contained in $\mathrm{mult}\,(l^p(\mathbb{Z}_+, w'))$.

THEOREM 7.2.2. *If $l^p(\mathbb{Z}_+, w)$ is a Beurling-Sobolev algebra, then*

$$l^p(\mathbb{Z}_+, w) \subset \mathrm{mult}\,(l^p(\mathbb{Z}_+, w'))$$

and the norm of this embedding does not exceed $C_{p,w}$.

It is easy to see that for the group case, that is, for $l^p(\mathbb{Z}, w)$ in place of $l^p(\mathbb{Z}_+, w)$, some more growth restrictions on w are required for the inclusion $A = l^p(\mathbb{Z}, w) \subset \mathrm{mult}\,(A') = \mathrm{mult}\,(l^p(\mathbb{Z}, w'))$. Indeed, the condition $c = \sup_k(\|e_k\|_{\mathrm{mult}(A')}/\|e_k\|_{p,w}) < \infty$ is necessary for such an inclusion; it implies that $w(k)w(-k) \geqslant \mathrm{const} \cdot |k|_*$ for all $k \in \mathbb{Z}$.

Thus, for every $1 \leqslant p < 2$ there exist regularly growing weights w satisfying the condition (iii) of theorem 7.1.5 but $c = \infty$ (take $w(k) = |k|_*^\alpha$ with $1/p' < \alpha < 1/2$). Consequently, there are algebras $l^p(\mathbb{Z}, w)$ for which the embedding $l^p(\mathbb{Z}, w) \subset \mathrm{mult}\,(l^p(\mathbb{Z}, w'))$ fails. On the contrary, for $p \geqslant 2$, any algebra $l^p(\mathbb{Z}, w)$ with regularly growing weight seems to be contained in $\mathrm{mult}\,(l^p(\mathbb{Z}, w'))$.

On the other hand, we can guarantee the embedding in question (and even the compactness of this embedding), if we require a stronger algebra condition. Below, we present some results from [ENZ] obtained by using two different approaches.

First, we give a simple sufficient condition for the embedding $l_0^p(w) \subset \mathrm{mult}\,(l^p(w'))$ factorizing it through an auxiliary l^1-space.

THEOREM 7.2.3. *I. For any weighted space $A = l^p(w)$, or $A = l_0^p(w)$, we have*

$$\|e_n\|_{\mathrm{mult}\,(A')} = \sigma(n),$$

where

(7.2)
$$\sigma(n) = \sup_k \frac{|k|_* w(k+n)}{|k+n|_* w(k)},$$

$k, n \in \mathbb{Z}$, or $k, n \in \mathbb{Z}_+$, respectively. Hence, we have the contractive embedding $l^1(\sigma) \subset \mathrm{mult}\,(A')$.

II. The embedding $l^p(w) \subset l^1(\sigma)$ is equivalent to

(7.3)
$$\frac{\sigma}{w} \in l^{p'}.$$

Being valid, this embedding is automatically compact for $p > 1$, and it is compact for $p = 1$ if and only if

(7.4)
$$\lim_k \frac{\sigma(k)}{w(k)} = 0,$$

where $k \longrightarrow \infty$ in the case of $\mathbb{Z}_+$, and $|k| \longrightarrow \infty$ in the case of $\mathbb{Z}$.

III. For every weighted space $l^p(w)$, condition (7.3) for $p > 1$, and condition (7.4) for $p = 1$, imply that $l^p(w) \subset_c \text{mult}(l^p(w'))$.

IV. For a Beurling algebra $A = l^1(\mathbb{Z}_+, w)$ the following are equivalent

a) the embedding $A \subset B = \text{mult}(A')$ is compact,

b) $l^1(\mathbb{Z}_+, w) \subset l^1(\mathbb{Z}_+, \sigma)$ is compact,

c) $\lim_{n,k \longrightarrow \infty} \frac{w(n+k)}{w(n)w(k)} = 0$,

d) the multiplication in $l^1(\mathbb{Z}_+, w)$ is asymptotically compact.

Moreover, if $A_0 = \{x \in A : \|x\|_A \leqslant 1\}$ is the unit ball of $A = l^1(\mathbb{Z}_+, w)$, then $\epsilon_n(A_0, B) = \sup_{k>n}(\sigma(k)/w(k))$ where $\epsilon_n(A_0, B)$ stands for the best polynomial approximation of degree n as defined in 3.2.3.

"Asymptotically compact multiplication" mentioned above means that for every $\epsilon > 0$ there exists an integer N such that $\|x * y\| < \epsilon\|x\| \cdot \|y\|$ for every $x, y \in l^p(\mathbb{Z}_+, w)$ satisfying $x_k = y_k = 0$ for $0 \leqslant k < N$.

The conditions a), b) and d) are still equivalent (after obvious modifications) for Beurling algebras $l^1(\mathbb{Z}, w)$ on $\mathbb{Z}$, but the behaviour of the weight σ is quite different in the cases of $\mathbb{Z}$ and $\mathbb{Z}_+$. Examples of weights w, for which the rate of the best polynomial approximations $\epsilon_n(A_0, B)$ can be computed explicitly, can be provided using the known Hardy field of functions, see Bourbaki [Bou]. In particular such a weight satisfying $\lim_{x \longrightarrow \infty} x^{-1}\log(w(x)) = 0$ generates an algebra $l^1(\mathbb{Z}_+, w)$ and satisfies the following dichotomy:

(i) either $\lim_{x \longrightarrow \infty}(w(x)/x) = 0$, and then $\sigma(n, w)/w(n) \simeq w(n)^{-1}$;

(ii) or $\lim_{x \longrightarrow \infty}(w(x)/x) > 0$, and then $\sigma(n, w)/w(n) \simeq n^{-1}$.

7.2.4. The best polynomial approximations related to the embedding

$$l^p(w) \subset \text{mult}(l^p(w'))$$

It is the key point of our approach. Knowing $\epsilon_m(A_0, B)$, it remains only to apply the theory of Section 3. As before, the cases of $\mathbb{Z}_+$ and $\mathbb{Z}$ are slightly different.

THEOREM 7.2.5. Let $1 \leqslant p \leqslant \infty$; suppose that $v = \log(w)$ satisfies the conditions (a), (b), and (iii) of theorem 7.1.5. Then the space $l^p(\mathbb{Z}_+, w)$ is an algebra, the embedding $A = l^p(\mathbb{Z}_+, w) \subset B = \text{mult}(l^p(\mathbb{Z}_+, w'))$ is compact, and the following estimates are valid for the best approximations $\epsilon_m(A_0, B)$ of the unit ball $A_0 \subset A$.

If condition (a) is satisfied with an exponent $\alpha \leqslant 1$, then

$$\epsilon_m(A_0, B) \leqslant (\sum_{j>m} w(j)^{-p'})^{1/p'}.$$

If condition (b) is satisfied, then $\epsilon_m(A_0, B) \leqslant c/m$ for $\alpha > 1 + 1/p'$; $\epsilon_m(A_0, B) \leqslant c \cdot (\log(m))^{1/p'}/m$ for $\alpha = 1 + 1/p'$, and $\epsilon_m(A_0, B) \leqslant c/m^{\alpha-1/p'}$ for $1/p' < \alpha < 1 + 1/p'$, where c stands for a constant depending on α and p.

As is mentioned above, for the case of $\mathbb{Z}$ we have some extra constraints in order that the embeddings in question would be compact. For instance, the condition

$$\lim_{|n|\longrightarrow 0}(\|e_n\|_B/\|e_n\|_A)=0$$

is obviously necessary for $A = l^p(\mathbb{Z},w) \subset_c B = \text{mult}\,(l^p(\mathbb{Z},w'))$. Since $\|e_n\|_B = \sigma(n) \geqslant |n|_*w(0)/w(-n)$, we obtain that if $\underline{\lim}_{|n|\longrightarrow\infty}|n|w(n)w(-n) > 0$, the embedding $l^p(\mathbb{Z},w) \subset \text{mult}\,(l^p(\mathbb{Z},w'))$ cannot be compact.

THEOREM 7.2.6. *Let w be a weakly symmetric normalized weight, that is $r_+ = r_- = 1$. Each of the following conditions implies that $l^p(\mathbb{Z},w)$ is a Beurling-Sobolev algebra compactly embedded into the multiplier space, $A = l^p(\mathbb{Z},w) \subset_c B = \text{mult}\,(l^p(\mathbb{Z},w'))$, with the following upper bounds for the best polynomial approximations $\epsilon_m(A_0,B)$ of the unit ball $A_0 \subset A$. Here $\mathcal{E}_m(p,\alpha)$ stands for the right hand side of the corresponding inequality in theorem 7.2.5 if $\alpha > 1$, and $\mathcal{E}_m(p,\alpha) = c \cdot m^{1-2\alpha+1/p'}$ otherwise.*

If $C_{1,w_\alpha} < \infty$ for an exponent $\alpha > 2^{-1}(1+1/p')$, where $w_\alpha = (w(k)/|k|_*^\alpha)_{k\geqslant 0}$, then $\epsilon_m(A_0,B) \leqslant C_{1,w_\alpha}\mathcal{E}_m(p,\alpha)$.

If $C_{p,w_\alpha} < \infty$ and $\alpha > 1/2$, then $\epsilon_m(A_0,B) \leqslant C_{p,w_\alpha}\mathcal{E}_m(1,\alpha)$.

For slowly increasing weights, we refer to [ENZ] for a simple direct estimate for the best approximations $\epsilon_m(A_0,B)$.

7.3. *Some explicit estimates of the norm controlling constants $c_1(\delta, l^p(w))$*

Now we are ready to obtain explicit estimates of the inverses in the Beurling-Sobolev algebras. To this end, we combine theorems of Section 3 with the estimates of the rate of polynomial approximations $\epsilon_m(A_0,B)$ provided in Subsections 7.1–7.2. Recall that the majorant M of theorem 3.2.5 depends on the distribution function λ_0 of the sequence $\epsilon_m(A_0,B)$ and on the multiplier norms $\sigma(k)$ of the basis vectors of $l^p(w)$. In 7.3.1 below we supposing $r_+ = r_- = 1$; for the case where $r_- < r_+$, see the remark at the end of this Subsection.

THEOREM 7.3.1. *Let A be a Beurling-Sobolev algebra, $A = l_0^p(\mathbb{Z},w)$ or $l_0^p(\mathbb{Z}_+,w)$, compactly embedded into $B = \text{mult}\,(l^p(w'))$. Then $\delta_1(A,\mathfrak{M}(A)) = 0$, and for all $\delta > 0$ we have the estimate $c_1(\delta, A, \mathfrak{M}(A)) \leqslant w(0)\delta^{-1} + M_1(\delta)$, where M is given in theorem 3.2.5 and $\|e_j\|_B = \sigma(j)$ (see formula (7.2)), and the constants $\mathcal{E}$ and C depend on the constant (7.1) and on the norm of the embedding $l_0^p(\mathbb{Z}_+,w) \subset B = \text{mult}\,(l^p(w'))$.*

7.3.2. **Examples of estimates of inverses on $\mathbb{Z}_+$.** Here are some typical Beurling-Sobolev algebras $A = l^p(\mathbb{Z}_+,w)$, $1 \leqslant p \leqslant \infty$.

(i) $w(n) = n_*^\alpha$, $\alpha > 1 + 1/p'$. Then, $c_1(\delta, l^p(\mathbb{Z}_+, n_*^\alpha)) \leqslant c(\alpha,p)/\delta^{4\alpha+2}$.

(ii) $w(n) = n_*^\alpha$, $\alpha = 1 + 1/p'$. Then, $c_1(\delta, l^p(\mathbb{Z}_+, n_*^\alpha)) \leqslant c(\alpha,p)(\log(1/\delta))^{\alpha/p'}/\delta^{4\alpha+2}$.

(iii) $w(n) = n_*^\alpha$, $1/p' < \alpha < 1 + 1/p'$. Then, $c_1(\delta, l^p(\mathbb{Z}_+, n_*^\alpha)) \leqslant c(\alpha,p)/\delta^{2\beta+1}$, where $\beta = (\alpha p' - 1)^{-1}$.

(iv) $w(n) = e^{n^\alpha}, 0 < \alpha < 1$. Then, $c_1(\delta, l^p(\mathbb{Z}_+, w)) \leqslant c(\alpha, p) \cdot \exp\{d_\alpha \delta^{-2\alpha(1-\alpha)}\} \cdot \delta^{-2} \cdot \log\frac{1}{\delta}$.

7.3.3. Examples of estimates of inverses on $\mathbb{Z}$. Comparison of the cases of $\mathbb{Z}$ and $\mathbb{Z}_+$ given in Subsections 7.1 and 7.2 shows that the explicit estimates of the inverses for rapidly growing weights should be the same on $\mathbb{Z}$ and on $\mathbb{Z}_+$, up to the sharp values of constants. Here, "rapidly" means at least as fast as the linearly growing weight $w(n) = |n|_*, n \in \mathbb{Z}$. For slower weights, e.g., for $w(n) = |n|_*^\alpha, \alpha < 1$, our method gives faster growth of the constants $c_1(\delta, l_p(w))$ on $\mathbb{Z}$ than on $\mathbb{Z}_+$. This method stops completely at the exponent $\alpha = \frac{1}{2}(1 + \frac{1}{p'})$. For this case we refer to the recent paper [E], where a different approach based on a Björk's paper [B] is employed. We restrict ourselves to a few examples illustrating the above theory.

(i) $w(n) = |n|_*^\alpha, \alpha > 1 + 1/p'$ or $1 \leqslant \alpha \leqslant 1 + 1/p'$. Then one has the same results as in 7.3.2 *(i)*, *(ii)* and *(iii)*, with modified constants.

(ii) $w(n) = |n|_*^\alpha, 2^{-1}(1+1/p') < \alpha < 1$. Then $c_1(\delta, l^p(\mathbb{Z}, |n|_*^\alpha)) \leqslant c(\alpha, p)/\delta^{2+2(1+\beta)(2-\alpha)}$. Observe that the right hand side of the last inequality diverges to infinity for $\alpha \longrightarrow 2^{-1}(1 + 1/p') = \alpha_p$, because $\beta \longrightarrow \infty$. However, in order to get a finite majorant, we can consider a weight growing slightly faster than the critical weight $w(n) = |n|_*^{\alpha_p}$. This can be done using $w(n) = |n|_*^{\alpha_p}(\log(1 + |n|_*))^\gamma$.

References

[B]   Björk J.-E., On the spectral radius formula in Banach algebras, Pacific J. Math. 40, No 2 (1972), 279–284.

[Bou]   Bourbaki N., Fonctions d'une variable réelle (théorie élémentaire), Chaps. 1–7, Actualités Sci. Industr., No 1074, 1132, Paris, Hermann 1958, 1961.

[C]   Cohen P., A note on constructive methods in Banach algebras, Proc. Amer. Math. Soc. 12 (1961), No 1, 159–164.

[D]   Dyn'kin E. M., Theorems of Wiener-Lévy type and estimates for Wiener-Hopf operators, Matematicheskie Issledovania (Kishinev, ed. by I. Gohberg), 8:3(29) (1973) 14–25 (Russian).

[E]   El-Fallah O., Ezzaaraoui A., Majorations uniformes de normes d'inverses dans les algèbres de Beurling, Preprint Univ. of Rabat.

[ENZ]   El-Fallah O., Nikolski N. K., and Zarrabi M., Resolvent estimates in Beurling-Sobolev algebras, Algebra i Analiz, 6 (1998), 1-80 (Russian); English transl.: St. Petersburg Math. J., 6 (1999), 1-69.

[Gar]   Garnett J. B., *Bounded analytic functions*, NY, AP, 1981.

[GRS]   Gelfand I. M., Raikov D. A., and Shilov G. E., *Commutative normed rings*, (Russian) Moscow, Fizmatgiz, 1960; English transl.: NY, Chelsea 1964.

[GMG]   Graham C. C. and McGehee O. C., *Essays in Commutative Harmonic Analysis*, NY-Heidelberg, Springer, 1979.

[HKKR] Helson H., Kahane J.-P., Katznelson Y., and Rudin W., The functions which operate on Fourier transforms, Acta Math. 102 (1959), 135–157.

[HR]   Hewitt E. and Ross K. A., *Abstract Harmonic Analysis, Vol. I and II*, NY-Heidelberg, Springer 1963 and 1970.

[K1]   Kahane J.-P., *Séries de Fourier absolument convergentes*, Heidelberg-NY, Springer 1970.

[K2]   Kahane J.-P., Sur le théorème de Beurling-Pollard, Math. Scand. 21 (1967), 71–79.

[KL]   Kerlin E. and Lambert A., Strictly cyclic shifts on l^p, Acta Math. Szeged, 35 (1973), 87–94.

[KS]   Kerman R. and Sawyer E., Convolution Algebras with Weighted Rearrangement—Invariant Norm, Studia Math., 108 (1994), 103–126.

[L]   Leblanc N., Les fonctions qui opèrent dans certaines algèbres à poids, Math. Scand. 25 (1969), 190–194.

[Lev]   Levina N. B., On the inverse Wiener-Lévy theorem, Funktsion. Analiz i ego Prilozhenia, 7:3 (1973), 84–85 (Russian); English transl.: Funct. Anal. Appl. (Plenum Publ., NY), 7:3 (1974), 241–242.

[New]   Newman D. J., A simple proof of Wiener's $1/f$ theorem, Proc. Amer. Math. Soc. 48 (1975), 264–265.

[NX]   Nicolau A. and Xiao J., Bounded functions in Möbius invariant Dirichlet spaces, J. Funct. Anal., 150:2 (1997), 383-425.

[N1]   Nikolski N., *Treatise on the shift operator*, NY-Heidelberg, Springer, 1986.

[N2]   Nikolski N., Norm control of inverses in radical and operator Banach algebras, to appear.

[N3]   Nikolski N., *Selected problems of weighted approximation and spectral analysis*, Trudy Mat. Inst. Steklov, vol.120 (Russian), Leningrad, Nauka 1974; English transl.: Series Proc. Steklov Inst. Math., 120, Providence, Amer. Math. Soc. 1976.

[N4]   Nikolski N., In search of the invisible spectrum, Ann. Inst. Fourier, 49:6 (1999), 1925-1998.

[Ru1]   Rudin W., *Fourier Analysis on Groups*, Interscience Tracts No.12, NY, Wiley 1962.

[S]   Shamoyan F. A., Applications of Dzhrbashyan integral representations to certain problems of analysis, Doklady AN SSSR, 261:3 (1981), 557–561 (Russian); English transl.: Soviet Math. Doklady 24:3 (1981), 563–567.

[Sh1]   Shapiro H. S., A counterexample in harmonic analysis. In: Approximation Theory, Banach Center Publications, Warsaw, 1979, Vol.4, 233–236 (submitted 1975).

[Sh2]   Shapiro H. S., Bounding the norm of the inverse elements in the Banach algebra of absolutely convergent Taylor series. Abstracts of the Sixth Summer St. Petersburg Meeting in Mathematical Analysis, June 22–24, 1997, St. Petersburg.

[Sr]   Sreider Yu. A., The structure of maximal ideals in rings of measures with involution, Matem. Sbornik 27(69) (1950), 297–318 (Russian); English transl.: AMS Transl. No 81 (1953), 28pp.

[St]   Stafney J. D., An unbounded inverse property in the algebra of absolutely convergent Fourier series, Proc. Amer. Math. Soc. 18:1 (1967), 497–498.

[SW]   Sundberg C. and Wolff T., Interpolating sequences for QA_B, Trans. Amer. Math. Soc., 276 (1983), 551-581.

[SzNF]   Szökefalvi-Nagy B. and Fois C., *Analyse harmonique des opérateurs de l'espace de Hilbert*, Budapest, Akadémiai Kiado, 1967.

[T]   Taylor J. L., *Measure algebras*, CBMS Conf. No 16, Amer. Math. Soc., Providence, R.I., 1972.

[To1]   Tolokonnikov V. A., Estimates in Carleson's corona theorem and finitely generated ideals in the algebra H_∞, Functional. Anal. i ego Prilozh. 14 (1980) 85–86 (Russian); English transl.: Funct. Anal. Appl. 14:4 (1980), 320–321.

[To2]   Tolokonnikov V. A., Corona theorems in algebras of bounded analytic functions. Manuscript deposed in VINITI, Moscow, No 251-84 DEP, 1984 (Russian).

[VP]   Vinogradov S. A. and Petrov A. N., The converse to the theorem on operation of analytic functions, and multiplicative properties of some subclasses of the Hardy space H^∞, Zapiski Nauchnyh Semin. St. Petersburg Steklov Institute, 232 (1996), 50–72 (Russian).

[W]   Wermer J., On a class of normed rings, Arkiv for Mat. 2:28 (1953), 537–551.

[Z]   Zygmund A., *Trigonometric series, Vol.1*, Cambridge, The University Press 1959.

Harmonic Analysis as found in Analytic Number Theory

Hugh L. Montgomery

Department of Mathematics
University of Michigan
Ann Arbor, MI 48109–1109 USA
phone: 734–763–3269
`hlm@math.lsa.umich.edu`

ABSTRACT. A wide variety of questions of Harmonic Analysis arise naturally in various contexts of Analytic Number Theory; in what follows we consider a number of examples of this type.

The author is grateful to Dr. Ulrike Vorhauer for advice and assistance at all stages of preparation of this paper.

1. Uniform Distribution

The definition of uniform distribution is fairly intuitive:

DEFINITION 1. *A sequence* $\{u_n\} \in \mathbb{T}$ *is* uniformly distributed *if for any* α, $0 \leqslant \alpha < 1$, *we have*

$$\lim_{N \to \infty} \frac{1}{N} \operatorname{card}\{1 \leqslant n \leqslant N : u_n \in [0, \alpha) \pmod 1\} = \alpha.$$

Let U_N be a measure with unit masses at the points u_n for $1 \leqslant n \leqslant N$. Then the Fourier transform of U_N is the exponential sum

$$\widehat{U}_N(k) = \sum_{n=1}^{N} e(-ku_n)$$

where $e(\theta) = e^{2\pi i\theta}$. (This notation was introduced by I. M. Vinogradov.) H. Weyl [36, 37] introduced an important criterion for uniform distribution in terms of the size of the U_N, namely that the following are equivalent statements concerning a sequence $\{u_n\}$:

Research supported in part by NSF Grant DMS 0070720.

271

J.S. Byrnes (ed.), Twentieth Century Harmonic Analysis - A Celebration, 271–293.

(a) The sequence $\{u_n\}$ is uniformly distributed;

(b) For each integer $k \neq 0$, $\widehat{U}_N(k) = o(N)$ as $N \to \infty$;

(c) If F is properly Riemann-integable on $\mathbb{T}$ then

$$\lim_{N \to \infty} \frac{1}{N} \sum_{n=1}^{N} F(u_n) = \int_{\mathbb{T}} F(x)\, dx .$$

From the formula for the value of a geometric series it is immediate that

$$\left| \sum_{n=1}^{N} e(n\theta) \right| \leqslant \frac{1}{|\sin \pi\theta|} \leqslant \frac{1}{2\|\theta\|}$$

where $\|\theta\|$ is the distance from θ to the nearest integer, $\|\theta\| = \min_{n\in\mathbb{Z}} |\theta - n|$. If θ is irrational and k is a non-zero integer then $k\theta$ is not an integer, and hence by the above with θ replaced by $k\theta$ we see that $\widehat{U}_N(k) = O_k(1)$ when $u_n = n\theta$. In this way we see easily that the sequence $n\theta$ is uniformly distributed if θ is irrational.

The proof of Weyl's Criterion depends on the existence of one-sided approximations to the characteristic function χ_I of an interval by trigonometric polynomials; these approximations should be close in the L^1 norm. Of course the existence of such trigonometric polynomials follows easily from the uniform approximation to continuous functions by trigonometric polynomials, but it is also useful to put this in a quantitative form. Erdős and Turán [9] showed that there exist trigonometric polynomials T_- and T_+ of degree at most K such that $T_-(x) \leqslant \chi_I(x) \leqslant T_+(x)$ for all x and such that $\int_{\mathbb{T}} T_\pm = \alpha + O(1/K)$, and thus that

$$\left| \operatorname{card}\{1 \leqslant n \leqslant N : u_n \in [0, \alpha) \pmod 1\} - N\alpha \right| \leqslant C\frac{N}{K} + C \sum_{k=1}^{K} \frac{|\widehat{U}(k)|}{k} .$$

In the 1970's Selberg considered how the large sieve could be refined, and in doing so discovered more natural functions that yield very sharp constants (see Selberg [32], pp. 213–226). Indeed, Beurling [5] defined the entire function

$$B(z) = \left(\frac{\sin \pi z}{\pi} \right)^2 \left(\frac{2}{z} + \sum_{n=0}^{\infty} \frac{1}{(z-n)^2} - \sum_{n=1}^{\infty} \frac{1}{(z+n)^2} \right),$$

and observed that $B(z) = O(e^{2\pi|\Im z|})$, that $B(x) \geqslant \operatorname{sgn}(x)$ for all real x, and that

$$\int_{\mathbb{R}} B(x) - \operatorname{sgn}(x)\, dx = 1 .$$

Indeed, Beurling showed that among functions with the prior properties, this function uniquely minimizes the above integral.

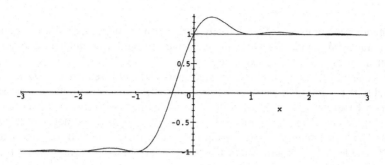

FIGURE 1. Beurling's function $B(x)$.

If $I = [a, b]$ is an interval of the real line and $\delta > 0$ then the Selberg majorant and minorant are

$$S_+(z) = \frac{1}{2}B(\delta(z - a)) + \frac{1}{2}B(\delta(b - z)),$$

$$S_-(z) = -\frac{1}{2}B(\delta(a - z)) - \frac{1}{2}B(\delta(z - b)).$$

Then $S_-(x) \leqslant \chi_I(x) \leqslant S_+(x)$ for all real x, $\widehat{S}(t) = 0$ for $|t| \geqslant \delta$, and $\int_{\mathbb{R}} S_\pm(x)\,dx = b - a \pm 1/\delta$. This approximation is optimal when $\delta(b - a)$ is an integer, and in any case is quite good. To obtain corresponding approximations for the circle group we apply the Poisson summation formula. Suppose that $b - a < 1$, so that the interval $[a, b]$ defines an arc of $\mathbb{T}$. We take $\delta = K + 1$ and set $T_\pm(x) = \sum_n S_\pm(n + x)$. Then $T_\pm$ is a trigonometric polynomial of degree not exceeding K, $T_-(x) \leqslant \chi_I(x) \leqslant T_+(x)$ for all x, and $\int_{\mathbb{T}} T_\pm(x)\,dx = b - a \pm 1/(K + 1)$. This is optimal when $(b - a)(K + 1)$ is an integer, and is close to optimal in any case.

Weyl's criterion has since been vastly generalized to describe the weak convergence of a sequence of measures μ_n to a limiting measure μ in terms of the convergence of the Fourier transforms of these measures. One fruitful generalization is to $\mathbb{T}^d$. Suppose that $\{u_n\}$ is a sequence of points in $\mathbb{T}^d$ and let $\mathcal{B} = [a_1, b_1] \times \cdots \times [a_d, b_d]$ denote a box in $\mathbb{T}^d$. Then the following are equivalent:

(a) $\lim_{N \to \infty} \frac{1}{N} \operatorname{card}\{1 \leqslant n \leqslant N : u_n \in \mathcal{B}\} = \operatorname{vol} \mathcal{B}$ for every box $\mathcal{B}$ in $\mathbb{T}^d$;

(b) If $k \in \mathbb{Z}^d$, $k \neq 0$, then $\sum_{n=1}^{N} e(k \cdot u_n) = o(N)$ as $N \to \infty$;

(c) If F is properly Riemann-integrable on $\mathbb{T}^d$ then $\lim_{N \to \infty} \frac{1}{N} \sum_{n=1}^{N} F(u_n) = \int_{\mathbb{T}^d} F(x)\,dx$.

Quantitative majorants in $\mathbb{T}^d$ are easily obtained by forming a product of one-dimensional majorants. Minorants are a little more elusive, but Barton, Vaaler and Montgomery [2] have given a construction that works pretty well.

In the same way that we used Weyl's Criterion to see that the sequence $\{n\theta\}$ is uniformly distributed if θ is irrational, we can use Weyl's Criterion for $\mathbb{T}^d$ to obtain a sharpening of Kronecker's Theorem. Suppose that $1, \theta_1, \ldots, \theta_d$ are linearly independent over the field $\mathbb{Q}$ of rational numbers. Then the points $n\boldsymbol{\theta} = (n\theta_1, n\theta_2, \ldots, n\theta_d)$ are not only dense in $\mathbb{T}^d$ (Kronecker's Theorem) but are actually uniformly distributed. Indeed, suppose that $\mathcal{B}$ is a box in $\mathbb{T}^d$, and let $T(\boldsymbol{x})$ be a trigonometric polynomial that minorizes $\chi_{\mathcal{B}}$ with $\int_{\mathbb{T}^d} T > 0$. Then

$$\text{card}\{M + 1 \leqslant n \leqslant M + N : n\boldsymbol{\theta} \in \mathcal{B}\} \geqslant \sum_{n=M+1}^{M+N} T(n\boldsymbol{\theta})$$

$$= \sum_{\boldsymbol{k}} \widehat{T}(\boldsymbol{k}) \sum_{n=M+1}^{M+N} e(n\boldsymbol{k} \cdot \boldsymbol{\theta})$$

$$\geqslant N\widehat{T}(0) - \frac{1}{2} \sum_{\boldsymbol{k} \neq 0} \frac{|\widehat{T}(\boldsymbol{k})|}{\|\boldsymbol{k} \cdot \boldsymbol{\theta}\|}.$$

Here the last sum is finite because there are only finitely many $\boldsymbol{k}$ for which $\widehat{T}(\boldsymbol{k}) \neq 0$ and because $\boldsymbol{k} \cdot \boldsymbol{\theta}$ is never an integer. Since $\widehat{T}(0) > 0$, the above is positive if N is sufficiently large. The remarkable feature here is that the expression above is independent of M. That is, if $n_1 < n_2 < n_3 < \cdots$ are the n for which $n\boldsymbol{\alpha} \in \mathcal{B}$ then the gaps $n_{i+1} - n_i$ are uniformly bounded above. This insight is critical to Bohr's definition of almost periodicity.

DEFINITION 2. *Let $f : \mathbb{R} \to \mathbb{C}$ be continuous. We say that a real number t is an ε-almost period of f if $|f(x + t) - f(x)| < \varepsilon$ for all real x. The function f is almost periodic if for every $\varepsilon > 0$ there is a number $C = C(\varepsilon)$ such that any interval of length at least C contains an ε-almost period.*

It is by the strengthened form of Kronecker's Theorem that we see that the *almost periodic polynomial*

$$T(x) = \sum_{n=1}^{N} a_n e(\lambda_n x)$$

is indeed an almost periodic function. It can also be shown that the almost periodic polynomials are dense in the space of almost periodic functions. Let $\zeta(s) = \sum_{n=1}^{\infty} n^{-s}$ be the Riemann zeta function. We write $s = \sigma + it$. If σ is fixed, $\sigma > 1$, then $\zeta(\sigma + it)$ is an almost periodic function of t. The concept of almost periodicity can be generalized to other norms. For example, when σ is fixed, $1/2 < \sigma < 1$ the function $\zeta(\sigma + it)$ is a mean-square almost periodic function, even though it is not an almost periodic function in the uniform norm. One might expect that by almost periodicity one could show that if the zeta function has a zero with real part $> 1/2$ then it would have many such zeros, all with approximately the same real part. However, attempts to prove such an assertion have so far been unsuccessful. If one could prove such a thing then the Riemann Hypothesis would likely follow, since we have fairly good upper bounds on the number of zeros of the zeta function with large real part. Let $N(\alpha, T)$ denote the number of zeros of $\zeta(s)$ in the rectangle $\alpha \leqslant \sigma \leqslant 1$, $0 \leqslant t \leqslant T$. Then $N(\alpha, T) \ll T^{\phi(\sigma)} \log T$ where $\phi(\sigma) < 1$ if $\sigma > 1/2$. (Following Vinogradov, we say that $f \ll g$ if $f = O(g)$. This notation saves a set of parentheses when there is no main term.) Although we have not succeeded so far to use almost periodicity to produce many zeros from one, we do know that translates of the zeta function are universal among non-zero analytic functions, in the following sense: Let $\mathcal{R}$ be a rectangle, $\mathcal{R} = \{s : \sigma_1 \leqslant \sigma \leqslant \sigma_2, t_1 \leqslant t \leqslant t_2\}$ with $1/2 < \sigma_1 < \sigma_2 < 1$. Let f be analytic and non-zero on a domain containing $\mathcal{R}$. Then for any $\varepsilon > 0$ there exists a real number τ such that $|f(s) - \zeta(s + i\tau)| < \varepsilon$ uniformly for $s \in \mathcal{R}$.

almost periodicity also arises in the error term in the Prime Number Theorem. Let $\Lambda(n)$ be von Mangoldt's lambda function, which is to say that $\Lambda(n) = \log p$ when $n = p^k$ for some positive integer k, and $\Lambda(n) = 0$ otherwise. We put $\psi(x) = \sum_{n \leqslant x} \Lambda(n)$. By integration by parts we see that the Prime Number Theorem in the form $\pi(x) \sim \mathrm{li}(x)$ is equivalent to the assertion that $\psi(x) \sim x$. Since $-\zeta'(s)/\zeta(s) = \sum_{n=1}^{\infty} \Lambda(n) n^{-s}$ for $\sigma > 1$, we can recover $\psi(x)$ from the logarithmic derivative of the zeta function by an inverse Mellin transform:

$$\psi(x) = \frac{-1}{2\pi i} \int_{c-i\infty}^{c+i\infty} \frac{\zeta'}{\zeta}(s) \frac{x^s}{s} \, ds \, .$$

By moving the contour to the left we see that this is

$$= x - \sum_{\rho} \frac{x^{\rho}}{\rho} - \frac{\zeta'}{\zeta}(0) + \frac{1}{2} \sum_{k=1}^{\infty} \frac{x^{-2k}}{k} \, .$$

Here ρ runs over all the non-trivial zeros of the zeta function, which is to say all those zeros with positive real part. This explicit formula for the error term in the Prime Number Theorem is essentially one that Riemann stated and von Mangoldt later proved. Write $\rho = \beta + i\gamma$. In the quantity $x^{\rho} = x^{\beta} x^{i\gamma}$, the second factor oscillates, but more slowly as x increases. To make it periodic we make an exponential change of variables. Suppose also that the Riemann Hypothesis is true, which is to say that $\beta = 1/2$ for all ρ. Then

$$\frac{\psi(e^y) - e^y}{e^{y/2}} = -\sum_{\rho} \frac{e^{i\gamma y}}{\rho} + o(1) \, .$$

This expression is mean-square almost periodic, and the sum on the right is its Fourier expansion. It is generally believed that the $\gamma > 0$ are linearly independent over $\mathbb{Q}$, so that the terms $e^{i\gamma y}$ behave like independent random variables.

2. Exponential Sums

In his seminal work [36, 37], Hermann Weyl not only gave his criterion for uniform distribution, but also a useful method for estimating exponential sums of the form $\sum_{n=1}^{N} e(P(n))$ where P is a polynomial with real coefficients. Indeed, today such an exponential sum is called a *Weyl sum*. Weyl's basic observation was that

$$
\left| \sum_{n=1}^{N} e(P(n)) \right|^2 = \sum_{n=1}^{N} \sum_{m=1}^{N} e(P(m) - P(n))
$$

$$
= \sum_{n=1}^{N} \sum_{h=1-n}^{N-n} e\big(P(n+h) - P(n)\big)
$$

$$
= \sum_{h=-N+1}^{N-1} \sum_{\substack{1 \leqslant n \leqslant N \\ 1-h \leqslant n \leqslant N-h}} e\big(P(n+h) - P(n)\big)
$$

$$
= N + 2\Re \sum_{h=1}^{N-1} \sum_{n=1}^{N-h} e\big(P(n+h) - P(n)\big).
$$

This manipulation is known as 'Weyl differencing'. If $P(x)$ has degree d then $P(x+h) - P(x)$ has degree $d - 1$ when $h \neq 0$. Hence if we perform this differencing $d - 1$ times then we are left with a geometric series, which we know how to estimate. In this way, Weyl showed that if $P(x) = \sum_{i=0}^{d} a_i x^i$ is a polynomial with real coefficients, and if at least one of the numbers $a_1, a_2, \ldots, a_d$ is irrational, then the sequence $\{P(n)\}$ is uniformly distributed modulo 1.

In the Weyl differencing, it is somewhat a disadvantage that the parameter h runs all the way from 1 to $N - 1$. Later, van der Corput found that h can be restricted. For $1 \leqslant n \leqslant N$ let y_n be a complex number, and suppose that $y_n = 0$ if $n < 1$ or $n > N$. Then

$$
(H+1)^2 \left| \sum_{n} y_n \right|^2 = \left| \sum_{h=0}^{H} \sum_{n} y_{n+h} \right|^2
$$

$$
= \left| \sum_{n} \sum_{h=0}^{H} y_{n+h} \right|^2
$$

By Cauchy's inequality this is

$$\leqslant (N+H) \sum_n \left| \sum_{h=0}^H y_{n+h} \right|^2$$

$$= (N+H) \sum_{h=0}^H \sum_{k=0}^H \sum_n y_{n+h} \overline{y_{n+k}}$$

$$= (N+H)(H+1) \sum_n |y_n|^2$$

$$+ 2(N+H)\Re \sum_{r=1}^H (H+1-r) \sum_n y_{n+r}\overline{y_n}.$$

Thus we obtain van der Corput's inequality, which asserts that

$$\left| \sum_{n=1}^N y_n \right|^2 \leqslant \frac{N+H}{H+1} \sum_{n=1}^N |y_n|^2 + \frac{2(N+H)}{H+1} \sum_{h=1}^H \left(1 - \frac{h}{H+1}\right) \left| \sum_{n=1}^{N-h} y_{n+h}\overline{y_n} \right|.$$

On taking $y_n = e(ku_n)$ and applying Weyl's Criterion twice, we obtain

THEOREM 1. (van der Corput) *If for each positive integer* h *the sequence* $\{u_{n+h} - u_n\}$ *is uniformly distributed* (mod 1), *then the sequence* $\{u_n\}$ *is uniformly distributed* (mod 1).

More recently it has been recognized that the above remains true even when h is restricted to lie in certain subsets of positive integers; we say that $\mathcal{H}$ is a *van der Corput set* if the above is true when h is restricted to lie in $\mathcal{H}$. This is equivalent to the existence of non-negative cosine polynomials

$$T(x) = a_0 + \sum_{\substack{1\leqslant h\leqslant H \\ h\in\mathcal{H}}} a_h \cos 2\pi h x$$

with $T(0) = 1$ and a_0 arbitrarily small. In this context it is no accident that we see the coefficients of the Fejér kernel in van der Corput's inequality. Since the set of perfect squares constitute a van der Corput set, there exist non-negative cosine polynomials of the form

$$T(x) = a_0 + \sum_{h=1}^H a_h \cos 2\pi h^2 x$$

with $T(0) = 1$ and a_0 arbitrarily small, but it is not known how rapidly a_0 tends to 0 as $H \to \infty$.

van der Corput devised a general method for estimating exponential sums of the form $\sum_{a \leqslant n \leqslant b} e(f(n))$ where f is a sufficiently smooth real valued function. As an example of a simple first result of this type, we mention that if $0 < \lambda_2 \leqslant f''(x) \leqslant A\lambda_2$ for $a \leqslant x \leqslant b$ then

$$\sum_{a \leqslant n \leqslant b} e(f(n)) \ll_A \lambda_2^{1/2}(b-a) + \lambda_2^{-1/2}.$$

In Process A of van der Corput, one exponential sum is made to give rise to another by a suitable application of the van der Corput inequality. This method is destructive in the sense that usually some cancellation is lost. In van der Corput's Process B, one takes $r(x) = e(f(x))$ for $a \leqslant x \leqslant b$, and $r(x) = 0$ otherwise. Then by the Poisson summation formula,

$$\sum_{a \leqslant n \leqslant b} e(f(n)) = \sum_n r(n) = \sum_\nu \widehat{r}(\nu) = \sum_\nu \int_a^b e(f(x) - \nu x)\, dx.$$

If f' is strictly increasing with $f'(a) = \alpha$, $f'(b) = \beta$, then for $\alpha \leqslant \nu \leqslant \beta$ we obtain a stationary phase at x_ν where $f'(x_\nu) = \nu$. If $f''(x) > 0$ for $a \leqslant x \leqslant b$ then the above is approximately

$$\sum_{\alpha \leqslant \nu \leqslant \beta} \frac{e(f(x_\nu) - \nu x_\nu + 1/8)}{\sqrt{f''(x_\nu)}}.$$

Thus the problem of estimating one sum is reduced to that of estimating another. Process B is non-destructive, since a second application of it takes us back to our initial sum. For a certain class of functions f, these two processes lead to estimates of the form

$$\sum_{a \leqslant n \leqslant b} e(f(n)) \ll \left(\max |f'| \right)^k (b-a)^\ell$$

for certain pairs (k, ℓ) in the square $0 \leqslant k \leqslant 1/2 \leqslant \ell \leqslant 1$. If (k, ℓ) is an exponent pair then Process A gives the exponent pair $\left(\frac{k}{2k+2}, \frac{k+\ell+1}{2k+2} \right)$, and Process B gives the exponent pair $(\ell - 1/2, k + 1/2)$. It is trivial that $(0, 1)$ is an exponent pair. By Process B it follows that $(1/2, 1/2)$ is an exponent pair, and by Process A this yields the further pair $(1/6, 2/3)$. The collection of exponent pairs that can be obtained in this way is indicated in Figure 2 below.

Recently, Huxley [17], building on work of Bombieri and Iwaniec [6], has slightly enlarged the region of known exponent pairs, but we are still far from proving the conjecture that (k, ℓ) is an exponent pair if $k > 0$ and $\ell > 1/2$. This is a quite deep conjecture, since the special case $f(x) = t \log x$ yields the Lindelöf Hypothesis, which asserts that $\zeta(1/2 + it) \ll t^\varepsilon$ for every $\varepsilon > 0$. Some useful exponent pairs are given in Table 1.

Quantitative estimates can also be derived for the Weyl sum $\sum_{n=1}^N e(P(n))$ in terms of the rational approximations to the coefficients of P. For example, by Weyl's method we find

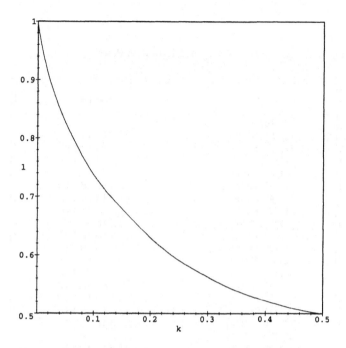

FIGURE 2. Exponent Pairs derived by van der Corput's processes.

that if $P(x) = \sum_{i=0}^{d} \alpha_i x^i$ and $|\alpha_d - a/q| \leqslant 1/q^2$ where $(a, q) = 1$, then

$$\sum_{n=1}^{N} e\big(P(n)\big) \ll_d N^{1+\varepsilon} \left(\frac{1}{q} + \frac{1}{N} + \frac{q}{N^d}\right)^{2^{1-d}}.$$

Here the most favorable circumstance is when $N \leqslant q \leqslant N^{d-1}$, in which case the upper bound is of the order $N^{1-2^{1-d}+\varepsilon}$. This is only slightly better than the trivial bound N if d is large, and it falls far short of what we conjecture, which is that

$$\sum_{n=1}^{N} e\big(P(n)\big) \ll_d N^{1+\varepsilon} \left(\frac{1}{q} + \frac{q}{N^d}\right)^{1/d}.$$

Here the most favorable situation arises when $q \approx N^{d/2}$, and then the upper bound is of the order $N^{1/2+\varepsilon}$.

This leads to a non-trivial estimate for $|S(\alpha)|$.

TABLE 1. Some Exponent Pairs.

(k,ℓ)	Operation
$(0,1)$	
$(1/254,247/254)$	$AAAAAAB$
$(1/126,20/21)$	$AAAAAB$
$(1/86,161/172)$	$AAAABAAB$
$(1/62,57/62)$	$AAAAAB$
$(1/50,181/200)$	$AAABABAAB$
$(1/42,25/28)$	$AAABAAB$
$(2/53,181/212)$	$AABABAAB$
$(1/24,27/32)$	$AABABAAB$
$(1/22,101/121)$	$AABABABAAB$
$(1/20,33/40)$	$AABAAB$
$(13/238,97/119)$	$AABAABAAB$
$(11/186,25/31)$	$AABAAAB$
$(4/49,75/98)$	$ABABAAAB$
$(11/128,97/128)$	$ABABAABAAB$
$(1/11,3/4)$	$ABABAAB$
$(1/10,81/110)$	$ABABABAAB$
$(13/106,75/106)$	$ABAABAAB$
$(11/86,181/258)$	$ABAABABAAB$
$(11/78,161/234)$	$ABAAABAAB$
$(22/117,25/39)$	$BABAAABAAB$
$(26/129,27/43)$	$BABAABABAAB$
$(11/53,33/53)$	$BABAABAAB$
$(13/55,3/5)$	$BABABABAAB$
$(1/4,13/22)$	$BABABAAB$
$(33/128,75/128)$	$BABABAABAAB$
$(13/49,57/98)$	$BABABAAAB$
$(19/62,52/93)$	$BAABAAAB$
$(75/238,66/119)$	$BAABAABAAB$
$(13/40,11/20)$	$BAABAAB$
$(81/242,6/11)$	$BAABABABAAB$
$(11/32,13/24)$	$BAABABAAB$
$(75/212,57/106)$	$BAABABAAAB$
$(11/28,11/21)$	$BAAABAAB$
$(81/200,13/25)$	$BAAABABAAB$
$(13/31,16/31)$	$BAAAAB$
$(75/172,22/43)$	$BAAAABAAB$
$(19/42,32/63)$	$BAAAAAB$
$(60/127,64/127)$	$BAAAAAAB$
$(1/2,1/2)$	B

3. Power Sums

Although we spend a lot of effort to estimate exponential sums, in the opposite direction it is sometimes useful to show that a cancelling sum is not always too cancelling. Turán's method of power sums provides tools of exactly this sort. Since error terms in analytic number theory are often expressed as a sum of oscillatory terms (as we have already seen in the case of the error term in the Prime Number Theorem), Turán's method assists us in proving that such error terms are sometimes large. To exemplify the method, we describe Turán's First Main Theorem. Let

$$s_\nu = \sum_{n=1}^{N} b_n z_n^\nu$$

and suppose that $|z_n| \geq 1$ for all n. Suppose that M is a given non-negative integer. We wish to show that there is a ν, $M + 1 \leq \nu \leq M + N$, such that $|s_\nu|$ is not too small compared with $|s_0|$. To this end we employ a simple duality argument, which is typical of Turán's method. Suppose that numbers a_ν have been determined so that

(3.1)
$$s_0 = \sum_{\nu=0}^{N-1} a_\nu s_{M+1+\nu}.$$

Then

$$|s_0| \leq \left(\sum_{\nu=0}^{N-1} |a_\nu| \right) \max_{0 \leq \nu \leq N-1} |s_{M+1+\nu}|.$$

By the definition of s_ν we see that (3.1) asserts that

$$\sum_{n=1}^{N} b_n = \sum_{n=1}^{N} b_n z_n^{M+1} \sum_{\nu=0}^{N-1} a_\nu z_n^\nu.$$

This identity certainly holds for arbitrary b_n provided that

$$1 = z_n^{M+1} \sum_{\nu=0}^{N-1} a_\nu z_n^\nu$$

for $1 \leq n \leq N$. That is, $P(z) = \sum_{\nu=0}^{N-1} a_\nu z^\nu$ should be a polynomial of degree at most $N - 1$ that satisfies the N conditions $P(z_n) = z_n^{-M-1}$. Without loss of generality the z_n are distinct, and hence $P(z)$ is uniquely determined. It can be shown that $\sum_{\nu=0}^{N-1} |a_\nu| \leq \sum_{k=0}^{N-1} \binom{M+k}{k} 2^k$, and thus we find that

$$\max_{M+1 \leq \nu \leq M+N} |s_\nu| \geq c(M, N) |s_0|$$

where

$$c(M, N) = \left(\sum_{k=0}^{N-1} \binom{M+k}{k} 2^k \right)^{-1}.$$

The constant here is best-possible, but it is disappointingly small, since it is only a little larger than

$$\left(\frac{N}{2e(M+N)} \right)^{N-1}.$$

Suppose that $T(x)$ is an exponential polynomial of N terms and period 1, say

$$T(x) = \sum_{n=1}^{N} b_n e(\lambda_n x)$$

where the λ_n are integers. Let I be a closed arc of the circle group $\mathbb{T}$, and let L denote the length of I. Then by Turán's First Main Theorem, it is easy to show that

$$\max_{x \in I} |T(x)| \geqslant \left(\frac{L}{2e} \right)^{N-1} \max_{x \in \mathbb{T}} |T(x)|.$$

Although the constant here depends on the number N of terms in $T(x)$, it is noteworthy that it is independent of the size of the frequencies λ_n. This inequality makes it possible to give a simple and motivated proof of the Fabry Gap Theorem.

The small constant $c(M, N)$ can be replaced by a larger constant if one is prepared to allow ν to run over a range longer than N. In more restricted situations the lower bound can be very good indeed. For example, suppose that

(3.2)
$$s_\nu = \sum_{n=1}^{N} e(\nu \theta_n).$$

Then

$$\sum_{\nu=1}^{K} \left(1 - \frac{\nu}{K+1} \right) |s_\nu|^2 = \sum_{m=1}^{N} \sum_{n=1}^{N} \sum_{\nu=1}^{K} \left(1 - \frac{\nu}{K+1} \right) e(\nu(\theta_m - \theta_n)).$$

Since the expression is real, we may take real parts to see that the above is

$$= \sum_{m=1}^{N} \sum_{n=1}^{N} \sum_{\nu=1}^{K} \left(1 - \frac{\nu}{K+1} \right) \cos 2\pi \nu (\theta_m - \theta_n) = \sum_{m=1}^{N} \sum_{n=1}^{N} \frac{1}{2} (\Delta_{K+1}(\theta_m - \theta_n) - 1)$$

where $\Delta_{K+1}(\theta)$ denotes the Fejér kernel. Since $\Delta_{K+1}(\theta) \geqslant 0$ for all θ, and $\Delta_{K+1}(0) = K+1$, it follows that the above is

$$\geqslant \frac{1}{2}(K+1)N - \frac{1}{2}N^2.$$

Thus if $K \geqslant (1+\varepsilon)N$ then $\max_{1\leqslant\nu\leqslant K}|s_\nu| \geqslant C(\varepsilon)\sqrt{N}$, and in particular

$$\max_{1\leqslant\nu\leqslant 2N}|s_\nu| \geqslant \sqrt{N/2}.$$

By working similarly with higher moments one can obtain still better lower bounds over longer ranges of ν: If s_ν is given by (3.2) and $1 \leqslant m \leqslant N/2$ then there is a ν, $1 \leqslant \nu \leqslant (12N/m)^m$ such that $|s_\nu| \geqslant \frac{1}{4}\sqrt{mN}$. It would be useful to have examples to show that this is close to best-possible.

For a more extensive survey of Turán's method one may consult Montgomery [21], pp. 85–107. For a detailed account of the subject and important applications, one should see the book of Turán [33].

4. Irregularities of Distribution

We now consider how well-distributed N points can be. If the points are to fall in $[0,1]$ then we could take $u_n = n/N$. These points are extremely well-distributed in the sense that the number of them in a subinterval $[0,\alpha]$ is $N\alpha + O(1)$. However, we find that it is not so easy to distribute points well in $\mathbb{T}^2$. Suppose that our points are $u_n = (u_1, u_2)$, let $\mathcal{R}(\alpha)$ be the rectangle $\mathcal{R}(\alpha) = [0,\alpha_1] \times [0,\alpha_2]$, and let $D(\alpha)$ be the discrepancy function

$$D(\alpha) = \mathrm{card}\{1 \leqslant n \leqslant N : u_n \in \mathcal{R}(\alpha)\} - N\alpha_1\alpha_2.$$

Roth [30] used a construction suggestive of wavelets to show that

$$\int_{\mathbb{T}^2} D(\alpha)^2\, d\alpha \gg \log N.$$

Since it is also possible to construct points that are this well-distributed, this solves the problem as to distribution in mean-square. Ostrowski [27] had observed that $D(\alpha) \ll \log N$ when $u_n = (n/N, n\sqrt{2})$. The problem of showing that in any case $\|D\|_\infty \gg \log N$ was solved by Schmidt [31] by means of a complicated induction. However, a curious difference arises here. Roth's argument generalizes easily to $\mathbb{T}^k$ to show that $\|D\|_2 \gg_k (\log N)^{(k-1)/2}$ for any $k > 1$. However, Schmidt's approach has not been extended to $k > 2$. It has been conjectured that $\|D\|_\infty \gg_k (\log N)^{k-1}$. This would be best possible, in view of constructions of Halton [14]. On the other hand, Pollington has recently mounted a wavelet approach to this problem that has led him to conclude that one should be able to show that $\|D\|_\infty \gg_k (\log N)^{k/2}$, and that this sup norm need not be larger. Hence one should regard the question of how large $\|D\|_\infty$ need be to be a wide open unsolved problem when $k > 2$. Halász [13] devised a variant of Roth's method that gives Schmidt's Theorem concerning $\|D\|_\infty$ when $k = 2$, and also gives the lower bound $\|D\|_1 \gg \sqrt{\log N}$ when $k = 2$. This is best possible, since Chen [8] has shown that if $0 < p < \infty$ and k is given, $k \geqslant 2$, then there exists a configuration of N points in $\mathbb{T}^k$ for which $\|D\|_p \ll_{p,k} (\log N)^{(k-1)/2}$.

Roth obtained his lower bound by the Cauchy–Schwarz inequality,

$$\int_{\mathbb{T}^k} D(\alpha) F(\alpha)\, d\alpha \leqslant \left(\int_{\mathbb{T}^k} D(\alpha)^2\, d\alpha \right)^{1/2} \left(\int_{\mathbb{T}^k} F(\alpha)^2\, d\alpha \right)^{1/2}$$

where F is a test function defined to be a sum, $F = \sum F_r$, of more basic orthogonal functions. Halász similarly used the inequality

$$\int_{\mathbb{T}^2} D(\alpha) F(\alpha)\, d\alpha \leqslant \|D\|_\infty \|F\|_1$$

where $F = \prod_r (1 + cF_r)$. This bears a strong resemblance to a Riesz product, as occurs in the theory of lacunary trigonometric series. For the lower bound of $\|D\|_1$, he wrote

$$\int_{\mathbb{T}^2} D(\alpha) F(\alpha)\, d\alpha \leqslant \|D\|_1 \|F\|_\infty$$

where

$$F = \prod_{r=1}^{R} \left(1 + \frac{i}{\sqrt{R}} F_r \right).$$

Since $-1 \leqslant F_r \leqslant 1$, it follows that $|F| \leqslant (1 + 1/R)^R \ll 1$.

Let $e(x_1), e(x_2), \ldots$ be an infinite sequence of unimodular complex numbers, and put $P_N(z) = \prod_{n=1}^{N}(z - e(x_n))$. Erdős asked whether it is possible to choose the x_n in such a way that the numbers $\max_{|z| \leqslant 1} |P_N(z)|$ are bounded as $N \to \infty$. Note that $\Im \log P_N(e(x))$ is just the discrepancy of the first N points, while the problem now being considered involves the harmonic conjugate $\Re \log P_N(e(x))$, but with the important difference that we need this quantity to be large and positive, not just large in absolute value. That this sequence can not remain bounded was first proved by Wagner [35], by means of a modified form of Schmidt's method. Later Beck [3] used Halász's modified form of Roth's method to obtain this in the following sharp quantitative form: There is an absolute constant $\delta > 0$ such that

$$\max_{|z| \leqslant 1} |P_N(z)| > N^\delta$$

for infinitely many N.

Measuring the distribution of points in $\mathbb{T}^k$ relative to rectangles with sides parallel to the coordinate axes is only one of many possibilities. If $\mathcal{S}$ is a measurable set then the quantity

$$D(\mathcal{S}) = \text{card}\{1 \leqslant n \leqslant N : u_n \in \mathcal{S}\} - N \operatorname{vol} \mathcal{S}$$

provides a measure of the distribution of the u_n. But when we consider $\mathcal{S}$ we would also include its translates, so we put $d(\alpha) = D(\mathcal{S} + \alpha)$. Then $\hat{d}(0) = 0$, but for $k \neq 0$ we have $\hat{d}(k) = \hat{\chi}_{\mathcal{S}}(-k)\hat{U}_N(k)$ where $\hat{U}_N(k) = \sum_{n=1}^{N} e(-k \cdot u_n)$. Hence by Parseval's identity,

$$\int_{\mathbb{T}^k} d(\alpha)^2\, d\alpha = \sum_{k \neq 0} |\hat{\chi}_{\mathcal{S}}(k)|^2 |\hat{U}_N(k)|^2.$$

The rate that $\widehat{\chi}_{\mathcal{S}}(\boldsymbol{k})$ tends to 0 as $\boldsymbol{k}$ tends to infinity in a particular direction depends on how much of the boundary of $\mathcal{S}$ is orthogonal to the direction in question. Thus in the case of a rectangle the Fourier coefficients decay slowly in the direction of the coordinate axes, but comparatively rapidly in other directions. Thus one may expect that points may be found so that $\widehat{U}_N(\boldsymbol{k})$ is small when $\widehat{\chi}_{\mathcal{S}}(\boldsymbol{k})$ is large, and vice versa. By using the Fejér kernel as in our discussion of power sums we can show that

$$\sum_{\substack{|k_1|\leqslant X_1 \\ |k_2|\leqslant X_2 \\ k\neq 0}} |\widehat{U}_N(\boldsymbol{k})|^2 \geqslant NX_1X_2 - N^2$$

for any positive real numbers X_1, X_2. In this way we can recover Roth's lower bound for $\|D\|_2$. By averaging over disks (and averaging over the radius as well) we find that there is a disk $\mathcal{D}$ for which $D(\mathcal{D}) \gg N^{1/4}$. More generally, if we start with a set $\mathcal{S}$, and are allowed to shrink, translate, and rotate $\mathcal{S}$, then we obtain this larger order of magnitude, in view of a general principle governing the mean square decay of the Fourier transform of the characteristic function of a set: If $\mathcal{C}$ is a simple, closed, piecewise C^1 curve in $\mathbb{R}^2$, and $\mathcal{S}$ is its interior, then

$$\int_{|t|\geqslant R} |\widehat{\chi}_{\mathcal{S}}(t)|^2 \, dt \sim \frac{|\mathcal{C}|}{2\pi^2 R}$$

as $R \to \infty$. This is due to Montgomery [19], [20] pp. 114–119; see also Herz [15, 16].

5. The Large Sieve

The large sieve was originated by Linnik [18] in a somewhat obscure form. It gained new life in the hands of Rényi [28], who viewed it as a statement about almost independent vectors. Today we usually think of this as an extension of Bessel's inequality for vectors in an inner product space: Let $\phi_1,\ldots,\phi_R$ be arbitrary vectors in an inner product space. Then the following three assertions concerning the constant C are equivalent:

(a) For any vector ξ in the space,

$$\sum_{r=1}^{R} |(\xi, \phi_r)|^2 \leqslant C\|\xi\|^2;$$

(b) For any complex numbers u_r we have

$$\left| \sum_{1\leqslant r,s\leqslant R} u_r\overline{u}_s(\phi_r, \phi_s) \right| \leqslant C \sum_{r=1}^{R} |c_r|^2;$$

(c) $C = \rho\big([(\phi_r, \phi_s)]\big)$.

(Here $\rho(A)$ denotes the spectral radius of the matrix A.) Rényi took the coordinates of his vectors to depend on arithmetic progressions or on Dirichlet characters, but the vectors he obtained in doing so were not very close to orthogonal, so the estimates he obtained were imperfect. Roth [29] had the excellent idea of taking $\phi_r = (e(n\alpha_r))$ where the α_r are well-spaced in $\mathbb{T}$. These vectors are quite close to being orthogonal, and we now think of the most basic form of the large sieve as being a statement about the mean square of a trigonometric polynomial at well-spaced points. That is, we take

$$S(\alpha) = \sum_{n=M+1}^{M+N} a_n e(n\alpha),$$

and we consider inequalities of the shape

$$\sum_{r=1}^{R} |S(\alpha_r)|^2 \leqslant \Delta \sum_{n=M+1}^{M+N} |a_n|^2.$$

We suppose that $\|\alpha_r - \alpha_s\| \geqslant \delta$ for $r \neq s$, and want Δ to depend on N and δ. If $a_n = e(-n\alpha_1)$ then $S(\alpha_1) = N$, and thus we must have $\Delta \geqslant N$. By averaging over translations of the α_r we can also show that $\Delta \geqslant \delta^{-1} - 1$. We find that Δ does not have to be much larger than is required by these considerations.

Gallagher [10] used an inequality of the Sobolev type,

$$|f(\alpha)| \leqslant \frac{1}{\delta} \int_{\alpha-\delta/2}^{\alpha+\delta/2} |f(x)|\,dx + \frac{1}{2} \int_{\alpha-\delta/2}^{\alpha+\delta/2} |f'(x)|\,dx,$$

to show that one can take $\Delta = 1/\delta + \pi N$. This is the best constant with respect to δ, but in arithmetic settings the coefficient of N is more important. The main advantage of Gallagher's approach is that it generalizes readily to other families of functions. To obtain good dependence on N we note that by duality the stated inequality is equivalent to the inequality

$$\sum_{n=M+1}^{M+N} \left| \sum_{r=1}^{R} y_r e(n\alpha_r) \right|^2 \leqslant \Delta \sum_{r=1}^{R} |y_r|^2.$$

On the left hand side we square out and take the sum over n inside. The diagonal terms give $N \sum_r |y_r|^2$, so it remains to consider the non-diagonal terms. This brings us to Hilbert's Inequality, which asserts that

$$\left| \sum_{\substack{r,s \\ r \neq s}} \frac{y_r \overline{y_s}}{r - s} \right| \leqslant \pi \sum_r |y_r|^2.$$

Montgomery and Vaughan [23], with some assistance from Selberg, found that this can be generalized, so that

$$\left| \sum_{\substack{r,s \\ r \neq s}} \frac{y_r \overline{y_s}}{\lambda_r - \lambda_s} \right| \leq \frac{\pi}{\delta} \sum_r |y_r|^2$$

provided that $|\lambda_r - \lambda_s| \geq \delta$ whenever $r \neq s$. Moreover for the circle group we have, correspondingly, the inequality

$$\left| \sum_{\substack{r,s \\ r \neq s}} \frac{y_r \overline{y_s}}{\sin \pi(\alpha_r - \alpha_s)} \right| \leq \frac{1}{\delta} \sum_r |y_r|^2$$

where $\|\alpha_r - \alpha_s\| \geq \delta$ for $r \neq s$. This gives the large sieve with the factor $\Delta = N + 1/\delta$. With a little more care one can obtain $\Delta = N + 1/\delta - 1$, and with this constant there are situations in which equality can occur.

In arithmetic situations the α_r are usually taken to be the Farey fractions a/q, in which $(a, q) = 1$ and $q \leq Q$. Since $\|a/q - a'/q'\| \geq 1/(qq') \geq 1/Q^2$ when a/q and a'/q' are distinct modulo 1, we find that

$$\sum_{q=1}^{Q} \sum_{\substack{a=1 \\ (a,q)=1}} |S(a/q)|^2 \leq (N + Q^2) \sum_{n=M+1}^{M+N} |a_n|^2.$$

The generalized Hilbert Inequality can also be established in a weighted form, in which we find that

$$\left| \sum_{\substack{r,s \\ r \neq s}} \frac{y_r \overline{y_s}}{\lambda_r - \lambda_s} \right| \leq \frac{3}{2} \pi \sum_r \frac{|y_r|^2}{\delta_r}$$

where $|\lambda_r - \lambda_s| \geq \delta_r$ when $s \neq r$. Here the constant $\frac{3}{2}\pi$ is certainly not best possible. The above also has a counterpart for the circle group, and hence we have a weighted form of the large sieve,

$$\sum_{r=1}^{R} \frac{|S(\alpha_r)|^2}{N + \frac{3}{2\delta_r}} \leq \sum_{n=M+1}^{M+N} |a_n|^2.$$

Many other variants of the large sieve have been derived, involving, for example, maximal partial sums (via the Carleson–Hunt Theorem), or the Hardy–Littlewood Maximal Theorem. As for further generalizations of Hilbert's Inequality, Montgomery and Vaaler [22] have shown that if $\rho_r = \beta_r + i\gamma_r$ with $\beta_r \geq 0$ for all r and $|\gamma_r - \gamma_s| \geq \delta_r$ for $s \neq r$, then

$$\left| \sum_{\substack{r,s \\ r \neq s}} \frac{y_r \overline{y_s}}{\rho_r + \overline{\rho_s}} \right| \leq 84 \sum_r \frac{|y_r|^2}{\delta_r}.$$

The proof depends on the theory of H^2 functions that are analytic in a half-plane.

6. Dirichlet Series

Let $D(s) = \sum_{n=1}^{N} a_n n^{-s}$. By Hilbert's Inequality we see that

$$\int_0^T |D(it)|^2 \, dt = (T + O(N)) \sum_{n=1}^{N} |a_n|^2,$$

and the weighted Hilbert Inequality gives

$$\int_0^T |D(it)|^2 \, dt = \sum_{n=1}^{N} |a_n|^2 (T + O(n)).$$

Thus we have some limitation on the amount of time that $|D|$ is large. Suppose that $|a_n| \leqslant 1$ for all n. Then by applying the above to D^2 we find that

$$\int_0^T |D(it)|^4 \, dt \ll (T + N^2) N^{2+\varepsilon}.$$

It would be very useful if we could interpolate between these estimates, in the sense that

$$\int_0^T |D(it)|^q \, dt \ll (T + N^{q/2}) N^{q/2+\varepsilon}$$

for real q, $2 \leqslant q \leqslant 4$. Indeed, this implies the Density Hypothesis concerning the Riemann zeta function. Many years ago, Hardy and Littlewood had conjectured that if $|\hat{f}(k)| \leqslant \hat{F}(k)$ for all k and $q \geqslant 2$ then $\|f\|_q \ll_q \|F\|_q$, and it can be shown that this majorant conjecture would imply the conjecture above. However, Bachelis [1] used a method of Katznelson to show that this is true only when q is an even integer (in which case it holds with constant 1).

In any case there is more that can be said about the number of times a Dirichlet polynomial can be large than follows from moment estimates. For example, if $0 \leqslant t_1 < t_2 < \ldots < t_R \leqslant T$ and $t_{r+1} - t_r \geqslant 1$ for all r, if $|D(it_r)| \geqslant V$ for all r, and if $|a_n| \leqslant 1$ for all n, then it is known that $R \ll N^2 V^{-2} T^\varepsilon$ provided that $V^2 > NT^{1/2+\varepsilon}$. It has been conjectured that this estimate for R holds when the last condition is weakened to read $V^2 \geqslant NT^\varepsilon$. However, Bourgain [7] has shown that if this is so then every Kakeya set in $\mathbb{R}^d$, $d \geqslant 2$, has Hausdorff dimension d. Thus there are those that doubt such a strong conjecture.

7. Spectral Characteristics of Zeros of the Zeta Function

Let $h(d)$ denote the class number of the quadratic number field with discriminant d. In an effort to derive a useful lower bound for $h(d)$ when d is negative, it was recognized that it would suffice to have a good supply of pairs of zeros of the Riemann zeta function that are $< c\frac{2\pi}{\log t}$ apart where $c < 1/2$. With this motivation, an attempt was made in 1971 to determine the distribution of $\gamma - \gamma'$ as γ and γ' run over nearby ordinates of zeros of the zeta function. One begins with a generalization of the explicit formula noted earlier. We observe that

$$\sum_{n \leqslant x} \Lambda(n) n^{-s} = \frac{-1}{2\pi i} \int_{c-i\infty}^{c+i\infty} \frac{\zeta'}{\zeta}(s+w) \frac{x^w}{w} \, dw$$

$$= -\frac{\zeta'}{\zeta}(s) + \frac{x^{1-s}}{1-s} - \sum_{\rho} \frac{x^{\rho-s}}{\rho - s} + \sum_{n=1}^{\infty} \frac{x^{-2n-s}}{2n+s}.$$

We assume the Riemann Hypothesis and combine two such explicit formulæ to see that

$$2 \sum_{\gamma} \frac{x^{i\gamma}}{1+(t-\gamma)^2} = -x^{-1/2} \left(\sum_{n \leqslant x} \Lambda(n) \left(\frac{x}{n}\right)^{-1/2+it} + \sum_{n>x} \Lambda(n) \left(\frac{x}{n}\right)^{3/2+it} \right)$$

$$+ x^{-1+it} \left(\log t + O(1) \right) + O(x^{1/2}/t).$$

We take the modulus squared of both sides, and integrate over $0 \leqslant t \leqslant T$. We write $x = T^\alpha$ for $0 \leqslant \alpha \leqslant 1$, and note that the expression on the right hand side can be asymptotically evaluated by using our mean square estimate for Dirichlet polynomials. Let

$$F(\alpha) = \left(\frac{T}{2\pi} \log T\right)^{-1} \sum_{\substack{0 < \gamma \leqslant T \\ 0 < \gamma' \leqslant T}} T^{i\alpha(\gamma-\gamma')} w(\gamma - \gamma')$$

where $w(u) = 4/(4+u^2)$. Then we find that F is real, even, non-negative, and $F(\alpha) = (1 + o(1)) T^{-2\alpha} \log T + \alpha + o(1)$ as $T \to \infty$, uniformly for $0 \leqslant \alpha \leqslant 1$. When $\alpha > 1$ the method fails because the Dirichlet polynomial is too long. But only the terms near the diagonal contribute, and one can use the Hardy–Littlewood quantitative form of the Twin Prime Conjecture to estimate that those terms contribute. In this way one is led to guess that $F(\alpha) = 1 + o(1)$ uniformly for $1 \leqslant \alpha \leqslant A$, for any $A > 1$. This is known as the Strong Pair Correlation Conjecture. By taking Fourier transforms, we are led to a conjecture concerning the distribution of the frequencies $\gamma - \gamma'$: The number of pairs γ, γ' of ordinates of zeros, $0 \leqslant \gamma \leqslant T$, $0 \leqslant \gamma' \leqslant T$, for which $2\pi\alpha/\log T \leqslant \gamma - \gamma' \leqslant 2\pi\beta/\log T$ is asymptotic to

$$\left(\delta + \int_\alpha^\beta 1 - \left(\frac{\sin \pi u}{\pi u}\right)^2 du \right) \frac{T}{2\pi} \log T.$$

Here $\delta = 1$ if $0 \in [\alpha, \beta]$, and $\delta = 0$ otherwise. That is, δ is a Dirac point mass at 0. This arises because of the possibility that $\gamma = \gamma'$ when $\alpha < 0 < \beta$. This is the Weak Pair Correlation Conjecture. Freeman Dyson observed that the density function here is that of a random hermitian matrix of unitary type, and thus we take the above, although only a conjecture, as evidence that the zeros of the zeta function are spectral in nature.

Goldston and Montgomery [12] showed that if the Riemann Hypothesis is true then the Strong Pair Correlation Conjecture is equivalent to the estimate

$$\int_0^X (\psi(x+h) - \psi(x) - h)^2\, dx \sim hX \log X/h$$

for $x^\varepsilon \leqslant h \leqslant x^{1-\varepsilon}$. A heuristic argument in favor of this can be obtained by expanding out and using the Hardy–Littlewood Conjecture concerning the number of d-twin primes not exceeding x.

Recently Montgomery and Soundararajan [26] considered the higher moments

$$\mu_k(X, h) = \frac{1}{X} \int_0^X (\psi(x+h) - \psi(x) - h)^k\, dx.$$

On expanding, one encounters enumerations of prime k-tuples. The Hardy–Littlewood Prime k-tuple Conjecture asserts that if $d_1, d_2, \ldots d_k$ are distinct integers then

$$\sum_{n \leqslant X} \prod_{i=1}^{k} \Lambda(n + d_i) = \mathfrak{S}(\mathcal{D})X + E(X, \mathcal{D})$$

where $\mathfrak{S}(\mathcal{D})$ is the 'singular series'

$$\mathfrak{S}(\mathcal{D}) = \sum_{\substack{1 \leqslant q_i < \infty \\ (1 \leqslant i \leqslant k)}} \prod_{i=1}^{k} \frac{\mu(q_i)}{\phi(q_i)} \sum_{\substack{1 \leqslant a_i \leqslant q_i \\ (a_i, q_i) = 1 \\ \sum a_i/q_i \in \mathbb{Z}}} e\left(\sum_{i=1}^{k} \frac{a_i d_i}{q_i}\right).$$

Gallagher [11] considered the moments $\mu_k(X, h)$ for smaller h of the form $h = c \log X$, and for that purpose showed that

$$\sum_{\substack{0 \leqslant d_i \leqslant X \\ d_i \text{ distinct} \\ (1 \leqslant i \leqslant k)}} \mathfrak{S}(\mathcal{D}) \sim X^k$$

as $X \to \infty$. For larger h, the mean value h is much larger than the usual size of the difference $\psi(x + h) - \psi(x) - h$, and so it is useful to consider the arithmetic function $\Lambda_0(n) = \Lambda(n) - 1$, whose mean value is asymptotically zero. For this function we have an alternative formulation of the prime k-tuple Conjecture, which asserts that

$$\sum_{n \leqslant X} \prod_{i=1}^{k} \Lambda_0(n + d_i) = \mathfrak{S}_0(\mathcal{D})X + E_0(X, \mathcal{D})$$

where now

$$\mathfrak{S}_0(\mathcal{D}) = \sum_{\substack{1<q_i<\infty \\ (1\leqslant i\leqslant k)}} \prod_{i=1}^{k} \frac{\mu(q_i)}{\phi(q_i)} \sum_{\substack{1\leqslant a_i\leqslant q_i \\ (a_i,q_i)=1 \\ \sum a_i/q_i \in \mathbb{Z}}} e\Big(\sum_{i=1}^{k} \frac{a_i d_i}{q_i}\Big).$$

Thus $\mathfrak{S}_0$ is the same as $\mathfrak{S}$ except that the possibility that $q_i = 1$ is now excluded. The mean value of $\mathfrak{S}_0$ is of course smaller, and hence more difficult to determine, but by elaborating on work of Montgomery and Vaughan [24] concerning the distribution of reduced residues modulo q in short intervals it can be shown that

$$\sum_{\substack{0\leqslant d_i\leqslant X \\ d_i \text{ distinct} \\ (1\leqslant i\leqslant k)}} \mathfrak{S}_0(\mathcal{D}) = \begin{cases} \dfrac{k!}{(k/2)!2^{k/2}}(-X\log X)^{k/2} + O\big(X^{k/2}(\log X)^{k/4}\big) & \text{if } k \text{ is even,} \\[2ex] O\big(X^{k/2-1/(7k)+\varepsilon}\big) & \text{if } k \text{ is odd.} \end{cases}$$

This is established unconditionally; when combined with plausible hypotheses concerning the size and behavior of the error terms $E(X,h)$ we are led to expect that

$$\mu_k(X,h) = \big(c_k + o(1)\big)Xh^{k/2}(\log X/h)^{k/2}$$

where the c_k are the moments of the normalized normal variable,

$$c_k = \begin{cases} \dfrac{k!}{(k/2)!2^{k/2}} & \text{if } k \text{ is even,} \\[2ex] 0 & \text{if } k \text{ is odd.} \end{cases}$$

Since these moments occur uniquely in the case of normal distribution, we are led to expect that the distribution of $\psi(x+h) - \psi(x)$, for $0 \leqslant x \leqslant X$, is approximately normal with mean h and variance $h\log X/h$.

Suppose that $X > T$. In terms of zeros of the zeta function, the Strong Pair Correlation Conjecture seems to be telling us that the mean square size of the sum

$$\sum_{0<\gamma\leqslant T} \cos\gamma\log x \qquad (0\leqslant x\leqslant X)$$

is the same as if the terms were uncorrelated random variables. (We recall from the theory of probability that if X_i are uncorrelated variables then $\mathrm{Var}\left(\sum X_i\right) = \sum \mathrm{Var}(X_i)$.) It seems that our new speculations concerning the μ_k can similarly be interpreted as asserting that the above sum is distributed as if the terms are independent random variables (as in the Central Limit Theorem). How this relates to the spectral nature of the zeros, or any possible underlying operators remains to be seen.

References

[1] G. F. Bachelis, *On the upper and lower majorant properties in* $L^p(G)$, Quart. J. Math. (Oxford) (2) **24** (1973), 119–128.

[2] J. T. Barton, H. L. Montgomery and J. D. Vaaler, *Note on a Diophantine inequality in several variables*, to appear.

[3] J. Beck, *The modulus of polynomials with zeros on the unit circle: a problem of Erdős*, Ann. of Math. (2) **134** (1991), 609–651.

[4] J. Beck and W. W. L. Chen, *Irregularities of distribution*, Cambridge Tract 89, Cambridge University Press, Cambridge, 1987.

[5] A. Beurling, *Sur les intégrales de Fourier absolument convergentes et leur application à une transformation fonctionelle*, Neuvième congrès des mathématiciens scandinaves, Helsingfors, 1938

[6] E. Bombieri and H. Iwaniec, *On the order of* $\zeta(\frac{1}{2} + it)$, Ann. Scuola Norm. Sup. Pisa Cl. Sci. (4) **13** (1986), 449–472.

[7] J. Bourgain, *On the distribution of Dirichlet sums*, J. Anal. Math. **60** (1993), 21–32.

[8] W. W. L. Chen, *On irregularities of distribution, I*, Mathematika **27** (1980), 153–170.

[9] P. Erdős and P. Turán, *On a problem in the theory of uniform distribution* I, Nederl. Akad. Wetensch. Proc. **51** (1948), 1146–1154; II, 1262–1269, (= Indag. Math. **10**, 370–378; 406–413).

[10] P. X. Gallagher, *The large sieve*, Mathematika **14** (1967), 14–20.

[11] ————, *On the distribution of primes in short intervals*, Mathematika **23** (1976), 4–9.

[12] D. A. Goldston and H. L. Montgomery, *On pair correlations of zeros and primes in short intervals*, Analytic number theory and Diophantine problems (Stillwater, 1984), Birkhäuser Verlag, Boston–Basel–Berlin, 1987, 183–203.

[13] G. Halász, *On Roth's method in the theory of irregularities of point distributions*, Recent Progress in Analytic Number Theory (Durham, 1979), Vol. 2, Academic Press, London, 1981, 79–84.

[14] J. H. Halton, *On the efficiency of certain quasirandom sequences of points in evaluating multidimensional integrals*, Num. Math. **2** (1960), 84–90.

[15] C. S. Herz, *Fourier transforms related to convex sets*, Ann. of Math. (2) **75** (1961), 81–92.

[16] C. S. Herz, *On the number of lattice points in a convex set*, Amer. J. Math. **84** (1962), 126–133.

[17] M. N. Huxley, *Area, Lattice Points and Exponential Sums*, Clarendon Press, Oxford, 1996.

[18] Ju. V. Linnik, *The large sieve*, Dokl. Akad. Nauk SSSR **30** (1941), 292–294.

[19] H. L. Montgomery, *The analytic principle of the large sieve*, Bull. Amer. Math. Soc. **84** (1978), 547–567.

[20] H. L. Montgomery, *Irregularities of distribution*, Congress of Number Theory (Zarautz, 1984), Universidad del Pais Vasco, Bilbao, 1989, 11–27.

[21] H. L. Montgomery, *Ten lectures on the interface between analytic number theory and harmonic analysis*, CBMS No. 84, Amer. Math. Soc., Providence, 1994.

[22] H. L. Montgomery and J. D. Vaaler, *A further generalization of Hilbert's inequality*, Mathematika **45** (1999), 35–39.

[23] H. L. Montgomery and R. C. Vaughan, *Hilbert's inequality*, J. London Math. Soc. (2) **8** (1974), 73–82.

[24] ————, *On the distribution of reduced residues*, Annals of Math. **123** (1986), 311–333.

[25] H. L. Montgomery and K. Soundararajan, *Beyond pair correlation*, to appear.

[26] H. L. Montgomery and K. Soundararajan, *Primes in short intervals*, to appear.

[27] A. Ostrowski, *Bemerkungen zur Theorie der Diophantischen Approximationen* I, Abh. Math. Sem. Hamburg **1** (1922), 77–98; II, **1** (1922), 250–251; III, **41** (1926), 224.

[28] A. Rényi, *Un nouveau théorème concernant les fonctions indépendantes et ses applications à la théorie des nombres*, J. Math. Pures Appl. **28** (1949), 137–149.

[29] K. F. Roth, *On irregularities of distribution*, Mathematika **1** (1954), 73–79.

[30] K. F. Roth, *On the large sieves of Linnik and Rényi*, Mathematika **12** (1965), 1–9.

[31] W. M. Schmidt, *Irregularities of distribution, VII*, Acta Arith. **21** (1972), 49–50.

[32] A. Selberg, *Collected Papers*, Volume II, Springer-Verlag, Berlin, 1991.

[33] P. Turán, *On a new method of analysis and its applications*, Wiley-Interscience, New York, 1984.

[34] J. D. Vaaler, *Some extremal functions in Fourier analysis*, Bull. Amer. Math. Soc. **12** (1985), 183–216.

[35] G. S. Wagner, *On a problem of Erdős in Diophantine approximation*, Bull. London Math. Soc. **12** (1980), 81–88.

[36] H. Weyl, *Über ein Problem aus dem Gebiete der diophantischen Approximationen*, Nachr. Ges. Wiss. Göttingen, Math.-phys. Kl. (1914), 234–244; Gesammelte Abhandlungen, Band I, Springer-Verlag, Berlin–Heidelberg–New York, 1968, 487–497.

[37] *Über die Gleichverteilung von Zahlen mod. Eins*, Math. Ann. **77** (1916), 313–352; Gesammelte Abhandlungen, Band I, Springer-Verlag, Berlin–Heidelberg–New York, 1968, 563–599; Selecta Hermann Weyl, Birkhäuser Verlag, Basel–Stuttgart, 1956, 111–147.

Mathematics of Radar

Bill Moran

1. Radar Fundamentals

1.1. *Introduction*

Radar is now used in many applications — meteorology, mapping, air traffic control, ship and aircraft navigation, altimeters on aircraft, police speeding control, etc. It is now being used in the form of ground penetrating radar for mineral exploration and delineation and for land-mine detection. Of course, its primary role is in defence. The theory of radar is well developed and has many interesting and difficult mathematical problems.

The aim of these notes is to provide a description of radar and its theory accessible to a mathematical audience with the hope of stimulating interest in the problems. Often mathematical treatments of radar ignore the problems of implementation. I intend to go as far into the engneering of a radar system as is necesssary to provide an understanding of these issues. In the first section we shall describe the operation of a radar system at a relatively detailed level. Later sections will cover some of the mathematics arising in radar design and use.

1.2. *A Simple Radar System*

The key references for this section are [3, 14, 16]. Typically radar systems comprise a transmitter and receiver, though recently passive radar systems relying on the ambient HF and VHF radiation have been built and made operational. We shall not consider such systems. We shall also assume that the radar transmitter and receiver are collocated (*monostatic*), again the usual situation. Nonetheless there is much current interest in *bistatic* and *multistatic* radars

The most important issues in designing a radar system are as follows.

1. What is its purpose? How it will be used significantly changes the method of processing the data. Some radars are used to detect targets, others to track them, others to produce images. Radars used to check the speeds of vehicles, for example, need to be able to measure the doppler accurately in a small range of velocities.
2. The next most serious issue is that of noise. Returns from relatively distant targets have very small amounts of energy which can be swamped by the noise generated

J.S. Byrnes (ed.), Twentieth Century Harmonic Analysis - A Celebration, 295–328.

within the receiver. Methods of maximizing the post processing signal-to-noise ratio are crucially important in radar systems.

3. As well as noise, a radar system will receive returns from objects which are not important for the operational purpose of the radar. For example, shipborn radars, typically receive much of the return energy from the sea surface, particularly if it is rough. This can mask a more important return. Such artefacts are collectively called "clutter". Much work has been done on extraction of signals of importance from the clutter.

4. The power available to the transmitter is a key factor, particularly in reducing noise problems. Evidently this dictates much about the size of the transmitter and the type of electronics to be used.

5. The resolution needed in both doppler and range and the range of values of these parameters for the targets of importance are significant factors in the design of a radar.

6. Of increasing importance in a defence environment are issues around jamming and other electronic counter-measures to obscure or mask a target, or otherwise render a radar system ineffective. Radar system developers have to be ready to counter such interventions.

A simple radar system is sketched in Figure 1.

At a general level, the various parts of a radar system are:

1. The antenna — usually serving the dual role of transmitting and receiving signals. The shape of the antenna dictates the shape of the beam that is transmitted. Focussing allows radar systems to put power where it is most needed. Inevitably all beams have some sidelobes. Shaping of these is an interesting mathematical problem which we do not have time to cover here. Many modern radars use phased array antennas which comprise many small antennas each of which can be assigned a separate (complex) weighting. Such a system allows the radar beam to be steered electronically rather than mechanically as is the case for conventional parabolic dish antennas.

2. The duplexer — this switches the antenna between transmission and reception. Both cannot happen at the same time usually. Typicallya radar will transmit for no more than 10% of time. During most of the remainder of the time it will be in receive mode.

3. The radio frequency (RF) oscillator — provides the carrier signal for the radar. This will usually be at least 10 times the bandwidth of the (intermediate frequency coded) waveform it carries.

4. The RF mixers — one in each of the transmitter and receiver. These serve to put the waveform onto the carrier signal in the transmitter and to remove it in the receiver by mixing it with the carrier and then passing it to the low pass filters.

5. Filters and amplifiers — these amplify the various signals and filter out unwanted frequencies. Filtering is a significant operation for the removal of noise. Noise is typically "broadband" in that its spectrum is wide. By filtering out many regions of the spectrum such noise is reduced. Amplifiers increase the power in the signals (and

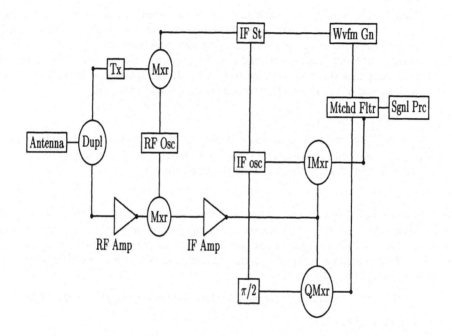

FIGURE 1. A simple radar system

the noise which is present) to make them less susceptible to noise in the later parts of
the receiver. Filtering also serves to remove high frequency components after mixing.
6. Components corresponding to the ones in 3, 4, 5 for the intermediate frequency (IF)
   stages of the system; in particular the demodulation mixers and filters which produce

the I and Q channels. These are obtained by mixing with two IF sinusoids $\pi/2$ radians out of phase with each other.

7. Waveform generator — produces the desired waveform for modulation onto the IF carrier and then onto the RF carrier for transmission. The waveform is also used to " matched filter" the return. We shall spend some time discussing waveform design in section 3.

8. Signal processing block — this serves to detect, identify, estimate the range and/or doppler of targets, to remove clutter and noise not removed earlier in the processing, etc. As analogue to digital (A/D) converters become faster and more accurate, signal processing is increasingly accomplished by converting the signals to digital form and then using digital circuitry and even software. In fact technology already at the (low) IF stage and I/Q channel separation can be accomplished digitally.

1.3. *Doppler and Range*

We begin here to build the mathematical model of radar processing, by and large following Borden's treatment [3]. Usually a transmitted signal comprises a slowly varying *waveform* superimposed on a rapidly oscillating sinusoid. Thus

$$(1.1) \qquad\qquad s(t) = w(t) . \cos(2\pi(f_c t + \phi(t)))$$

where $w(t)$ is the amplitude modulation waveform, $\phi(t)$ represents the frequency or phase modulation and f_c is the carrier frequency. It is important to make the rather obvious observation at this stage that all signals transmitted and received are real-valued. Of course we can represent the signal as the result of amplitude modulating two sinusoids with *opposite phase* (that is, differing by $\pi/2$), thus

$$s(t) = w(t) . \cos(2\pi\phi(t))cos(2\pi f_c t) - w(t) \sin(2\pi\phi(t)) \sin(2\pi f_c t),$$

which in *complex form* is

$$(1.2) \qquad\qquad s(t) = \Re\big(w(t) . e^{2\pi\phi(t)} . e^{2\pi f_c t}\big),$$

provided the waveform $w(t)$ is real. We write

$$(1.3) \qquad\qquad s_0(t) = w(t) e^{2\pi\phi(t)}$$

for the complexification of the slowly varying part.

Much of the theory of radar processing takes place in the complex domain as we shall see later. To give an idea of the relative variability of the two components of (1.1), the carrier frequency will often be in the range 1–10GHz ($= 1 - 10 \times 10^9$ cycles per second). The variability of the waveform of course depends on its form. Often phase coded pulses are used which, in theory, switch instantaneously, but limitations in the practical implementation usually restrict the highest frequency components of the waveform to be no more than $1/10$ of the carrier frequency.

In the complex domain we can write the signal as

$$(1.4) \qquad s(t) = \Re\left(w(t).e^{2\pi i\phi(t)}.e^{2\pi i f_c t}\right) = \Re\left(s_0(t).e^{2\pi i f_c t}\right)$$

The transmitted signal hits a *target* whose distance from the (collocated) transmitter and receiver is R. Let us assume for the moment that the target is stationary relative to the radar system and that it comprises a single scatterer. Then the return signal will, to a good approximation (see Section [deleted -Ed.] for a more accurate view of radar scattering), be a delayed version of the original signal which depends on the range R of the target. Specifically the signal voltage at the antenna of the receiver is

$$(1.5) \qquad s_u(t) = As\left(t - \frac{2R}{c}\right)$$

where c is the speed of light and A represents the attentuation of the signal by the reflection. As I have already said, this is an approximation to the true situation, but a good one for most purposes. In fact, even if the target is totally unchanging the transmitted waveform really excites the electrons within the material of the target which then reradiate a modulated version of the signal. This effect is slight, but its use is being developed for the purposes of target identification. More importantly the phase of the waveform will change on reflection from the target. This can be accommodated within the complex form of the signal (that is, the complex signal whose real part is given in (1.2)) by allowing A to be complex. Otherwise an extra phase shift is imposed on the carrier.

When the waveform returns to the receiver some noise is added ("receiver noise") which arises from the thermal activity generated within the components of the receiver. It is important to realise here that the fall-off of the signal strength (ie power) as a function of distance of the target is as the fourth power of the range R, so that typically the received signal power is very small (for a distant target 10^{-18} watts is not unusual — there is a saying among radar engineers that the total signal energy received by all radars ever is not enough to turn a page) and in this context receiver noise can be significant. To a good approximation, receiver noise is Gaussian. That is we can write

$$s_r(t) = s_u(t) + N(t),$$

where $N(t)$ is a Gaussian process, for the signal after the initial stages of the receiver.

I have mentioned the issue of noise several times and its importance cannot be over-emphasised. At any moment it will often be the most significant part of the signal in the system prior to the matched filter. Extraction of signal from noise is a major issue. In radar literature the noise process is assumed typically to be "white Gaussian noise". From a mathematical perspective there are problems with this assumption. Such a continuous time process cannot exist since it would have infinite power. To be more mathematically precise, we assume that the power spectral density of the noise is flat across the interval $[-B, B]$ where B represents the bandwidth of the receiver; that is, the maximum frequency it will handle, as would be the case if the signal had been passed through a (perfect) lowpass filter prior to further processing. The treatment of noise in this situation involves interesting ideas in the field of mathematical statistics — specifically estimation theory and hypothesis testing, but we do not have time to address them here. Accordingly we refer the reader to [13, 18].

Now we insert the possibility that the target is moving. This has the effect of modifying the transmitted waveform other than just by delay. It imposes the doppler effect on it. If this is done correctly (ie relativistically) it results in a " time dilation" of the return signal, so that, if the target has a radial velocity v, the return signal $s_u(t)$ becomes

$$s_u(t) = As(\alpha t - \frac{2R}{c}),$$

where

$$\alpha = \frac{(1 - \frac{v}{c})}{(1 + \frac{v}{c})}.$$

When v is much smaller than c this is approximated by $\alpha = (1 - 2v/c)$. A further approximation is possible if, as is usually the case, the signal is " narrow band"; that is, if its (Fourier) spectrum is essentially in a range $(f_c - \delta, f_c + \delta)$ where δ is small compared to f_c. For most radar applications this is a reasonable assumption since the signal modulating the carrier will have relatively low bandwidth. In this case, the return signal is approximated by shifting the frequency of the return from a stationary target at the same range by $f_d = (2v/c)f_c$. This is best written in terms of the complex signal

$$s_u(t) = \Re\left(s_0(t - \frac{2R}{c}).e^{2\pi i f_c(1 - 2v/c)(t - \frac{2R}{c})}\right)$$

This equation is the standard one used in most radar calculations.

1.4. *I and Q channels*

As we have already said, processing of radar signals is done largely in the complex domain. Of course the signal transmitted from and returned to the radar antenna is real. However in a very natural way as we have seen (1.2) the transmitted signal is treated as the real part of a complex signal which is simpler to understand. When the return signal is received it is turned into a complex signal — that is, it is made into two real signals. This is usually done in two steps. First the signal is " demodulated" to one whose carrier frequency (the "intermediate frequency" or IF) is much lower and so in practical terms easier to work with. Demodulation is done by mixing (that is, multiplying) the return signal with a signal (pure tone) whose frequency differs from the carrier frequency by the IF. In other words we form

$$s_m(t) = s_r(t)cos2\pi(f_c + f_{IF})t$$

where f_{IF} is the IF. The resulting signal is then passed through a low-pass filter which removes all frequencies greater than the carrier frequency and leaves intact signals whose spectrum lies well below the carrier frequency. Typically for an S-band radar (ie $f_c \sim 3 \times 10^9$

GHz) the IF will be around 100MHz or less and may be accomplished by two stages of de-modulation. Using the product formula for sine and cosine we see that the resulting signal is approximately

$$s_{IF}(t) = w(t) \cdot \cos 2\pi\phi(\tau) \cos 2\pi((f_{IF} - f_d)\tau)) \\ -w(\tau) \sin 2\pi\phi(\tau) \sin 2\pi(f_{IF} - f_d)\tau),$$

where $\tau = t - \frac{2R}{c}$. It needs to be realised that when I say "approximately" the degree of approximation in this game is very large. Because of the need to detect very small signals the filters used will often attenuate the power in unwanted frequencies by 10^{-8}.

At this point the conventional radar system will mix the signal with both $\cos 2\pi f_{IF}t$ and $\sin 2\pi f_{IF}t$ and again low pass filter to form the I and Q channels respectively. The radar engineer really would like to obtain an analytic signal so the correct thing to do is to form the Hilbert transform of the real signal $s_r(t)$. This is impractical, and so it is approximated by this method. The approximation is very good for narrow band signals. The complex signal which results from these transformations is

$$(1.6) \qquad s_c(t) = s_0(t - \frac{2R}{c})e^{-2\pi i f_d t}.$$

1.5. Ambiguity

Let us return to the case when there is no doppler component; that is, when the target is stationary. The return signal is as in (1.5) together with some noise. What happens next in the radar receiver depends on the information we want to extract from the signal. However at this point there are just two types of information we might want to obtain:

1. detection of the target — that is to decide on the basis of a statistical hypothesis test whether we indeed have a signal or just noise;
2. estimation of the range R of the target.

It turns out that the optimal detector (in the sense of the Neymann-Pearson Lemma) and the maximum likelihood range estimator both require the same operation — to take the correlation of the received signal with a copy of the transmitted signal. This also has the effect of maximizing the post-processing signal-to-noise ratio. In fact it makes sense to do this in the complex domain. Thus we form the function

$$Q(x) = \int s_c(t)\overline{s_0(t - x)} \, dt$$

where the bar over the signal indicates (for when we treat complex signals) complex conjugation. The maximum absolute value of this signal tells us the best estimate of the range in Gaussian noise for this choice of transmitted signal. Note that we will usually have some control over the shape of the latter and we shall spend some time later on issues concerned with its design.

When there is doppler present the radar behaves in a similar way. The return signal is correlated ("matched filtered") with the transmitted signal, so that the resulting signal looks like this:

$$(1.7) \qquad \chi(x, f) = \int_{\mathbb{R}} s_c(t)\overline{s_0(t - x)e^{2\pi i f t}}\, dt$$

where now the return signal $s_c(t)$ is as in (1.6). The effect of the matched filter is to maximise the signal power relative to the noise power in the post-processing signal.

We call this the *radar ambiguity function* of the signal s_0 and write it as $\chi_{s_0}(x, f)$. It is the output we would obtain from our receiver for a signal delay $x = 2R/c$ and velocity $v = f.c/2f_c$. An example for a rectangular pulse is given in Figure 2. We shall deal with

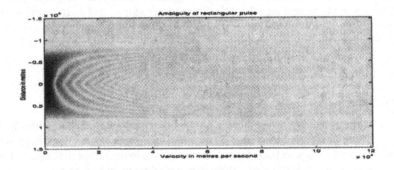

FIGURE 2. Ambiguity of rectangular waveform

the ambiguity function in Section 2. We shall return to the choice of waveforms in Section 3.

1.6. *Extended targets*

When the transmitted signal is reflected from an extended scatterer or from multiple scatterers which are not time varying, the resulting signal is obtained as a convolution of the so-called *reflectivity kernel* of the target(s) with the transmitted signal. Thus if $k(t)$ is this kernel, where now t is a measure of range (which is the same as time in this context), the reflected signal is

$$s_u(t) = \int_{\mathbb{R}} s(t - \tau)k(\tau)\, d\tau.$$

Normally velocity does not vary from one part of the target to another so that this subjected to a constant phase shift, though it is straightforward to derive the formula for varying doppler.

2. Ambiguity Functions

This section discusses the mathematical structures lying behind the ambiguity function (particularly the narrow band case) as we have described it in Section 1.5. We shall do so in terms of the unitary representation theory of the Heisenberg group and we shall deal with the theoretical ideas needed first.

2.1. *Representations of the Heisenberg Group*

The narrow band ambiguity function can be described in several different ways at varying levels of abstraction. In an attempt to use some of the power of the abstract setting without demanding too much of the reader, I have attempted to steer a middle course between papers of Schempp [19–22] and the "naive" viewpoint of Wilcox [4]. In many respects the point of view adoped here is that of Auslander and Tolimieri [12]; this couches the theory in terms of the infinite dimensional irreducible representations of the Heisenberg group. As these are probably the most sophisticated ideas discussed here, we present them in detail.

2.1.1. *The Heisenberg Group.* The three dimensional Heisenberg group is the group of 3×3 upper triangular matrices with 1's on the diagonal, thus:

$$(2.1) \qquad u = \begin{matrix} 1 & x & z \\ 0 & 1 & f \\ 0 & 0 & 1 \end{matrix} \ .$$

We denote the collection of all such matrices with real entries by $\mathcal{H}$ and note that they form a group under matrix multiplication. Observe that the inverse of u above is

$$u^{-1} = \begin{matrix} 1 & -x & -z + xf \\ 0 & 1 & -f \\ 0 & 0 & 1 \end{matrix} \ .$$

This group has the topological structure of $\mathbf{R}^3$, so that we can talk about continuity of functions on $\mathcal{H}$. Lebesgue measure m on $\mathbf{R}^3$ is invariant in the sense that

$$(2.2) \qquad \int_{\mathbf{R}^3} f(\mathbf{u}\mathbf{x}) \, dm(\mathbf{x}) = \int_{\mathbf{R}^3} f(\mathbf{x}) \, dm(\mathbf{x}),$$

where the variable x represents an element of $\mathcal{H}$ and u is an arbitrary element of $\mathcal{H}$. Here f is any suitable function for which the integrals make sense (such as continuous complex-valued functions with compact support). In fact, for this group, the measure is both left invariant and right invariant, so that equation (2.2) also holds when u multiplies x on the right in the argument of f.

2.1.2. *Unitary Representations.* A *unitary representation of* $\mathcal{H}$ is a map $x \to U_x$ from $\mathcal{H}$ into the group of unitary operators $\mathcal{U}(\mathfrak{H})$ of some (separable) Hilbert space $\mathfrak{H}$, which is a homomorphism in the sense that

$$U_{x_1} . U_{x_2} = U_{x_1 . x_2},$$

where the . on the left side indicates multiplication of operators and on the right multiplication of group elements in $\mathcal{H}$. We shall also assume that this map is continuous into the weak-operator topology on $\mathcal{U}(\mathfrak{H})$; that is,

$$x \mapsto \langle \xi, U_x \eta \rangle$$

is continuous for every ξ and η in the Hilbert space.

There are some simple such representations of $\mathcal{H}$ for which $\mathfrak{H}$ is one dimensional (and so its unitary group can be identified with the group of complex numbers of absolute value 1 under multiplication). An example is the *trivial representation* **1** where

$$\mathbf{1}_x = 1$$

for all $x \in H$. A slightly less trivial example is of the form

$$(2.3) \qquad \begin{pmatrix} 1 & x & z \\ 0 & 1 & f \\ 0 & 0 & 1 \end{pmatrix} \mapsto e^{2\pi i(\alpha x + \beta f)}$$

where α and β are any real numbers. This gives a two-dimensional family, parametrized by α and β, of one-dimensional representations, but these are not important for our purposes.

2.1.3. *Irreducibility and Equivalence.* We need to make two more general definitions from representation theory. First we need to say when two representations, say $x \to U_x^{(1)}$ on $\mathfrak{H}_1$, and $x \to U_x^{(2)}$ on $\mathfrak{H}_2$, are *equivalent.* By this we mean that they are the same up to a change of Hilbert basis, or equivalently, that there is a unitary operator V from $\mathfrak{H}_1$ onto $\mathfrak{H}_2$ such that

$$V U_x^{(1)} = U_x^{(2)} V$$

for all $x \in \mathcal{H}$.

A representation of $\mathcal{H}$ ($x \mapsto U_x$ on $\mathfrak{H}$, say) is *irreducible* if it cannot be broken up into sub-representations, that is, if there are not two subspaces $\mathfrak{H}_1$ and $\mathfrak{H}_2$ each invariant under all of the operators U_x ($x \in H$), such that $\mathfrak{H} = \mathfrak{H}_1 \oplus \mathfrak{H}_2$ in the sense of Hilbert spaces. The restriction of U to these subspaces produces representations $U^{(1)}$ and $U^{(2)}$ on $\mathfrak{H}_1$ and $\mathfrak{H}_2$ respectively, such that, for all $x \in H$,

$$U_x(\xi_1 + \xi_2) = U_x^{(1)} \xi_1 + U_x^{(2)} \xi_2,$$

where $\xi_j \in \mathfrak{H}_j$ ($j = 1, 2$). While it may not be obvious, this is equivalent to the non-existence of a non-trivial subspace of $\mathfrak{H}$ which is left invariant by each of the operators U_x, ($x \in H$). This is a consequence of the unitary property of the operators. Equally it is equivalent by Schur's Lemma to the non-existence of a non-trivial bounded operator on $\mathfrak{H}$

which commmutes with all of the U_x. Evidently all of the one-dimensional representations described in (2.3) are irreducible.

2.1.4. *The Schrödinger Representation.* Now we define a representation of $\mathcal{H}$ on the Hilbert space $L^2(\mathrm{R})$ of complex square integrable functions on R (or equivalently complex-valued finite energy signals on R) by letting the matrix (2.3) act on a signal s as a point target with doppler shift f and delay x would. Specifically we write

$$(2.4) \qquad\qquad U_x^\gamma(s)(t) = e^{2\pi i \gamma(ft-z)} s(t - x).$$

Note that the variable z acts as a scalar multiplier (namely the scalar $e^{2\pi i \gamma z}$) which makes the homomorphism property work out correctly. Here γ is a non-zero real number. Now the Stone-von Neumann theorem says that, up to equivalence of representations, the representations of the form U^γ are the only irreducible representations which are not one dimensional, so that (2.3) and (2.4) are descriptions of all of the irreducible unitary representations of the Heisenberg group up to equivalence. The particular forms of the infinite dimensional irreducible representations of the Heisenberg group given in (2.4) are called the *Schrödinger representation.*

In fact the representation in (2.4) is characterized via the Stone–von Neumann theorem as the unique (up to equivalence) irreducible representation whose restriction to the centre of $\mathcal{H}$ (that is, to the matrices as in (2.1) for which $f = x = 0$) is just $z \to e^{2\pi i \gamma z} I$ where I is the identity operator on $L^2(\mathrm{R})$.

2.2. Ambiguity Functions

It is now easy to define the ambiguity function corresponding to a signal s in terms of the infinite dimensional irreducibles of the Heisenberg group. Writing $U = U^1$, we define

$$(2.5) \qquad\qquad \chi_s(x, f) = \langle s, U_x s \rangle$$

where

$$(2.6) \qquad\qquad \mathbf{x} = \begin{pmatrix} 1 & f & 0 \\ 0 & 1 & x \\ 0 & 0 & 1 \end{pmatrix}.$$

This produces the function we have defined in (1.7). This view of ambiguity functions can be found in the paper [12] of Auslander and Tolimieri. The choice of element of the group may seem arbitrary, but remember that the only effect of changing the top right hand corner of this matrix is to multiply the ambiguity function by a scalar of absolute value 1. In fact, it is really the absolute value of the ambiguity function which is important, since it represents the response of the system to a target of a given range and frequency, and this is unchanged by a change in the top right hand corner. In effect we are making a choice of cross section of the quotient of $\mathcal{H}$ by its centre Z.

Once we have made this definition, it makes sense to allow the *cross ambiguity function*

$$\chi_{s,s'}(x, f) = \langle s, U_x s' \rangle,$$

where x is as in (2.6).

2.2.1. *Properties.* The (cross-)ambiguity function has a number of remarkable properties, one of the most important of which is Moyal's identity:

THEOREM 1. *Let s and s' be in $L^2(R)$. Then*

$$\|\chi_{s,s'}\|_{L^2(R^2)} = \|s\| \|s'\|.$$

This is a simple consequence of the following theorem:

THEOREM 2. *Let s_1, s_2, s_3, s_4 be in $L^2(R)$. Then*

$$\langle \chi_{s_1,s_2} \cdot \chi_{s_3,s_4} \rangle_{L^2(R^2)} = \langle s_1, s_3 \rangle \langle s_2, s_4 \rangle.$$

One of the important consequences of Moyal's identity is that the unitary representation $x \mapsto U_x$ is a *square integrable* representation modulo the centre of $\mathcal{H}$. Such representations have been widely studied.

Another remarkable property of this quadratic form on $L^2(R)$ is its variation under the Fourier transform operator.

THEOREM 3. *Let $s \in L^2(R)$. Then, for any $(x, f) \in R^2$,*

(2.7) $$\chi_{\mathcal{F}(s)}(-f, x) = e^{2\pi i f x} \chi_s(x, f).$$

In fact this arises from an interesting intertwining property of the Schrödinger representation. The map

$$(x, f, z) \mapsto (f, -x, -z)$$

is an anti-automorphism of the Heisenberg group which we call κ and so $x \mapsto \kappa(x)^{-1}$ is an automorphism. Call this ν. Then

$$\nu(x, f, z) = (-f, x, z - fx).$$

Moreover

$$\mathcal{F}(U_x(s)) = U_{\nu(x)}(\mathcal{F}(s)),$$

so that the Fourier transform operator intertwines the two representations U and D where

(2.8) $$D_x = U_{\nu(x)}.$$

Now, for $s \in L^2(R)$,

$$e^{2\pi i z} \chi_s(x, f) = \langle s, U_x s \rangle = \langle \mathcal{F}(s), \mathcal{F}(U_x s) \rangle,$$

by the Plancherel Theorem, and in view of (2.8) this equals

$$\langle \mathcal{F}(s), D_x \mathcal{F}(s) \rangle = \langle \mathcal{F}(s), U_{\nu(x)} \mathcal{F}(s) \rangle = e^{2\pi i f x} \chi_{\mathcal{F}(s)}(-f, x),$$

which proves equation (2.7). A corresponding equality holds for the bilinear form.

2.3. Weil-Brezin Formula (Zak Transform)

There is another way of realising the unitary representations we have described and used above. It will simplify formulae if we deal only with the case $\gamma = 1$, though of course the theory extends in an obvious way to all other non-zero values of γ. Instead of using the Hilbert space $L^2(\mathbb{R})$ we consider a Hilbert space $\mathfrak{K}$ of functions F on the Heisenberg group itself. Let Γ be the subgroup of $\mathcal{H}$ consisting of matrices of the form

$$(2.9) \qquad \begin{pmatrix} 1 & n & z \\ 0 & 1 & m \\ 0 & 0 & 1 \end{pmatrix}$$

where $n, m \in \mathbb{Z}$ and $z \in \mathbb{R}$. Consider the space $\mathfrak{K}$ of functions F on $\mathcal{H}$ which satisfy

1. $F(\gamma x) = e^{2\pi i z} F(x)$ for all $\gamma \in \Gamma$ as in (2.9);
2. $\int_{\mathcal{H}/\Gamma} |F(x)|^2 \, dx < \infty$,

with the inner product

$$(2.10) \qquad \langle F, G \rangle_{\mathfrak{K}} = \int_{\mathcal{H}/\Gamma} F(x)\overline{G(x)} \, dx.$$

Next we define a representation V of $\mathcal{H}$ just by the simple formula

$$(2.11) \qquad V_x(F)(y) = F(yx).$$

We need to make a few remarks about these formulae. First note that 1. makes $|F|$ constant on the right cosets of Γ, so that the integral in 2., which is over the right coset space $\Gamma_0 \backslash H$, makes sense. This applies equally to the integral in (2.10). Secondly, the space of functions $\mathfrak{K}$ may be realised as the space of L^2 sections of a line bundle on the torus $\mathbf{T}^2$. Thirdly, those familiar with induced representations will recognize that this is just the representation τ, induced from the subgroup Γ to $\mathcal{H}$, where τ is given by

$$\begin{pmatrix} 1 & n & z \\ 0 & 1 & m \\ 0 & 0 & 1 \end{pmatrix} \longrightarrow e^{2\pi i z},$$

that is, $V = \mathrm{ind}_{\Gamma}^{\mathcal{H}} \tau$, in the terminology of Mackey. We note that Γ is a normal subgroup of $\mathcal{H}$ and so we may apply standard Mackey theory. The stabilizer of this representation is just Γ itself and so the induced representation is irreducible. Now the Stone–von Neumann Theorem tells us that the only irreducible representation whose restriction to the centre of $\mathcal{H}$, the subgroup of matrices with x and y both zero, is just $e^{2\pi i z}$ times the identity operator, is the one we have called U^γ earlier with $\gamma = 1$. It follows that the representation V and $U = U^1$ are equivalent, so there is a unitary operator W intertwining U and V.

It is not difficult to calculate that the intertwining operator is $W : L^2(\mathrm{R}) \mapsto \mathfrak{K}$ given by

$$(2.12) \qquad W(s)(f, x, z) = e^{2\pi i z} \sum_{k \in \mathbf{Z}} s(x + k) e^{2\pi i k f}.$$

This formula is called the *Weil-Brezin formula* or (usually without the term $e^{2\pi i z}$) the *Zak transform*. In essence it is a map between functions on R and functions on the unit square (the cross sections of the line bundle) and maps $L^2(\mathrm{R})$ onto $L^2(I^2)$ where I represents the unit interval. In fact the map is unitary. Since it intertwines two irreducible representations it has to be unitary up to a scalar multiple.

This gives us another way of representing ambiguity functions, since for $s \in L^2(\mathrm{R})$,

$$(2.13) \qquad e^{2\pi i z} \chi_s(x, f) = \langle s, U_x \rangle = \langle W(s), V_x W(s) \rangle.$$

This alternative view is a useful device in realising theoretical features of χ.

2.4. Resolution and Hermite functions

This section is taken largely from the work of Wilcox [4]. He uses a slightly modified form of the ambiguity function. Indeed in many papers the ambiguity of signal s is defined in a *symmetrised* form:

$$(2.14) \qquad A_{\mathrm{S}}(x, f) = \int_{\mathrm{R}} s(t + \frac{x}{2}) \overline{s(t - \frac{x}{2})} e^{2\pi i f t} \, dt.$$

This formalism corresponds to a slightly different description of the Heisenberg group as the group on R^3 whose multiplication is given by

$$(2.15) \quad (x_1, f_1, z_1) * (x_2, f_2, z_2) = (x_1 + x_2, f_1 + f_2, z_1 + z_2 + B(x_1, f_1; x_2, f_2))$$

where

$$(2.16) \qquad B(x_1, f_1; x_2, f_2) = \frac{1}{2}(x_1 f_2 - f_1 x_2).$$

It is important to be aware that this does not change the significant properties of the ambiguity function. In fact

$$A_{\mathrm{S}}(x, f) = e^{\pi i f x} \chi_{\mathrm{S}}(x, f).$$

The Hermite functions play a central role in the theory of this object. For our purposes we define them by follows:

$$W_n(t) = \frac{2^{1/4}}{\sqrt{n!}} H_n(2\sqrt{\pi} t) e^{-\pi t^2} \qquad (n = 0, 1, 2, 3, \ldots)$$

where H_n is the nth Hermite polynomial:

$$H_n(t) = (-1)^n e^{x^2/2} \left(\frac{d}{dx}\right)^n e^{-\frac{x^2}{2}} \qquad (n = 0, 1, 2, 3, \ldots).$$

The Hermite functions form an orthogonal basis of $L^2(\mathbb{R})$. Their ambiguities are easily calculated in terms of known functions:

$$A_{W_n}(x,y) = e^{-\frac{\pi}{2}(x^2+y^2)} L_n(\frac{1}{2}(\pi(x^2+y^2)))$$

where L_n is the nth Laguerre function

$$L_n(x) = \frac{1}{n!} e^x \left(\frac{d}{dx}\right)^n (x^n e^{-x}) \text{ for } x > 0.$$

Evidently such functions are radially symmetric. In fact this characterizes Hermite waveforms, as Wilcox observes in [4].

THEOREM 4. *An ambiguity function $A_s(x,y)$ is radially symmetric, that is,*

$$A_s(x,y) = f(x^2+y^2)$$

if and only if $s = cW_n$ for some n.

We mention here too the many papers of Schempp on this subject and on the ambiguity function in general (see, for example, [19–22]).

The sharpness of the peak of the ambiguity function (at $(0,0)$, of necessity) is a measure of the waveform's capacity to resolve nearby (in both doppler and range) targets. One way of measuring resolution capabilities is as follows. Fix a minimum response resolution ϵ and consider the level curve

$$|A_s(x,y)|^2 = 1 - 4\pi^2\epsilon^2.$$

This is approximately an ellipse, which Wilcox calls the *resolution ellipse*. Write its equation as

$$\beta^2 x^2 + 2\gamma xy + \tau^2 y^2 = \epsilon^2.$$

In fact

$$\beta = \int_{\mathbb{R}} f^2 |\mathcal{F}(s)(f)|^2 \, df \qquad \tau = \int_{\mathbb{R}} t^2 |s(t)|^2 \, dt$$

are respectively measures of the *bandwidth* and the time duration of the signal. Two close point targets will produce a response which comprises two copies of the ambiguity function with slightly different centres. These two targets will be most difficult to resolve if their resolution ellipses share a common major axis. The two targets are said to be *resolvable* if the centres of their resolution ellipses are one major semi-axis apart. This number — the *smallest resolvable separation* — is given by

$$R(\epsilon)^2 = \frac{2\epsilon^2}{\tau^2 + \beta^2 - \sqrt{(\tau^2-\beta^2)^2 + 4\gamma^2}}.$$

Wilcox calls $R(\epsilon)/\epsilon$ the *resolution factor* ρ, but perhaps for mathematical precision the resolution factor should be the derivative of $R(\epsilon)$ at 0. Wilcox has shown that the best resolution factor obtainable subject to constraints $\beta \leqslant \beta_0$ and $\tau \leqslant \tau_0$ on the bandwidth and time duration is

$$\rho = \frac{1}{\min(\beta_0, \tau_0)}.$$

The Hermite functions arise in this context as follows. Let Ω_n be the subspace of $L^2(R)$ consisting of all functions of the form $P(t)e^{-\pi t^2}$ where P is a polynomial of degree at most n. The following theorem of Hardy characterizes Ω_n.

THEOREM 5. *Let $s \in L^2(R)$ satisfy*

$$s(t) = O(t^n e^{-\pi t^2}), \qquad |t| \to \infty,$$
$$\mathcal{F}s(f) = O(f^n e^{-\pi f^2}), \qquad |f| \to \infty.$$

Then $s \in \Omega_n$

Now we have the following theorem of Wilcox.

THEOREM 6. *The Hermite waveform W_n has the smallest resolution factor of all wave-forms in Ω_n.*

2.5. Characterization

We discuss here the problem of how to describe the class $\mathcal{A}$ of ambiguity functions χ_s as s ranges over all functions in $L^2(R)$. One easy characterization arises from the observation that the Fourier transform of the ambiguity function with respect to the doppler variable is

$$\mathcal{F}_2^{-1}(\chi_s)(x, \tau) = s(\tau)\overline{s(\tau - x)},$$

where we use the notation $\mathcal{F}_2$ to mean the Fourier transform with respect to the second variable. Making the change of variable

$$\tau - x = u, \qquad \tau = \tau,$$

we obtain the function

$$H(u, \tau) = s(\tau)\overline{s(u)}.$$

We write $H = \mathcal{R}(\chi_s)$ to denote this transformation; that is,

$$\mathcal{R}(F)(s, t) = \mathcal{F}_2^{-1}(F)(t - s, t). \tag{2.17}$$

To describe ambiguity functions then it is enough to describe these functions intrinsically and this is easily accomplished. The following is a theorem of Wilcox [4], but see also [12].

THEOREM 7. *Let $H \in L^2(R^2)$ satisfy*
1. $\overline{H(u, \tau)} = H(\tau, u)$
2. $H(x, x) \geqslant 0$
3. $H(v, v)H(x, \tau) = H(x, v)H(v, \tau)$

Then there is a signal $s \in L^2(R)$ such that $\mathcal{R}(\chi_S) = H$; that is,

$$\chi_S(x, f) = \mathcal{F}_2(K)(x, f)$$

where $K(u, \tau) = H(\tau - u, \tau)$.

A somewhat similar theorem, also due to Wilcox, characterizes ambiguity functions in terms of their expansions in tensor product bases of the Hilbert space

$$L^2(R^2) = L^2(R) \otimes L^2(R).$$

A *tensor* orthonormal basis of $L^2(R^2)$ is a basis $\psi_{m,n}(x, y) = \phi_m(x)\overline{\phi_n(y)}$ where (ϕ_n) is an orthonormal basis of $L^2(R)$. Any member F of $L^2(R^2)$ has an expansion

(2.18)
$$F(x, y) = \sum_{m,n} c_{m,n} \phi_m(x)\overline{\phi_n(y)}$$

which converges in $L^2(R^2)$. The following simple theorem characterizes ambiguity functions in terms of their expansion in such a basis.

THEOREM 8. *A function $F \in L^2(R^2)$ is an ambiguity function if, in equation (2.18),*

$$c_{m,n} = a_n\overline{a_m}$$

for some sequence (a_n).

This has the following immediate corollary

COROLLARY 1. *A function $F \in L^2(R^2)$ is an ambiguity function if, in equation (2.18),*

$$c_{k,k}c_{m,n} = c_{m,k}c_{k,n} \quad and \quad c_{k,k} \geqslant 0 \text{ for all } k, m, n.$$

Another characterization of ambiguity functions arises from their description in terms of the representation of the Heisenberg group. To this end, we recall that a complex-valued function p on $\mathcal{H}$ is *positive definite* if, for any choice of N complex numbers c_n and elements of the group x_n,

$$\sum_{m,n=1}^{N} c_m\overline{c_n}p(x_m x_n^{-1}) \geqslant 0.$$

positive definite functions arise as functions of the form

$$p(x) = \langle \xi, R(x)\xi \rangle,$$

where R is a representation of the group and ξ and is an element of the Hilbert space on which it acts. Among the positive definite functions, the ones which arise from irreducible representations are characterized by the following property: we call a positive definite function p *extremal* if, for all other positive definite functions q for which $p - \alpha q$ is also positive definite for some $\alpha > 0$ q is a multiple of p.

Since we know the irreducible representations of the Heisenberg group, we also know all extremal positive definite functions. An ambiguity function is now a function $\chi(x, f)$ in $L^2(R)$ for which the function

$$p(x, f, z) = e^{2\pi i z} \chi(x, f)$$

is an extremal positive definite function. While this characterization is not as easily checked as the earlier ones it is of theoretical importance.

2.6. *The Abiguity Problem*

The simple abiguity problem is that of determining a signal s from its ambiguity χ_s. To make it mathematically meaningful we restrict the signal to belong to $L^2(R)$, so that for such signals we are interested in the question: is $s \mapsto \chi_s$ one to one? The answer to this question is yes.

It is relatively simple to show the following theorem (see, for example, [12]),

THEOREM 9. *Let s_1 and s_2 be two $L^2(R)$ functions for which $\chi_{s_1} = \chi_{s_2}$ then $s_1 = s_2$.*

However this is not really an answer to the radar engineer's problem for several reasons, one of which is that the ambiguity function is not what is important. It is the absolute value of the ambiguity function which is really of interest. In this context, the problem arises of determining all possible waveforms s whose ambiguity functions have the same absolute value. Much of this section is adapted from work of Philippe Jaming indexJaming, Philippe and his collaborators (see [7, 10, 15]). Following Bueckner ([9]) we state here the following:

DEFINITION 1 (General Abiguity Problem). *Given $s_1, s_2 \in L^2(R)$ what is the set of all pairs s_1', s_2' such that*

(2.19) $$|\chi_{s_1, s_2}(x, f)| = |\chi_{s_1', s_2'}(x, f)|$$

for all $(x, f) \in R^2$?

This problem is unsolved in general, but the following simple observation shows that equation (2.19) does not force $(s_1, s_2) = (s_1', s_2')$. Let

(2.20) $$s_j' = U_{x'} s_j \text{ where } x' \in \mathcal{H}.$$

Then , with $x = (x, f, 0) \in \mathcal{H}$,

$$|\chi_{s_1', s_2'}(x, f)| = |\langle s_1', U_x s_2' \rangle| = |\langle U_{x'} s_1, U_x U_{x'} s_2 \rangle|$$
$$= |\langle s_1, U_{x'}^* U_x U_{x'} s_2 \rangle| = |\langle s_1, U_x U_z s_2 \rangle|$$

where $z = x^{-1} x'^{-1} x x'$. Now z is a commutator and so (since the Heisenberg group is two step nilpotent) belongs to the centre of $\mathcal{H}$. It follows from the irreducibility of U that U_z is just a scalar times the identity operator, say $U_z = e^{2\pi i z} I$, and hence that

$$|\langle s_1, U_x U_z s_2 \rangle| = |\langle s_1, U_z U_x s_2 \rangle| = |\langle e^{2\pi i z} s_1, U_x s_2 \rangle| = |\langle s_1, U_x s_2 \rangle| = |\chi_{s_1, s_2}(x, f)|.$$

We will say that s'_1, s'_2 are *Heisenbery related* to s_1 and s_2 if a relationship holds as in (2.20). Two such related pairs have the same absolute values for their ambiguity functions. The problem is now restated as: Does (2.19) force s'_1, s'_2 to be Heisenberg related to s_1, s_2?.

Unfortunately this is not the case. For instance if

$$s_1(t) = s_2(t) = \left(\frac{\sin t}{t}\right)^n \sin(2\pi nt)$$

$$s'_1(t) = s'_2(t) = \left(\frac{\sin t}{t}\right)^n \cos(2\pi nt)$$

then

$$|\chi_{s_1, s_1}(x, f)| = |\chi_{s'_1, s'_1}(x, f)|,$$

for all $(x, f) \in \mathbf{R}^2$, though s_1 is not Heisenbery related to s'_1. The proof of this is relatively easy if one works in the Fourier domain and uses the identity (2.7). The Fourier transforms of these functions comprise two non-zero identical "pieces" around $\pm 2n\pi$; the difference between the two being in the signs of these pieces. The ambiguity therefore is non-zero only when translated of these pieces overlap. It is now clear that the ambiguities can differ only is sign.

For special classes of functions it is the case that all s'_1, s'_2 are obtained by such transformations. For instance, Bueckner [9] and De Buda [5] have proved that, if s_1 and s_2 are both of the form $P(t)\exp(-\frac{t^2}{2})$ with P a polynomial, then s'_1 and s'_2 are Heisenbery related to s_1 and s_2.

Jaming indexJaming, Philippe [10] has obtained some results on the abiguity problem for compactly supported functions. His results are stated in terms of the symmetrised form, though obviously simple modifications convert to the χ-form of the ambiguity function. Let $u \in L^2(\mathbf{R})$ be a compactly supported function and suppose v satisfies $|A_v| = |A_u|$. Then it is relatively easy to see that v is also compactly supported. Moreover if the support of u is contained in an interval of length $2a$ then the support of v is also contained in an interval of length $2a$. In fact Jaming shows that the interval of support must have the same length for both u and v.

We may now assume that supports of both u and v are contained in $[-a, a]$ and no smaller interval, in particular, u and v are compactly supported. The Paley-Wiener theorem ensures that $A_u(x, y)$ and $A_v(x, y)$ are both entire functions of exponential type in the y variable. Now (cf : [10]) the following holds.

$$(2.21) \qquad A_u(x, y)\overline{A_v(x, \bar{y})} = A_v(x, y)\overline{A_u(x, \bar{y})} \quad \text{for all } x \in \mathbf{R}, y \in \mathbf{C}.$$

On the other hand, by the Hadamard factorisation theorem, an entire function $f(z)$ of exponential type is entirely determined by its zeros, up to a factor $\lambda e^{\mu z}$ with $\lambda, \mu \in \mathbf{C}$. Unfortunately (2.21) only tells us that, for fixed x, if z is a zero of $A_u(x, .)$ then either z or $\bar{z}$ is a zero of $A_v(x, .)$.

Several cases occur, for instance, A_u may only have real zeros (e.g. if $u = A_{[a,b]}$), then A_u and A_v have the same zeroes.

There are some functions u for which every ambiguity partner v is such that either A_u and A_v have the same zeroes, or A_u and A_{Zv} have the same zeroes, where $Zu(t) = u(-t)$. The final alternative is that $A(u)$ and $A(v)$ may have some common non-real zeroes and some conjugate zeroes.

In what follows, after replacing u by Zu or by some function Heisenbery related to u, we shall assume that $A(u)$ and $A(v)$ have the same zeroes. In other words we now consider the following restricted radar abiguity problem :

PROBLEM 1 (Restricted Radar Abiguity Problem). *Given a compactly supported $u \in L^2(R)$, what is the set of ambiguity partners v of u, such that for every $x \in R$, $A(u)(x, .)$ and $A(v)(x, .)$ have the same zeroes in the complex plane ?*

Jaming calls such ambiguity partners *restricted ambiguity partners* and shows that there exist compactly supported functions u which have ambiguity partners that are not restricted ambiguity partners either of u or of Zu. He also shows the following result.

THEOREM 10. *Let $u \in L^2(R)$ be a compactly supported function and let v be a restricted ambiguity partner of u. If Ω is the open set of all x such that $A(u)(x, .)$ is not identically 0, there exists a locally constant function ϕ on Ω such that, for every t_0, t_1, t_2 belonging to the support of u,*

$$\phi(t_2 - t_1) + \phi(t_1 - t_0) \equiv \phi(t_2 - t_0) \quad (2\pi)$$

and

$$v(x) = ce^{i\phi(x-a-x_0)}e^{i\omega x}u(x - a)$$

for some $a \in R, \omega \in R, c \in \mathrm{T}$ and some x_0 belonging to the support of u. Conversely, every function v of that form is a (restricted) ambiguity partner of u.

This theorem essentially states that if u is "simple" (in particular, the support is an interval) then the solutions of the abiguity problem are "simple", whereas for complicated u (for example, when the support has big gaps) the solutions are also complicated. In [7], the discrete abiguity problem is considered and various interesting results are obtained.

2.7. *Approximation*

An important problem for the radar engineer is to invent waveforms with specific abiguity properties. The abiguity problem of the previous section is one part of that problem — it examines the possibility of uniquely specifying a waveform in terms of its ambiguity (or its absolute value), and have we have seen without further constraints this is not successful. Nonetheless the radar engineer needs despite the non-uniqueness to be able to find waveforms with appropriate ambiguity fucntions. One way to do this is first to invent a function $F \in L^2(\mathrm{R}^2)$ with the appropriate properties and then find the ambiguity function which most closely approximates it. In other words, the problem is to find the signal $s \in L^2(\mathrm{R})$ which satisfies

$$s = \mathrm{argmin}||A_s - F||_{L^2(\mathrm{R}^2)}.$$

Using the unitary operator $\mathcal{R}$ described in equation (2.17), we obtain

$$||A_s - F||^2 = ||s(t)\overline{s(\tau)} - \mathcal{R}(F)||^2 = 2(1 - \langle \mathcal{R}(F), s \otimes \bar{s} \rangle)$$

if both F and s are normalized. The problem now is one of maximizing

$$\langle \mathcal{R}(F), s \otimes \bar{s} \rangle)$$

which corresponds to finding the eigenvector whose eigenvalue has the largest absolute value of the integral operator with kernel $\mathcal{H}$, that is,

$$\mathcal{H}(u)(t) = \int_{\mathrm{R}} H(t, \tau)u(\tau) \, d\tau.$$

This operator is compact and so its eigenvalues $\mu_1, \mu_2, \ldots$ decrease in absolute value and tend to zero. We choose for our signal s the eigenvector corresponding to μ_1.

2.8. *The Wide-Band Ambiguity Function*

The classical ambiguity function of Woodward is a result of making the narrow band approximation. There are situations where this is not valid. For example in sonar work where the speed of sound through water is not sufficiently high compared to velocities of targets that the approximation is valid. Also in radar the approximation breaks down in two ways:

1. the velocity is a substantial proportion of the velocity of light. This is an unlikely scenario in practice.
2. the radar is "wide-band"; that is, it uses a broad spectrum. This is an area of increasing interest.

The *wide band cross ambiguity* function is defined by

$$W_{S_1, S_2}(x, \alpha) = \alpha^{1/2} \int_{\mathrm{R}} s_1(t)\overline{s_2(\alpha(t + x))} \, dt,$$

where $\alpha > 0$ represents the scaling due to doppler and x corresponds to range as before; the bar over the second term in the integral corresponds to complex conjugation. The (auto)-ambiguity function of a signal s is $W_s(x, \alpha)$.

This formula too corresponds to a representation of a group. This time the group in question is the so called $ax + b$ group G of 2×2 matrices

$$\begin{pmatrix} a & b \\ 0 & 1 \end{pmatrix}$$

where $a > 0$ and $b \in \mathrm{R}$. The representation is again on $L^2(\mathrm{R})$ and is given by

$$U_{a,b}(s)(t) = a^{1/2}s(at + b)$$

for $f \in L^2(\mathbb{R})$. The wide-band ambiguity function is then just

$$W_{s_1,s_2}(x, \alpha) = \langle s_1, U_{\alpha,x}(s_2) \rangle_{L^2(\mathbb{R})}.$$

Several properties of the narrow band ambiguity function carry over to the wide band case. However it is necessary to restrict to a smaller class of functions to make the theory closely approximate that of the narrow band ambiguity. We write $H^2(\mathbb{R})$ for the Hardy class of functions whose Fourier transforms vanish on the negative half-line. Note that this subspace is invariant under the action of $U_{a,b}$ for all $a > 0$ and $b \in \mathbb{R}$. The original representation therefore cannot have been irreducible, though the restriction to $H^2(\mathbb{R})$ is. We need to restrict further to the those functions s which satisfy

$$\int_{\mathbb{R}^+} |\mathcal{F}(s)(f)|^2 \, \frac{df}{f} < \infty.$$

This class is designated $\mathfrak{H}$ and we write

$$\langle s_1, s_2 \rangle = \int_{\mathbb{R}^+} \mathcal{F}(s_1)(f) \overline{\mathcal{F}(s_2)(f)} \, \frac{df}{f}.$$

It is now possible to obtain a Moyal type identity. For u in $H^2(\mathbb{R})$ and $v \in \mathfrak{H}$, $W_{u,v}$ is in $\mathfrak{K} = L^2(\mathbb{R}^+ \times \mathbb{R}; a^{-1} \, da \, db)$, and

$$\langle W_{u,v}, W_{u',v'} \rangle_{\mathfrak{K}} = \langle v, v' \rangle_{L^2(\mathbb{R})} \langle u, u' \rangle_{\mathfrak{H}}.$$

There are more aspects of the narrow band theory which carry over to the wide band theory with appropriate modifications. For example the approximation results of Section 2.7 have a counterpart in the wide band theory. The $ax + b$ group is an example (the simplest) of a non-unimodular group — its left and right invariant measures are different. It is this fact which produces the need for the two separate Hilbert spaces $H^2(\mathbb{R})$ and $\mathfrak{H}$. For more information on this issue, the reader is referred to [6]. For more on the general theory of the wide-band ambiguity, we refer the reader to [1, 2, 11, 17, 23]. Jaming [10] has considered the ambiguity problem for wide-band ambiguity.

3. Waveform Design and Processing

For many purposes the ideal radar waveform would produce an ambiguity function which was the so-called "thumbtack" — that is, zero everywhere except at the origin. This would have ideal range and doppler discrimination. However Moyal's identity tells us that no (finite energy) signal gives rise to that waveform since the result has to be in $L^2(\mathbb{R}^2)$ and have norm equal to the square of the norm of the signal. A very short pulse would seem to have many advantages in this sense. It has a good thumbtack-like ambiguity at least in the range direction. However it has many disadvantages. If it is not to have extremely high power (which of course means that the electronics have to be able to deliver this amount of power to the antenna) then the total electromagnetic energy hitting the target is very small and the resulting energy returned to the receiver orders of magnitude smaller. The received energy falls off as the fourth power of the distance of the target so total *energy on target* is a significant factor in radar detection, particularly of distant targets. From this perspective longer waveforms are better. Another disadvantage of a short pulse is that it is a very wideband signal, and so not so good in the doppler direction. Moreover, to produce the shape of siuch a pulse and effectively deal with it on reception requires electronics capable of handling such signals. Maintaining linear responses over such a frequency spread is difficult. Radar engineers then aim to have long waveforms whose auto- correlation (matched filter) produces something approximating a thumb-tack. This process is called *pulse compression*. A purely random signal of infinite length (and therefore energy) — that is one comprising a bi-infinite sequence of Gaussian random variables of zero mean and constant variance has an expected ambiguity which is a thumbtack. Figure 3 gives a the ambiguity of a finite waveform obtained using a random number generator. As a result pseudo-random codes (which are finite and deterministic approximations to random signals) are often used as waveforms in radar. Many papers have been written examining the properties of different waveforms. We give a small sample of these here.

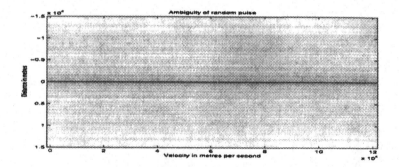

FIGURE 3. Ambiguity of random waveform

Before we do, however, we note that for some purposes a thumbtack is not the best waveform. For example there are circumstances in which a waveform which is extremely doppler tolerant (that is its ambiguity is a ridge along the doppler axis) and in others we may require range tolerance.

3.1. *Conventional waveforms*

We shall first talk about waveforms for a convventional radar.

3.1.1. *Rectangular and Gaussian pulses.*
The simplest waveform used in radar is just a simple pulse, either rectangular or Gaussian. Such pulses are shown in Figure 4 and their ambiguity functions in Figure 5. In fact, of course, we plot the absolute value of the ambiguity

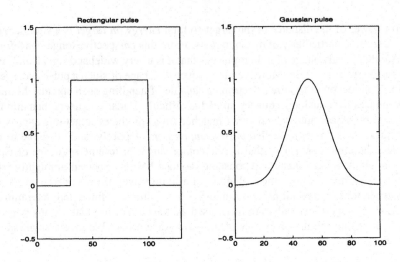

FIGURE 4. Rectangular and Gaussian waveforms

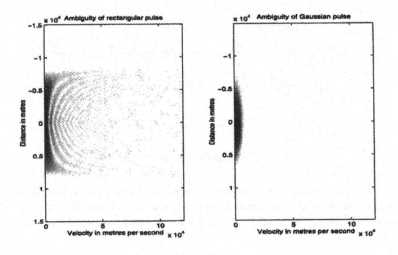

FIGURE 5. Ambiguities of rectangular and Gaussian waveforms

since in general it is a complex function. Notice the "sinc" behaviour on the doppler direction of the rectangular pulse. It has a relatively sharp peak in the doppler direction but then has "sidelobes". The effect of these sidelobes is to mix targets of differing dopplers into the same doppler bin. This is an issue for shipborne radars, in particular, where the reflection from the sea close to the ship might obscure a distant (and therefore fainter) airplane whose doppler should separate it.

A glance at the doppler axes in Figure 5 shows that I have plotted doppler shifts corresponding to quite unrealistic velocities. I should point out that the values assume a 100 microsecond pulse (a long one) at 10 GHz. A more realistic ambiguity plot is shown in Figure 6 — it gives the ambiguities of the two waveforms for velocities up to mach 3 — about 1km metres per second. This too is an unrealistic representation since the grey scales do not adequately convey the significance of the pixels in the image. Again from an engineering viewpoint it is better to plot the logarithm of the values of the ambiguity — to be precise we need a dB plot where the value in decibels is equal to $20 \log_{10} |\chi|$. These are shown in Figure 7. As can be seen, the main problems with these waveforms are

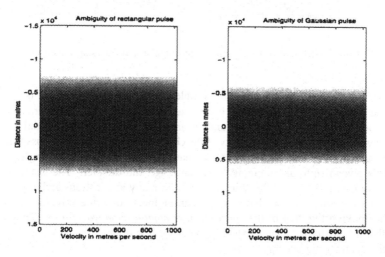

FIGURE 6. Ambiguities of rectangular and Gaussian waveforms out to mach 3

- significant doppler tolerance;
- very low range resolution.

To reiterate, a long pulse is used here and better range resolution could be obtained with a shorter pulse. However, compared to many other waveforms both of these have significant weaknesses.

3.1.2. *Chirp waveforms.* Perhaps the next most common waveform is called a "chirp" or "linear chirp". It is a signal with a linear increase or decrease in frequency over time. Using

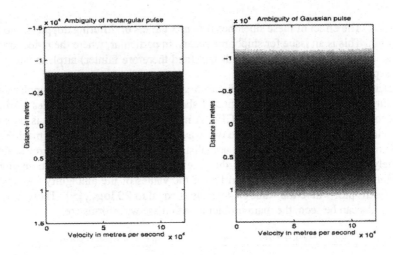

FIGURE 7. dB plots of ambiguities of rectangular and Gaussian waveforms

the fact that we have effectively real and complex (or I and Q channels) it is possible to transmit the signal $e^{-\alpha i t^2}$ (rather than just its real part, say). The real part is shown in Figure 8, along with its auto- correlation and the ambiguity of the complex signal is shown in Figure 8. It will be seen that the ambiguity of the chirp has a ridge running across the diagram which has quite a sharp peak, however, it does have a slight fall from left to right across the image. This gives rise to an ambiguity between range and doppler. The sharpness of the peak is a very useful feature of the chirp It is used in many applications and in particular in synthetic aperture radar. As we have said, radar engineers describe waveforms with auto-correlations looking like that of the chirp as *pulse compression waveforms*. They have the property that they behave to some extent (at least for low doppler values) like a very sharp pulse while retaining high energy.

3.1.3. *Barker sequences.* The fourth favourite waveform is based on Barker codes. Such a code is a sequence of ± 1s with the property that the auto correlation takes just the values $0, \pm 1$ and the length of the code. For example the code $[1 - 1111]$ has auto correlation

$$[101050101].$$

Barker codes have almost perfect auto correlation properties and quite doppler intolerant ambiguity properties. The ambiguity is shown in Figure 10. Unfortunately they are only known to exist in lengths 3, 5, 7, 11 and 13. Much mathematical effort has gone into trying to show that these are the only lengths of Barker codes and much computing effort into finding longer ones without success in both cases.

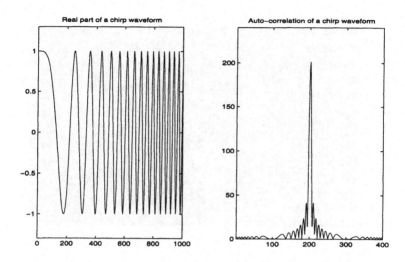

FIGURE 8. Chirp waveform and auto- correlation

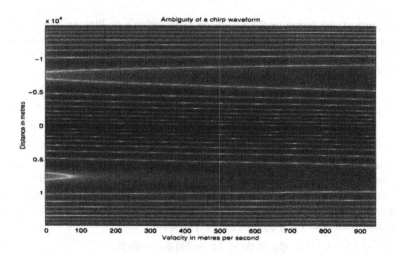

FIGURE 9. Ambiguity of Chirp waveform

One way in which radar engineers have sought to overcome the lack of longer Barker sequences is to use so called "Barker on Barker" sequences. These are formed by choosing

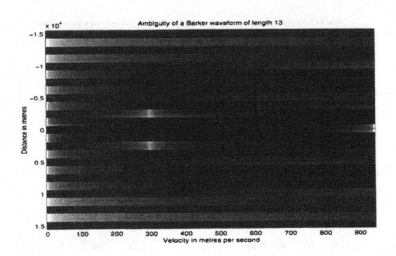

FIGURE 10. Ambiguity of Barker waveform of length 13

two Barker codes (which may be the same) and substituting a multiple of one for each entry in the other using that entry to determine the mulltiplicative factor. Thus a Barker of length 3 is $[1 - 1 - 1]$ and using it together with the Barker of length 5 given above we obtain the following code of length 15:

$$[\;1 -1 -1 -1\quad 1\quad 1\quad 1 -1 -1\quad 1 -1 -1\quad 1 -1 -1].$$

Repeated substitution of this kind allows us to obtain arbitrarily long codes. Unfortunately their auto correlation and ambiguity properties are considerably less than ideal. Figure 11 illustrates this for a Barker 7 on a Barker 13.

3.1.4. *Costas Arrays.* Costas arrays are used in the design of stepped frequency radars; that is, radars in which the waveform comprises a sequence of pure tones at differing frequencies. Thus a chosen collection of N frequencies $\mathbf{f} = (f_1, f_2, \ldots, f_N)^T$ are transmitted in consecutive time intervals of equal duration in this order. The frequencies are chosen from equally spaced ones $\boldsymbol{\phi} = (\phi_1, \phi_2, \ldots, \phi_N)^T$, so that

$$(3.1) \qquad\qquad \phi_j = \phi_1 + (j - 1)(\phi_2 - \phi_1).$$

The choice of order can be described bya permutation matrix P so that $= P\boldsymbol{\phi}$. In order to obtain good ambiguity performance, the matrix P is chosen so that for any shift of the matrix horizontally or vertically the non-zero terms overlap in at most one place. Such a matrix is called a *Costas array*. This can be re-expressed in terms of the set S:

$$(3.2) \qquad\qquad S = \{(i, j) | P_{ij} = 1\}.$$

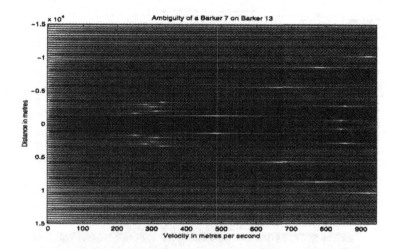

FIGURE 11. Ambiguity of a "Barker on Barker" waveform

Now P is a Costas array if, for $\mathbf{r, s, r', s'}$ distinct elements of S, no equation of the form

(3.3) $$\mathbf{s - r = s' - r'}$$

can hold.

There are several ways to construct Costas arrays (see Golomb [8]). Here is one due to Golomb. Let α and β be two distinct primitive elements in the Galois field $GF(p^n)$, and let P be the matrix whose entries are 0 except at the elements of the set

(3.4) $$A = \{(i,j) : \alpha^i + \beta^j = 1 \qquad (1 \leqslant i \leqslant q - 2)\}.$$

Then P is a permutation matrix and indeed a Costas array. Figure 3.1.4 is an example of a Costas array of size **31** and Figure 3.1.4 is its ambiguity.

3.2. *Complementary Waveforms*

Following the view that the ideal waveform is a thumb-tack, we should require of such a waveform that at least its auto- correlation be a spike at the origin and zero elsewhere. As we have seen, the chirp approximates this quite well. No single waveform can achieve this perfect auto correlation, though Barker waveforms approach the ideal. It is possible however to find a pair of waveforms which have the property that if transmitted separately, each correlated against a copy of itself and the results added then we do obtain a perfect spike. Such waveforms are called *complementary pairs*. We require of a pair of waveforms $w_1(t)$

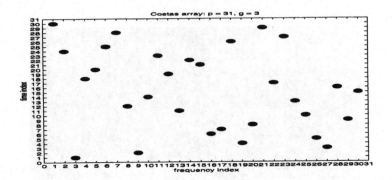

FIGURE 12. Costas array

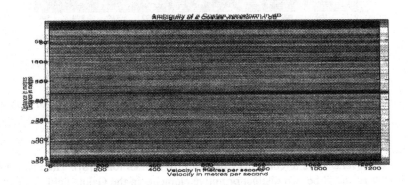

FIGURE 13. Costas array

and $w_2(t)$ that

$$\text{corr}(w_1, w_1)(t) + \text{corr}(w_2, w_2)(t) = \Delta_T(t),$$

where Δ_T is a short (triangular) spike at 0 of width T. This is equally expressed in terms of convolutions:

$$(3.5) \qquad w_1 * \widetilde{w_1}(t) + w_2 * \widetilde{w_2}(t) = \Delta_T(t).$$

where $\widetilde{w}(t) = \overline{w(-t)}$. As we have said, it is possible to find such pairs.

In fact it is possible to find a pair of discrete codes p_1 and p_2 which are finite sequences of ± 1s satisfing

$$(3.6) \qquad \text{corr}(p_1, p_1)(k) + \text{corr}(p_2, p_2)(k) = 0 \text{ except when } k = 0,$$

and from these easily construct waveforms satisfying (3.5). The classical construction of these is due to Golay and independently, Shapiro. It is an inductive construction starting with the two codes

$$p_1^{(1)} = [1, \quad 1]$$
$$p_2^{(1)} = [1, \ -1].$$

Then longer codes are constructed by the formula:

$$p_1^{(k)} = [p_1^{(k-1)}, \quad p_2^{(k-1)}]$$
$$p_2^{(k)} = [p_1^{(k-1)}, \ -p_2^{(k-1)}].$$

Our intention is to transmit the two waveforms in separate channels and on reception keep them separate and matched filter each against the corresponding transmitted form. The appropriate ambiguity function for such a pair of codes is the sum of the ambiguities of each code separately. By Moyal's identity, since p_1 and p_2 are orthogonal, so are χ_{p_1} and χ_{p_2}. After normalisation so that the sum of the energies in p_1 and p_2 equals 1, we have the following:

$$(3.7) \qquad ||\chi_{p_1} + \chi_{p_2}||^2 = ||\chi_{p_1}||^2 + ||\chi_{p_2}||^2 = ||p_1||^4 + ||p_2||^4 = \frac{1}{2}.$$

This shows that the ambiguities of complementary waveforms already have some chance of being more thumbtack-like than single channel waveforms, since the corresponding answer for a single channel waveform is 1, and the height at the centre $((0, 0))$ is the same in each case. The ambiguity of a complementary pair is illustrated in Figure 14. In fact at dopplers of interest in typical applications (up to mach 3, say) complementary pairs are quite doppler tolerant. What is remarkable is that at all dopplers the range ambiguity is zero outside half the range of the waveform.

It may seem that complementary sequences offer ambiguity properties superior to single waveforms. However there are serious issues involved in the implementation of these wave-forms. How does one maintain a significant separation of the two complementary sequences? If they are separated in time, then the target must remain coherent over the time span of the transmission for the advantages to be worthwhile. Separation in frequency, while feasible, also carries with it disadvantages. The responses of targets are frequency dependent as is attentuation through the atmosphere. Moreover, for extended targets, the effects of doppler change with different frequencies. These result in degradation from the ideal situation described here. The complexity of the electronics is also increased. One final observation: ultimately there is really only one signal transmitted, whether it be spread over a long time period (as time separation would require) or have a high bandwidth (in the case of frequency separation). The use of complementary waveforms is merely a device to manipulate the ambiguity function of a single waveform to push the lobes into more desired parts of the range-doppler plane.

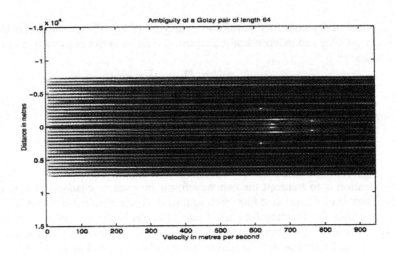

FIGURE 14. Ambiguity of a Golay pair of length 64

It is possible to go beyond just two waveforms. We recall the construction of the PONS matrix. This works by analogy with the Golay pairs but produces 2^n waveforms of length 2^n which are complementary in pairs, so that one could use any number of complementary sequence pairs separated by frequency or time or a mixture of both. The recursive method of construction is to take two "parent" codes s_1 and s_2 of length l, say, and use them to construct four "children" of double the length:

$$s_1 \cdot s_2$$
$$s_1 \cdot -s_2$$
$$s_2 \cdot s_1$$
$$-s_2 \cdot s_1,$$

where $\cdot$ indicates concatenation. It is relatively straightforward to see that, if the two parent codes are orthogonal, so are the four children. Classical PONS is now formed by starting with the two codes

$$\begin{matrix} 1 & 1 \\ 1 & -1 \end{matrix}$$

and repeatedly applying the above construction.

Of course the issues raised in the preceding paragraph are correspondingly exacerbated, but for N waveforms the power in the ambiguity is now (by a repeat of calculation (3.7)) just $1/N$. As Figure 15 illustrates, when we use half of the PONS matrix of length 64 the improvement in ambiguity is remarkable. We have perfect range sidelobes at all dopplers.

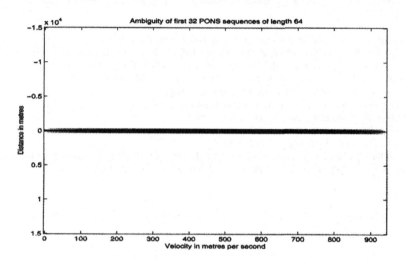

FIGURE 15. Ambiguity of the first 32 rows of the PONS matrix of length 64

References

[1] R.A. Altes. *Methods of wide-band signal design for radar and sonar systems*. PhD thesis, Univ. of Rcohester, Rochester, N.Y., 1970.

[2] R.A. Altes. Wide-band, proportional bandwidth wigner-ville analysis. *IEEE Trans. Acoustics, Speech, Signal Processing*, 38(6):1005–1012, June 1990.

[3] B. Borden. *Radar Imaging of Airborne Targets*. Institute of Physics Publishing, Bristol and Philadelphia, 1999.

[4] Wilcox C.H. The synthesis problem for radar ambiguity functions. Technical Report 157, MRC Tech. Summary Report, 1960. republished in Radar and Sonar part I vol 32 I.M.A. vol in Math. and its Appl,1991, 229–260.

[5] de Buda R. Signals that can be calculated from their ambiguity function. *IEEE Trans. Information Theory*, IT16:195–202, 1970.

[6] H. Dym and H.P. McKean. *Fourier Series and Integrals*. Academic Press, New York and London, 1972.

[7] G. Garrigós, Ph. Jaming, and J.-B. Poly. Zéros de fonctions holomorphes et contre-exemples en théorie des radars. To appear.

[8] S. Golomb. Costas arrays.

[9] Bueckner H.F. Signals having the same ambiguity functions. Technical Report 67-C-456, General Electric, Research and Development Center, Schnectady, N.Y., 1967.

[10] Ph. Jaming. Phase retrieval techniques for radar ambiguity functions. to appear.

[11] Auslander L. and Gretner I. Wide-band ambiguity functions and the $ax + b$ group. In *Signal processing part I: Signal Processing Theory*, volume 22 of *I.M.A. vol in Math and its Appl*, pages 1–12, 1990.

[12] Auslander L. and Tolimieri R. radar ambiguity functions and group theory. *SIAM J. Math Anal*, 16:577–6, 1985.

[13] Scharf L. *Statistical Signal Processing*.

[14] F.E. Nathanson. *Radar design principles*. McGraw-Hill, New York, 1969.

[15] Jaming Ph. *Trois problèmes d'analyse harmonique*. PhD thesis, Université d'Orléans, 1998.

[16] M.I. Skolnik. *Introduction to Radar Systems*. McGraw-Hill, New York, 2nd edition, 1981.

[17] J. Speiser. Wide-band ambiguity functions. *IEEE trans. Inform. Theory*, 13:122–3, 1967.

[18] Poor V. *Detection and Estimation Theory*. Springer-Verlag.

[19] Schempp W. *R. Math. Rep.Acad. Sci.Canada*, 4, 1984.

[20] Schempp W. *R. Math. Rep.Acad. Sci.Canada*, 5, 1984.

[21] Schempp W. *R. Math. Rep.Acad. Sci.Canada*, 6, 1984.

[22] Schempp W. *Math. Appl.*, pages 217–260, 1984.

[23] L.B. White. The wide-band ambiguity function and altes' q-distribution: constrained synthesis and time-scale filtering. *IEEE Trans. Inform. Theory*, 38(2):886–892, March 1992.

The Mathematical Theory of Wavelets

G. Weiss and E.N. Wilson

Washington University
Department of Mathematics
St. Louis, MO 63130

ABSTRACT. We present an overview of some aspects of the mathematical theory of wavelets. These notes are addressed to an audience of mathematicians familiar with only the most basic elements of Fourier Analysis. The material discussed is quite broad and covers several topics involving wavelets. Though most of the larger and more involved proofs are not included, complete references to them are provided. We do, however, present complete proofs for results that are new (in particular, this applies to a recently obtained characterization of "all" wavelets in section 4).

1. Introduction

A *wavelet* is a function ψ in $L^2(\mathbb{R})$ such that the system

$$(1.1) \qquad \psi_{jk}(x) \equiv 2^{j/2}\,\psi(2^j x - k)$$

j, k ϵ Z, is an orthonormal basis for $L^2(\mathbb{R})$. Observe that if τ_k is the translation operator mapping ψ into $(\tau_k\psi)(x) = \psi(x-k)$, $k \epsilon Z$, and D^j is the dilation operator defined by $(D^j\psi)(x) = 2^{j/2}\,\psi(2^j x)$, then the system $\{\psi_{jk}\}$ is obtained by *first* applying the translation τ_k to ψ and, *secondly*, the dilation D^j to the function $\tau_k\psi$. As we shall see later on, it is important to respect this order of applying these operators: the translation operator is applied before the dilation operator.

Two examples of wavelets were known for a long time: the *Haar wavelet* and the *Shannon wavelet*. The former is the function

$$(1.2) \qquad \psi(x) = \begin{cases} 1, & \text{if } 0 \leqslant x < 1/2 \\ -1, & \text{if } 1/2 \leqslant x < 1 \\ 0, & \text{elsewhere} \end{cases}.$$

329

J.S. Byrnes (ed.), Twentieth Century Harmonic Analysis - A Celebration, 329–366.
© 2001 *Kluwer Academic Publishers. Printed in the Netherlands.*

The latter is the function ψ whose Fourier transform is

(1.3)
$$\widehat{\psi}(\xi) = \begin{cases} 1, & \text{if } \xi \in (-1, -1/2] \cup [1/2, 1) \\ 0, & \text{elsewhere} \end{cases}.$$

The Fourier transform we shall use is given by the equality

(1.4)
$$\widehat{f}(\xi) = \int_{-\infty}^{\infty} f(x) e^{-2\pi i \xi x}\, dx$$

whenever $f \in L^1(\mathbb{R})$. We assume that the reader is acquainted with the basic L^2- theory of the Fourier transform. In particular, $(\tau_k f)^\wedge(\xi) = e^{-2\pi i k \xi} \widehat{f}(\xi)$ and $(D^j f)^\wedge(\xi) = 2^{-j/2} \widehat{f}(2^{-j} \xi)$. Thus, translations by k are converted by the Fourier transform into *modulations* by $-k$ (multiplication by $e^{-2\pi i k \xi}$); the dilations D^j become the dilations D^{-j} after taking the Fourier transform. The Plancherel theorem, the fact that $\{2^j S\}$, $j \in \mathbb{Z}$, is a partition of $\mathbb{R} - \{0\}$ when $S = (-1, -1/2] \cup [1/2, 1)$, and the completeness of the system $\{e^{2\pi i k \xi}\}$, $k \in \mathbb{Z}$, in $L^2(S)$, immediately imply that the Shannon function ψ in (1.3) is a wavelet. That the Haar function defined in (1.2) is a wavelet has been well known since it was introduced in 1910 [Ha]. In any case, this is an easy application of the characterizations of wavelets we shall present.

In the early eighties many different constructions of wavelets were discovered. This included several other similar methods of reproducing functions. For example, pairs of systems $\{\phi_{jk}\}$ and $\{\psi_{jk}\}$, $j, k \in \mathbb{Z}$, were introduced so that for any $f \in L^2(\mathbb{R})$ we have the reproducing formula

(1.5)
$$f = \sum_{j,k \in \mathbb{Z}} \langle f, \phi_{jk} \rangle \psi_{jk}$$

for all $f \in L^2(\mathbb{R})$.

We will present a careful accounting of who produced the results we describe throughout the text as well as in an appendix at the end of this exposition. Soon after the "new wavelets" were introduced it became apparent that they had important applications in various different areas. This attracted many investigators whose principal interest was in these applications. Perhaps this detracted attention from the mathematical theory that is associated with wavelets and similar concepts. Our purpose is to present some of this theory. It is our belief that it is a beautiful subject connected to many areas of mathematics.

It is clear from the little that has been presented so far that the Fourier transform must play an important role in the study of bases and similar systems that are constructed by applying translations, dilations and modulations to a specific function. Let us illustrate this by presenting a characterization of those $\psi \in L^2(\mathbb{R})$ such that $\{\psi_{j,k}\}$, $j, k \in \mathbb{Z}$, is an orthonormal system:

PROPOSITION I. *Suppose $\psi \in L^2(\mathbb{R})$. Then $\{\psi (. - k) : k \in \mathbb{Z}\}$ is an orthonormal system if and only if*

(A)
$$\sum_{k \in \mathbb{Z}} | \widehat{\psi} (\xi + k)|^2 = 1 \ \text{for a.e. } \xi \in \mathbb{R}.$$

The proof of this fact is very simple. The orthonormality condition is $\langle \psi (. - j), \psi (. - l) \rangle = \delta_{jl}$, which, by the Plancherel theorem, is equivalent to

$$\delta_{jl} = \int_{\mathbb{R}} \widehat{\psi} (\xi) \overline{\widehat{\psi} (\xi)} e^{-2\pi i (j-l)\xi} \, d\xi.$$

We can then "periodize" this integral so that it takes the form

$$\int_0^1 \sum_{k \in \mathbb{Z}} | \widehat{\psi} (\xi + k)|^2 e^{-2\pi i (j-l)\xi} \, d\xi$$

and we see that the orthonormality condition is equivalent to the statement that the 1-periodic function $\sum_{k \in \mathbb{Z}} | \widehat{\psi} (\xi + k)|^2$, which clearly belongs to $L^2 ([0, 1))$, has Fourier coefficients 0 corresponding to all non-zero frequencies and the zero-coefficient is 1. But this is equality (A).

PROPOSITION II. *The systems $\{\psi_{j_1,k}\}$ and $\{\psi_{j_2,k}\}$, $k \in \mathbb{Z}$, are orthogonal to each other whenever $j_1 \neq j_2$ if and only if*

(B)
$$\sum_{k \in \mathbb{Z}} \widehat{\psi} (\xi + k) \overline{\widehat{\psi} (2^j(\xi + k))} = 0 \ \ \text{for a.e. } \xi \in \mathbb{R} \text{ whenever } j \geqslant 1.$$

By a change of variable the orthogonality condition can be reduced to the case $j_1 = 0$ and $j_2 \geqslant 1$. A periodization argument, just like the one we just described then gives us equality (B).

Thus, we see that the characterization of *all* wavelets is reduced to finding a condition that implies the completeness of the system $\{\psi_{jk}\}$, $j, k \in \mathbb{Z}$. It turns out that, again, a simple equality, involving the Fourier transform of ψ, provides us with such a characterization of completeness:

PROPOSITION III. *(The characterization of all wavelets in $L^2(\mathbb{R})$). A function $\psi \in L^2(\mathbb{R})$ is a wavelet if and only if the system $\{\psi_{jk}\}$ is orthonormal and*

(C) $$\sum_{j\in\mathbb{Z}} |\hat{\psi}(2^j \xi)|^2 = 1 \quad for\ a.e\ \xi \in \mathbb{R}.$$

Unlike Proposition I, this result is not immediate. It is also quite new. We shall discuss its proof in the sequel. For the moment, let us make some observations.

The characterization of orthonormality involved averaging (summing) over the group of integral translations. Since the group of dyadic dilations also plays a basic role in the definition of a wavelet it is natural to expect that averaging over this last group plays a part in this characterization. In fact, this is precisely what is the case in equality (C). What is surprising, however, is that there is a characterization of all wavelets that involves only sums over dilations:

PROPOSITION IV. *(Another characterization of all wavelets) Suppose* $\psi \in L^2(\mathbb{R})$, *then* ψ *is a wavelet if and only if* $\|\psi\|_2 \geqslant 1$, *equality* (C) *is satisfied, and*

(D) $$t_q(\xi) = \sum_{j\geqslant 0} \hat{\psi}(2^j \xi) \overline{\hat{\psi}(2^j(\xi+q))} = 0 \ for\ a.e.\ \xi \in \mathbb{R},$$

whenever q is an odd integer.

Let us explain the role played by the hypothesis $\|\psi\|_2 \geqslant 1$. Let H be a separable Hilbert space and $\mathcal{E} = \{e_\alpha : \alpha \in \mathcal{A}\}$ a countable collection of vectors in H ($\mathcal{A}$ can be N or $\{(j,k) : j, k \in \mathbb{Z}\}$) such that $\|u\|^2 = \sum_{\alpha\in\mathcal{A}} |(u, e_\alpha)|^2$ for each $u \in H$. Such a collection $\mathcal{E}$ is then called a *tight frame* (of constant 1) for H. If $\|e_\alpha\| \geqslant 1$ for all $\alpha \in \mathcal{A}$ then, letting $u = e_{\alpha_0}$, we have

$$\|e_{\alpha_0}\|^2 = \sum_{\alpha \in \mathcal{A}} |(e_{\alpha_0}, e_\alpha)|^2 = \|e_{\alpha_0}\|^4 + \sum_{\alpha\neq\alpha_0} |(e_{\alpha_0}, e_\alpha)|^2.$$

Hence,

$$\|e_{\alpha_0}\|^2 (1 - \|e_{\alpha_0}\|^2) = \sum_{\alpha \neq \alpha_0} |(e_{\alpha_0}, e_\alpha)|^2.$$

Because of our assumption that $\|e_{\alpha_0}\| \geqslant 1$, the left side of this equality cannot be strictly bigger than 0, while the right side cannot be negative. It follows that $\|e_{\alpha_0}\| = 1$ and $(e_{\alpha_0}, e_\alpha) = 0$ for all α_0 and $\alpha \neq \alpha_0$. That is, $\mathcal{E}$ is an orthonormal basis (see pages 336-7 of [HW] for a more complete account of these matters).

The two equalities (C) and (D) characterize those $\psi \in L^2(\mathbb{R})$ for which the system $\{\psi_{jk}\}$, j, k in $\mathbb{Z}$, is a tight frame of constant 1 for $L^2(\mathbb{R})$. The condition $\|\psi\|_2 \geqslant 1$ assures us that this system is an orthonormal basis; that is, that ψ is a wavelet.

The four equations (A), (B), (C) and (D) not only provide us with a rather simple characterization of all wavelets, but they are most useful for constructing large classes of wavelets.

For example, it follows immediately from (A) or (C) that if ψ is a wavelet then $|\widehat{\psi}(\xi)| \leqslant 1$ a.e. Since $\|\psi\|_2 = \|\widehat{\psi}\|_2 = 1$ this means that $\{\xi : \widehat{\psi}(\xi) \neq 0\}$ must have measure at least 1. The Shannon wavelet is an example for which this set has measure precisely 1. It is natural to consider the class of all wavelets ψ such that $\mathbf{W} = \mathbf{W}_\psi = \{\xi : \widehat{\psi}(\xi) \neq 0\}$ has measure 1. For such ψ we clearly must have $|\widehat{\psi}| = \chi_W$. It is natural to call the class of such wavelets the collection of *Minimally Supported Frequency* (MSF) wavelets. It is an easy exercise to show that the MSF wavelets are characterized as the class of all $\psi \epsilon L^2(\mathbb{R})$ such that $|\widehat{\psi}(\xi)|$ assumes only the values 0 or 1 a.e. and equations (A) and (C) are satisfied. The sets $\mathbf{W} = \mathbf{W}_\psi$ on which the Fourier transform of MSF wavelets is not zero are called *wavelet sets*. (A) and (C) are equivalent to the statement:

THEOREM (1.1). *W is a wavelet set if and only if each of the collections* $\{W - k\}, k$ *in* $\mathbb{Z}$, *and* $\{2^j W\}, j$ *in* $\mathbb{Z}$ *is a partition of* $\mathbb{R}$.

In the course of this exposition the reader will find many examples of wavelets constructed by the use of these 4 equations and in the appendix we will give a still larger class of wavelets obtained by these means.

We shall also consider the subject of wavelets $\psi \epsilon L^2(\mathbb{R}^n)$. Not only will we show many of the various properties they enjoy, but we will generalize the concept by showing how other dilations and translations can be used for obtaining orthonormal bases or tight frames from a particular function (or a collection of functions); moreover, we will extend all these matters to higher dimensions. In order to do this most efficiently it is useful to discuss "continuous wavelets" associated with $\mathbb{R}^n$. For many considerations the theory of these wavelets is simpler.

2. Continuous Wavelets in One and More Dimensions

Let G be the *affine group* associated with $\mathbb{R}$ consisting of all

$$(a, b) \epsilon \mathbb{R} \times \mathbb{R}, \ a \neq 0, \quad \text{with the group operation}$$

$$(c, d) \circ (a, b) = (ac, b + \frac{d}{a}).$$

This operation is consistent with the action of $g = (a, b) \epsilon G$ on x in $\mathbb{R}$ given by $g(x) = a(x + b)$. Observe that $g^{-1} = (a, b)^{-1} = (a^{-1}, -ab)$. For $\psi \epsilon L^2(\mathbb{R})$ let

$$(2.1) \qquad (T_g \psi)(x) = \frac{1}{\sqrt{|a|}} \psi(\frac{x}{a} - b) = \frac{1}{\sqrt{|a|}} \psi(g^{-1}(x)) \equiv \psi_{a,b}(x).$$

Then $g \longrightarrow T_g$ is a unitary representation of G acting on $L^2(\mathbb{R})$.

The mapping W_ψ taking $f \epsilon L^2(\mathbb{R})$ into the function

$$(W_\psi f)(g) = \int_{\mathbb{R}} f(x) \, \overline{(T_g \psi)(x)} \, dx = \langle f, \psi_{a,b} \rangle$$

on G is the (*continuous*) *wavelet transform of f*. A goal in wavelet theory is to find a condition on ψ that allows us to reconstruct f from its wavelet transform via the reproduction formula

$$(2.2) \qquad f(x) = \int_G \langle f, \psi_g \rangle \, \psi_g(x) d\lambda(g) = \int_G (W_\psi f)(g) \, T_g \psi(x) \, d\lambda(g),$$

where λ is (left) Haar measure on G ($d\lambda(a, b) = \frac{da\,db}{|a|}$). One can consider (1.5) as a discrete version of this reproduction formula when $\varphi = \psi$.

This condition, the *admissibility condition* for ψ, was discovered by Calderón in 1964 [C] and can be expressed in the form

$$(2.3) \qquad 1 = \int_{\mathbb{R}-\{0\}} |\widehat{\psi}(a\,\xi)|^2 \frac{da}{|a|}$$

for a.e. ξ. We will show the equivalence of (2.2) and (2.3) in a considerably more general context.

It is clear that (2.3) is the "continuous" analog of equality (C). In fact it is much more than an analog. Suppose that $\psi \in L^2(\mathbb{R})$ satisfies (C), then

$$\log 2 = \int_1^2 \frac{da}{a} = \int_1^2 \sum_{j \in \mathbb{Z}} |\widehat{\psi}(2^j a)|^2 \frac{da}{a} = \sum_{j \in \mathbb{Z}} \int_{2^j}^{2^{j+1}} |\widehat{\psi}(a)|^2 \frac{da}{a}$$

$$= \int_0^\infty |\widehat{\psi}(a)|^2 \frac{da}{a} = \int_0^\infty |\widehat{\psi}(a\,\xi)|^2 \frac{da}{a}.$$

for $\xi > 0$. Thus, it follows that, after a renormalization, ψ satisfies (2.3).

This shows, essentially, that each wavelet is also a continuous wavelet. On the other hand, it is clear that the converse is not true; being a wavelet is more restrictive than being a continuous wavelet.

Let us also observe that, in the continuous case, the order of the operation of translation (by $-b$) followed by dilation (by $\frac{1}{a}$), as performed in the definition of $\psi_{a,b}$, can be reversed and we would, again, have that the same admissibility condition (2.3) is equivalent to the reproducing formula (2.2). More precisely, if G is endowed with the operation $(a, b) \cdot (c, d) = (ac, ad + b)$ (that corresponds to the action $x \longrightarrow ax + b$ on $\mathbb{R}$) and

$$(S_g \psi)(x) = \frac{1}{\sqrt{|a|}} \psi(g^{-1} x) = \frac{1}{\sqrt{|a|}} \psi\left(\frac{x - b}{a}\right) \equiv \widetilde{\psi}_{a,b}(x),$$

we have

(2.4) $$\int_G \langle f, \psi_{a,b} \rangle \, \psi_{a,b}\,(x)\, \frac{dadb}{|a|} = \int_{\widetilde{G}} \langle f, \widetilde{\psi}_{a,b} \rangle \, \widetilde{\psi}_{a,b}\,(x)\, \frac{dadb}{a^2}$$

for all $f \in L^2(\mathbb{R})$ ($\widetilde{G}$ is the "new" version of the affine group with this last multiplication, so that its left Haar measure is $\frac{dadb}{a^2}$).

Let us now pass to the extensions of these notions and results to n dimensions. The *Full Affine Group of Motions on* $\mathbb{R}^n$, $G^\#$, consists of all pairs $(a, b) \in GL(n, \mathbb{R}) \times \mathbb{R}^n$ (endowed with the product topology) together with the operation

$$(\alpha, \beta) \cdot (a, b) = (\alpha a, b + a^{-1}\beta).$$

This operation is associated with the action $x \longrightarrow a(x + b)$ on $\mathbb{R}^n$. The subgroup

$$\mathcal{N} = \{(a, b) \in G^\# : a = I,\ b \in \mathbb{R}^n\}$$

is clearly a normal subgroup of $G^\#$.

We consider a class of subgroups, $\{G\}$, of $G^\#$ of the form

$$G = \{(a, b) \in G^\# : a \in D, b \in \mathbb{R}^n\},$$

where D is a closed subgroup of $GL(n, \mathbb{R})$. We can identify D with the subgroup $\{(a, b) \in G : a \in D, b = 0\}$) of G. We refer to D as the *dilation subgroup* and $\mathcal{N}$ will be called the *translation subgroup of* G.

If μ is left Haar measure for D, then $d\lambda(a, b) = d\mu(a)db$ is the element of left Haar measure for G.

Let T be the unitary representation of G acting on $L^2(\mathbb{R}^n)$ defined by

(2.5) $$(T_{(a,b)}\psi)(x) = |\det a|^{-1/2}\, \psi(a^{-1}x - b) \equiv \psi_{a,b}(x)$$

for $(a, b) \in G$ and $\psi \in L^2(\mathbb{R}^n)$. Observe that $(a, b)^{-1} = (a^{-1}, -ab)$. We then have

(2.6) $$(T_{(a,b)}\psi)^\wedge(\xi) = |\det a|^{\frac{1}{2}}\, \widehat{\psi}(a^*\xi)e^{-2\pi i\xi \cdot ab}$$

where a^* is the transpose of a.

The *wavelet transform* W_ψ *associated with* ψ is now defined by

$$(W_\psi f)(a, b) = \langle f, \psi_{a,b} \rangle = \int_{\mathbb{R}^n} f(y)\, \overline{\psi(a^{-1}y - b)}\, \frac{dy}{\sqrt{|\det a|}}$$

whenever $f \in L^2(\mathbb{R}^n)$ and $(a, b) \in G$. Our first goal is to find an admissibility condition for ψ that guarantees the general version of the Calderón reproducing formula

(2.7)
$$f(x) = \int_G \langle f, \psi_{a,b} \rangle \, \psi_{a,b}(x) d\lambda(a,b)$$

for all $f \in L^2(\mathbb{R}^n)$. The analog of (2.4), involving the operation $(\alpha, \beta) \circ (a, b) = (\alpha a, \alpha b + \beta)$, is valid in this general case; hence, the same admissibility condition applies to both "versions" of G. This condition is

THEOREM (2.1). *Equality(2.7) is valid for all $f \in L^2(\mathbb{R}^n)$ if and only if for a.e. $\xi \neq 0$*

(2.8)
$$\Delta_\psi(\xi) = \int_D |\widehat{\psi}(a^* \xi)|^2 d\mu(a) = 1.$$

(compare with (2.3)).

The following argument also provides a (weak) meaning for (2.7).

W_ψ obviously maps $L^2(\mathbb{R}^n)$ into $L^\infty(G)$. We claim that, if (2.8) is satisfied, then W_ψ is an isometry from $L^2(\mathbb{R}^n)$ into $L^2(G, \lambda)$:

$$\|W_\psi f\|^2_{L^2(G,\lambda)} = \int_D \int_{\mathbb{R}^n} |\langle f, \psi_{a,b} \rangle|^2 db d\mu(a) =$$

$$\int_D \int_{\mathbb{R}^n} |\int_{\mathbb{R}^n} \widehat{f}(\xi) \overline{\widehat{\psi}(a^* \xi)} e^{i2\pi \xi \cdot ab} d\xi|^2 |\det a| db d\mu(a)$$

$$= \int_D [\int_{\mathbb{R}^n} |\{\widehat{f} \overline{\widehat{\psi}(a^* \cdot)}\}^\vee (ab)|^2 |\det a| db] d\mu(a)$$

$$= \int_D [\int_{\mathbb{R}^n} |\{\widehat{f} \overline{\widehat{\psi}(a^* \cdot)}\}^\vee (b)|^2 db] d\mu(a)$$

$$= \int_D \int_{\mathbb{R}^n} |\widehat{f}(\xi)|^2 |\widehat{\psi}(a^* \xi)|^2 d\xi d\mu(a) = \int_{\mathbb{R}^n} |\widehat{f}(\xi)|^2 \Delta_\psi(\xi) d\xi$$

$$= \|\widehat{f}\|^2_2 = \|f\|^2_2.$$

By polarization, therefore, we have

(2.9)
$$\langle W_\psi f, W_\psi h \rangle_{L^2(G)} = \langle f, h \rangle_{L^2(\mathbb{R}^n)}$$

for all f and $g \in L^2(\mathbb{R}^n)$. In particular, the adjoint, W_ψ^*, of W_ψ is a left inverse of W_ψ : $W_\psi^* W_\psi = I$. We also have shown that the reproducing formula is valid in the weak sense. We refer the reader to [S] for more general versions of this reproducing formula.

Suppose, on the other hand, W_ψ satisfies (2.9) (so that (2.7) is valid in the weak sense) and ξ_0 is a point of differentiability for the integral of Δ_ψ. Let $|\widehat{f}(\xi)|^2 = |B_r(\xi_0)|^{-1} X_{B_r(\xi_0)}(\xi)$, where $B_r(\xi_0)$ is the ball of radius $r > 0$ centered at ξ_0. Reversing the equality chain used to obtain (2.9) we see that

$$\frac{1}{|B_r(\xi_0)|} \int_{B_r(\xi_0)} \Delta_\psi(\xi)\, d\xi = 1$$

for all $r > 0$. Letting $r \longrightarrow 0$ we obtain $\Delta_\psi(\xi_0) = 1$. Since a.e. $\xi \in \mathbb{R}^n$ is such a point of differentiability, the admissibility condition (2.8) is true and the theorem is established.

It is natural to ask: for what dilation groups D does there exist a $\psi \in L^2(\mathbb{R}^n)$ satisfying the admissibility condition (2.8)? We shall call such groups *admissible*. When $n = 1$ and D is the group of non-zero numbers (or positive numbers) with multiplication being the group operation, it is clear that the admissibility condition is verified by any ψ such that $\hat{\psi}$ is a bounded function supported in a compact set in $\mathbb{R} - \{0\}$ (appropriately scaled). Thus, there exists $\psi \in L^2(\mathbb{R}^n)$ satisfying the admissibility condition . For the same reason, in $\mathbb{R}^n$, the group $D = \{aI : a \in \mathbb{R} - \{0\}\}$ is also admissible. The group $SO(2)$, acting on $\mathbb{R}^2$, however, is not admissible. For if this group were admissible, then there exists $\psi \in L^2(\mathbb{R}^2)$ such that

$$1 = \int_0^{2\pi} |\hat{\psi}(e^{i\theta}\rho e^{i\varphi})|^2\, d\theta = \int_0^{2\pi} |\hat{\psi}(\rho e^{i\theta})|^2\, d\theta.$$

Thus,

$$\int_0^\infty \rho\, d\rho = \int_0^\infty \rho \int_0^{2\pi} |\hat{\psi}(\rho e^{i\theta})|^2\, d\theta\, d\rho = ||\hat{\psi}||^2_{L^2(\mathbb{R}^2)} < \infty$$

which is clearly impossible.

However, the 1-parameter groups

(2.10)
$$D = \left\{ a = e^{tL} : t \in \mathbb{R},\ L = \begin{pmatrix} 1 & -1 \\ 1 & 1 \end{pmatrix} \right\}$$

and

(2.11)
$$D = \left\{ \begin{pmatrix} x & y \\ 0 & 1 \end{pmatrix} : x \neq 0, x, y \in \mathbb{R} \right\}$$

are admissible.

The following (almost) characterization of the admissible groups D can be applied to see the validity of our claim about the last two examples.

THEOREM (2.2). *Let D be a closed subgroup of $DL(n, \mathbb{R})$ and consider the right action $\xi \longrightarrow a^* \xi$ of G on $\mathbb{R}^n$ for $\xi \in \mathbb{R}^n$. Let*

$$D_\xi^\epsilon = \{a \in D : ||a^* \xi - \xi|| \leqslant \epsilon\}$$

be the ϵ-stabilizer of ξ, for $\epsilon \geqslant 0$, and let $D_\xi = D_\xi^0$ be the stabilizer of ξ. If either

   i. $|\det a| = \Delta(a)$ *for all* $a \in D$, *or*
   ii. $\{\xi \in \mathbb{R}^n : D_\xi$ *is noncompact*$\}$ *has positive Lebesgue measure*

holds, then D *is not admissible. If both* i *and*

   iii. $\{\xi \in \mathbb{R}^n : D_\xi^\epsilon$ *is non-compact for all* $\epsilon > 0\}$ *has positive Lebesgue measure*

fail, then D *is admissible.*

(Δ is the modular function of D : the Radon-Nikodym derivative $d\mu_l/d\mu_r$, where μ_l is left Haar measure and μ_r is right Haar measure on D, normalized so that $\Delta(I) = 1$, where I is the identity element of D.) The proof is by no means immediate and will appear in [LWWW].
When D is given by (2.11) and $\xi = (\xi_1, \xi_2)$ with $\xi_1 \neq 0$ we have

$$D_\xi^\epsilon = \left\{ \begin{pmatrix} x & y \\ 0 & 1 \end{pmatrix} : (x-1)^2 + y^2 \leqslant \epsilon^2/\xi_1^2 \right\};$$

This is compact for any positive ϵ, so condition 2 fails. Furthermore, for $a = \begin{pmatrix} x & y \\ 0 & 1 \end{pmatrix}$, $d\mu_l(a) = dxdy/x^2$ and $d\mu_r(a) = dxdy/|x|$ so $\Delta(a) = d\mu_l(a)/d\mu_r(a) = 1/|\det a|$ and so condition i is also invalid. Hence D is admissible. When D is a general 1-parameter subgroup of $GL(n; \mathbb{R})$ (i.e., $D = \{e^{tL} : t \in \mathbb{R}\}$ for some $n \times n$ matrix L), D is unimodular and $\det(e^{tL}) = e^{t\,tr(L)}$ so i holds $\Leftrightarrow tr(L) = 0$. In this case D is not admissible. When $tr(L) \neq 0$, i fails and it is easy to check that 2 also fails, so D is admissible.

A *homogeneous Galilei group* is a subgroup D of $GL(n + 1, \mathbb{R})$ which is of the form (2.11) with y replaced by an $n \times 1$ column vector (i.e., a member of $\mathbb{R}^n$) and x replaced by an invertible $n \times n$ matrix satisfying various stipulations; e.g., x is an orthogonal matrix or the product of an orthogonal matrix and a non-zero scalar. The subgroup G of the affine group on $\mathbb{R}^{n+1}$ whose dilation group is D and whose translation group includes all $\mathbb{R}^{n+1}$ translations is then an *inhomogeneous Galilei group.*

Using Theorem (2.2), it follows easily that D is not admissible if we allow x to be orthogonal while D is admissible if we allow x to be the product of an orthogonal matrix and a non-zero scalar. In the second case, the family $\{T_g \psi : g \in G\}$ determined by any continuous wavelet ψ and the reproducing formula associated with this family provide examples of what are known in physics as *coherent states*. The admissibility conditions for certain classes of these groups have been obtained by several authors. In general these derivations are rather complicated (see [Co]); theorem (2.2) does provide a simpler and more unified method for solving these problems.

As an introduction to the notion of "discretizing" continuous wavelets let us observe that the proof of Theorem (2.1) in the special case $n = 1$ and D the group $\{2^j : j \in \mathbb{Z}\}$ shows that (C) in Proposition III is equivalent to the reproducing formula

(2.12)
$$f = \sum_{j \in \mathbb{Z}} \int_{\mathbb{R}} \langle f, \psi_{jt} \rangle \, \psi_{jt} \, dt,$$

where $\psi_{jt}(x) \equiv 2^{j/2} \psi(2^j x - t)$ for $j \in \mathbb{Z}$ and $t \in \mathbb{R}$. That is, (2.12) is a "discretization" of (2.2) with the sum $\sum_{j \in \mathbb{Z}}$ replacing the integral $\int_{\mathbb{R}} \cdot \dfrac{da}{|a|}$. The replacement of the integral over $\mathbb{R}$ in (2.12) by the sum $\sum_{k \in \mathbb{Z}} \langle f, \psi_{jk} \rangle \psi_{jk}$ gives us the orthonormal wavelet expansion of f if ψ is such a wavelet. It turns out that there are other discretizations of (2.12). For example, if n is an odd integer and ψ is an o.n. wavelet, then

$$f = \sum_{j \in \mathbb{Z}} \sum_{k \in \mathbb{Z}} \frac{1}{n} \langle f, \psi_{j\frac{k}{n}} \rangle \psi_{j\frac{k}{n}}$$

with convergence in $L^2(\mathbb{R})$ (see [CS]). This is an example of a phenomenon known as *oversampling*. Letting n tend to ∞ we obtain (at least formally) the equality (2.12) which can be thought of as "the ultimate oversampling property" of a ψ satisfying (C). There are several examples of discretizations of the continuous wavelet properties. In order to appreciate the complexity of "discrete" over "continuous" wavelets we now present generalizations to non-dyadic wavelets, wavelet systems, and related families in n-dimensions.

Let $\Gamma = P\mathbb{Z}^n$ be a lattice in $\mathbb{R}^n$ (P any invertible $n \times n$ matrix) and A an $n \times n$ *dilation* matrix (each eigenvalue λ of A satisfies $|\lambda| > 1$) for which $A\Gamma \subset \Gamma$. For $\psi \in L^2(\mathbb{R}^n)$ let $\mathcal{E} = \{\psi_{j,\gamma} : j \in \mathbb{Z}, \gamma \in \Gamma\}$, where

(2.13)
$$\psi_{j,\gamma}(x) = |\det A|^{j/2} \psi(A^j x - \gamma).$$

$\mathcal{E}$ is the *affine system* generated by ψ, the lattice Γ, and the dilation matrix A. ψ is a *wavelet relative to Γ and A* if and only if $\mathcal{E}$ is an orthonormal basis of $L^2(\mathbb{R}^n)$. We shall also use the further notations: $\Gamma^* = \{\gamma' \in \mathbb{R}^n : \langle \gamma', \gamma \rangle \in \mathbb{Z} \text{ for all } \gamma \in \Gamma\}$ and $B = A^*$, the transpose of A. Then $B\Gamma^* \subset \Gamma^*$. Let S be the set difference $\Gamma^* \setminus B\Gamma^*$. We then have the following generalizations of Proposition III and Proposition IV.

PROPOSITION III'. *ψ is a wavelet relative to Γ and A if and only if its affine system $\mathcal{E}$ is an orthonormal set in $L^2(\mathbb{R}^n)$ and*

(C')
$$\sum_{j \in \mathbb{Z}} |\widehat{\psi}(B^j \xi)|^2 = |\det P| \quad \text{for a.e. } \xi \in \mathbb{R}^n.$$

PROPOSITION IV'. *The affine system $\mathcal{E}$ generated by ψ is a tight frame of constant 1 if and only if equality (C') is satisfied and, for each $s \in S$,*

(D')

$$\sum_{j \geqslant 0} \widehat{\psi}\,(B^j\,\xi)\,\overline{\widehat{\psi}\,(B^j(\xi+s))} = 0 \ \text{ for a.e. } \xi \in \mathbb{R}^n.$$

ψ is a wavelet if and only if (C') and (D') are satisfied with $\|\psi\|_2 \geqslant 1$.

Now suppose A is of the form e^L for some $n \times n$ real matrix L; this is a very mild extra condition on A, e.g., it is automatically satisfied if A has no negative eigenvalues. Write A^t for e^{tL} and B^t for $(A^t)^* = e^{tL^*}$. Then $D = \{A^t : t \in \mathbb{R}\}$ is a one parameter subgroup of $GL(n, \mathbb{R})$ with Haar measure $d\mu\,(A^t) = dt$. If ψ is a wavelet relative to $\Gamma = \mathbb{Z}^n$ and A, then equation (C') implies

$$\int_D |\widehat{\psi}\,((A^t)^*\,\xi)|^2\,d\mu\,(A^t) = \int_{-\infty}^{\infty} |\widehat{\psi}\,(B^t\,\xi)|^2\,dt$$

$$= \int_0^1 \sum_{j \in \mathbb{Z}} |\widehat{\psi}\,(B^t(B^j\,\xi))|^2\,dt = 1$$

for *a.e.* $\xi \in \mathbb{R}^n$. So ψ is a continuous wavelet relative to D. It is easy to construct examples where A belongs to larger subgroups (non 1-parameter) (D') and ψ remains a continuous wavelet relative to (D'). In view of Proposition IV', we again conclude that it is relatively easy for a function $\psi \in L^2(\mathbb{R}^n)$ to be a continuous wavelet but that far more structure is required for ψ to be a discrete wavelet. We are also led to pose the question of determining, for a given admissible group $D \subset GL(n, \mathbb{R})$, which discrete subgroups of D and which $\mathbb{R}^n$ lattices give rise to discrete systems analogous to $\mathcal{E}$ which are either orthonormal bases or tight frames for $L^2(\mathbb{R}^n)$.

Another observation is in order. In many situations it is appropriate to generate a wavelet basis with more than one funtion ψ. In the fourth section, for example, we shall see that $L = 2^n - 1$ functions $\psi^1, \ldots, \psi^L$ are needed to obtain MRA wavelets in n-dimensions. This is reflected in what follows.

With Γ and A as above and with L an integer $\geqslant 1$, we can associate with each family $\Psi = \{\psi^1, \psi^2, \ldots, \psi^L\} \subset L^2(\mathbb{R}^n)$ an affine system $\mathcal{E}_\Psi = \{\psi^l_{j\,\gamma} : j \in \mathbb{Z}, \gamma \in \Gamma, 1 \leqslant l \leqslant L\}$ where $\psi^l_{j\,\gamma}$ is defined for ψ^l by (2.13). In addition, Ψ generates a *quasi-affine system* $\widetilde{\xi}_\Psi = \{\widetilde{\psi}^l_{j\,\gamma} : j \in \mathbb{Z}, \gamma \in \Gamma, 1 \leqslant l \leqslant L\}$ where

$$\widetilde{\psi}^l_{j\,\gamma}\,(x) = \left\{ \begin{array}{ll} \psi^l_{j\,\gamma}\,(x) & \text{if } j \geqslant 0 \\ |\det A|^j\,\psi^l\,(A^j(x-\gamma)) & \text{if } j < 0 \end{array} \right\}$$

Recall that an arbitrary collection $\{e_\alpha : \alpha \in \mathcal{A}\}$ in $L^2(\mathbb{R}^n)$ is a *Bessel family* if there is a constant $C > 0$ such that

$$\sum_{\alpha \in A} |\langle f, e_\alpha \rangle|^2 \leqslant C \, ||f||_2 \text{ for all } f \in L^2(\mathbb{R}^n).$$

Suppose $\Psi = \{\psi^1, \ldots, \psi^L\}$ and $\Phi = \{\varphi^1, \varphi^2, \ldots, \varphi^L\}$ are two families in $L^2(\mathbb{R}^n)$ for which the affine systems $\mathcal{E}_\Psi$ and $\mathcal{E}_\Phi$ are Bessel families. Then Φ is an *affine dual* of Ψ if $\mathcal{E}_\Phi$ is dual to $\mathcal{E}_\Psi$ in the sense that for all $f, g \in L^2(\mathbb{R}^n)$,

$$(2.14) \qquad \langle f, g \rangle = \sum_{l=1}^{L} \sum_{j \in \mathbb{Z}} \sum_{\gamma \in \Gamma} \langle f, \psi^l_{j\gamma} \rangle \, \langle \varphi^l_{j\gamma}, g \rangle.$$

Φ is a *quasi-affine dual* of Φ if (2.14) holds when $\psi^l_{j\gamma}$ and $\varphi^l_{j\gamma}$ are replaced by $\widetilde{\psi}^l_{j\gamma}$ and $\widetilde{\varphi}^l_{j\gamma}$.

THEOREM (2.3). *Using the above notation, suppose $\mathcal{E}_\Psi$ and $\mathcal{E}_\Phi$ are Bessel families. Then Φ is an affine dual of Ψ if and only if*

$$(2.15) \qquad \sum_{l=1}^{L} \sum_{j \in \mathbb{Z}} \widehat{\psi^l}(B^j \xi) \, \overline{\widehat{\varphi^l}(B^j \xi)} = |det P|$$

for a.e. $\xi \in \mathbb{R}^n$ and, for each $s \in \mathcal{S} = \Gamma^* \setminus B\Gamma^*$,

$$(2.16) \qquad t_s(\xi) = \sum_{l=1}^{L} \sum_{j \geqslant 0} \widehat{\psi^l}(B^j \xi) \, \overline{\widehat{\varphi^l}(B^j(\xi + s))} = 0$$

for a.e. $\xi \in \mathbb{R}^N$.
Moreover, these two equations also characterize the relation that Φ is a quasi-affine dual of Ψ.

Note that (2.15) and (2.16) reduce to (C') and (D') when $L = 1$ and $\varphi^1 = \psi = \psi^1$. $\Psi = \{\psi^1, .., \psi^L\}$ is a *wavelet system* (relative to Γ and A) if $\mathcal{E}_\Psi$ is an *orthonormal basis* of $L^2(\mathbb{R}^n)$; in particular, (2.15) and (2.16) must hold with $\varphi^i = \psi^i$ for $1 \leqslant i \leqslant L$.

The generalization to wavelet systems and the reversal of the order of dilation and translation in passing from affine systems to quasi-affine systems raise further questions, and there is reason to hope this may be elucidated by the less technically formidable investigation of continuous wavelets for subgroups of $GL(n, \mathbb{R})$. This is an area of active research which we shall not comment upon further in these notes. Instead, we turn to the techniques needed to prove the new characterization of dyadic wavelets announced in Proposition III in the first section and the generalized Proposition III'.

3. Shift Invariant Subspaces and a New Characterization of Wavelets

Proposition III was stated as a conjecture by the first author in a seminar. Two students, M. Bownik and Z. Rzeszotnik, proved it independently. We present the n-dimensional extension of an argument in the Ph.D. thesis of the latter [R]; for a different approach see [B₂].

Suppose φ is a non-zero function in $L^2(\mathbb{R}^n)$. Let $\mathcal{A}_\varphi$ denote the algebraic span of the translates $\varphi(\cdot - k) = \tau_k \varphi$ where $k \in \mathbb{Z}^n$. That is, $\mathcal{A}_\varphi$ is the linear space of all finite linear combinations of the translates $\tau_k \varphi$. Let

$$(3.1) \qquad \omega_\varphi(\xi) = \sum_{l \in \mathbb{Z}^n} |\widehat{\varphi}(\xi + l)|^2.$$

It is clear that ω_φ is a 1-periodic function that is integrable on $\mathbb{T}^n = \{\xi = (\xi_1, \xi_2, ..., \xi_n) : 0 \leqslant \xi_j < 1, j = 1, 2, ..., n\}$ (by "1-periodic" we mean that it is 1-periodic in each variable). Moreover, if $f = \sum_{finite} a_k \tau_k \varphi$ is the general element of $\mathcal{A}_\varphi$, then

$$\widehat{f}(\xi) = \left\{ \sum_{finite} a_k e^{-2\pi i k \cdot \xi} \right\} \widehat{\varphi}(\xi) = t(\xi) \widehat{\varphi}(\xi).$$

Conversely, if $\widehat{f}(\xi) = t(\xi) \widehat{\varphi}(\xi)$, with t a trigonometric polynomial, then $f \in \mathcal{A}_\varphi$. Thus, for such an f we have

$$\|f\|_2^2 = \|\widehat{f}\|_2^2 = \int_{\mathbb{R}^n} |t(\xi)|^2 |\widehat{\varphi}(\xi)|^2 \, d\xi = \sum_{k \in \mathbb{Z}^n} \int_{\tau_k \mathbb{T}^n} |t(\xi)|^2 |\widehat{\varphi}(\xi)|^2 \, d\xi =$$

$$\sum_{k \in \mathbb{Z}^n} \int_{\mathbb{T}^n} |t(\xi + k)|^2 |\widehat{\varphi}(\xi + k)|^2 \, d\xi = \int_{\mathbb{T}^n} |t(\xi)|^2 \omega_\varphi(\xi) d\xi.$$

This shows that the mapping $\widetilde{U}_\varphi$ that assigns to $f \in \mathcal{A}_\varphi$ the unique trigonometric polynomial t such that $\widehat{f} = t\widehat{\varphi}$ is an isometry between $\mathcal{A}_\varphi$ and the space P_φ of all trigonometric polynomials endowed with the norm

$$\|t\|_{L^2(\mathbb{T}^n, \omega_\varphi)} = \left(\int_{\mathbb{T}^n} |t(\xi)|^2 \omega_\varphi(\xi) \, d\xi \right)^{1/2}.$$

Thus, $\widetilde{U}_\varphi$ has a unique extension to an isometry U_φ between V_φ, the closure of $\mathcal{A}_\varphi$ in $L^2(\mathbb{R}^n)$, and the space $L^2(\mathbb{T}^n, \omega_\varphi)$ consisting of all 1-periodic functions s satisfying $\|s\|_{L^2(\mathbb{T}^n, \omega_\varphi)} < \infty$. Observe that, as functions on $\mathbb{T}^n$ (or $\mathbb{R}^n$ and 1-periodic), two elements that are equal on $\Omega_\varphi = \{\xi : \omega_\varphi(\xi) \neq 0\}$ represent the same element of $L^2(\mathbb{T}^n, \omega_\varphi)$.

When $f \in V_\varphi$, then $U_\varphi f = s$ with $s \in L^2(\mathbb{T}^n, \omega_\varphi)$ and

(3.2) $$\|f\|_2 = \|s\|_{L^2(\mathbb{T}, \omega_\varphi)},$$

where $\widehat{f} = s \widehat{\varphi}$. Conversely, any $s \in L^2(\mathbb{T}^n, \omega_\varphi)$ gives rise to an $f \in V_\varphi$ via the last equality.

The translates of $\varphi, \tau_k \varphi, k \in \mathbb{Z}^n$ generate V_φ in the manner just described. In view of the notions we have been discussing, it is natural to ask if V_φ, which is clearly shift invariant, contains a θ which also generates V_φ and $\{\tau_k \theta\}, k \in \mathbb{Z}^n$, is a tight frame for this subspace? The answer is "yes" and is easily obtained: Let $s(\xi) = \omega_\varphi(\xi)^{-1/2}$ for $\xi \in \Omega_\varphi$ and, say, $s(\xi) = 0$ outside Ω_φ. It is easily seen that $\widehat{\theta}(\xi) = s(\xi)\widehat{\varphi}(\xi)$ gives a function $\theta \in L^2(\mathbb{R}^n)$ having these properties: obviously $s \in L^2(\mathbb{T}^n, \omega_\varphi)$; in fact,

$$\|s\|^2_{L^2(\mathbb{T}^n, \omega_\varphi)} = \int_{\mathbb{T}^n} X_{\Omega_\varphi}(\xi)d\xi = |\Omega_\varphi| \leqslant 1.$$

If $f \in V_\varphi$ so that $\widehat{f} = t\widehat{\varphi}$ and $\widehat{\theta} = s\widehat{\varphi}$, then $\{\theta_k\} = \{\tau_k \theta\}$ is a tight frame for V_φ if and only if

(3.3) $$\sum_{k \in \mathbb{Z}^n} |\langle f, \theta_k \rangle|^2 = \|f\|^2_2.$$

But, using Plancherel's theorem, and "periodizing" the integral over $\mathbb{R}^n$,

$$\langle f, \theta_k \rangle = \int_{\mathbb{R}_n} t(\xi)\widehat{\varphi}(\xi)e^{2\pi ik\cdot\xi}\overline{s(\xi)\widehat{\varphi}(\xi)}d\xi =$$

$$\int_{\mathbb{T}^n} t(\xi)\overline{s(\xi)}e^{2\pi ik\cdot\xi}\sum_{l\in\mathbb{Z}^n} |\widehat{\varphi}(\xi+l)|^2 \, d\xi = \int_{\mathbb{T}^n} t(\xi)\overline{s(\xi)}\omega_\varphi(\xi)e^{2\pi ik\cdot\xi}\, d\xi.$$

Thus, $\{\langle f, \theta_k \rangle\}, k \in \mathbb{Z}^n$, is the sequence of Fourier coefficients of the function $t(\xi)\overline{s(\xi)}\omega_\varphi(\xi)$. Thus, by this calculation and (3.2),

(3.4) $$\sum_{k \in \mathbb{Z}^n} |\langle f, \theta_k \rangle|^2 = \int_{\mathbb{T}^n} |t(\xi)|^2 |s(\xi)|^2 (\omega_\varphi(\xi))^2 \, d\xi = \|f\|^2_2.$$

This proves (3.3)

On the other hand, by (3.2) (with s replaced by t),

$$\|f\|^2_2 = \int_{\mathbb{T}^n} |t(\xi)|^2 \omega_\varphi(\xi)d\xi.$$

From this equality, (3.3) and (3.4) we have

(3.5) $$0 = \int_{\mathbb{T}^n} |t(\xi)|^2 \omega_\varphi(\xi)[1 - |s(\xi)|^2 \omega_\varphi(\xi)] \, d\xi$$

for all $t \in L^2(\mathbb{T}^n, \omega_\varphi)$. Choosing $t = \chi_E$ where E is either $\{\xi \in \Omega_\varphi : 1 > |s(\xi)|^2 \omega_\varphi(\xi)\}$ or $\{\xi \in \Omega_\varphi : 1 < |s(\xi)|^2 \omega_\varphi(\xi)\}$ we see that (3.5) is equivalent to

$$1 - |s(\xi)|^2 \omega_\varphi(\xi) = 0$$

for a.e. $\xi \in \Omega_\varphi$. We have proved

LEMMA (3.6). *For each space* V_φ, $\varphi \in L^2(\mathbb{R}^n)$, *we can find* $\theta \in V_\varphi$ *such that*

$$\{\theta(\cdot - k)\}, k \in \mathbb{Z}^n,$$

is a tight frame (of constant 1) for V_φ. *All such* θ *are characterized by having Fourier transforms* $\widehat{\theta}(\xi) = \nu(\xi) \omega_\varphi(\xi)^{-1/2} \widehat{\varphi}(\xi)$, *where* ν *is a 1-periodic unimodular function. Moreover, the tight frame property of these* θ *is characterized by the equality*

(3.7)
$$\sum_{k \in \mathbb{Z}^n} |\widehat{\theta}(\xi + k)|^2$$

(3.8)
$$= \chi_{\Omega_\varphi}(\xi) \text{ a.e. in } \mathbb{R}^n.$$

The elements of $V_\varphi = V_\theta$ *are precisely those whose Fourier transform is of the form*

$$t(\xi) \widehat{\theta}(\xi), t \in L^2(\mathbb{T}^n, d\xi).$$

Remark. Equality (3.7) is, clearly, a more general version of equality (A) in Proposition I that characterizes the orthonormality of the system

$$\{\psi(\cdot - k)\}, k \in \mathbb{Z}^n.$$

From this lemma we obtain the following characterization of shift invariant subspaces:

THEOREM (3.1). *Suppose* V *is a closed subspace of* $L^2(\mathbb{R}^n)$. V *is shift invariant if and only if there exists a sequence of functions* $\{\theta^j\}$, $1 \leq j$, *belonging to* V *that are mutually orthogonal such that each* θ^j *generates a tight frame (of constant 1),* $\{\theta^j(\cdot - k)\}, k \in \mathbb{Z}^n$ *for the space* V_{θ^j} *and*

(3.9)
$$V = \oplus_{j=1}^\infty V_{\theta^j}{}^1$$

Remark. All but a finite number of the θ^j can be the zero function; in this case $V_{\theta^j} = \{0\}$. Unless V is the trivial space $\{0\}$, let us order the θ^j so that the non-zero ones are listed at the beginning.

---

[1]The symbol $\oplus_{j=1}^\infty$ denotes the orthogonal direct sum of the sequence of subspaces that follows it.

Proof. It is clear that if V satisfies (3.9) then it is shift invariant. Thus, we only need to show that a shift invariant closed subspace V satisifes (3.9). Toward this end we choose a non-zero $\varphi \in V$ (if such φ exists) and apply Lemma (3.6) to obtain a $\theta \in V_\varphi$ satisfying (3.7). We let $\theta^1 = \theta$ and consider the orthogonal complement of V_{θ^1} in V. Applying the same argument to $V \cap V_{\theta^1}^\perp$ we obtain θ^2 in this orthogonal complement. Continuing in this fashion we obtain (3.9) (if we wish to be completely rigorous, we invoke the separability of V and Zorn's lemma).

Now suppose V is shift invariant and, thus, equals a direct sum as in (3.9). Fix $j \geqslant 1$ and $\xi \in \mathbb{R}^n$. Let $\Theta^j(\xi)$ be the vector in $l^2(\mathbb{Z}^n)$ whose k^{th} coordinate is $\widehat{\theta^j}(\xi + k), k \in \mathbb{Z}^n$. Since $\{\theta^j(\cdot - k)\}_{k \in \mathbb{Z}^n}$ is a tight frame, equality (3.7) tells us that

$$(3.10) \qquad \|\Theta^j(\xi)\|_{l^2(\mathbb{Z}^n)}^2 = \sum_{k \in \mathbb{Z}^n} |\widehat{\theta^j}(\xi + k)|^2 = \mathcal{X}_{\Omega_{\theta^j}}(\xi) = 0 \text{ or } 1$$

(in order to avoid having to repeatedly add the expression "a.e." we tacitly assume that we only choose ξ in a subset of $\mathbb{R}^n$ whose complement has measure 0 and, for all such ξ, (3.7) and the other related properties we invoke are valid).

The orthogonality of the spaces V_{θ^j}, and a periodization argument like the one that gives us equalities (A) and (B), yields

$$(3.11) \qquad \langle \Theta^j(\xi), \Theta^{j'}(\xi) \rangle_{l^2(\mathbb{Z}^n)} = 0$$

if $j \neq j'$.

Let $\mathcal{L}(\xi)$ be the closure in $l^2(\mathbb{Z}^n)$ of the linear space generated by the vectors $\Theta^j(\xi), 1 \leqslant j$. It is an immediate consequence of (3.10) and (3.11) that the sequence $\{\Theta^j(\xi)\}, 1 \leqslant j$ is a tight frame (of constant 1) for $\mathcal{L}(\xi)$ (even if $\mathcal{L}(\xi) = \{0\}$). Let $P(\xi)$ be the orthogonal projection of $l^2(\mathbb{Z}^n)$ onto $\mathcal{L}(\xi)$ and $e_0 \in l^2(\mathbb{Z}^n)$ the vector all of whose coordinates are 0 except for the coordinate corresponding to $0 \in \mathbb{Z}^n$, which has value $1 : e_0(k) = 0$ if $k \neq 0$ and $e_0(0) = 1$. Then, using the tight frame property for $\{\Theta^j(\xi)\}$ we have

$$1 \geqslant \|P(\xi)e_0\|_{l^2(\mathbb{Z}^n)}^2 = \sum_{j \geqslant 1} |\langle P(\xi)e_0, \Theta^j(\xi) \rangle|^2 =$$

$$\sum_{j \geqslant 1} |\langle e_0, P(\xi)\Theta^j(\xi) \rangle|^2 = \sum_{j \geqslant 1} |\langle e_0, \Theta^j(\xi) \rangle|^2 =$$

$$\sum_{j \geqslant 1} |\widehat{\theta^j}(\xi)|^2 \equiv \sigma(\xi) \ (= \sigma_V(\xi)).$$

This shows

LEMMA (3.11). *If V is a closed shift invariant subspace of $L^2(\mathbb{R}^n)$ and we represent V as the orthogonal direct sum (3.9), then*

(3.12) $$\sigma_V(\xi) = \sum_{j \geq 1} |\widehat{\theta}^j(\xi)|^2 \leq 1.$$

The following result will be an essential tool we shall use in the proof of the principal result of this section. An interesting feature is that the dyadic dilation operator D arises naturally in this study of the properties associated with the translation operators τ_k.

LEMMA (3.13). *Suppose we have the same hypothesis as in the previous lemma. If* $\sigma_V \in L^1(\mathbb{R}^n)$, *then*

(3.14) $$\cap_{j \in \mathbb{Z}} D^j V = \{0\}.$$

Proof. Suppose there exists a non-zero $f \in \cap_{j \in \mathbb{Z}} D^j V$; we might as well assume $||f||_2 = 1$. Since

$$f \in \cap_{j \geq 0} D^{-j} V$$

we must have $D^j f \in V$ for $j \geq 0$.

If $g \in V$ then, by Lemma (3.6) and equality (3.8),

$$\widehat{g}(\xi) = \sum_{j=1}^{\infty} m^j(\xi) \widehat{\theta}^j(\xi)$$

(convergence is in $L^2(\mathbb{R}^n)$, where each m^j is a 1-periodic function in $L^2(\mathbb{T}^n \cap \Omega_{\theta^j})$ that is uniquely determined on Ω_{θ^j}). Moreover,

(3.15) $$||\widehat{g}||_2^2 = \sum_{j=1}^{\infty} ||m^j||_{L^2(\mathbb{T}^n \cap \Omega_{\theta^j})}^2,$$

(by (3.2) since, in this case, $\omega_{\varphi} = \mathcal{X}_{\Omega_{\theta^j}}$). In particular, applying the above equality for $\widehat{g}(\xi)$ and (3.15) to $g = D^l f, l \geq 0$,

$$2^{-nl/2} \widehat{f}(2^{-l}\xi) = \sum_{j=1}^{\infty} m_l^j(\xi) \widehat{\theta}^j(\xi),$$

where m_l^j is 1-periodic, supported in Ω_{θ^j} and

(3.16) $$1 = ||D^l f||_2^2 = \sum_{j=1}^{\infty} ||m_l^j||_{L^2(\mathbb{T}^n \cap \Omega_{\theta^j})}^2 = \sum_{j=1}^{\infty} ||m_l^j||_{L^2(\mathbb{T}^n)}^2.$$

Therefore, if $l \geq 0$ we have

(3.17)
$$\hat{f}(\xi) = 2^{nl/2} \sum_{j=1}^{\infty} m_l^j(2^l \xi) \, \hat{\theta}^j(2^l \xi).$$

Let Q be the translate of $\mathbb{T}^n = \{\xi = (\xi_1, \xi_2, \ldots, \xi_n) : 0 \leqslant \xi_j < 1, j = 1, 2, \ldots, n\}$ by $k = (k_1, k_2, \ldots, k_n) \in \mathbb{Z}^n$ satisfying $k_1, k_2, \ldots, k_n \geqslant 1$. We claim that $\hat{f}(\xi) = 0$ for $\xi \in Q$. To see this we first observe that

(3.18)
$$\int_Q \sum_{j=1}^{\infty} |m_l^j(2^l \xi)|^2 \, d\xi = 2^{-nl} \int_{2^l Q} \sum_{j=1}^{\infty} |m_l^j(\xi)|^2 \, d\xi =$$
$$2^{-nl} 2^{nl} \int_{\mathbb{T}^n} \sum_{j=1}^{\infty} |m_l^j(\xi)|^2 \, d\xi = 1.$$

This is a consequence of the fact that $2^l Q$ is the disjoint union of 2^{nl} lattice point translates of $\mathbb{T}^n$, the 1-periodicity of m_l^j, and (3.16). Hence, using (3.17) and (3.18),

$$\int_Q |\hat{f}(\xi)| \, d\xi \leqslant 2^{nl/2} \sum_{j=1}^{\infty} \int_Q |m_l^j(2^l \xi)| \, |\hat{\theta}^j(2^l \xi)| \, d\xi \leqslant$$

$$2^{nl/2} \int_Q \left(\sum_{j=1}^{\infty} |\hat{\theta}^j(2^l \xi)|^2 \right)^{1/2} \left(\sum_{j=1}^{\infty} |m_l^j(2^l \xi)|^2 d\xi \right)^{1/2} d\xi \leqslant$$

$$2^{nl/2} \left(2^{-nl} \int_{2^l Q} \sum_{j=1}^{\infty} |\hat{\theta}^j(\eta)|^2 d\eta \right)^{1/2} \left(2^{-nl} \int_{2^l Q} \sum_{j=1}^{\infty} m_l^j(\eta)|^2 d\eta \right)^{1/2} \leqslant$$

$$\left(\int_{2^l Q} \sigma_V(\eta) d\eta \right)^{1/2} \cdot 1.$$

But the last expression tends to 0 as $l \longrightarrow \infty$ since σ_V is integrable and the points η of $2^l Q$ satisfy $2^l \sqrt{n} \leqslant 2^l |k| \leqslant |\eta|$. This shows that $f(\xi) = 0$ when $\xi = (\xi_1, \xi_2, \ldots, \xi_n)$ with $\xi_j \geqslant 1$ for $j = 1, 2, \ldots, n$.

Since $D^l f \in V$ for $l \geqslant 1$ we can apply this argument to $\hat{f}(2^{-l} \xi)$ to obtain the fact that $\hat{f}(\xi)$ vanishes in the entire first quadrant. It is also easy to see that this proof, with obvious changes, shows that $\hat{f}$ vanishes in the remaining quadrants. Hence, $\hat{f}$ and f vanish a.e., contradicting the assumption $f \neq 0$.

We are now ready to establish the main result of this section:

THEOREM (3.2). *Suppose $\psi \in L^2(\mathbb{R}^n)$ and the system $\{\psi_{jk}\}$, $j \in \mathbb{Z}$, $k \in \mathbb{Z}^n$ is orthonormal, then this system is an orthonormal basis for $L^2(\mathbb{R}^n)$ if and only if ψ satisfies equality (C).*

This means that equalities (A), (B) and (C) in section 1 completely characterize all wavelets.

Proof. For each $j \in \mathbb{Z}$ let

$$W_j = \overline{span\{\psi_{jk} : k \in \mathbb{Z}^n\}},$$

where we assume that the system $\{\psi_{jk}\}, j \in \mathbb{Z}, k \in \mathbb{Z}^n$ is orthonormal. As we explained above (see Proposition III) if ψ is a wavelet, then ψ satisfies (C). Thus, all we need to do is to show that, under our assumptions, if (C) is satisfied then ψ is a wavelet.

Clearly, $W_j \perp W_{j'}$ when $j \neq j'$. Thus,

$$L^2(\mathbb{R}^n) = \{\oplus_{j \in \mathbb{Z}} W_j\} \oplus \{\oplus_{j \in \mathbb{Z}} W_j\}^\perp$$

and we must show that $\{\oplus_{j \in \mathbb{Z}} W_j\}^\perp = \{0\}$. Let

$$V = \{\oplus_{j \geqslant 0} W_j\}^\perp.$$

Since each $W_j, j \geqslant 0$, is shift invariant, it follows that V is shift invariant. We claim that

(3.19) $$\{\oplus_{j \in \mathbb{Z}} W_j\}^\perp \subset \cap_{l \in \mathbb{Z}} D^l V.$$

This follows from the fact that $D^l V = \{\oplus_{j \geqslant l} W_j\}^\perp \supset \{\oplus_{j \in \mathbb{Z}} W_j\}^\perp$. Thus, it is natural to see if Lemma (3.13) can be applied in order to conclude that $\cap_{l \in \mathbb{Z}} D^l V = \{0\}$. We have, by (3.9),

(3.20) $$V = \oplus_{j=1}^\infty V_{\theta^j},$$

where the functions θ^j satisfy the properties described in Theorem (3.1). We want to show that $\sigma_V(\xi) = \sum_{j=1}^\infty |\widehat{\theta^j}(\xi)|^2$ is integrable. In fact, we shall apply (3.9) to

(3.21) $$L^2(\mathbb{R}^n) = V \oplus \{\oplus_{l \geqslant 0} W_l\}$$

which is clearly shift invariant. Given $q \in \mathbb{Z}^n$ and $l \geqslant 0$ there exist unique $k, p \in \mathbb{Z}^n$ with $p = (p_1, p_2, \ldots, p_n)$ satisfying $0 \leqslant p_j \leqslant 2^l - 1, j = 1, 2, \ldots, n$, such that $q = 2^l k + p$. There exist precisely 2^{nl} such p and each of these lattice points determines the orthonormal system

$$\{\psi_{lq}(x)\} = \{2^{nl/2} \psi(2^l(x - k) - p)\}, k \in \mathbb{Z}^n,$$

which is generated by the lattice point translations of the function $\varphi_p^l(x) = 2^{nl/2} \psi(2^l x - p)$. That is, each space $W_l, l \geqslant 0$, which is shift invariant, has the special (3.8) decomposition

(3.22) $$W_l = \oplus_p V_{\varphi_p^l}.$$

Putting together (3.20), (3.21) and (3.22), therefore, we have

$$L^2(\mathbb{R}^n) = \{\oplus_{j=1}^\infty V_{\theta^j}\} \oplus \{\oplus_{l \geqslant 0}\{\oplus_p V_{\varphi_p^l}\}\}.$$

Applying Lemma (3.11) we then must have

$$\sum_{j=1}^{\infty} |\widehat{\theta^j}(\xi)|^2 + \sum_{l=0}^{\infty} \sum_p |\widehat{\varphi_p^l}(\xi)|^2 =$$

$$\sigma_V(\xi) + \sum_{l=0}^{\infty} \sum_p |\widehat{\psi_{lp}}(\xi)|^2 \leqslant 1.$$

But $\widehat{\psi_{lp}}(\xi) = 2^{-nl/2} e^{-2\pi i 2^{-l}\xi \cdot p} \widehat{\psi}(2^{-l}\xi)$ and, thus,

$$|\widehat{\psi_{lp}}(\xi)| = 2^{-nl/2} |\widehat{\psi}(2^{-l}\xi)|$$

Consequently, the sum $\sum_p |\widehat{\psi_{lp}}(\xi)|^2$ consists of 2^{nl} equal terms and, thus, must be equal to $|\widehat{\psi}(2^{-l}\xi)|^2$. We conclude that

$$1 \geqslant \sigma_V(\xi) + \sum_{l=0}^{\infty} \sum_p |\widehat{\psi_{lp}}(\xi)|^2 = \sigma_V(\xi) + \sum_{l=0}^{\infty} |\widehat{\psi}(2^{-l}\xi)|^2.$$

Because of (C), therefore, $\sigma_V(\xi) \leqslant 1 - \sum_{l=0}^{\infty} |\widehat{\psi}(2^{-l}\xi)|^2 = \sum_{l=1}^{\infty} |\widehat{\psi}(2^l\xi)|^2$. Hence,

$$\int_{\mathbb{R}^n} \sigma_V(\xi)d\xi \leqslant \int_{\mathbb{R}^n} \sum_{l=1}^{\infty} |\widehat{\psi}(2^l\xi)|^2 d\xi = \left(\sum_{l=1}^{\infty} 2^{-nl}\right) ||\psi||_2^2 = \frac{1}{2^n - 1} < \infty.$$

This shows that σ_V is integrable and Theorem (3.2) is established.

Before ending this section let us make some observations. We introduced the characterization of systems $\Psi = \{\psi^1, \ldots, \psi^L\}$ that generate tight frames in the discussion following Proposition IV. Theorem (3.2) was stated and proved for the case L=1. As we stated before, we did this for simplicity. If we assume the orthonormality of the system $\{\psi_{jk}^l\}, l = 1, \ldots, L, 1 \leqslant j, k \in \mathbb{Z}^n$, it is easy to see that our proof goes through when equality (C) is satisfied. It is a curious fact that if we assume that the functions ψ^l have norm at least 1 then (C) and (D) are equivalent to the property that the system $\{\psi_{jk}^l\}$ is an orthonormal basis for L^2. We need not assume that the L functions $\psi^l, l = 1, \ldots, L$, are mutually orthogonal. This orthogonality is a consequence of (C) and (D).

One can extend Theorem (3.2) to tight frames that are semiorthogonal (that is the subspaces W_j are mutually orthogonal and the system $\{\psi_{jk}^l\}$ is a tight frame for the subspace it generates).

The characterization of shift invariant subspaces, at least in the case of finitely many subspaces V_θ, was introduced in [RS]. The general case was also obtained by M. Bownik [B$_2$] using a different approach.

4. Multiresolution Analyses in $\mathbb{R}^n$

A *multiresolution analysis* (MRA) is a sequence of subspaces $\{V_j\}, j \epsilon \mathbb{Z}$, of $L^2(\mathbb{R}^n)$ satisfying the following conditions

$$(4.1) \qquad\qquad V_j \subset V_{j+1}, j \epsilon \mathbb{Z},$$

$$(4.2) \qquad\qquad f \epsilon V_j \text{ if and only if } f(2\cdot) \epsilon V_{j+1}, j \epsilon \mathbb{Z},$$

$$(4.3) \qquad\qquad \overline{\cup_{j \epsilon \mathbf{z}} V_j} = L^2(\mathbb{R}^n),$$

$$(4.4) \qquad\qquad \text{there exists } \varphi \epsilon V_0 \text{ such that } \{\varphi(\cdot - k)\}, k \epsilon \mathbb{Z}^n$$
$$\text{is an orthonormal basis for } V_0.$$

The function φ in (4.4) is called a *scaling function* for this MRA.

It is not hard to show that (4.1), (4.2) and (4.4) imply

$$(4.5) \qquad\qquad \cap_{j \epsilon \mathbf{z}} V_j = \{0\}.$$

The proof of the one-dimensional version of this implication can be found on page 45 of [HW]. The argument given there is very similar to the one we presented in Lemma (3.13). In fact, we adapted the argument in [HW] to provide the proof of (3.13).

The construction of a wavelet basis from an MRA can be described in the following way: For each $j \epsilon \mathbb{Z}$ let $W_j = V_{j+1} \cap V_j^\perp$. A consequence of the above hypotheses is that these spaces are mutually orthogonal with

$$(4.6) \qquad\qquad V_j \oplus W_j = V_{j+1}, j \epsilon \mathbb{Z},$$

and

$$(4.7) \qquad\qquad L^2(\mathbb{R}^n) = \oplus_{j \epsilon \mathbf{z}} W_j.$$

If we can find a system $\Psi = \{\psi^1, \psi^2, \ldots, \psi^L\} \subset W_0$ such that

$$\{\psi^l(\cdot - k)\}, l = 1, 2, \ldots, L, k \epsilon \mathbb{Z}^n$$

is an orthonormal basis for W_0, then (4.2), (4.6) and (4.7) imply that

$$\{\psi_{jk}^l\}, l = 1, 2, \ldots, L, j \epsilon \mathbb{Z}, k \epsilon \mathbb{Z}^n,$$

is an orthonormal basis for $L^2(\mathbb{R}^n)$. That is, Ψ generates a wavelet basis for $L^2(\mathbb{R}^n)$.

It is convenient to express these properties in terms of the Fourier transform. By doing so we shall see that the construction of appropriate systems Ψ raises some interesting problems and, in particular, we will discover that L must equal $2^n - 1$. It is clear that

$$\widehat{V}_0 = \{\hat{f} : f \in V_0\} = \{\hat{f}(\xi) = m(\xi)\widehat{\varphi}(\xi) : m \in L^2(\mathbb{T}^n)\}$$

(where m is 1-periodic and its Fourier coefficients are the coefficients in the expansion $f = \sum_{k \in \mathbb{Z}^n} \alpha_k \varphi(\cdot - k)$, that is provided by property (4.4)).

The elements of $\widehat{V}_0$, by (4.2), also provide us with those of $\widehat{V}_{-1}$ and $\widehat{W}_{-1}$. In each case they are of the form $\hat{\theta}(2\,\xi)$ with $\theta \in V_0$; moreover, we must have

(4.8) $$\hat{\theta}(2\,\xi) = m(\xi)\widehat{\varphi}(\xi)$$

and, since φ satisfies (A),

$$2^{-n}\|\hat{\theta}\|^2_{L^2(\mathbb{R}^n)} = \int_{\mathbb{R}^n} |m(\xi)|^2\,|\widehat{\varphi}(\xi)|^2 d\xi = \int_{\mathbb{T}^n} |m(\xi)|^2 \sum_{k \in \mathbb{Z}^n} |\widehat{\varphi}(\xi + k)|^2\,d\xi$$

$$= \int_{\mathbb{T}^n} |m(\xi)|^2\,d\xi.$$

By polarization, therefore,

(4.9) $$2^{-n}\langle\hat{\theta}, \widehat{\varphi}\rangle_{L^2(\mathbb{R}^n)} = \langle t, s\rangle_{L^2(\mathbb{T}^n)}$$

whenever $\hat{\theta}(2\,\xi) = t(\xi)\widehat{\varphi}(\xi)$ and $\widehat{\psi}(2\,\xi) = s(\xi)\widehat{\varphi}(\xi)$ are representations of $\theta, \psi \in V_0$ via equality (4.8). We shall examine some consequences of (4.8) and (4.9) when $\theta = \varphi$ and $\psi \in W_0$ is such that the system $\{\psi(\cdot - k)\}, k \in \mathbb{Z}^n$, is orthonormal. It will be useful to consider the set of vertices of $\mathbb{T}^n$:

$$\vartheta_n = \{\varepsilon = (\varepsilon_1, \varepsilon_2, \cdots, \varepsilon_n) : \varepsilon_j = 0 \text{ or } 1, 1 \leqslant j \leqslant n\}.$$

By (4.8) we have $\widehat{\varphi}(2\,\xi) = m_0(\xi)\widehat{\varphi}(\xi)$ and since φ satisfies equality (A) (because the translates by $k \in \mathbb{Z}^n$ of φ form an orthonormal system) we have

$$1 = \sum_{k \in \mathbb{Z}^n} |\widehat{\varphi}(2\,\xi + k)|^2 = \sum_{k \in \mathbb{Z}} |m_0(\xi + \frac{1}{2}k)|^2\,|\widehat{\varphi}(\xi + \frac{1}{2}k)|^2$$

$$= \sum_{\varepsilon \in \vartheta_n} |m_0(\xi + \frac{1}{2}\varepsilon)|^2 \sum_{l \in \mathbb{Z}^n} |\widehat{\varphi}(\xi + \frac{\varepsilon}{2} + l)|^2 = \sum_{\varepsilon \in \vartheta_n} |m_0(\xi + \frac{1}{2}\varepsilon)|^2.$$

We have used the 1-periodicity on m_0 and the fact that to each $k \in \mathbb{Z}^n$ there exists a unique $l \in \mathbb{Z}^n$ and $\varepsilon \in \vartheta_n$ such that $(1/2)k = l + (1/2)\varepsilon$. This shows that the 2^n-dimensional vector with components $m_0(\xi + (1/2)\varepsilon)$ has norm 1 (for a.e. $\xi \in \mathbb{R}^n$):

(4.10)
$$1 = \sum_{\varepsilon \in \vartheta_n} |m_0(\xi + \frac{1}{2}\varepsilon)|^2.$$

Since we are assuming $\psi \in W_0$ also satisfies (A) this shows that the "filter" m defined by the equality $\widehat{\psi}(2\xi) = m(\xi)\widehat{\varphi}(\xi)$ also satisfies (4.10). [2]

We are also assuming that φ and $\psi(\cdot - k)$ are orthogonal for all $k \in \mathbb{Z}^n$. Hence,

$$0 = \int_{\mathbb{R}^n} \widehat{\varphi}(\xi)\overline{\widehat{\psi}(\xi)} e^{2\pi i k \cdot \xi} d\xi = \int_{\mathbb{T}^n} \left\{ \sum_{l \in \mathbb{Z}^n} \widehat{\varphi}(\xi + l)\overline{\widehat{\psi}(\xi + l)} \right\} e^{2\pi i k \cdot \xi} d\xi.$$

That is, all Fourier coefficients of the 1-periodic function $\sum_{l \in \mathbb{Z}^n} \widehat{\varphi}(\xi + l)\overline{\widehat{\psi}(\xi + l)}$ are 0. Thus for a.e. $\xi \in \mathbb{R}^n$

(4.11)
$$0 = \sum_{k \in \mathbb{Z}^n} \widehat{\varphi}(\xi + k)\overline{\widehat{\psi}(\xi + k)}.$$

Consequently,

$$0 = \sum_{k \in \mathbb{Z}^n} \widehat{\varphi}(2\xi + k)\overline{\widehat{\psi}(2\xi + k)} =$$

$$\sum_{k \in \mathbb{Z}^n} m_0(\xi + \frac{1}{2}k)\overline{m(\xi + \frac{1}{2}k)} \, |\widehat{\varphi}(\xi + \frac{1}{2}k)|^2 =$$

$$\sum_{\varepsilon \in \vartheta_n} m_0(\xi + \frac{1}{2}\varepsilon)\overline{m(\xi + \frac{1}{2}\varepsilon)} \sum_{l \in \mathbb{Z}^n} |\widehat{\varphi}(\xi + \frac{1}{2}\varepsilon + l)|^2 =$$

$$\sum_{\varepsilon \in \vartheta_n} m_0(\xi + \frac{1}{2}\varepsilon)\overline{m(\xi + \frac{1}{2}\varepsilon)}.$$

This shows that the 2^n-dimensional vectors $\{m_0(\xi + \frac{1}{2}\varepsilon)\}$ and $\{m(\xi + \frac{1}{2}\varepsilon)\}, \varepsilon \in \vartheta_n$, are orthogonal to each other:

(4.12)
$$0 = \sum_{\varepsilon \in \vartheta_n} m_0(\xi + \frac{1}{2}\varepsilon)\overline{m(\xi + \frac{1}{2}\varepsilon)}.$$

In fact, these properties characterize the "wavelets" $\psi \in W_0$: $\psi \in W_0$ is such that the system $\{\psi(\cdot - k)\}, k \in \mathbb{Z}_n$, is orthonormal if and only if the vector $\{m(\xi + \frac{1}{2}\varepsilon)\}, \varepsilon \in \vartheta_n$, has norm 1,

[2]In general, $f \in V_0$ iff $\widehat{f}(\xi) = t(\xi)\widehat{\varphi}(\xi)$ with t 1-periodic and in $L^2(\mathbb{T}^n)$. We shall call t the *filter* associated with f. It is unique.

as in (4.10), and satisfies (4.12) (just reverse the order of the sequence in the equalities that established (4.10) and (4.12)).

Let us use the notation $\psi^0 = \varphi$ and $\psi^1 = \psi$ (for a $\psi \in W_0$ having the properties we just discussed). Suppose we find $\psi^l, l = 2, 3, \ldots, L$, in W_0 with filters m_l (that is, $\widehat{\psi^l}(2\xi) = m_l(\xi)\widehat{\varphi}(\xi) \in L^2(\mathbb{T}^n), m_l$ 1-periodic) such that

(4.13)
$$\sum_{\varepsilon \in \vartheta_n} m_{l_1}(\xi + \tfrac{1}{2}\varepsilon)\overline{m_{l_2}(\xi + \tfrac{1}{2}\varepsilon)} = \delta_{l_1 l_2}$$

a.e., then the collection $\Psi = \{\psi^1, \psi^2, \ldots \psi^L\}$ generates the orthonormal system $\{\psi^l_{j,k}\}, l = 1, 2, \ldots, L, j \in \mathbb{Z}, k \in \mathbb{Z}^n$. It is clear that L cannot exceed $2^n - 1$ since the $(L+1) \times 2^n$ matrix has row vectors $\{m_l(\xi + \tfrac{1}{2}\varepsilon)\}, \varepsilon \in \vartheta_n, 0 \leqslant l \leqslant L$, that satisfy (4.13); that is, the $L+1$ row vectors form an orthonormal system for a.e. ξ (as the last calculation in this section shows, we must have $L = 2^n - 1$). Consequently, we are presented with the following question: given the MRA $\{V_j\}, j \in \mathbb{Z}$, with the scaling function φ, how do we find $2^n - 1$ filters $m_l(\xi)$ such that the $2^n \times 2^n$ matrix $\mathcal{M}(\xi)$ having rows $\{m_l(\xi + \tfrac{1}{2}\varepsilon)\}, \varepsilon \in \vartheta_n$, is unitary for a.e. ξ? It is clear from our discussion that once this is achieved, then the collection $\Psi = \{\psi^1, \psi^2, \ldots, \psi^{2^n-1}\}$ generates a wavelet basis for $L^2(\mathbb{R}^n)$. In this connection let us make the following observation. Since $\mathcal{M}(\xi)$ is unitary a.e., its "first column";

$$\begin{bmatrix} m_0(\xi) \\ m_1(\xi) \\ \cdots \\ m_{2^n-1}(\xi) \end{bmatrix}$$

is a vector of norm 1:

$$\sum_{l=0}^{2^n-1} m_l(\xi)\overline{m_l(\xi)} = 1.$$

Thus, $\sum_{l=0}^{2^n-1} \overline{m_l(\xi)}\, \widehat{\psi^l}(2\xi) = \sum_{l=0}^{2^n-1} |m_l(\xi)|^2 \widehat{\varphi}(\xi) = \widehat{\varphi}(\xi)$. This shows that

$$\varphi \in \overline{\operatorname{span}\{\psi^l_{-1,k} : l = 0, 1, 2, \ldots, 2^n - 1, k \in \mathbb{Z}^n\}} \subset V_{-1} \oplus W_{-1}.$$

Since this last space equals V_0 which is shift invariant and is spanned by $\{\varphi(\cdot - k)\}, k \in \mathbb{Z}^n$, it follows that the span of the functions $\psi^l_{-1,k}, 1 \leqslant l \leqslant 2^n - 1$, just considered, must be W_{-1}. We can conclude, therefore, that Ψ generates a wavelet basis.

It is natural to consider the problem of finding the filters $m_l, 1 \leqslant l \leqslant 2^n - 1$, given the filter m_0 associated with the scaling function m_0. We begin by looking for a filter $m_1(\xi)$ such that the vector $\{m_1(\xi + \tfrac{1}{2}\varepsilon)\}, \varepsilon \in \vartheta_n$, has norm one and is orthogonal to $\{m_0(\xi + \tfrac{1}{2}\varepsilon)\}, \varepsilon \in \vartheta_n$, for a.e. ξ. This can be done in several ways. Here is a simple construction: Let $\mathbf{1} = (1, 1, \ldots, 1) \in \mathbb{R}^n$ and $\varepsilon_0 = (1, 0, \ldots, 0) \in \mathbb{R}^n$. If we define $m_1(\xi) = \overline{e^{i\xi \cdot \mathbf{1}} m_0(\xi + \tfrac{\varepsilon_0}{2})}$, then the two vectors $\mathcal{M}_l(\xi) = \{m_l(\xi + \frac{1}{2}\varepsilon)\}, \varepsilon \in \vartheta_n, l = 0, 1$ are orthogonal to each other and of unit length a.e.. This provides us with one of the desired wavelet basis generators ψ^1 if we define it by the equality $\widehat{\psi}^1(2\xi) = m_1(\xi)\widehat{\varphi}(\xi)$. The search for $\psi^2, \ldots, \psi^{2^n-1}$ is more involved. Perhaps a few observations in the case $n = 2$ provide us with an insight into this situation. If m_0 is real-valued let $\mathcal{M}_0(\xi) = (m_0(\xi_1, \xi_2), m_0(\xi_1, \xi_2 + \frac{1}{2}), m_0(\xi_1 + \frac{1}{2}, \xi_2), m_0(\xi_1 + \frac{1}{2}, \xi_2 + \frac{1}{2}))$, and define the row vectors

$$\mathcal{M}_l(\xi) = \mathcal{M}_0(\xi)L_l,$$

$l = 1, 2, 3$, where

$$L_1 = \begin{pmatrix} 0 & 0 & 1 & 0 \\ 0 & 0 & 0 & -1 \\ -1 & 0 & 0 & 0 \\ 0 & 1 & 0 & 0 \end{pmatrix}, \; L_2 = \begin{pmatrix} 0 & -1 & 0 & 0 \\ 1 & 0 & 0 & 0 \\ 0 & 0 & 0 & -1 \\ 0 & 0 & 1 & 0 \end{pmatrix}, \; L_3 = \begin{pmatrix} 0 & 0 & 0 & -1 \\ 0 & 0 & -1 & 0 \\ 0 & 1 & 0 & 0 \\ 1 & 0 & 0 & 0 \end{pmatrix}.$$

These provide us with the unitary matrix

$$U(\xi) = \begin{bmatrix} \mathcal{M}_0(\xi) \\ \mathcal{M}_1(\xi) \\ \mathcal{M}_2(\xi) \\ \mathcal{M}_3(\xi) \end{bmatrix} = \begin{bmatrix} m_0(\xi) \ldots & m_0(\xi + \frac{1}{2}(1,1)) \\ m_1(\xi) \ldots & m_1(\xi + \frac{1}{2}(1,1)) \\ m_2(\xi) \ldots & m_2(\xi + \frac{1}{2}(1,1)) \\ m_3(\xi) \ldots & m_3(\xi + \frac{1}{2}(1,1)) \end{bmatrix}$$

such that the first column gives us the filters determining the collection $\Psi = \{\psi_1, \psi_2, \psi_3\}$ that generates a wavelet basis associated with the MRA we are considering: $\widehat{\psi}^l(2\xi) = m_l(\xi)\widehat{\varphi}(\xi), l = 0, 1, 2, 3$. If the components of $\mathcal{M}_0(\xi)$ are complex-valued let

$$(s_0(\xi)m_0(\xi), \; s_1(\xi)m_0(\xi + \frac{1}{2}(0,1)),$$

$$s_2(\xi)m_0(\xi + \frac{1}{2}(1,0)), \; s_3(\xi)m_0(\xi + \frac{1}{2}(1,1,)))$$

$$= (|m_0(\xi)|, |m_0(\xi + \frac{1}{2}(0,1))|, |m_0(\xi + \frac{1}{2}(1,0))|, |m_0(\xi + \frac{1}{2}(1,1))|)$$

and

$$S(\xi) = \begin{bmatrix} s_0(\xi) & 0 & 0 & 0 \\ 0 & s_1(\xi) & 0 & 0 \\ 0 & 0 & s_2(\xi) & 0 \\ 0 & 0 & 0 & s_3(\xi) \end{bmatrix},$$

where each $s_l(\xi)$ is unimodular, $l = 0, 1, 2, 3$. $S(\xi)$ is, then, a unitary matrix and, letting

(4.14) $$\mathcal{M}_l(\xi) = \mathcal{M}_0(\xi)S(\xi)L_lS^*(\xi),$$

$l = 0, 1, 2, 3$, we, again, obtain the desired filters by selecting the first components of these vectors. The matrices I, L_1, L_2, L_3 can be considered to represent the generators $1, i, j, k$ of the quaternions (observe that $L_l^2 = -I, l = 1, 2, 3$, $L_1L_2 = L_3$, $L_2L_3 = 1$ and $L_3L_1 = L_2$). We can try to extend the idea of this construction by using the Cayley numbers and, for higher dimensions, the Clifford algebras. One encounters, however, some difficulties by following this path. Even in the two-dimensional case, the vectors defined by (4.14) may lack desired smoothness due to the discontinuities of the signum functions. Thus, even if the scaling function is compactly supported and has other desirable properties, we cannot expect the wavelets so obtained to be compactly supported.

One can obtain compactly supported complex-valued wavelets from an MRA in $\mathbb{R}^2$, however, by taking tensor products of 1-dimensional MRA's. The general case can be easily understood once we present the following two-dimensional case. Let θ be a scaling function in $L^2(\mathbb{R})$ that is compactly supported with an accompanying low pass filter s such that the wavelet ζ satisfying $\widehat{\zeta}(2\xi) = e^{2\pi i \xi}\overline{s(\xi)}\,\widehat{\theta}(\xi)$ is compactly supported (see chapter 2 of [HW] where the Daubechies wavelets are constructed; $s(\xi)$ in this case is a trigonometric polynomial). Then $\varphi(x, y) = \theta(x)\theta(y)$ is a scaling function for an MRA in $L^2(\mathbb{R}^n)$ and the polynomials $m_0(\xi) = m_0(\xi_1, \xi_2) = s(\xi_1)s(\xi_2)$, $m_1(\xi) = s(\xi_1)e^{2\pi i \xi_2}\overline{s(\xi_2 + \frac{1}{2})}$, $m_2(\xi) = e^{2\pi i \xi_1}\overline{s(\xi_1 + \frac{1}{2})}s(\xi_2)$, $m_3(\xi) = e^{2\pi i(\xi_1 + \xi_2)}\overline{s(\xi_1 + \frac{1}{2})}\,\overline{s(\xi_2 + \frac{1}{2})}$ will then provide us a unitary matrix $U(\xi)$, as in equality (4.14), from which we obtain the system of compactly supported wavelets $\Psi(x, y) = \{\psi^1(x, y), \psi^2(x, y), \psi^3(x, y)\}$ satisfying $\widehat{\psi^l}(2\xi) = m_l(\xi)\widehat{\varphi}(\xi), l = 1, 2, 3$. It is clear that this construction extends to n-dimensions and it gives us a compactly supported scaling function $\varphi(x) = \varphi(x_1, \ldots, x_n)$ that will produce a system $\Psi(x) = \{\psi^1(x), \ldots, \psi^{2n-1}(x)\}$ of compactly supported wavelets; furthermore, the 1-dimensional scaling functions whose product is φ can be different scaling functions of the variables $x_1, x_2, \ldots, x_n$.

Wavelets such as the ones we just described, as well as more general tensor products obtained by partitioning the variables $x_1, x_2, \ldots, x_n$ into m subsets, are sometimes called *separable*. Construction of wavelets having smooth filters is challenging even in the separable case, if we require that they not be tensor products of one dimensional functions (see [A] for an elegant, but not simple, such construction).

There exist wavelet bases for $L^2(\mathbb{R}^n)$ that are generated by single functions. They can be constructed directly by using the basic equations (A), (B), (C), and (D) (see [SoW]) or their existence can be established by employing operator theoretic methods (see [DL]). Clearly, these cannot be MRA wavelets. The wavelets obtained in [SoW] are MSF wavelets; that is, the absolute value of their Fourier transforms are characteristic functions of a set $W \subset \mathbb{R}^n$, a *wavelet* set. Such sets are "fractal" and enjoy various interesting properties that are described in [SoW].

It is natural to ask if there is a characterization of MRA wavelets. The answer is that there exists a simply stated condition that determines whether a function (or a collection of functions) is a wavelet obtained from an MRA. Let us first consider the one dimensional case and suppose ψ is an MRA wavelet; that is, $\widehat{\psi}(2\xi) = m_1(\xi)\widehat{\varphi}(\xi), \widehat{\varphi}(2\xi) = m_0(\xi)\widehat{\varphi}(\xi)$ and m_0, m_1 are 1-periodic functions satisfying, in particular,

$$|m_0(\xi)|^2 + |m_1(\xi)|^2 = 1,$$

for a.e. ξ (this follows from the 1-dimensional versions of the fact that $\mathcal{M}(\xi)$ is unitary). Consequently,

$$|\widehat{\psi}(2\xi)|^2 + |\widehat{\varphi}(2\xi)|^2 = \{|m(\xi)|^2 + |m_0(\xi)|^2\}|\widehat{\varphi}(\xi)|^2 = 1 \cdot |\widehat{\varphi}(\xi)|^2$$

a.e. Iterating this argument we obtain

$$|\widehat{\varphi}(\xi)|^2 = |\widehat{\varphi}(2^N \xi)|^2 + \sum_{j=1}^{N} |\widehat{\psi}(2^j \xi)|^2$$

a.e. for each positive integer N. It is clear that the limits

$$\lim_{N \to \infty} \sum_{j=1}^{N} |\widehat{\psi}(2^j \xi)|^2 \text{ and } \lim_{N \to \infty} |\widehat{\varphi}(2^N \xi)|^2$$

exist (observe that these sums are bounded and increasing) and the integrability of $|\widehat{\varphi}|^2$, together with Fatou's lemma, imply that the last limit is 0 a.e.. Thus,

(4.15)
$$|\widehat{\varphi}(\xi)|^2 = \sum_{j=1}^{\infty} |\widehat{\psi}(2^j \xi)|^2 \text{ a.e.}$$

Since $\{\varphi(\cdot - k) : k \in \mathbb{Z}\}$ is an orthonormal system, we have, by Proposition I,

(4.16)
$$D_\psi(\xi) \equiv \sum_{j=1}^{\infty} \sum_{k \in \mathbb{Z}} |\widehat{\psi}(2^j(\xi + k))|^2 = 1$$

a.e.. The 1-periodic function D_ψ (clearly integrable on $[0, 1)$ whenever $\psi \in L^2(\mathbb{R})$) is known as the *dimension function*. Equality (4.16) characterizes the 1-dimensional MRA wavelets:

Theorem. *Suppose* $\psi \in L^1(\mathbb{R})$ *is a wavelet. Then* ψ *is an MRA wavelet if and only if* $D_\psi(\xi) = 1$ *a.e..*

See chapter seven of [HW] for a discussion and appropriate credits for this result.

The dimension function D_ψ can be defined by equality (4.16) for any $\psi \in L^2(\mathbb{R}^n)$ (the only change is that k now ranges throughout $\mathbb{Z}^n$). Essentially the same argument we have just given shows that if

$$\Psi = \{\psi^1, \psi^2, \dots \psi^L\}, L = 2^n - 1$$

is an MRA wavelet system, then

$$(4.17) \qquad D_\Psi(\xi) = \sum_{l=1}^{L} \sum_{j=1}^{\infty} \sum_{k \in \mathbf{Z}^n} |\widehat{\psi}^l(2^j(\xi + k))|^2 = 1$$

a.e.. We also observe that this equality can only hold if $L = 2^n - 1$. Indeed, if (4.17) is true, then

$$1 = \int_{\mathbf{T}^n} D_\Psi(\xi) d\xi = \sum_{l=1}^{L} \sum_{j=1}^{\infty} \sum_{k \in \mathbf{Z}^n} \int_{\mathbf{T}^n} |\widehat{\psi}(2^j(\xi + k))|^2 d\xi$$

$$= \sum_{l=1}^{L} \sum_{j=1}^{\infty} \int_{\mathbf{R}^n} |\widehat{\psi}(2^j \xi)|^2 d\xi = \sum_{l=1}^{L} \sum_{j=1}^{\infty} 2^{-jn} ||\psi^l||_2^2 = L\left(\frac{1}{2^n - 1}\right).$$

5. The connectivity of wavelets.

Let us return to the "classical" 1-dimensional wavelets. We shall discuss the connectivity of this class. We begin by showing that the set of MRA wavelets is an arcwise connected set:

THEOREM (5.1). *If ψ_0 and ψ_1 are two MRA wavelets, then there exists a continuous map $A : [0,1] \longrightarrow L^2(\mathbb{R})$ such that $A(0) = \psi_0, A(1) = \psi_1$ and $A(t)$ is an MRA wavelet for all $t \in [0,1]$.*

This result is due to Xingde Dai and Rufeng Liang . Their proof is presented in [Wu] (they are members of the Wutam Consortium). We shall present the basic ideas of their argument. In order to do so we will use some of the notions introduced in [Wu]. We begin by introducing three "multipliers" that play important roles in the theory of wavelets:

Definition. A measurable function ν on $\mathbb{R}$ is a *wavelet multiplier* if and only if $(\nu \widehat{\psi})^\vee$ is an o.n. wavelet whenever ψ is an o.n. wavelet. ν is a *scaling function multiplier* if and only if $(\nu \widehat{\varphi})^\vee$ is a scaling function whenever φ is a scaling function. A measurable function μ is a *low pass filter multiplier* if and only if μm is a low pass filter whenever m is a low pass filter.

If $(\nu \widehat{\psi})^\vee$ is an MRA wavelet whenever ψ is an MRA wavelet we say that ν is an *MRA wavelet multiplier*.

These multipliers have been completely characterized in [Wu]:

THEOREM (5.2). *ν is a wavelet, MRA wavelet or scaling function multiplier if and only if it is unimodular and $\nu(2\xi)/\nu(\xi)$ is a.e. equal to a 1-periodic function. μ is a low pass filter multiplier if and only if it is a unimodular function that is equal a.e. to a 1-periodic function.*

(The term "unimodular function" means that the function in question has absolute value 1 a.e.).

A few observations are in order. If ψ_0 is an MRA wavelet then the other wavelets belonging to the same MRA are precisely those functions ψ such that $\widehat{\psi} = \nu\,\widehat{\psi}_0$ where ν is unimodular and a.e. equal to a 1-periodic function. If ν is a wavelet multiplier, $\psi = (\nu\,\widehat{\psi}_0)^\vee$ may very well belong to a different MRA (by the last theorem in section 4 and Theorem (5.2), however, we do know that such a ψ is an MRA wavelet). We shall be interested in the classes

$$\mathcal{W}_{\psi_0} = \{\text{all wavelets } \psi : |\widehat{\psi}(\xi)| = |\widehat{\psi}_0(\xi)| \text{ a.e.}\}$$

whenever ψ_0 is a wavelet. It follows from the above observations that if ψ_0 is an MRA wavelet then each $\psi \in \mathcal{W}_{\psi_0}$ is an MRA wavelet and $\mathcal{W}_{\psi_0}$ contains all the wavelets generated by the MRA that produced ψ_0. Furthermore, if $\psi = (\nu\,\widehat{\psi}_0)^\vee$, where ν is a wavelet multiplier, then $\psi \in \mathcal{W}_{\psi_0}$. Thus, if, for a wavelet ψ_0, we let

$$\mathcal{M}_{\psi_0} = \{\psi = (\nu\,\widehat{\psi}_0)^\vee : \nu \text{ a wavelet multiplier}\},$$

then

(5.3) $$\mathcal{M}_{\psi_0} \subset \mathcal{W}_{\psi_0}$$

We shall consider a third class of wavelets generated by an MRA wavelet ψ_0. In order to introduce this class it is helpful to use the following characterization of the scaling functions (see chapter 7 of [HW]):

THEOREM (5.4). *A function* $\varphi \in L^2(\mathbb{R})$ *is a scaling function for an MRA if and only if*

1. $\displaystyle\sum_{k \in \mathbb{Z}} |\widehat{\varphi}(\xi + k)|^2 = 1$ *a.e.*;
2. $\displaystyle\lim_{j \to \infty} |\widehat{\varphi}(2^{-j}\xi)| = 1$ *a.e.*;
3. *there exists a 1-periodic* $m \in L^2([-1/2, 1/2))$ *such that*
$$\widehat{\varphi}(2\xi) = m(\xi)\widehat{\varphi}(\xi).$$

An immediate consequence of this theorem is that $|\widehat{\varphi}|$ is the Fourier transform of a scaling function whenever φ is a scaling function. The function m is, of course, the low pass filter associated with the scaling function φ via the equality in 3 of Theorem (5.4); furthermore, m is uniquely determined by φ and satisfies the Smith-Barnwell equality

(5.5) $$|m(\xi)|^2 + |m(\xi + \tfrac{1}{2})|^2 = 1$$

for a.e. ξ. The most general wavelet belonging to the MRA generated by φ satisfies

(5.6) $$\widehat{\psi}(2\xi) = e^{2\pi i \xi}\, s(2\xi)\overline{m(\xi + \tfrac{1}{2})}\,\widehat{\varphi}(\xi)$$

where s is a unimodular 1-periodic function. An immediate consequence of (5.5) and (5.6) is that if φ^1 and φ^2 are two scaling functions such that $|\widehat{\varphi}^1(\xi)| = |\widehat{\varphi}^2(\xi)|$ a.e. and ψ^1, ψ^2 are MRA wavelets associated with φ^1 and φ^2, then $|\widehat{\psi}^1(\xi)| = |\widehat{\psi}^2(\xi)|$ a.e. It follows that the class $\mathcal{S}_{\psi_0}$ of all MRA wavelets associated with a scaling function φ satisfying $|\widehat{\varphi}(\xi)| = |\widehat{\varphi}_0(\xi)|$ a.e. (ψ_0 being an MRA wavelet associated with φ_0) is a subset of $\mathcal{W}_{\psi_0}$. On the other hand, equality (4.15) shows that $\mathcal{W}_{\psi_0} \subset \mathcal{S}_{\psi_0}$ Thus,

$$(5.7) \qquad\qquad \mathcal{W}_{\psi_0} = \mathcal{S}_{\psi_0}$$

whenever ψ_0 is an MRA wavelet. In fact we have

THEOREM (5.8). *If ψ is an MRA wavelet, then*

$$\mathcal{M}_\psi = \mathcal{W}_\psi = \mathcal{S}_\psi.$$

This result is proved in [Wu].

The sets $\mathcal{M}_\psi$ and $\mathcal{W}_\psi$ can be defined for any wavelet ψ, not necessarily MRA. In view of Theorem (5.8), it is natural to ask if the equality $\mathcal{M}_\psi = \mathcal{W}_\psi$ is true for *all* wavelets ψ. The answer is "No;" Q. Gu constructed a clever counterexample.

Let us describe the two basic ideas of the proof of this theorem. First one shows that each of the classes $\mathcal{W}_\psi, \psi$ an MRA wavelet, is connected. Second, one shows that each MRA wavelet ψ_1 can be connected to the Shannon wavelet ψ_0. As we shall see, this part of the argument is facilitated by the fact that we can choose ψ_1, by Theorem (5.8), so that its associated scaling function φ_1 satisfies $\widehat{\varphi}_1(\xi) \geqslant 0$ a.e. (since $\mathcal{W}_{\psi_1} = \mathcal{S}_{\psi_1}$ this clearly can be done).

For the first part of this proof, which establishes that $\mathcal{W}_\psi$ is connected, one chooses $\psi_1 \in \mathcal{W}_\psi = \mathcal{M}_\psi$ (by Theorem (5.8)) so that $\widehat{\psi}_1 = \nu \widehat{\psi}$ for an appropriate wavelet multiplier. Since ν is unimodular (Theorem (5.2)) we can write $\nu(\xi) = e^{i\lambda(\xi)}$; $\lambda(\xi)$, however, is not unique, but it is easy to construct an appropriate λ so that $\nu_t(\xi) = e^{it\lambda(\xi)}$ is a wavelet multiplier ($\nu_t(2\xi)\overline{\nu_t(\xi)}$ is 1-periodic) for $t \in [0,1]$. We then obtain a continuous map $\theta: t \longrightarrow \psi_t \equiv (\nu_t\widehat{\psi})^\vee$ on $[0,1]$ such that for $t = 0$ we have $\theta(0) = \psi_0 = \psi$ and, for $t = 1$, we have $\theta(1) = \psi_1$.

The second part of the proof is somewhat more involved. We present the basic features of the argument. As explained above, we can assume that we have a wavelet ψ_1 constructed from a scaling function φ_1 such that $\widehat{\varphi}_1(\xi) \geqslant 0$ a.e.. It suffices to connect ψ_1 to the Shannon wavelet ψ_0 whose scaling function φ_0 satisfies $\widehat{\varphi}_0(\xi) = \chi_{[-\frac{1}{2},\frac{1}{2}]}(\xi)$. It follows that the corresponding filters, m_0 and m_1, are non-negative. It is tempting to define

$$(5.9) \qquad\qquad m_t(\xi) = \sqrt{(1-t)m_0(\xi)^2 + tm_1(\xi)^2}$$

for $t \in [0, 1]$ (observe that the Smith-Barnwell equality, $m_t(\xi)^2 + m_t(\xi + \frac{1}{2})^2 = 1$ is true). This provides us with a continuous path of low pass filters. The corresponding scaling functions satisfying

$$\widehat{\varphi}_t(\xi) = \Pi_{j=1}^{\infty} m_t(2^{-j}\xi)$$

form a continuous path of scaling functions. The desired path of wavelets is, then, obtained by letting

$$\widehat{\psi}_t(2\xi) = e^{2\pi i \xi} \overline{m_t(\xi + \pi)} \widehat{\varphi}_t(\xi)$$

for $t \in [0, 1]$. This scheme "almost works". The main modification needed is that the intermediate filters be obtained by an equality that is technically more complicated than (5.9) (see [Wu] page 588).

The consideration of the connectivity of wavelets is very natural. The "first wavelets" (besides the Haar and Shannon wavelets), constructed in the early eighties by Lemarié and Meyer, are, in a real sense, obtained by a continuous "smoothing" (on the Fourier transform side) of the Shannon wavelet. A general result on connectivity was obtained in [BDW]. The authors in this work concerned themselves with wavelets produced by very smooth filters. The paths obtained were continuous with respect to a topology that is considerably stronger than that produced by the $L^2(\mathbb{R})$-norm. There is a topological impediment that prevents one from connecting two wavelets in general when this stronger topology is used. For example, if ψ_0 is the Haar wavelet and $\psi_1(x) = \psi_0(x - 1)$ is its translate by 1, then $\widehat{\psi}_1(\xi) = s(\xi)\widehat{\psi}_0(\xi) = e^{-i\xi}\widehat{\psi}_0(\xi)$. The fact that the function that is identically 1 and $s(\xi)$ are not in the same homotopy class prevents the existence of a path joining these two wavelets that is continuous in the topology used in [BDW]. These questions, as well as extensions of the results in this last work, are discussed by G. Garrigós in [Ga].

An interesting connectivity result was obtained by D. Speegle [SP] who showed that the collection of MSF wavelets is connected. Let us say a few words to put this result in some perspective. One of the first examples of a wavelet that is not an MRA wavelet was obtained by J-L Journée. His wavelet is an MSF wavelet (see [HW], page 64). It follows that the union of the MRA and MSF wavelets is an arcwise connected subset of the surface of the unit sphere in $L^2(\mathbb{R})$. Auscher [A] and Lemarié [Le] have, independently, shown that if one makes rather mild assumptions about the Fourier transform of a wavelet (continuity and a decrease at ∞), the wavelet arises from an MRA. We see therefore, that "most" wavelets are either MRA or MSF wavelets. It is not unreasonable to conjecture, therefore, that the collection of all wavelets in $L^2(\mathbb{R})$ is connected. The answer to this conjecture, however, is not yet known.

If we consider the question of connectivity and, more generally, multipliers, in connection with higher dimensional wavelets produced by more general dilations we find various factors that alter the form of the results we seek. We have seen, for example, that the MRA wavelets in higher dimensions require more than one generator. This, however, depends on the dilation that is used. This is not the case for those wavelets on $\mathbb{R}^2$ obtained by the translations by the points of the lattice $\mathbb{Z}^2 \subset \mathbb{R}^2$ and the dilations obtained from the integral powers of the matrix

$$M = \begin{pmatrix} 1 & -1 \\ 1 & 1 \end{pmatrix}.$$

That is, if $\psi \in L^2(\mathbb{R}^2)$, the system $\{\psi_{jk}\}, j \in \mathbb{Z}, k \in \mathbb{Z}^2$ is defined by

(5.10) $$\psi_{jk}(x) = (\det M)^{j/2} \psi(M^j x - k) = 2^{j/2} \psi(M^j x - k)$$

for $x \in \mathbb{R}^2$ ($M^j x$ denoting the vector obtained by multiplying the matrix M^j by the column vector

$$x = \begin{pmatrix} x_1 \\ x_2 \end{pmatrix} \in \mathbb{R}^2)$$

A function ψ such that $\{\psi_{jk}\}, j \in \mathbb{Z}, k \in \mathbb{Z}^2$, is an orthonormal basis for $L^2(\mathbb{R}^2)$ is called a *Quincunx wavelet*. There exist MRA wavelets, ψ, of this type producing orthonormal bases of the form $\{\psi_{jk}\}$, just as in the one-dimensional case. That is, even though the dimension of the underlying space, $\mathbb{R}^2$, is 2, we do not need $3 (= 2^2 - 1)$ generators to obtain a "wavelet basis" for $L^2(\mathbb{R}^2)$ as is the case described in section 4.

The quincunx wavelets share many other properties with the "classical" wavelets on $\mathbb{R}$ obtained by integral translations and dyadic dilations. In particular, the Wutam program, for the quincunx wavelets, is just like the one we just presented.

Let $\psi \in L^2(\mathbb{R}^2)$ and $\{\psi_{jk}\}$ be the system defined by equality (5.10). Then, letting M_1 be the transpose of M, we have

(5.11) $$\widehat{\psi}_{jk}(\xi) = 2^{-j/2} \widehat{\psi}(M_1^{-j} \xi) e^{-2\pi i k \cdot M_1^{-j} \xi}$$

for $\xi \in \mathbb{R}^2, j \in \mathbb{R}$ and $k \in \mathbb{Z}^2$. Until further notice, we shall use the term "wavelet" for an orthonormal quincunx wavelet (that is, a function $\psi \in L^2(\mathbb{R}^2)$ such that the system defined by (5.10) is an o.n. basis for $L^2(\mathbb{R}^2)$). The characterization of wavelets obtained by general dilations, Theorem (2.3), reduced to the situation we are now studying, becomes

PROPOSITION (5.12).. *Let* $\psi \in L^2(\mathbb{R}^2)$ *such that* $\|\psi\|_2 \geqslant 1$. *Then* ψ *is an o.n. wavelet if and only if*

(I) $$\sum_{j \in \mathbb{Z}} |\widehat{\psi}(M_1^j \xi)|^2 = 1 \text{ for a.e. } \xi \in \mathbb{R}^2,$$

(II) $$\sum_{j \geqslant 0} \widehat{\psi}(M_1^j \xi) \overline{\widehat{\psi}(M_1^j(\xi + q))} = 0 \text{ for a.e. } \xi \in \mathbb{R}^2$$

whenever $q = \begin{pmatrix} q_1 \\ q_2 \end{pmatrix} \in \mathbb{Z}^2$ *with* q_1, q_2 *having different parity.*

Let $\mathbb{T} = \mathbb{T}^2 = \{\xi = (\xi_1, \xi_2) : -\frac{1}{2} \leqslant \xi_j < \frac{1}{2}, j = 1, 2\}$ and

$$S = (M\mathbb{T})\backslash\mathbb{T}$$

($M\,\mathbb{T}$ is the square with vertices $(0,1)$, $(-1,0)$, $(0,-1)$ and $(1,0)$). Letting ψ be defined by having $\widehat{\psi}(\xi) = \mathcal{X}_s(\xi)$, it is easy to check that ψ is an orthonormal wavelet that is also an MRA wavelet. A scaling function ψ for the MRA is obtained by letting $\widehat{\varphi}(\xi) = \mathcal{X}_{\mathbb{T}}(\xi)$ and its associated low pass filter m is the characteristic function of $\mathcal{X}_M^{-1}\,\mathbb{T}$, restricted to $\mathbb{T}$, and then extended to be 1-periodic in each variable. In fact, this provides us with an example of an MRA, MSF wavelet in the quincunx case (the analogue of the Shannon wavelet).

The notion of multiplier (wavelet, scaling function or filter multiplier) as well as the definition of the sets W_ψ, $\mathcal{M}_\psi$ and S_ψ extends in an obvious way, and so does Theorem (5.8). In fact, it is not hard to obtain the version of Theorem (5.1) that asserts that the *MRA quincunx wavelets are path-connected.* The details of this "Wutam program" adapted to this situation were presented in a Ph.D. qualifying oral exam by one of our students, L. Zhang.

6. Summary and Bibliographical comments.

As we stated at the beginning, one of our motivations was to present the "Mathematical Theory of Wavelets." Obviously, we cannot do this in any exhaustive way in this rather short article. Our aim was to describe some of the beauty of this theory and, if possible, elaborate topics that are new. We hope that this write-up does have some of these properties.

We have not considered many important uses of wavelet theory in mathematics. For example, their application to the construction and deriving properties of a large collection of important function spaces (the Besov and Triebel-Lizorkim spaces) is not a topic we discussed. Chapters 5 and 6 in [HW] and [FJW] deal with this topic.

We described various kinds of wavelets, scaling functions, filters and we gave characterizations of almost all these function with the exception of low pass filters. One of us was involved in the characterizations of *all* low pass filters in 1-dimension and for dyadic dilations [PSW]. The characterization of the dual systems (Φ, Ψ) announced by Theorem (2.3) really pertains to the case where the dilation matrix A has integer entries and the translations are obtained from the lattice $\mathbb{Z}^n$ (in terms of the notation used in section 2, the matrix $P^{-1}AP$ must have integer entries and this similarity reduces the problem to the lattice $\mathbb{Z}^n$). There are, however, wavelet systems that involve more general dilations and translations; in [CCMW] the problem of the characterization of such systems is solved.

Though we concentrated our attention to wavelets, we indicated that other systems, obtained by applying, say, modulations and translations to a fixed function, are also of interest. For example, systems of the form

(6.1)
$$g_{mn}(x) = e^{2\pi i m x} g(x - n),$$

$m, n \in \mathbb{Z}$, the *Gabor systems*, have been studied extensively. A very general context that produces the "continuous" versions of such systems (as well as the continuous wavelets) can be described by the collection of $(n + 2) \times (n + 2)$ matrices of the form

(6.2)
$$g = \begin{pmatrix} 1 & b & z \\ 0 & A & c \\ 0 & 0 & 1 \end{pmatrix},$$

where b is an n-dimensional row vector, c an n-dimensional column vector, $A \in GL(n, \mathbb{R})$ and $z \in \mathbb{R}$. If we let the matrix (6.2) act on the column vector (on the right)

$$\begin{pmatrix} u \\ v \\ y \end{pmatrix},$$

where $u, y \in \mathbb{R}$ and $v \in \mathbb{R}^n$, we obtain the action

$$\begin{pmatrix} u \\ v \\ y \end{pmatrix} \longrightarrow \begin{pmatrix} u + b \cdot v + zy \\ Av + cy \\ y \end{pmatrix}$$

This induces the following mapping

$$(T_g \psi)(u, v, y) = |\det A|^{-\frac{1}{2}} e^{2\pi i (u + b \cdot v + zy)} \psi(Av + cy)$$

defined on a function $\psi(v)$, $v \in \mathbb{R}^n$. If $u = 0, b = 0$ and $y = 1$, we obtain the map (2.1) (strictly speaking, we should use g^{-1} instead of g; the variable z is not important and its main function is to make sure that the matrices (6.2) form a group). If $A = I, y = 1, u = 0$ we obtain the "continuous Gabor system," which is also referred to as the *Weyl-Heisenberg* transform. It is natural to consider the various themes treated in the previous questions in this general setting. For example, we can try to find admissibility conditions, such as Theorem (2.1), when A belongs to a subgroup of $GL(n, \mathbb{R})$ and $b, c \in \mathbb{R}^n$. In the Weyl-Heisenberg case this is considered in [LP]; see, also, the discussion in [DGM] that is relevant to this case. The authors of [LWWW] are engaged in an investigation of these problems in the general case. Particularly challenging is the question of the discretizations of the continuous transforms. The Gabor system

$$\psi_{mn}(x) = e^{2\pi i m b \cdot x} \psi(x + nc)$$

is a particular discretization of this type that is well known; various conditions guaranteeing the orthonormality and frame properties of these functions have been considered by many authors (see, in particular, [Cz]).

We have included some items in the bibliography that are not referred to directly. In general these pertain to presentations of continuous wavelets and their discretizations that should be compared with our presentation; there are, also, certain aspects of what we describe here that are, formally, quite similar to work done in representations theory. We suggest that the reader examine [BGZ], [BT], [Car], [DM], [GM], [GMP₁], [GMP₂], [GP], [K], [LP], [M], and [ST] for the work done that is related to section 2.

One last word about the "characterization" equalities we talked about. Clearly it is a good thing to find out descriptions of the class of all functions having certain properties (wavelets, scaling functions, tight frame wavelets, low pass filters, etc.) We claimed, at the end of section 1, that the equalities we are considering have been very useful for many constructions. The book [HW] presents many examples. We cite [BGRW] as another example that attracted some attention.

References

[A]   Ayache, A. *Construction of Non-separable Dyadic Compactly Supported Orthonormal Wavelet Bases for* $L^2(\mathbb{R}^2)$ *of arbitrary high regularity*. Rev. Mat. Iberoamericana 15, no. 1: 37-58, 1999.

[AAG₁] Ali, S. T., Antoine, J-P, and Gazeau, J-P. *Square-integrability of Group Representatives on Homogenous Spaces I: Reproducing Triples and Frames; II: Generalized Square-integrability and Equivalent Families of Coherent States*. Ann. Inst. H. Poincaré, 55: 829-890, 1991.

[AAG₂] Ali, S.T., Antoine, J-P, and Gazeau. J-P. *Coherent States, Wavelets and Their Generalizations*. Springer-Verlag New York, Inc.: 1-418, plus xiii, 2000.

[B₁]   Bownik, M. *A Characterization of Affine Dual Frames in* $L^2(R^n)$. Vol. 8, No. 2, 3, Appl. & Comp. Harm. Anal.: 203-20, 2000.

[B₂]   Bownik, M. *Characterization of Mulitwavelets in* $L^2(\mathbb{R}^n)$. preprint.

[BDW]   Bonami, A., Durand, S., and Weiss, Guido. *Wavelets Obtained by Continuous Deformations of the Haar Wavelet*. Revista Matemática Iberoamericana, Vol. 12, No. 1 (1996) pp 1-18.

[BGRW] Brandolini, L., Garrigós, G., Rzeszotnik, Z., and Weiss, Guido. *The Behaviour at the Origin of a class of Band-Limited Wavelets*. Contemporary Mathematics, volume 247, AMS: 75-92, 1999.

[BGZ]   Bacry, H., Grossman, A., and Zak, J. *Geometry of Generalized Coherent States*. Group Theoretical methods in physics (Fourth International Colloquium, Nijmegen, 1975), Springer Verlag, Berlin, Lecture Notes in Physics, 50: 249-268, 1976.

[BT]   Bernier, D., and Taylor, K. F. *Wavelets from Square-integrable Representations*. SIAM J. Math. Anal., 27(2): 594-608, 1996.

[C]   Calogero, Andrea. *A Characterization of Wavelets on General Lattices*. J. Geom. Anal. (to appear)

[Ca]   Calderón, A. P. *Intermediate Spaces and Interpolation, the Complex Method*. Stud. Math., 24: 113-190, 1964.

[Car]   Carey, A. L. *Square Integrable Representations of Non-unimodular Groups*. Bull. Austral. Math. Soc., 15: 1-12, 1976.

[CCMW] Chui, C. K., Czaja, W., Maggioni, M., and Weiss, Guido. *Characterization of Tight Wavelets Frames With Arbitrary Dilation and General Tightness Preserving Oversampling*. To appear.

[Co] Corbett, J. *Coherent States on Kinematic Groups: The Study of Spatio-Temporal Wavelets and Their Application to Motion Estimation.* Ph.D. Thesis, Washington University, 1999.

[CS] Chui, C.K. and Shi, X.L. $n\times$ *Oversampling Preserves any Tight Affine Frame for Odd n.* Proc. of the AMS 121(2): 511-517, 1994.

[Cz] Czaja, W. *Characterization of Gabor Systems via the Fourier Transform.* Collectanea Matematica Vol 51 (2): 205-224, 2000.

[D] Daubechies, I. *Ten Lectures on Wavelets.* Society for Industrial and Applied Mathematics, Philadelphia, 1992.

[DGM] Daubechies, I., Grossman, A., and Meyer, Y. *Painless Nonorthogonal Expansions.* J. Math. Phys. 27(5): 1271-1283, 1986.

[DL] Dai, X. and Larson, D. *Wandering Vectors for Unitary Systems and Orthogonal Wavelets.* Memoirs No. 640 of the AMS: 1-68, 1998.

[DM] Duflo, M., and Moore, C. C. *On the Regular Representation of a Non-unimodular Locally Compact Group.* Journal of Functional Analysis, 21: 209-243, 1976.

[FJW] Frazier, M. and Jawerth, B. and Weiss, Guido. *Littlewood-Paley Theory and the Study of Function Spaces.* Published for the Conference Board of the Mathematical Sciences, Washington D.C., 1991.

[Ga] Garrigós, G. *The Characterization of Wavelets and Related Functions and the Connectivity of Alpha-localized Wavelets in R.* Ph.D. Thesis, Washington University, 1998.

[GHHLWW] Gilbert, J. E., Han, Y-S, Hogan, J.A., Lakely, J. D., Weiland, D., and Weiss, Guido. *Smooth Molecular Decomposition of Functions and Singular Integral Operators.* To appear as a Memoir of the AMS.

[GM] Grossman, A., and Morlet, J. *Decomposition of Hardy Functions into Square Integrable Wavelets of Constant Shape.* SIAM J. Math. Anal., 15: 723-736, 1984.

[GMP$_1$] Grossman, A., Morlet, J., and Paul, T. *Transforms Associated to Square Integrable Group Representations. I. General Results.* Journal of Math. Physics, 26 (10): 2473-2479, 1985.

[GMP$_2$] Grossman, A., Morlet, J., and Paul, T. *Transforms Associated to Square Integrable Group Representations II. Examples.* Ann Inst. H. Poincaré Phys. Théor, 45(3): 293-309, 1986.

[GP] Grossman, A., and Paul, T. *Wave Functions on Subgroups of the Group of Affine Canonical Transformations.* Resonances-models and phenomena (Bielefeld, 1984), Lecture Notes in Physics, 211: 128-138. Springer, Berlin, 1984.

[Ha] Haar, A., *Zür Theorie der Orthogonalen Funktionen Systems.* Math. Ann. 69: 331-371, 1910.

[HW] Hernández, E., and Weiss, Guido. *A First Course in Wavelets.* CRC Press, Boca Raton, Florida, 1996.

[HWW] Hernández, E., Wang, X., and Weiss, Guido. *Characterization of Wavelets, Scaling Functions and Wavelets associated with Multiresolution Analyses.* Israel Math. Conf. Proceedings (conference held in the Technion, Haifa, 1996), Vol. 13, AMS publications: 51-87, 1999.

[HeWa] Heil, C. E., and Walnut, D. F. *Continuous and Discrete Wavelet Transforms.* SIAM Review, 31: 628-666, 1989.

[K] Kalisa, C. *Etats Cohérents Affines: Canoniques, Galileéns et Relativistes.* Pd.D. Thesis, Université de Louvain, 1993.

[Le] Lemarié-Rieusset, P.G. *Sur l'Existence des Analyse Multi Resolutions en Théorie des Ondelettes.* rev. Mat. Iberoamericano, Vol 8, no. 3: 457-474, 1992.

[LWWW] Laugesen, R. S., Wilson, E. N., Weaver, N., and Weiss, Guido. *A Generalized Calderón Reproducing Formula and its Associated Continuous Wavelets.* In preparation.

[LP] Liu, H., and Peng, L. *Admissible Wavelets Associated with the Heisenberg Group.* Pacific Jour. of Math., vol 180, No. 1: 101-123, 1997.

[M]   Meyer, Y. *Principe d'incertitude, Bases Hilbertiennes et Algèbres d'Opérateurs.* Seminaire Bourbaki, 38 (662): 1-15, 1986.

[PSW]   Papadakis, M., Sikić, H., and Weiss, Guido. *The Characterization of Low Pass Filters and Some Basic Properties of Wavelets, Scaling Functions and Related Concepts.* Jour. of Fourier Analysis & Applications, Vol. 5, Issue 5: 495-521, 1999.

[PSWX]   Paluszyński, M., Sikić, H., Weiss, G., and S. Xiao. *Generalized Low Pass Filters and MRA Frame Wavelets.* To appear in the Jour. of Geometric Analysis (2000).

[R]   Rzeszotnik, Z. *Characterization Theorems in the Theory of Wavelets.* Ph.D. Thesis, Washington University, 2000.

[RS]   Ron, Amos and Shen, Zuowei. *Affine Systems in $L^2(\mathbb{R}^d)$. II. Dual systems.* J. Fourier Anal. Appl. 3(1997), 627-637.

[S]   Saeki, S. *On the Reproducing Formula of Calderón.* JFAA, Vol. II, No. 1, 1995.

[SoW]   Soardi, P. M. and Weiland, D. *Single Wavelets in n-dimensions.* J. Fourier Anal. Appl. 4, No. 3: 299-315, 1998.

[SP]   Speegle, D. *The S-Elementary Wavelets are Path Connected.* Proc. Amer. Math. Soc. 127, no. 1: 223-233, 1999.

[ST]   Schulz, E., and Taylor, K. F. *Extensions of the Heisenberg Group and Wavelet Analysis in the Plane.* CRM Proceedings and Lecture Notes, Serge Dubuc, Gilles Deslauriers, editors, 18: 99-107, 1999.

[SW]   Stein, E.M. and Weiss, Guido. *Introduction to Fourier Analysis on Euclidean Spaces.* Princeton Univ. Press, Princeton, NJ.: 1-489, 1970.

[W]   Wickerhauser, M. V. *Adapted Wavelet Analysis from Theory to Software.* AK Peters, Lts., Wellesley, Massachusetts, 1994.

[Wu]   Wutam Consortium Publication. *Basic Properties of Wavelets.* Jour. of Fourier Analysis and Applications 4, Issue 4: 285-320, 1998.

Part 2

Problems

Assorted Problems

Various Authors

1. Preservation of bandlimitedness under non-affine time warping for multi-dimensional functions

S. Azizi, J.N. McDonald, and D. Cochran, Arizona State University

If $f \in L^2(\mathbf{R}^d)$ has compactly supported Fourier transform and $\gamma : \mathbf{R}^d \to \mathbf{R}^d$ has the form $\gamma(x) = Ax + b$ where $A \in GL_d(\mathbf{R})$ and $b \in \mathbf{R}^d$, then $h = f \circ \gamma$ also has compactly supported Fourier transform. It is possible to construct specific f and γ so that f and $f \circ \gamma$ both have compactly supported Fourier transforms, f is not the zero function, and γ is a continuous and invertible function that is not of the affine form just given. This construction can be accomplished, for example, by considering a d-fold cartesian product of known one-dimensional examples such as described in [1].

Under the assumption that $\gamma : \mathbf{R}^d \to \mathbf{R}^d$ is continuous and invertible, is it true that $f \circ \gamma$ has compactly supported Fourier transform for all $f \in L^2(\mathbf{R}^d)$ with compactly supported Fourier transform if and only if $\gamma(x) = Ax + b$?

[1] D. Cochran, S. Azizi, and J.N. McDonald, "Harmonic analysis and sampling in warped spaces," in *20th Century Harmonic Analysis — A Celebration*, J.S. Byrnes, ed.

2. Window pairs that determine spectral phase

S. Shetty, J.N. McDonald, and D. Cochran, Arizona State University

Consider a finite sequence

$$(2.1) \qquad a(0), \ a(1), \ ..., \ a(n)$$

of complex numbers and the associated polynomial

$$(2.2) \qquad p(z) = a(0) + a(1)z + \cdots + a(n)z^n.$$

The *spectral magnitude* of the sequence (2.1) or its associated polynomial (2.2) is the function $s : \mathbf{R} \to \mathbf{C}$ defined by $s(\omega) = |p(e^{i\omega})|$. In general, it is clear that two different polynomials can have identical spectral magnitude functions. If, however, two polynomials $p(z)$ and $q(z)$ have identical spectral magnitude functions *and* their respective derivatives $p'(z)$ and $q'(z)$

J.S. Byrnes (ed.), Twentieth Century Harmonic Analysis - A Celebration, 369–386.

also have identical spectral magnitude functions (i.e., $|p'(z)| = |q'(z)|$ for all z with $|z| = 1$), then it can be shown that $p(z) = cq(z)$ for some complex constant c with $|c| = 1$.

Define finite sequences $w_1(k) = 1$ and $w_2(k) = k$ for $k = 0, ..., n$ and consider the sequences $b_1(k) = a(k)w_1(k)$ and $b_2(k) = a(k)w_2(k)$ obtained by "windowing" the original sequence $a(0), ..., a(n)$ with each of these new sequences. The polynomial associated with $b_2(0), ..., b_2(n)$ is $r(z) = zp'(z)$, which has spectral magnitude function identical to that of $p'(z)$. Hence, the spectral magnitude functions of the two windowed sequences are sufficient to determine $p(z)$ up to a unimodular constant factor.

The problem posed here is to characterize all pairs of window sequences $w_1(0), ...w_1(n)$ and $w_2(0), ...w_2(n)$ with the property that the spectral magnitude functions of the sequences $b_1(k) = a(k)w_1(k)$ and $b_2(k) = a(k)w_2(k)$ are sufficient to determine any polynomial $p(z)$ of degree n up to a unimodular constant factor. The corresponding problem with spectral phase in place of spectral magnitude is also of interest.

3. Questions on Riesz Products

J. P. Kahane, Université Paris-Sud

Riesz products, identified with positive measures, are of the form

$$\mu_a = \prod_{n=1}^{\infty} \left(1 + \Re(a_n e^{i\lambda_n x})\right)$$

$$\mu_b = \prod_{n=1}^{\infty} \left(1 + \Re(b_n e^{i\lambda_n x})\right)$$

where (λ_n) is a given Hadamard sequence ($\lambda_1 > 0$, $\frac{\lambda_{n+1}}{\lambda_n} > q > 1$), $a = \{a_n\}$ and $b = \{b_n\}$, $|a_n| < 1$ and $|b_n| < 1$ for all n.

3.1. *Question 1*

Give a necessary and sufficient condition on (a, b) for $\mu_a \sim \mu_b$.

Partial answers are:

$$\left.\begin{array}{l} \sum_1^{\infty} |a_n - b_n|^2 < \infty \\ \sup |a_n| < 1 \end{array}\right\} \Rightarrow \mu_a \sim \mu_b \quad \text{(Peyzière, Brown-Moran)}$$

$$\left.\begin{array}{l} \sum_1^{\infty} |a_n - b_n|^2 < \infty \\ |a_n| = |b_n| \end{array}\right\} \Rightarrow \mu_a \sim \mu_b$$

(Parreau, Ann. Inst. Fourier 40, 2 (1990), 391–405).

Let us write

$$\mu_{a,\varphi} = \prod_{n=1}^{\infty} \left(1 + \Re(a_n e^{i(\lambda_n x + \varphi_n)})\right)$$

3.2. *Question 2*

$\mu_a \ll \mu_b \Leftrightarrow \mu_{a,\varphi} \ll \mu_{b,\varphi}$.

Fan Ai-Lua is the author of this question. A positive answer would also answer question 1, using the following theorem of Klimer and Saeki (Ann. Inst. Fourier 38, 2 (1988), 63–93) (*cf*. also Fan, Studia Math. (1991) 249–266).

If the φ_n are random with the usual probability,

$$\left.\begin{array}{c} \sum_1^\infty |a_n - b_n|^2 \left(1 + \frac{\cos^2(t_n - s_n)}{\sqrt{2 - |a_n - b_n|}}\right) < \infty \\ t_n = \arg(a_n + b_n) \\ s_n = \arg(a_n - b_n) \end{array}\right\} \Leftrightarrow \mu_{a,\varphi} \sim \mu_{b,\varphi} a.s.$$

Motivations and hints can be found in the article of Fan in Bull. Sc. Math. 2^e s., 117 (1993), 421–439.

4. Completeness of sets of complex exponentials in convex sets: open problems

B.N. Khabibullin, Department of Mathematics, Frunze str. 32, Bashkir State University, Ufa, Bashkortostan, 450074, Russia,
algeom@bsu.bushedu.ru *and/or* khabib-bulat@mail.ru

Let $\Lambda = \{\lambda_n\}$, $n \in \mathbb{N}$, be a sequence of pairwise different complex numbers (points) in the complex plane $\mathbb{C}$, $0 \notin \Lambda$, and $\lambda_n \to \infty$ as $n \to +\infty$.

Let X be a *convex* open or closed set on $\mathbb{C}$. By $H(X)$ denote the space of all continuous complex-valued functions on X which are holomorpic in the interior Int X of X (if Int $X \neq \emptyset$) *with the topology of uniform convergence on compact subsets of* X. An exponential system Exp $\Lambda = \{\exp(\lambda_n z)\}$ is complete in X if the closure of the linear span of Exp Λ in $H(X)$ coincides with $H(X)$.

The completeness problem for a set X is to determine *all* such sequences of exponents Λ for which the system Exp Λ is complete in X. On the background of known results it is natural to look for the solution of the problem in terms of the density of the distribution of the points of Λ and in terms of the geometric characteristics of X. It is natural to subdivide the completeness problem into six cases according to the type of X:

1) $X = [-a, a]$ is a segment on real axis $\mathbb{R}$, $a > 0$, and $H(X) = C[-a, a]$;
2) $X = (-a, a)$, where $a \in (0, +\infty]$, or $X = [0, +\infty)$, or $X = (-\infty, 0]$;
3) $X = G$ is an *unbounded convex* domain in $\mathbb{C}$, $[0, +\infty) \subset G$;
4) $X = \text{Clos } G$, where G is the same as above, Clos G is the closure of G;
5) $X = G$ is a *bounded convex* domain in $\mathbb{C}$;
6) $X = K$, where K is a *compact convex* set in $\mathbb{C}$, Int $K \neq \emptyset$;

Below, every item (k-!) or (k-?) is a continuation of item k), and (k-!) (respectively (k-?)) denotes a solved (respectively unsolved) problem.

(1-?) No precise condition is known for the completeness of systems Exp Λ in $[-a, a]$, and finding such a condition is a very difficult problem. In this case we present a test for the completeness of systems Exp Λ in $[-a, a]$ to within a single exponential (see [5, Theorem 7] together with recent results of B. Cole and T. Ransford [2]).

Here and below we denote by $\mathcal{G}$ *the class of all extended Green functions* $g_D(\zeta, 0)$ for the point 0 and arbitrary bounded regular (for the Dirichlet problem) domains $D \subset \mathbb{C}$, $0 \in D$, $g_D(\zeta, 0) \equiv 0$ for $\zeta \notin D$.

THEOREM 1. *If*

$$\sup_{g \in \mathcal{G}} \left(\sum_n g(\lambda_n) - \frac{a}{\pi} \int_{-\infty}^{+\infty} g(iy)\, dy \right)$$

equals $+\infty$, *then the system* Exp Λ *is complete in* $[-a, a]$. *Conversely, if this quantity is bounded above, then by deleting a single (arbitrary) exponential from the system* Exp Λ *we obtain a system that is not complete in* $[-a, a]$.

(2-!) In this case the completeness problem was solved completely by the profound Beurling-Malliavin theorem on the radius of completeness [1].

(3-!) When $X = G$ is an unbounded convex domain in $\mathbb{C}$, the completeness problem was solved completely in [3, Theorem 2].

(4-?) The *width* of an unbounded convex domain G, $[0, +\infty) \subset G$, is defined to be the quantity $d_G = \lim_{x \to +\infty} d_G(x)$, where $d_G(x) = \sup\{|z - z'| : z, z' \in G, \Re z = \Re z' = x\}$, $x \in [0, +\infty)$.

If either $d_G = +\infty$ or, for all $x \in \mathbb{R}$ we have $d_G(x) \neq d_G < +\infty$, then sharp conditions of completeness of Exp Λ in Clos G are the same as in (3-!). In the opposite case, there is a test for the completeness of a system Exp Λ in Clos G only under conditions that the sequence Λ is separated from the imaginary axis (see [4, Corollary 4.2]):

THEOREM 2. *Let G be an unbounded convex domain of width $2\pi d < +\infty$, $[0, +\infty) \subset G$, and suppose there exists one value of $x \in \mathbb{R}$ such that $d_G(x) = d_G$. Suppose that a sequence Λ satisfies the condition $|\Re \lambda_n| \geqslant \delta |\lambda_n|$, $n \geqslant n_0$, for a certain number $\delta > 0$. The system* Exp Λ *is complete in* Clos G *if and only if*

$$\sup_{1 \leqslant r < R < +\infty} \left(\max\left\{ \sum_{\substack{r \leqslant |\lambda_n| < R \\ \Re \lambda_n < 0}} -\Re \frac{1}{\lambda_n}, \sum_{\substack{r \leqslant |\lambda_n| < R \\ \Re \lambda_n > 0}} \Re \frac{1}{\lambda_n} \right\} - d \log \frac{R}{r} \right) = +\infty.$$

(5-?) At present the completeness problem has not been solved satisfactorily for any bounded convex domain G. In our article [5, Theorem 5] we give a complete solution of the completeness problem in a bounded convex domain G in terms of so-called Jensen functions. After [2] this result can be formulated in the following way.

Let G^* be symmetric to G with respect to $\mathbb{R}$, and let $s_G(\theta)$ be the arc length of the boundary of Clos G^* from a given point to the next (counter-clockwise) point of contact of the supporting line that is orthogonal to the direction θ.

Let $g \in \mathcal{G}$. The function $k_g(\theta) = \int_0^{+\infty} g(te^{i\theta}) \, dt$ is called the *indicator of g*. Always the indicator k_g is the support function of a convex compact set K_g such that $0 \in \text{Int } K_g \neq \emptyset$. We set $S(K_g, G) = \frac{1}{2} \int_0^{2\pi} k_g(\theta) \, ds_G(\theta)$. This quantity is the mixed area of the convex sets K_g and G^*.

THEOREM 3. *The system is complete in the convex bounded domain G if and only if* $\inf d \geqslant 1$, *where the infimum is taken over the numbers d for which* $\sup_{g \in \mathcal{G}} \left(\sum_n g(\lambda_n) - \frac{d}{\pi} S(K_g, G) \right) < +\infty$.

In this theorem, the class $\mathcal{G}$ of Green functions plays the role of test functions. However, the class $\mathcal{G}$ is too wide. Some more transparent sufficient conditions can be extracted from the Theorem 3. For example, they are in [6, Theorem A]. It is possible that these conditions are necessary.

(6-?) No precise condition is known for the completeness of systems Exp Λ in K. The completeness problem in K is apparently a very difficult problem (cf. with $X = [-a, a]$). We present a test for the completeness of systems Exp Λ in K to within two exponentials (see [5, § 7, 3.] together with [2]).

THEOREM 4. *If the quantity* $\sup_{g \in \mathcal{G}} \left(\sum_n g(\lambda_n) - \frac{1}{\pi} S(K_g, K) \right)$ *equals* $+\infty$, *then the system* Exp Λ *is complete in K. Conversely, if this quantity is bounded above by a finite number, then by deleting two (arbitrary) exponentials from the system* Exp Λ *we obtain a system that is not complete in K.*

References

[1] Arne Beurling and Paul Malliavin *On the closure of characters and the zeros of entire functions.*, Acta Math. 1967. Vol. 118. P. 79–93.

[2] Brian J. Cole and Thomas J. Ransford *Jensen measures and harmonic measures.*, Submitted to Crelle's Journal, June 28, 2000 (to appear).

[3] Bulat N. Khabibullin *On the growth of entire functions of exponential type along the imaginary axis.*, Mat. Sb. 1989. Vol. 180. No. 5. P. 706–719; English transl. in Math. USSR Sb. 1990. Vol. 67. No. 1.

[4] Bulat N. Khabibullin *On the growth of entire functions of exponential type with given zeros along the line.*, Analysis Mathematica. 1991. Vol. 17. No. 3. P. 239–256. (Russian)

[5] Bulat N. Khabibullin *Sets of uniqueness in spaces of entire functions of a single variable.*, Izv. Akad. Nauk. SSSR, Ser. Mat. 1991. Vol. 55. No. 5. P. 1401–1423; English transl. in Math. USSR Izv. 1992. Vol. 39. No. 2.

[6] Bulat N. Khabibullin *Completeness of systems of entire functions in spaces of holomorphic functions.*, Mat. Zametki. 1999. Vol. 66. No. 4. P. 603–616; English transl. in Math. Notes. 1999. Vol. 66. No. 4.

5. Density of Domain of the Weighted Hilbert Transform

Alexander Kheifets, kheifets@mail.ru

5.1. *Problem 1*

Let Δ be a nonnegative measurable function on the unit circle T. Assume that $\frac{1}{\Delta} \in L^1$ and for definiteness $\|\frac{1}{\Delta}\|_1 = 1$. In this case $f \in L^2(\Delta)$ implies $f \in L^1$, which allows one to define the Hilbert transform Hf for $f \in L^2(\Delta)$:

$$(Hf)(t) = \frac{1}{2\pi i} \oint_T \frac{f(\tau)d\tau}{\tau - t_+} + \frac{1}{2\pi i} \oint_T \frac{f(\tau)d\tau}{\tau - t_-} , \; t \in T,$$

where the first and the second integrals are understood as the radial limits from inside and outside of the unit disk D respectively. But Hf need not be in $L^2(\Delta)$. We say that f is in the domain of H on $L^2(\Delta)$ if both f and Hf are in $L^2(\Delta)$. Assume that the domain of H is dense in $L^2(\Delta)$. The question is: can this property be characterized in terms of Δ?

One can show that if $\frac{1}{\Delta} = |g|$, where g is in the Hardy space H^1_+, then the domain of H is dense in $L^2(\Delta)$ if and only if g is an exposed point of the unit ball in the space H^1_+. A point b_0 in a complex or real Banach space B is said to be an exposed point of the unit ball if $\|b_0\| = 1$ and there exists a real continuous linear functional F on B such that $F(b_0) = 1$ and $F(b) < 1$ for all other b in the unit ball. Regarding extreme and exposed points of the unit ball in H^1 see [1], [2], [3] and [4].

Obviously integrability of Δ is a sufficient condition for density of the domain but it is not a necessary one. The interesting case is the one when Δ is not integrable.

5.2. *Problem 2*

Let w be a measurable unimodular function on the unit circle T: $|w(t)| = 1$ for a.a. $t \in T$. Assume that set $P_+ w | H^2_+$ is dense in H^2_+, where H^2_+ is the Hardy space of analytic functions and P_+ is the orthogonal projection from L^2 onto H^2_+. The question is: how this property can be characterized in terms of w?

One can show that the property holds true if and only if w is a canonical solution to a Nehari problem. The Nehari problem consists in finding all the functions w bounded in modulo by 1 on the unit circle T: $|w(t)| \leqslant 1$ for a.a. $t \in T$, with given $P_- w$, where P_- is the orthogonal projection from L^2 onto H^2_-. Details regarding the Nehari problem can be found in [5], [6], [7], [8], [1], [4].

It is also known that w is a canonical solution of a Nehari problem if and only if it is of the form $w = \frac{g}{|g|}$, where g is an exposed point of the unit ball in H^1_+. In this case the representation is unique. See [2], [4] .

References

[1] J. Garnett, *Bounded analytic functions*, Academic Press, (1981)

[2] D. Sarason, *Exposed points in H^1*, Operator Theory: Advances and Applications, Birkhäuser, Part I: Vol. 41, (1989), pp. 485-496; Part II: Vol 48 (1990), pp. 333-347

[3] A. Kheifets, *On regularization of γ-generating pairs*, Journal of Funtional Analysis, Vol. 130 (1995), pp. 310-333

[4] A. Kheifets, *Nehari's interpolation problem and exposed points of the unit ball in the Hardy space H^1*, Israel Mathematical conference proceedings, Vol. 11 (1997), pp. 145-151

[5] V.M. Adamjam, D.Z. Arov, M.G. Krein, *Infinite Hankel matrices and generalized problems of Carathéodory-Fejér and M. Riesz*, Funktional. Anal. Appl. 2 (1968), 1-19; English transl.: Funct. Anal. Appl. 2 (1968),1-18

[6] V.M. Adamjam, D.Z. Arov, M.G. Krein, *Infinite Hankel matrices and generalized Carathéodory-Fejér and I. Schur problems*, Funktional. Anal. Appl. 4 (1968), 1-17; English transl.: Funct. Anal. Appl. 2 (1968), 269-281

[7] D.Z. Arov, *γ-generating matrices, j-inner matrix-functions and related extrapolation problems*, Teor. Funktii Funktsional. Anal. i Prilozhen. 51 (1989), 61-67; 52 (1989), 57-65; 53 (1990), 57-65; English transl.: J. Soviet Math. 52 (1990), 3487-3491, 52 (1990), 3421-3425, 58 (1992), 532-537

[8] N.K. Nikol'skii, *Treatise on the shift operator*, Grundlehren der mathematishen Wissenshaften 273, Springer, Berlin, 1986 (Translation from Russian edition, Nauka, Moscow, 1980)

6. Harmonic Sliding Analysis Problem

Vladimir Ya.Krakovsky, UNESCO/IIP International Research-Training Centre for Information Technologies and Systems, 40 Academician Hlushkov Ave, Kyiv 03680 Ukraine, krakovsk@uasoiro.freenet.kiev.ua

Harmonic sliding analysis (HSA) is a dynamic spectrum analysis [1] in which the next analysis interval differs from the previous one by including the next signal sample and excluding the first one from the previous analysis interval. Such a harmonic analysis is necessary for time-frequency localization [2] of analysed signals with given peculiarities. Using the well-known Fast Fourier transform (FFT) is not effective in this context. More effective are known recursive algorithms which use only one complex multiplication for computing one harmonic during each analysis interval.

To yield an instant spectrum

$$(6.1) \qquad F_q(p) = \frac{1}{N} \sum_{k=q-N+1}^{q} f(k)W_N^{-pk}, \; p \in \overline{0, P-1}, \; q = 0, 1, 2, \ldots$$

it is possible to use a simple recursive algorithm, described in [3], [4]:

$$(6.2) \qquad F_q(p) = F_{q-1}(p) + \frac{1}{N}[f(q) - f(q-N)]W_N^{-pq}, \; p \in \overline{0, P-1}, \; q = 0, 1, 2, \ldots .$$

This algorithm has a remarkable peculiarity which permits one to organize HSA so that one complex multiplication may be used for computing two, four and even eight (for complex signals) spectrum harmonics at once [5], [6], [7], [8]. This may be done as follows. Let algorithm (2) be presented as follows:

(6.3) $$F_q(p) = F_{q-1}(p) + \Delta F_q(p), \; p \in \overline{0, P-1}, \; q = 0, 1, 2, \ldots ,$$

(6.4) $$\Delta F_q(p) = \frac{1}{N}[f(q) - f(q-N)] \exp[-j\frac{2\pi}{N}pq].$$

The spectrum increments $\Delta F_q(p)$ may be used not only for the spectrum harmonic p, but for spectrum harmonics

(6.5) $$p_i = iN/4 + p, \; i \in \overline{1, 3}$$

and

(6.6) $$p_k = kN/4 - p, \; k \in \overline{1, 4}$$

as well, using known properties of the complex exponential function. In a summarized (and simplified) view the algorithm (3) modification may be presented as follows:
a) for spectrum harmonics (5)

(6.7) $$F_q(p_i) = F_{q-1}(p_i) + (-j)^{iq}\Delta F_q(p), \; q = 0, 1, 2, \ldots ,$$

b) for spectrum harmonics (6)

(6.8) $$F_q(p_k) = F_{q-1}(p_k) + (-j)^{kq}\Delta F_q(-p), \; q = 0, 1, 2, \ldots ,$$

where $\Delta F_q(-p)$ are complex conjugated spectrum increments $\Delta F_q(p)$, if the signal samples are real, and if the signal samples are complex, $\Delta F_q(-p)$ are generated by inverting the signs of the products of the signal increments $\Delta f_q = \frac{1}{N}[f(q) - f(q-N)]$ with the imaginary part of the weighting function and then forming the appropriate algebraic sums.

In such a way it is possible to use one complex multiplier for computing up to four harmonics for a real signal, and up to eight harmonics for a complex signal, simultaneously.

We can now state the HSA problem: Is it possible to double the speed of the response by using an additional multiplier? If so, how?

References

[1] V.N.Plotnikov, A.V.Belinsky, V.A.Sukhanov and Yu.N.Zhigulevtsev, "Spectrum digital analyzers," Moscow: Radio i svjaz', 1990. - 184 p. (In Russian).

[2] I.Daubechies, "The wavelet transform, time-frequency localization and signal analysis," IEEE Trans.Inform. Theory, vol.36, No.5, pp.961-1005, 1990.

[3] R.D.Lejtes and V.N.Sobolev, "Synthetic telephone systems digital modeling." Moscow: Svjaz', 1969. (In Russian).

[4] J.W.Schmitt and D.L.Starkey, "Continuously updating Fourier coefficients every sampling interval." USA Patent 3778606 Dec.11, 1973.

[5] V.Ya.Krakovsky, "Algorithms for increase of speed of response for digital analyzers of the instant spectrum," Kibernetika, No.5, pp.113-115, 1990. (In Russian).

[6] V.Ya.Krakovskii, "Digital analyzers for dynamic spectral analysis." Meas. Tech. (USA), vol.36, no.12, p.1324-30. Translation of: Izmer.Tekh. (Russia), vol.36, no.12, p.13-16 (Dec.1993).

[7] V.Ya.Krakovskii. "Generalized representation and implementation of speed-improvement algorithms for instantaneous spectrum digital analyzers." Cybernetics and System Analysis, Vol.32, No.4, pp.592-597, March 1997.

[8] V.Ya.Krakovsky. "Spectrum Sliding Analysis Algorithms & Devices." Proceedings of the 1st International Conference on Digital Signal Processing and Its Application, June 30 - July 3, 1998, Moscow, Russia, Volume I, pp.104-107.

input montgomery-problems

7. Questions on Kazhdan's Property (T) on Hypergroups

Liliana Pavel, University of Bucharest, Faculty of Mathematics, Academiei 14, 70109 Bucharest, Romania, lpavel@@lan.unibuc.ro

Isolated points of the dual of a locally compact group (endowed with the hull-kernel topology) were first discussed by Diximier [1]. Kazhdan [5] discovered there exists a clear connection between the fact that the class of the one dimensional trivial representation is an isolated point in the dual of a locally compact group (in this case the group is said to have property (T)) and many interesting group properties.

In [6] we have initiated the study of property (T) on hypergroups (with Haar measure), obtaining only some introductory results. We have given an appropriate definition of property (T) for hypergroups: this definition is an extension of the corresponding one for locally compact groups of [3]. Consequently, we say that a hypergroup K has property (T) if each continuous representation of K which has almost invariant vectors has also invariant vectors. (We recall that a representation π of K on the Hilbert space $\mathcal{H}_\pi$ has almost invariant vectors if, for any $\varepsilon > 0$ and $C \subset K$, C compact, there exists $a \in \mathcal{H}_\pi$, with $\|a\| = 1$ such that $\|\pi_x a - a\| < \varepsilon$, $\forall x \in C$; π has invariant vectors if there exists $b \in \mathcal{H}_\pi$, with $\|b\| = 1$, such that $\pi_x b = b, \forall x \in K$.)

We note that in the particular case when K is a locally compact group this definition is equivalent with the one given by Kazhdan [5]. For arbitrary hypergroups, we have only obtained that if K is a hypergroup with property (T), then the class of the one dimensional trivial representation is an isolated point in the dual of K ([6, Theorem 2]), so we ask if the converse is still valid.

It is well known that any amenable non-compact locally compact group has no property (T). This is an immediate consequence of the fact that a locally compact group G is amenable if and only if it satisfies Reiter's condition (P_2): $\forall \varepsilon > 0$, $\forall C \subset G$, C compact, $\exists f \in L_2(G)$, $f \geqslant 0$, $\|f\|_2 = 1$, such that $\|\delta_x * f - f\|_2 < \varepsilon$, $\forall x \in C$, or, equivalently, the left regular representation λ_G of G on $L_2(G)$ has almost invariant vectors; λ_G obviously has no invariant vectors if G is not compact.

For non-compact hypergroups (with Haar measure), the amenability is not equivalent to (P_2). In [7, Example 4.6] an example is given of a non-compact amenable hypergroup that does not satify (P_2); it is also proved that (P_2) implies the amenability ([7, Theorem 4.1]). Consequently, for hypergroups, we can obtain, with similar arguments as for locally compact groups, that any non-compact hypergroup K satisfying condition (P_2) has no property (T).

For example, if K is a non-compact amenable hypergroup with a supernormal subhypergroup, or if K is a non-compact commutative hypergroup with the Plancherel measure on the dual such that its support contains the trivial character, then K satisfies (P_2) ([7, Theorem 4.7, Lemma 4.5]). In these cases, K has no property (T).

We can not yet give a general answer concerning the relation between amenability and property (T) for the hypergroups case. Thus, we close with the question: Does there exist an amenable non-compact hypergroup having property (T) (?).

References

[1] J. Dixmier, *Points isolés dans le dual d'un groupe localement compact.* Bull. Sc. Math. 85 (1961), 91-96.

[2] J.M.G. Fell, *Weak containment and induced representation of groups.* Canad. J. Math. 14, (1962), 237-268.

[3] P. de la Harpe, A. Valette, *La propriété (T) de Kazhdan pour les groupes localement compacts.* Astérisque 175 (1989).

[4] R. Jewett, *Spaces with an Abstract Convolution of Measures.* Advances in Math. 18 (1975), 1-101.

[5] D. A. Kazhdan, *Connection of the dual space of a group with the structure of its closed subgroups.* Functional Anal. i Prilozen 1 (1967), 71-74.

[6] L. Pavel, *On hypergroups with Kazhdan's property (T).* to appear in Math. Reports.

[7] M. Skantharajah, *Amenable hypergroups.* Illinois J. Math. 36 (1992), 15-46.

[8] M. Voit, *On the dual space of a commutative hypergroups.* Arch. Math. 56 (1991), 380-385.

[9] M. Voit, *Properties of Subhypergroups.* Semigroup Forum 56 (1997), 373-391.

[10] P.S. Wang, *On Isolated Points in the Dual Spaces of Locally Compact Groups.* Math. Ann. 218 (1975), 19-34.

8. Optimal Speech Signal Partition into One-Quasiperiodical Segments

Taras K. Vintsiuk, UNESCO/IIP International Research-Training Centre for Information Technologies and Systems, Kyjiv 03680 Ukraine,
vintsiuk@uasoiro.freenet.kiev.ua

8.1. *Introduction*

It is well known that analysis of such complicated signals as speech signals has to be carried out synchronically with a current pitch period (quasiperiod). Besides, for speech signal it is important to find current one-quasiperiod segment duration, beginnings and ends as well.

To solve this problem quasi-periodicity and non-periodicity signal models are proposed. Each hypothetical one-quasiperiodical signal segment is considered as a random distortion of previous or following one taken with the unknown multiplying factor. The problem of optimal current pitch period discrimination and speech signal partition into quasiperiodical and non-periodical segments consists in 1) the finding the best quasiperiod beginnings or the one-quasiperiod segments under restrictions on both value and changing of current quasiperiod duration and multiplying factor and 2) the association of optimal one-quasiperiod segment signals into large quasiperiodic and non-periodic segments. For this problem solving an effective algorithm based on dynamic programming have to be proposed.

8.2. *One-Quasiperiodicity Models*

Let the signal f_n, $n = 1 : N$ be observed where f_n is a signal value at the discrete uniform time $n\Delta t$ with step Δt, for example $\Delta t = 50$ μs for speech signal. If the $(s+1)$-th one-quasiperiod signal segment beginning is denoted by n_s, then $(n_s - 1)$ will be the end of the s-th one-quasiperiod. Further segment signal f_{n_s-1+j}, $j = 0 : (T_s - 1)$, $T_s = n_s - n_{s-1}$ will be called s-th one-quasiperiod segment with duration T_s, if it is approximated sufficiently well by neighbouring ones, $(s-1)$-th or $(s+1)$-th, which respectively precedes or follows the s-th one-quasiperiod segment. The latter is taken with the unknown multiplying number α_s^- or α_s^+:

(8.1)
$$f_{n_s-1+j}^- = \begin{cases} \alpha_s^- f_{n_s-2+j}, & j = 0 : (\min(T_s, T_{s-1}) - 1); \\ 0, & j = \min(T_s, T_{s-1}) : (T_s - 1), \end{cases}$$

(8.2)
$$f_{n_s-1+j}^+ = \begin{cases} \alpha_s^+ f_{n_s+j}, & j = 0 : (\min(T_s, T_{s+1}) - 1); \\ 0, & j = \min(T_s, T_{s+1}) : (T_s - 1). \end{cases}$$

Let us introduce a priori restrictions for multiplying number value α and both current quasiperiod duration value T_s and its changing $\Delta_s = T_s - T_{s-1}$:

(8.3) $\quad \{\alpha_s : 0 \leqslant \alpha_{\min} \leqslant \alpha_s \leqslant \alpha_{\max}\} = \mathcal{A}, \quad T_{\min} \leqslant T_s \leqslant T_{\max}, \quad |\Delta_s| \leqslant \Delta_{\max}.$

Let us fix the elementary quasi-periodicity (EQP) measure for the s-th one-quasiperiodic signal segment $f_{n_s-1+j} = f_{n_s-T_s+j}$, $j = 0 : (T_s - 1)$ as:

$$d\left((n_s, T_s), \Delta_s^-\right) = \overset{\alpha \in \mathcal{A}}{\min} \sum_{j=0}^{T_s-1} \left(f_{n_s-1+j} - f_{n_s-1+j}^-\right)^2 =$$

(8.4)
$$= \overset{\alpha \in \mathcal{A}}{\min} \sum_{j=0}^{\min(T_s-\Delta_s^-, T_s)-1} \left(f_{n_s-T_s+j} - \alpha f_{n_s-2T_s+\Delta_s^-+j}\right)^2 + \sum_{j=\min(T_s-\Delta_s^-, T_s)}^{T_s-1} f_{n_s-T_s+j}^2$$

or

$$d\left((n_s, T_s), \Delta_s^+\right) = \overset{|\Delta^+| \leqslant \Delta_{\max}}{\min} \overset{\alpha \in \mathcal{A}}{\min} \sum_{j=0}^{T_s-1} \left(f_{n_s-1+j} - f_{n_s-1+j}^+\right)^2 =$$

(8.5)
$$= \min_{0 \leqslant \Delta^+ \leqslant \Delta_{max}} \left(\min_{\alpha \in A} \sum_{j=0}^{\min(T_s - \Delta^+, T_s) - 1} \left(f_{n_s - T_s + j} - \alpha f_{n+j} \right)^2 + \sum_{j=\min(T_s - \Delta^+, T_s)}^{T_s - 1} f^2_{n_s - T_s + j} \right).$$

As it is followed from the expression (8.4-8.5) the signal segment

$$f_{n_{s-1}+j} = f_{n_s - T_s + j}, \quad j = 0 : (T_s - 1)$$

is tested on quasiperiodicity by comparison with previous segment

$$f_{n_s - 2T_s + \Delta_s^- + j}, \, j = 0 : \left(T_s - \Delta_s^- - 1 \right)$$

and all possible following ones f_{n_s+j}, $j = 0 : (T_s - \Delta^+ - 1)$, $|\Delta^+| \leqslant \Delta_{max}$ but only the best comparison result is associated with the quasiperiodicity measure value $d\left((n_s, T_s), \Delta_s^\pm \right)$.

Any permissible variant $\left((n_s, T_s), \Delta_s^\pm \right)$, $s = 0, 1, 2, ..., P$ of the signal f_n, $n = 1 : N$ segmentation on P one-quasiperiodic segments under restrictions (8.3) is characterised by the sum of respective EQP measure values:

(8.6)
$$G\left(n_s, s = 0 : P \right) = \sum_{s=0}^{P} d\left((n_s, T_s), \Delta_s^\pm \right)$$

To find for the signal f_n, $n = 1 : N$ the best segmentation onto unknown number P one-quasiperiod segments it is necessary to minimise the criteria (8.6) on all permissible sequences n_s, $s = 0 : P$.

8.3. *Problems to be Solved*

The following problems have to be solved.

1. To propose effective dynamic programming algorithm for optimal partition of signal into one-quasiperiodical segments.
2. To determine ways how to unite one-quasiperiodical segments into large quasiperiodic or non-periodic ones.
3. For each time n to formulate some necessary and sufficient conditions that a certain pair (n^*, T^*), $n^* < n$ is an optimal one-quasiperiodic segment beginning n^* and duration T^* independently on future signal after n.

References

[1] Taras K. Vintsiuk. Optimal Joint Procedure for Current Pitch Period Discrimination and Speech Signal Partition into Quasi-Periodic and Non-Periodic Segments. - In "Text, Speech, Dialogue", Proc. of the First Workshop on Text, Speech, Dialog - TSD'98, Brno, 1998, pp 135-140.

9. A question of "complexity"

J. A. Ward, Murdoch University

This problem arises in discrete-time worst-case system identification [9].

In *time-domain* identification we wish to identify the impulse response h of a time-invariant linear system within a prescribed tolerance, using a finite number of sampled values of the output y corresponding to a chosen input signal u. The output signal y is the sum of the system response $u * h$ and a noise term η, so that $y = u * h + \eta$. Both u and h are bounded real or complex valued one-sided sequences, and $*$ denotes the usual convolution. It is assumed that h belongs to some specified model set, such as one of the polynomial sequence sets $\mathcal{P}_n$, or the sets $\mathcal{V}(p, r)$ that are associated with linear systems whose transfer functions H are rational functions and whose poles (if any) occur outside the unit circle. Here H is the z-transform $\widehat{h}$ of the impulse response. In *frequency-domain* identification the aim is to identify H using a finite number of noisy sample values of H on the unit circle. The sample values are obtained by measuring the system's response to a sinusoidal input. It is also assumed that H belongs to some prescribed model set, such as the disc algebra $A(\mathbb{D})$, although additional conditions may be placed on H.

In *worst-case identification* the noise term η is simply assumed to be uniformly bounded and $||\eta||_\infty \leqslant \delta$, where $||\eta||_\infty$ denotes the supremum norm. The ℓ^1, ℓ^2 and H^∞-norms are most often used to measure the accuracy of model estimates (tolerance).

Both time- and frequency-domain worst-case identification problems can be recast within a more abstract framework [7], [2]. Suppose that $\mathcal{X}$ is a normed linear space,

$$\varphi_1, \varphi_2, \ldots, \varphi_N, \ldots$$

a uniformly bounded sequence of continuous linear functionals on $\mathcal{X}$, and that $\tau > 0$. Suppose also that

$$y_k = \varphi_k(h) + \eta_k \text{ and } |\eta_k| \leqslant \delta \text{ for each } 1 \leqslant k \leqslant n,$$

where h is an unknown element of a given subset $\mathcal{M}$ of $\mathcal{X}$. Then, using $y_1, y_2, \ldots, y_n$, we want to find $\widetilde{h}$ in $\mathcal{M}$ so that $\left\| h - \widetilde{h} \right\| < \tau$. If we can do this, then we say that $\{\varphi_1, \varphi_2, \ldots, \varphi_n\}$ is a (δ, τ)-*identifying set* for $\mathcal{M}$. The connection with time-domain identification is that each $\varphi_k(h) = (u * h)_k$, where u is the chosen input, while for frequency-domain identification $\varphi_k(h) = H(z_k) = \widehat{h}(z_k)$ where $z_0, z_1, \ldots, z_{n-1}$ are points on the unit circle.

The *complexity* of identification in $\mathcal{M}$ for a given tolerance τ and noise bound δ is defined to be the minimum number of 'observations' or functionals required to obtain an estimate within that tolerance for the given noise bound. It depends on $\mathcal{M}$ and on the type of functionals that are allowed. Since $||f||_2 = ||f||_{H^2} \leqslant ||f||_{H^\infty} \leqslant ||f||_1$ for any sequence f, the complexity of identification in $\mathcal{M}$ decreases as we replace the ℓ^1 norm on $\mathcal{M}$, by the H^∞ norm, and then by the ℓ^2 norm. It is known that ℓ^1 identification is typically 'exponential' in complexity, and so not very practical (see [1], for example). On the other hand, both H^∞ and ℓ^2 identification are typically 'polynomial' [3], [5], [8].

A drawback to using the conservative ℓ^∞ norm to measure the noise component is that a single outlier may make this very large, and this in turn may make it difficult to 'identify' the system using standard techniques. An obvious way to address this difficulty is to replace the ℓ^∞ norm by a norm that is less sensitive to outliers. More generally we work with a sequence $(\|\|_n)$ of seminorms on $\mathbf{R}^n$, assuming that each is dominated by the ℓ^∞ norm. The problem then is to determine the complexity for different choices of $\mathcal{M}$ and norms on the noise.

We are helped in this task by the observation [3], [2], [7], that when the ℓ^∞ norm is used to measure the noise, then linear functionals $\varphi_1, \varphi_2, \varphi_3, \ldots, \varphi_n$ form a (δ, τ)-identifying set for an absolutely convex set $\mathcal{M}$ if and only if

$$\max_{1 \leqslant k \leqslant n} |\varphi_k(h)| \geqslant \delta \text{ for each } h \in \mathcal{M}_\tau,$$

where $\mathcal{M}_\tau = \{ h \in \mathcal{M} : \|h\| = \tau \}$.

A related result [6] for the generalised noise measure case is that there is a 'robustly convergent' algorithm for estimating h on the basis of n test functional values if and only if there is a number $\varepsilon > 0$ such that

$$\lim_{n \to \infty} \inf \|(\varphi_1(h), \varphi_2(h), \ldots, \varphi_n(h))\|_n \geqslant \varepsilon \|h\| \text{ for each } h.$$

This is relevant because it can be shown that if there is such an algorithm then for any noise bound δ and tolerance level τ there is a finite subset of the φ_ks that form a (δ, τ)-identifying set for $\mathcal{M}$.

For a concrete version of the problem, what is the complexity of identification if we suppose that $\mathcal{M} = A(\mathbb{D})$ with its usual norm, and that $\|\|_n = \|\|_*$ for all n, where $\|\|_*$ is an *Orlicz* or *Lorenz* norm on $A(\mathbb{D})$? Of particular interest is the case of the Lorenz norm given by $\|\mathbf{x}\|^w = \|\mathbf{x}\|_w / \|(1, 1, \ldots, 1)\|_w$, where

$$\|\mathbf{x}\|_w = \sup_{\pi \in \Sigma_n} \sum_{j=1}^n |x_{\pi(j)}| \, w_j \text{ for each } \mathbf{x} = (x_1, x_2, \ldots, x_n) \in \mathbf{R}^n,$$

and Σ_n denotes the set of permutations of $\{1, 2, \ldots, n\}$, and where $\mathbf{w} = (1, 1, \ldots, 1, 0, 0, \ldots, 0)$, with the first K terms equal to 1.

References

[1] M.A. Dahleh, T. Theodosopoulos and J.N. Tsitsiklis, The sample complexity of worst-case identification of FIR linear systems, *Systems Control Lett.*, **20** (1993), 157-166.

[2] D.W.Hadwin, K.J.Harrison and J.A.Ward, Worst-case identification of linear systems: existence and complexity, in *Proc. Centre for Math. and its Applications*, **36** (1999), 53-65.

[3] Harrison, K.J., J.R. Partington and J.A. Ward, Complexity of identification of linear systems with rational transfer functions, *Math. Control Signals Systems*, (1998)

[4] Harrison, K.J. and J.A. Ward, Fractional covers for convolution products, *Result. Math.*, **30** (1996), 67-78.

[5] Harrison, K.J., J.A. Ward and D.K. Gamble, Sample complexity of worst-case H^∞-identification, *Systems Control Lett.*, **27** (1996), 255-260.

[6] Makila, P.M. and J.R. Partington, Robustness in H^∞ identification, preprint 1999.

[7] Partington, J.R., Interpolation in normed spaces from values of linear functionals, *Bull. London Math. Soc.*, **26** (1994), 165-170.

[8] Partington, J.R., Worst-case identification in ℓ^2: linear and non-linear algorithms, *Systems Control Lett.*, **22** (1994), 93-98.

[9] Partington, J.R., *Interpolation, identification and sampling*, Oxford University Press, 1997.

10. Rate of decay of convolution vs. frequency of sign changes

H.S. Shapiro

All references, as well as further background discussion concerning the present problem may be found in reference [S].

Let f be in $L^\infty(\mathbb{R}^+)$ (the class of real-valued bounded measurable functions on $(0, \infty)$). Then, the following, due to B.F. Logan, Jr., is known: If

$$(10.1) \qquad F(y) := \frac{2}{\pi} \int_0^\infty \left(\frac{y}{x^2 + y^2} \right) f(x)\, dx, \quad y > 0$$

satisfies

$$(10.2) \qquad F(y) = O(e^{-ay}) \quad \text{as } y \to \infty$$

for some $a > 0$, then $\tilde{f}$, *the even extension of f to $\mathbb{R}$* (which is in $L^\infty(\mathbb{R})$) *has spectrum disjoint from* $(-a, a)$. the converse is also true. (The *spectrum* here means the support of the distributional Fourier transform.) As an illustration, look at $f(x) = \cos ax$ (then, $\tilde{f}(x) = \cos ax$, $x \in \mathbb{R}$). Here $F(y)$ is the Poisson integral of $\cos ax$, evaluated at a point of the imaginary axis, so $F(y) = e^{-ay} \cos ax$. Observe that in this case, $f(x)$ *has an essential sign change on each interval in $\mathbb{R}^+$ of length greater than π/a* (that is, on each such interval it assumes positive values, as well as negative values, on a set of positive measure). In a 1965 doctoral thesis, B.F. Logan, Jr. raised the question whether (10.1) implies an asymptotic lower bound for the amount of oscillation of f. Namely, defining $\sigma(x)$ as the number (possibly infinite) of points of $(0, x)$, on each neighborhood of which f has an essential sign change, he asked whether

$$(10.3) \qquad \liminf_{x \to \infty} \frac{\sigma(x)}{x} \geq \frac{a}{\pi},$$

For each $f \in L^\infty(\mathbb{R})$, not identically zero and satisfying (10.2). He proved the answer is affirmative if f also is assumed to *be the restriction to $\mathbb{R}^+$ of an entire function of exponential type*. So far as I know, no one has proved (or disproved) this assertion without that last hypothesis.

In my paper [S] I remarked that, as a consequence of a recent theorem due to Baouendi and Rothschild, the weakening of (10.2) to

$$(10.4) \qquad\qquad F(y) = O(y^{-n}), \quad y \to \infty$$

for every positive integer n implies the (weak) oscillation result: f *has an essential sign change on* (b, ∞) *for every* $b > 0$.

Certain questions now almost pose themselves: Suppose $F(y)$ has, as $y \to \infty$, a rate of decrease intermediate between (10.2) and (10.4) (e.g., $F(y) = O(e^{-y^t})$ for some t with $0 < t < 1$.) Can one assert anything about the frequency of sign changes of f, beyond what already follows from (10.4)?

A further avenue of generalizations appears when we observe that (10.1) is a convolution on the group $\mathbb{R}^+$ (with respect to the Haar measure $\frac{dx}{x}$). After a logarithmic variable change it becomes a usual convolution on $\mathbb{R}$, with the "kernel" $K(x) := (2/\pi)(\cos x)^{-1}$ (details in [S]). Thus, all the questions we have raised so far can be put in the form: Deduce from the rate of decay at $+\infty$ of $g \star K$, for some $g \in L^\infty(\mathbb{R})$, lower bounds for the (asymptotic frequency of) essential sign changes of g. Once this standpoint is taken, one can raise these questions for other kernels, like $K(x) = e^{-x^2}$.

One final remark: It is a corollary of the result that (10.2) implies disjointness of spectrum $\hat{f}$ from $(-a, a)$, that: *If (10.2) holds for every* $a > 0$, *then* f *vanishes a.e.* This result predates Logan's work, and is a special case (after transformation from $\mathbb{R}^+$ to $\mathbb{R}$ by the logarithmic variable change) of general results due to I.I. Hirschmann, Jr. from 1951. We might express matters thusly: for some class of kernels in $L^1(\mathbb{R})$, Hirschmann found the *critical* rate at which a convolution of a nontrivial function in $L^\infty(\mathbb{R})$ with this kernel may decay at ∞ (i.e., faster decay is impossible). There is some evidence that *substantial* decay implies corresponding oscillatory behavior; this seems a promising and mostly unexplored area.

References

[S]　Shapiro, H.S., *Notes on a theorem of Baouendi and Rothschild*, Expositions Math. **13** (1995) 247–275.

[G. Tsoucari, K. Bethanis, P. Tzamalis, and A. Hountas] G. Tsoucari,[1] K. Bethanis,[2] P. Tzamalis,[2] and A. Hountas[2]

[Direct Methods and Quantum Mechanics] Direct Methods and Quantum Mechanics
[1]Laboratoire de recherche des musees de France, Paris, France
[2]Physics Laboratory, Agricultural University, Athens, Greece

11. The Problem

Direct Methods for crystal structure determination are based on mathematical relations, so called *phase relations*, between the moduli (observed by diffraction experiments) and the (unknown) phases of the Fourier coefficients E(H) of the periodic electron density ρ(r). The latter can be approximated by one electron quantum mechanical *wave functions* $\psi(r)$:

(11.1) $$\rho(r) = | \quad \psi(r)^2| \Leftarrow FT \Rightarrow E(p) = \Psi(p) * \Psi(-p)^*$$

$$\psi \quad \Leftarrow FT \Rightarrow \quad \Psi$$

A question arises then, whether it is possible to obtain *phase relations from fundamental Quantum Mechanics*. We show below that the Schrodinger equation in Fourier (momentum) space, written below in appropriate atomic units, provides a basis for such relations:

Direct space   Momentum space

(11.2) $-\Delta\psi(r)/2 + V(r)\psi(r) = \varepsilon_0\psi(r) \Leftarrow FT \Rightarrow p^2\Psi(p)/2 + W(p) * \Psi(p) = \varepsilon_0\Psi(p)$

with $V \Leftarrow FT \Rightarrow W$

For a periodic crystalline structure the convolution integral in momentum space is re-placed by a discrete summation and eq. (2) is re-arranged as (3). We see then (with an eigenvalue $\varepsilon_0 < 0$) *that the phase of the wave function $\Psi(p)$ is invariant under the potential (convolution) operator $W(p)*$* :

(11.3) $$(H^2/2 - \varepsilon_0)\Psi(H) = -\Sigma_K W(K) \quad \Psi(H - K)$$

The next key remark is that the Fourier Coefficients $W(K)$ of the potential function $V(r)$ are identical to the crystallographic expression $- Z \sqrt{N} E(K) / K^2$. For a unit cell containing N identical atoms of atomic number Z we have, as shown in Appendix:

(11.4) $$W(K) = -Z\sqrt{N}E(K)/K^2$$

Eq. (4) is the *key relation* linking the Quantum Mechanical potential function to the diffraction experiment for a crystal. Thus eq. (3) is written as (5), as shown in Appendix

(11.5) $$(H^2/2 - \varepsilon_0)\Psi(H) = Z\sqrt{N}\Sigma_K E(K)\Psi(H - K)/K^2$$

(11.6) $$E(H) = \Sigma_K\Psi(K)\Psi^*(K - H)$$

The pair of eq. (5) and (6) forms a system of self consistent equations, that is a usual procedure in Quantum Mechanics. They provide the physical basis of the TWIN algorithm developed for crystal structure determination (Hountas, A. and Tsoucaris, G. (1995) Acta Cryst. A51, 754-763; see also: J. Navaza and G. Tsoucaris, in Phys. Rev. (1981); G. Berthier et al. in J. Quantum Chemistry, 1996, p. 195-199). The final form (5) of the Schrodinger equation in momentum space has a strong similarity with one of the fundamental phase relations of Direct Methods:

(11.7) $$Phase \quad of \quad E(H) = Phase \quad of \quad [\Sigma_K E(K)E(H - K)]$$

Note that the positive factor $1/K^2$ in eq. (5) plays the role of a weighting factor for each contributor $E(K) \Psi(H-K)$ in the summation, that is a usual technique in Direct Methods.

The similarity becomes closer with a further approximation related to the classical Linear Combination of Atomic Orbitals LCAO method:

$\Psi(H) = c E(H)$ with a known constant $c > 0$.

In conclusion, we have shown a close relation between a fundamental formula in Direct Methods (obtained by Probability theory applied to the FT of the electron density of a crystal) and the Schrodinger equation in momentum (i.e. FT) space. We further note that the FT between position and momentum is a central notion in Quantum Mechanics. It corresponds to a kind of "built in" feature of Quantum Mechanics and constitutes the fundamental link between the position and the momentum of a quantum mechanical particle, well known as the Uncertainty Principle.

Note: The above presentation holds for a simplified model of electrons moving independently to each other (no interelectronic interaction) in the potential created by the nuclei of the crystal. Such an approximation is of course inappropriate for a correct quantum description of molecular orbitals, but it is sufficient for the determination by Direct Methods of approximate phases of the Fourier coefficients EH (with observed moduli).

12. Appendix

The electron-nuclei attractive potential for N atoms of atomic number Z_j at positions r_j is (in atomic units):

$V(r) = - \Sigma j \, Z_j \, / \, |r - r_j|$ with $r \in R^3$

If r plays the role of time and $K \in R^3$ that of frequency, then the FT at K of $V(r)$ is :

$W_K = - \Sigma_j \, [\, Z_j \exp (\, 2\pi \, i \, K. \, r_j \,) \,] \, / \, K^2$

reminding that in 3D we have:

$1/|r| \quad \Leftarrow FT \Rightarrow 1/|K|^2$

Note that for simplicity the above notation of modulus for the 3D vectors r and K has been omitted throughout the text.

The Fourier coefficients E_K (so called normalized structure factors) involved in diffraction experiments have a similar expression apart for the (positive) factor $1/K^2$. Indeed, the FT at K of N identical atoms -considered as Dirac masses $1/\sqrt{N}$ at positions r_j - is given by:

$E_K = \Sigma j \, [\exp (\, 2\pi \, i \, K. \, r_j \,) \,] \, / \, \sqrt{N}$

For atoms of atomic number Z, eq. (4) follows.

How to Use the Fourier Transform in Asymptotic Analysis

V. Gurarii and J. Steiner,[1] V. Katsnelson,[2] V. Matsaev[3]

[1] *Centre for Mathematical Modelling*
Swinburne University of Technology, Melbourne, Australia

[2]*The Weizmann Institute, Rehovot, Israel*

[3]*Tel Aviv University, Tel Aviv, Israel*

ABSTRACT. This introductory paper presents a method for the analysis of differential equations with polynomial coefficients which also provides a further insight into the Stokes Phenomenon. The method consists of a chain of steps based on the concept of the Stokes Structure and Fourier-like transforms adjusted to this Stokes Structure. Although the main object here is Bessel's equation our approach can be extended to more general matrix equations. It will be shown (i) how to derive the Stokes Structure directly from differential equations without any previous knowledge of Bessel or hypergeometric functions, (ii) how to adjust Fourier transforms to the Stokes Structure, (iii) how to answer questions on the interrelation between formal and actual solutions of Bessel's equation using Fourier Analysis, and finally (iv) how to evaluate the coefficients of the Stokes Structure, thus providing a new insight into the Stokes Phenomenon.

1. Introduction

In [4], [5] an approach for the study of a general class of matrix differential equations with polynomial coefficients was presented. However, this study does not cover many equations which require special attention. One such case is the classical Bessel's equation. It was explained in [3] how to derive properties of solutions of Bessel's equation from the Fourier-dual hypergeometric equations. In particular, it was shown how the monodromic properties of hypergeometric functions are transfered to solutions of Bessel's equation as algebraic relations.

J.S. Byrnes (ed.), Twentieth Century Harmonic Analysis - A Celebration, 387–401.

The Hankel functions $H_\nu^{(1)}(z)$ and $H_\nu^{(2)}(z)$ of order ν (or Bessel functions of the third kind) are unique solutions of Bessel's equation

$$(1.1) \qquad y'' + \frac{1}{z}y' + \left(1 - \frac{\nu^2}{z^2}\right)y = 0$$

satisfying the Hankel inequalities (or expansions)

$$(1.2) \qquad H_\nu^{(1)}(z) = \left(\frac{2}{\pi z}\right)^{1/2} e^{i(z-\nu\pi/2-\pi/4)}(1 + o(1))$$

$$(1.3) \qquad H_\nu^{(2)}(z) = \left(\frac{2}{\pi z}\right)^{1/2} e^{-i(z-\nu\pi/2-\pi/4)}(1 + o(1))$$

as $z \to +\infty$. They can be continued analytically as single-valued functions to the whole Riemann surface of $\log z : 0 < |z| < \infty$, $-\infty < \arg z < +\infty$.

The functions $P_1(z)$, $P_2(z)$ defined by

$$(1.4) \qquad H_\nu^{(1)}(z) \equiv \left(\frac{2}{\pi z}\right)^{1/2} e^{i(z-\nu\pi/2-\pi/4)} P_1(z)$$

$$(1.5) \qquad H_\nu^{(2)}(z) \equiv \left(\frac{2}{\pi z}\right)^{1/2} e^{-i(z-\nu\pi/2-\pi/4)} P_2(z)$$

are known as the *phase amplitudes* of the Hankel functions $H_\nu^{(1)}(z)$, $H_\nu^{(2)}(z)$ respectively. It follows that

$$(1.6) \qquad P_1(z) = 1 + o(1), \quad P_2(z) = 1 + o(1)$$

as $z \to +\infty$. They can also be extended as analytic single-valued functions to the whole Riemann surface of $\log z$. Moreover, they satisfy respectively the following pair of differential equations

$$(1.7) \qquad \mathcal{L}_1 P_1(z) = 0, \; \mathcal{L}_2 P_2(z) = 0$$

with the pair of differential operators $\mathcal{L}_1$, $\mathcal{L}_2$ defined by

$$(1.8) \qquad \mathcal{L}_1 = z^2 D_z^2 + 2iz^2 D_z - b$$

and

$$(1.9) \qquad \mathcal{L}_2 = z^2 D_z^2 - 2iz^2 D_z - b$$

where $b = \nu^2 - 1/4$ and $D_z \overset{\text{def}}{=} \frac{d}{dz}$.

On the other hand there exists a unique pair of factorially divergent power series

(1.10)
$$\hat{P}_1(z) = 1 + \sum_{m=1}^{\infty} \frac{a_{1,m}}{z^m}, \ \hat{P}_2(z) = 1 + \sum_{m=1}^{\infty} \frac{a_{2,m}}{z^m}$$

formally satisfying equations (1.7) respectively.

It is natural to introduce the Fourier-dual operators $\mathcal{L}_1^*$, $\mathcal{L}_2^*$ to $\mathcal{L}_1$, $\mathcal{L}_2$

(1.11)
$$\mathcal{L}_1^* \stackrel{\text{def}}{=} \xi\,(\xi - 2i)\,D_\xi^2 + 2\,(\xi - i)\,D_\xi - b$$

(1.12)
$$\mathcal{L}_2^* \stackrel{\text{def}}{=} \xi\,(\xi + 2i)\,D_\xi^2 - 2\,(\xi + i)\,D_\xi - b$$

where $D_\xi \stackrel{\text{def}}{=} \frac{d}{d\xi}$.

There exists a unique pair $F_1(\xi)$, $F_2(\xi)$ of solutions of $\mathcal{L}_1^* F_1(\xi) = 0$, $\mathcal{L}_2^* F_2(\xi) = 0$ respectively, analytic at the singular point $\xi = 0$. This pair is nothing but the pair of Gauss hypergeometric functions

(1.13)
$$F_1(\xi) = F\left(\frac{1}{2} - \nu, \frac{1}{2} + \nu, 1, \xi/2i\right)$$

(1.14)
$$F_2(\xi) = F\left(\frac{1}{2} - \nu, \frac{1}{2} + \nu, 1, -\xi/2i\right).$$

It is not difficult to check that the formal power series $\hat{P}_1(z)$, $\hat{P}_2(z)$ can be represented respectively as formal Laplace transforms of the formal hypergeometric series

(1.15)
$$\sum_{m=0}^{\infty} \frac{(\frac{1}{2} + \nu)_m (\frac{1}{2} - \nu)_m}{(2i)^m (1)_m m!} \xi^m$$

(1.16)
$$\sum_{m=0}^{\infty} (-1)^m \frac{(\frac{1}{2} + \nu)_m (\frac{1}{2} - \nu)_m}{(2i)^m (1)_m m!} \xi^m$$

where

(1.17)
$$(a)_m \stackrel{\text{def}}{=} a(a+1)\ldots(a+m-1) = \frac{\Gamma(a+m)}{\Gamma(a)},$$

while the phase amplitudes $P_1(z)$, $P_2(z)$ can be represented as classical Laplace transforms of (1.13), (1.14) respectively, see [3]. In other words, these formal series and the phase amplitudes are generated in the same manner by different branches of the same hypergeometric function.

Moreover, using (1.4), (1.5) we obtain the following integral representations of Hankel functions

$$(1.18) \qquad H_\nu^{(1)}(z) = \left(\frac{2z}{\pi}\right)^{1/2} e^{i\left(z - \frac{1}{2}\nu\pi - \frac{1}{4}\pi\right)} \int_0^{+\infty} e^{-z\xi} F\left(\frac{1}{2} - \nu, \frac{1}{2} + \nu, 1, \frac{\xi}{2i}\right) d\xi$$

$$(1.19) \qquad H_\nu^{(2)}(z) = \left(\frac{2z}{\pi}\right)^{1/2} e^{-i\left(z - \frac{1}{2}\nu\pi - \frac{1}{4}\pi\right)} \int_0^{+\infty} e^{-z\xi} F\left(\frac{1}{2} - \nu, \frac{1}{2} + \nu, 1, -\frac{\xi}{2i}\right) d\xi,$$

which, upon using the monodromic properties of hypergeometric functions, yield the following monodromic relation, see [3]

$$(1.20) \qquad \begin{pmatrix} 1 & 0 \\ -T_2 e^{2iz} & 1 \end{pmatrix} \begin{pmatrix} P_1\left(ze^{2\pi i}\right) \\ P_2\left(ze^{2\pi i}\right) \end{pmatrix} = \begin{pmatrix} 1 & T_1 e^{-2iz} \\ 0 & 1 \end{pmatrix} \begin{pmatrix} P_1(z) \\ P_2(z) \end{pmatrix}$$

where T_1, T_2 are complex constants.

This relation suggests an algebraic structure for the phase amplitudes $P_1(z)$, $P_2(z)$ on the Riemann surface of $\log z$, which will form the basis of our present investigation. The principal idea of this paper is to apply Fourier transforms to this algebraic structure rather than to the original differential equation. It should be noted in fact that our approach presented in Sections 2, 6, 7, 8, 9, 10 to follow does not depend on the original differential equation.

2. The Stokes Structure $\mathfrak{S}$

DEFINITION   *A pair of functions* $P_1(z)$, $P_2(z)$

(i) *analytic on the Riemann surface of* $\log z$ *with at most exponential growth at* $z = \infty$ *in every sector* $S_{\alpha,\beta} = \{z : -\infty < \alpha < \arg z < \beta < +\infty\}$

(ii) *satisfying inequalities*

$$(2.1) \qquad P_1(z) = 1 + o(1), \ z \to \infty, \ z \in S_c(1)$$

$$(2.2) \qquad P_2(z) = 1 + o(1), \ z \to \infty, \ z \in S_c(2)$$

in the closed subsectors

$$(2.3) \qquad S_c(1) \subset S(1) \stackrel{\text{def}}{=} \{z : -\pi < \arg z < 2\pi, \ 0 < |z| < \infty\}$$

$$(2.4) \qquad S_c(2) \subset S(2) \stackrel{\text{def}}{=} \{z : -2\pi < \arg z < \pi, \ 0 < |z| < \infty\}$$

(iii) *satisfying the monodromic relation (1.20) with complex constants* T_1, T_2 *written as*

$$(2.5) \qquad P_1(ze^{2\pi i}) = P_1(z) + T_1 P_2(z) e^{-2iz}$$

$$(2.6) \qquad P_2(ze^{2\pi i}) = P_2(z) + T_2 P_1(ze^{2\pi i}) e^{2iz}$$

are the elements of the Stokes Structure

(2.7) $$\mathfrak{S} = \{P_1(z), P_2(z)\}.$$

3. From Differential Equation to $\mathfrak{S}$

This technique does not require any previous knowledge or properties of the solutions of (1.1) nor of the hypergeometric functions.

DEFINITION The rays

(3.1) $$l = \{z : \ Re(iz) = 0\}$$

are called *separation rays* for (1.1).

Let us look for solutions y_1, y_2 of (1.1)

(3.2) $$y_1(z) = \left(\frac{2}{\pi z}\right)^{1/2} e^{i(z - \nu\pi/2 - \pi/4)} P_1(z)$$

(3.3) $$y_2(z) = \left(\frac{2}{\pi z}\right)^{1/2} e^{-i(z - \nu\pi/2 - \pi/4)} P_2(z)$$

such that

(3.4) $$P_1(z) = 1 + o(1), \ P_2(z) = 1 + o(1)$$

as $z \to \infty$ along a separation ray l on the Riemann surface of $\log z$.

In terms of $P_1(z)$, $P_2(z)$ the differential equations (1.7) together with conditions (3.4) can be equivalently rewritten respectively as

(3.5) $$P_1(z) = 1 - \frac{b}{2i} \int_z^{\infty_l} \frac{P_1(w)}{w^2} dw + \frac{b}{2i} \int_0^{\infty_l} e^{2iw} \frac{P_1(w + z)}{(w + z)^2} dw$$

(3.6) $$P_2(z) = 1 + \frac{b}{2i} \int_z^{\infty_l} \frac{P_2(w)}{w^2} dw - \frac{b}{2i} \int_0^{\infty_l} e^{-2iw} \frac{P_2(w + z)}{(w + z)^2} dw$$

with $\infty_l = \infty \cdot e^{i \arg l}$.

The integral equations (3.5), (3.6) can be analyzed using successive iterations to construct the unique solutions $P_1(z)$, $P_2(z)$ satisfying inequalities (3.4) respectively, see, for example [2]. Further analysis of these integral equations for a specially chosen l shows that $P_1(z)$, $P_2(z)$ form in fact the Stokes Structure $\mathfrak{S}$ defined above by (2.7).

4. Formal and Actual Solutions

Choosing the separation ray $\arg z = 0$ as the paths of integration in (3.5), (3.6) respectively to construct the solutions $P_1(z)$, $P_2(z)$ and using the uniqueness of this pair and that of $H_\nu^{(1)}(z)$, $H_\nu^{(2)}(z)$ also yield

$$(4.1) \qquad H_\nu^{(1)}(z) \equiv y_1(z) = \left(\frac{2}{\pi z}\right)^{1/2} e^{i(z-\nu\pi/2-\pi/4)} P_1(z)$$

$$(4.2) \qquad H_\nu^{(2)}(z) \equiv y_2(z) = \left(\frac{2}{\pi z}\right)^{1/2} e^{-i(z-\nu\pi/2-\pi/4)} P_2(z)$$

which are identical to (1.4), (1.5). Thus the solutions of (3.5), (3.6) for this chosen separation ray are nothing but the *phase amplitudes* $P_1(z)$, $P_2(z)$ of the Hankel functions $H_\nu^{(1)}(z)$, $H_\nu^{(2)}(z)$ respectively.

Another pair of linearly independent solutions of (1.1) is

$$(4.3) \qquad \hat{y}_1(z) = \left(\frac{2}{\pi z}\right)^{1/2} e^{i(z-\nu\pi/2-\pi/4)} \hat{P}_1(z)$$

$$(4.4) \qquad \hat{y}_2(z) = \left(\frac{2}{\pi z}\right)^{1/2} e^{-i(z-\nu\pi/2-\pi/4)} \hat{P}_2(z)$$

where

$$(4.5) \qquad \hat{P}_1(z) = \sum_{m=0}^{\infty} \frac{(\frac{1}{2}+\nu)_m(\frac{1}{2}-\nu)_m}{(2i)^m(1)_m} \frac{1}{z^m}$$

$$(4.6) \qquad \hat{P}_2(z) = \sum_{m=0}^{\infty}(-1)^m \frac{(\frac{1}{2}+\nu)_m(\frac{1}{2}-\nu)_m}{(2i)^m(1)_m} \frac{1}{z^m}.$$

Formal substitution of these solutions into (1.1) yield, after canceling the exponentials, power series in z^{-1} with zero coefficients. However, the above power series are clearly factorially divergent for any z if ν is not a half integer. Thus, these solutions can be regarded as *formal solutions* as opposed to *actual solutions*.

Three natural questions arise immediately:

(1) how to relate the pair of formal solutions one to another,

(2) how to relate the pair of formal solutions to actual solutions $H_\nu^{(1)}(z)$, $H_\nu^{(2)}(z)$,

(3) how to decode properly the symbol $o(1)$ in the expansions above.

Stokes (1857) was the first one to formulate and answer the first two questions for Airy's differential equation $y'' - zy = 0$ related to Bessel's equation for $\nu = \frac{1}{3}$. To answer the third question, Poincaré (1886) considered formal solutions as asymptotic representations of actual solutions. However, as discovered a century later, see [1], this approach is not satisfactory since it does not answer question (1) altogether, only answers partially question (2), and does not provide sufficient information about the remainder.

5. The Stokes Phenomenon

Using (4.1), (4.2) the monodromic relations (2.5), (2.6) can be rewritten in terms of $H_\nu^{(1)}(z), H_\nu^{(2)}(z)$ as

$$(5.1) \qquad H_\nu^{(1)}\left(ze^{2\pi i}\right) = -H_\nu^{(1)}(z) + ie^{-i\nu\pi}T_1 H_\nu^{(2)}(z)$$

$$(5.2) \qquad H_\nu^{(2)}\left(ze^{2\pi i}\right) = -H_\nu^{(2)}(z) + ie^{i\nu\pi}T_2 H_\nu^{(1)}\left(ze^{2\pi i}\right).$$

These, in turn, yield extended Hankel expansions valid outside the sectors in (2.3), (2.4).

All these Hankel expansions are of the form

$$(5.3) \qquad z^{-1/2}\left(A(\nu)e^{iz} + B(\nu)e^{-iz}\right).$$

Again, Stokes (1857) was the first to discover that the constants $A(\nu)$ and $B(\nu)$ are discontinuous as arg z changes continuously when crossing the separation rays. The existence of such discontinuities is called the *Stokes Phenomenon* and the corresponding values of the jumps in $A(\nu)$, $B(\nu)$ can be expressed in terms of *connection coefficients* T_1, T_2 very important in many applications. A modern insight into the Stokes Phenomenon can be found in [1].

A fourth question then arises:

(4) how to evaluate the connection coefficients T_1, T_2.

Starting with the Stokes Structure we will present a technique that answers questions (1)-(3). The culmination of our approach will be to answer question (4), obtaining explicit expressions for the connection coefficients independently of any knowledge of the actual solutions of the differential equation.

6. Fourier-Like Transforms Adjusted to $\mathfrak{S}$

Let $P_1(z)$, $P_2(z)$ be elements of $\mathfrak{S}$ with (unknown) T_1, T_2 in its monodromic relations (2.5), (2.6) and $S_c(1) \subset S(1)$, $S_c(2) \subset S(2)$ a pair of closed subsectors with angles greater than π.

Let

$$(6.1) \qquad H(z) = a_0 z\left(1 + o(1)\right), \ z \to \infty$$

be analytic on the Riemann surface of log z, and

$$(6.2) \qquad C(1/z) = c_0 + c_1 z + \dots$$

an entire function with complex $a_0 \neq 0$; $c_0, c_1, \dots$.

We define the general Fourier-like transforms of $P_j(z)$, $j = 1, 2$ as

$$(6.3) \qquad F_j(\xi) \overset{\text{def}}{=} \frac{1}{2\pi i} \int_{\gamma(j)} e^{H(z\xi)} C(z) P_j(z)\, dz/z, \ j = 1, 2$$

with paths of integration $\gamma(1)$, $\gamma(2)$ as boundaries of $S_c(1)$, $S_c(2)$ respectively, oriented so that $S_c(j)$ are to the right of $\gamma(j)$.

7. Main Result

THEOREM 1. *Let $P_1(z)$, $P_2(z)$ be the elements of the Stokes Structure $\mathfrak{G}$.*
Then for each $j = 1, 2$ in the dual complex ξ-plane
  (i) *there exists a ray l_j emanating from the origin such that $F_j(\xi)$ is continuous for $\xi \in l_j$, and $F_j(\xi)$ can be continued analytically to some open sector containing the ray l_j;*
 (ii) *there exists a neighborhood of the origin such that $F_j(\xi)$ can be further continued analytically to this neighborhood as a single-valued function;*
(iii) *moreover, $F_j(\xi)$ can be continued analytically to the whole ξ-plane along every path not crossing the point*

$$(7.1) \qquad \xi_0 \equiv \xi_{0,j} = \frac{2i}{a_0} (-1)^{j-1}.$$

8. From $\mathfrak{G}$ to Formal Power Series

Consider the special cases of Fourier-like transforms (6.3) for $H(z) = z$, $C(z) = 1$. These are nothing but the Borel transforms of $P_j(z)$

$$(8.1) \qquad F_j^{(0)}(\xi) \overset{\text{def}}{=} \frac{1}{2\pi i} \int_{\gamma(j)} e^{z\xi} P_j(z)\, dz/z, \ j = 1, 2.$$

Their inversion formulae are nothing but the Laplace transforms of $F_j^{(0)}(\xi)$

$$(8.2) \qquad P_j(z) = z \int_{l_j} e^{-z\xi} F_j^{(0)}(\xi)\, d\xi, \ j = 1, 2.$$

Due to (i), (ii) of Theorem 1 the integrals (8.1) are absolutely convergent for $\xi \in l_j$ and $F_j^{(0)}(\xi)$ can be represented by their Taylor series, which can be regarded as formal power series in ξ

$$(8.3) \qquad \hat{F}_j^{(0)}(\xi) \overset{\text{def}}{=} \sum_{k=0}^{\infty} f_{j,k}^{(0)} \xi^k.$$

Substituting $\hat{F}_j^{(0)}(\xi)$ for $F_j^{(0)}(\xi)$ in (8.2) and writing

$$(8.4) \qquad a_{j,k} \overset{\text{def}}{=} k!\, f_{j,k}^{(0)}$$

yield

$$(8.5) \qquad \hat{P}_j^{(0)}(z) \overset{\text{fps}}{=} z \int_{l_j} e^{-z\xi} \hat{F}_j^{(0)}(\xi)\, d\xi = \sum_{k=0}^{\infty} \frac{a_{j,k}}{z^k}, \quad j = 1, 2.$$

The symbol fps means that (8.5) should be perceived on the level of formal power series.

9. Formal Series as Strong Expansions

Although for an element P_j of the Stokes Structure (2.7)

$$(9.1) \qquad \lim_{\substack{z \to \infty\, z \in S_c(j)}} P_j(z) = 1,$$

it is not at all obvious that the Stokes Structure guarantees the next limits

$$(9.2) \qquad \lim_{\substack{z \to \infty\, z \in S_c(j)}} (P_j(z) - 1)z.$$

However, the formal series

$$(9.3) \qquad \sum_{k=0}^{\infty} \frac{a_{1,k}}{z^k}, \sum_{k=0}^{\infty} \frac{a_{2,k}}{z^k}$$

are Poincaré asymptotic expansions for $P_1(z)$, $P_2(z)$ in sectors $S(1)$, $S(2)$ respectively. This means that for any subsector $S_c(j)$ of $S(j)$ and for $z \in S_c(j)$ there exists $M_N > 0$ such that the following estimates are valid for $N = 1, 2, \ldots$

$$(9.4) \qquad \left| P_j(z) - \sum_{k=0}^{N-1} \frac{a_{j,k}}{z^k} \right| < \frac{M_N}{|z|^N}.$$

It should be noted, however, that these approximations are too rough to provide real information about the behavior of the remainders

$$(9.5) \qquad P_j(z) - \sum_{k=0}^{N-1} \frac{a_{j,k}}{z^k}$$

since we don't know how M depends on N.

In fact the formal series (9.3) are much better and more precise asymptotic expansions for $P_j(z)$ than the Poincaré expansions.

THEOREM 2.   For any subsector $S_c(j)$ of $S(j)$ and for $z \in S_c(j)$ there exists $a > 0$ depending only on $S_c(j)$ such that the following estimates are valid for $N = 1, 2, \ldots$

$$(9.6) \qquad \left| P_j(z) - \sum_{k=0}^{N-1} \frac{a_{j,k}}{z^k} \right| < \frac{M a^N N!}{|z|^N}.$$

These expansions are known as *strong asymptotic expansions*, see [7], [6]. In contrast to Poincaré expansions they have the following uniqueness property:

WATSON'STHEOREM. Watson's Theorem *If $P_1(z)$, $P_2(z)$ are analytic functions in a sector S with its angle not less than π, and $\sum_{k=0}^{\infty} \frac{a_k}{z^k}$ is their strong asymptotic expansion in S, then $P_1(z) \equiv P_2(z)$.*

The inequalities (9.6) answer our question (3).

10. Power Series Representation of $F_j(\xi)$

Now that we have $\sum_{k=0}^{\infty} \frac{a_{j,k}}{z^k}$ it is natural to formally substitute these for $P_j(z)$ into the general Fourier-like transforms (6.3) to yield the formal Fourier-like transforms

$$(10.1) \qquad \hat{F}_j(\xi) \stackrel{\text{def}}{=} \frac{1}{2\pi i} \int_{\gamma(j)} e^{H(z\xi)} C(z) \hat{P}_j^{(0)}(z)\, dz/z$$

resulting in the power series in ξ

$$(10.2) \qquad \hat{F}_j(\xi) = \sum_{k=0}^{\infty} f_{j,k}\xi^k \equiv \sum_{k=0}^{\infty} \xi^k s_k \left(\sum_{m=0}^{k} a_{j,m} c_{k-m} \right)$$

with

$$(10.3) \qquad s_k = \frac{1}{2\pi i} \int_{\gamma^*(j)} e^{H(z)} \frac{1}{z^{k+1}} dz, \; k = 0, 1, \ldots$$

and

$$(10.4) \qquad \gamma^*(j) = \gamma(j)\, e^{i \arg \xi}.$$

THEOREM 3. *The power series $\hat{F}_j(\xi)$ are absolutely convergent and thus represent the analytic functions*

$$(10.5) \qquad \widetilde{F}_j(\xi) = \sum_{k=0}^{\infty} \xi^k \left(s_k \left(\sum_{m=0}^{k} a_{j,m} c_{k-m} \right) \right)$$

inside the circle of radius $\frac{2}{|a_0|}$ with its center at $\xi = 0$. Moreover, if $\xi \in l_j$ and $|\xi| < \frac{2}{|a_0|}$ then

$$(10.6) \qquad \widetilde{F}_j(\xi) \equiv F_j(\xi).$$

Thus, the Fourier-like transforms $F_j(\xi)$ can be represented both by the integral transforms (6.3) and by the convergent Taylor series (10.5) for $\xi \in l_j$, $|\xi| < |\xi_0|$, where l_j and ξ_0 are defined in Theorem 1 (i) and (iii), (7.1) respectively.

11. Evaluation of Borel Transforms

Now let us return to Bessel's equation (1.1) and remember that the elements $P_1(z)$, $P_2(z)$ of the Stokes Structure (2.7) are the phase amplitudes of the Hankel functions $H_\nu^{(1)}(z)$, $H_\nu^{(2)}(z)$.

It follows from Theorem 2 that in particular the formal series (9.3) are Poincaré asymptotic expansions of $P_1(z)$, $P_2(z)$. On the other hand, one can derive from integral equations (3.5), (3.6) that the formal power series (4.5), (4.6) are also Poincaré asymptotic expansions of $P_1(z)$, $P_2(z)$. It should be noted, however, that it is a hard problem to derive from integral equations (3.5), (3.6) that the formal power series (4.5), (4.6) are strong asymptotic expansions for $P_1(z)$, $P_2(z)$.

The uniqueness property of Poincaré asymptotic expansions yields

$$(11.1) \qquad a_{1,k} = \frac{(\frac{1}{2}+\nu)_k(\frac{1}{2}-\nu)_k}{(2i)^k(1)_k}$$

$$(11.2) \qquad a_{2,k} = (-1)^k \frac{(\frac{1}{2}+\nu)_k(\frac{1}{2}-\nu)_k}{(2i)^k(1)_k}$$

that is

$$(11.3) \qquad \hat{P}_j(z) \equiv \hat{P}_j^{(0)}(z)$$

and the left-, right-hand sides of (11.3) are defined by (8.5) and by (4.5), (4.6) respectively.

It is worth noting that (11.3) is in fact the converse of an important principle that was named in [3] as the *Principle of Functional Closure: If a formal series satisfying a differential-difference-algebraic relation can be summed to an analytic function in a region of the complex plane, then this function satisfies exactly the same relation in this region.*

It follows from (8.4) and (8.3) that

$$(11.4) \qquad \hat{F}_j^{(0)}(\xi) = \hat{F}\left(\frac{1}{2}+\nu, \frac{1}{2}-\nu, 1, \pm\frac{\xi}{2i}\right)$$

where $\hat{F}\left(\frac{1}{2}+\nu, \frac{1}{2}-\nu, 1, \pm\frac{\xi}{2i}\right)$ are power series expansions in ξ of Gauss' hypergeometric function $F\left(\frac{1}{2}+\nu, \frac{1}{2}-\nu, 1, \pm\frac{\xi}{2i}\right)$, respectively.

12. Interrelation between Solutions

It follows from (11.4) and (8.1) that the Borel transforms of the phase amplitudes of the Hankel functions are in fact the hypergeometric functions, while the formal Borel transforms of the formal power series are the corresponding (formal) hypergeometric series.

The following relations are valid

(12.1)
$$F\left(\frac{1}{2}+\nu,\frac{1}{2}-\nu,1,\frac{\xi}{2i}\right) = \frac{1}{2\pi i}\int_{\gamma(1)} e^{\xi z} P_1(z)\, dz/z$$

(12.2)
$$F\left(\frac{1}{2}+\nu,\frac{1}{2}-\nu,1,-\frac{\xi}{2i}\right) = \frac{1}{2\pi i}\int_{\gamma(2)} e^{\xi z} P_2(z)\, dz/z$$

(12.3)
$$\hat{F}\left(\frac{1}{2}+\nu,\frac{1}{2}-\nu,1,\frac{\xi}{2i}\right) \overset{\text{def}}{=} \frac{1}{2\pi i}\int_{\gamma(1)} e^{\xi z} \hat{P}_1(z)\, dz/z$$

(12.4)
$$\hat{F}\left(\frac{1}{2}+\nu,\frac{1}{2}-\nu,1,-\frac{\xi}{2i}\right) \overset{\text{def}}{=} \frac{1}{2\pi i}\int_{\gamma(2)} e^{\xi z} \hat{P}_2(z)\, dz/z.$$

The representations (12.1)-(12.4) together with their respective inversion formulae

(12.5)
$$P_1(z) = z\int_0^\infty e^{-z\xi} F\left(\frac{1}{2}+\nu,\frac{1}{2}-\nu,1,\frac{\xi}{2i}\right) d\xi$$

(12.6)
$$P_2(z) = z\int_0^\infty e^{-z\xi} F\left(\frac{1}{2}+\nu,\frac{1}{2}-\nu,1,-\frac{\xi}{2i}\right) d\xi$$

(12.7)
$$\hat{P}_1(z) \overset{\text{def}}{=} z\int_0^\infty e^{-z\xi} \hat{F}\left(\frac{1}{2}+\nu,\frac{1}{2}-\nu,1,\frac{\xi}{2i}\right) d\xi$$

(12.8)
$$\hat{P}_2(z) \overset{\text{def}}{=} z\int_0^\infty e^{-z\xi} \hat{F}\left(\frac{1}{2}+\nu,\frac{1}{2}-\nu,1,-\frac{\xi}{2i}\right) d\xi$$

reveal the following one-to-one correspondences (denoted by the symbol $\leftrightarrow$) below

(12.9)
$$\hat{P}_j(z) \leftrightarrow \hat{F}_j(\xi) \equiv \hat{F}\left(\tfrac{1}{2}+\nu,\tfrac{1}{2}-\nu,1,\pm\tfrac{\xi}{2i}\right) \leftrightarrow$$
$$\leftrightarrow F\left(\tfrac{1}{2}+\nu,\tfrac{1}{2}-\nu,1,\pm\tfrac{\xi}{2i}\right) \leftrightarrow P_j(z).$$

These interrelations show that both formal series $\hat{P}_1(z)$, $\hat{P}_2(z)$ and actual functions $P_1(z)$, $P_2(z)$, are generated in the same manner by different branches of the same hypergeometric function, thus answering questions (1) and (2).

REMARK

Formulae (12.5), (12.6) together with (4.1), (4.2) yield again the integral representations (1.18), (1.19) for Hankel Functions. It is curious that we could not find these representations, the most principal in our context, in the classical literature on Bessel functions. In the literature, the Hankel expansions are commonly derived from the representations

(12.10)
$$H_\nu^{(1)}(z) = \frac{\Gamma\left(\frac{1}{2}-\nu\right)\left(\frac{z}{2}\right)^\nu}{\pi^{3/2}i}\int_{\gamma_1} e^{izt}\left(t^2-1\right)^{\nu-\frac{1}{2}} dt$$

$$(12.11) \qquad H_\nu^{(2)}(z) = \frac{\Gamma\left(\frac{1}{2} - \nu\right)\left(\frac{z}{2}\right)^\nu}{\pi^{3/2} i} \int_{\gamma_2} e^{-izt} \left(t^2 - 1\right)^{\nu - \frac{1}{2}} dt$$

with γ_1, γ_2 simple loops bypassing $t = \pm 1$ but not enclosing $t = \mp 1$, respectively, $|\arg z| < \frac{\pi}{2}$, and $\nu \ne \frac{1}{2}, \frac{3}{2}, \ldots$.

These are derived by reducing Bessel's equation to

$$(12.12) \qquad zw'' + (2\nu + 1)\, w' + zw = 0$$

for the variable $w = z^{-\nu} y$, and then applying the Laplace transform to this special equation with linear coefficients. Unfortunately, this approach is generally not possible for other differential equations.

13. The Connection Coefficients

Consider the Fourier-like transforms $F_1(\xi)$, $F_2(\xi)$ of $P_1(z)$, $P_2(z)$ defined by (6.3) with

$$(13.1) \qquad H(z) = 2iz, \ C(z) = -\frac{b\pi}{z}, \ b = \nu^2 - \frac{1}{4}.$$

$$(13.2) \qquad F_j(\xi) \overset{\text{def}}{=} -\frac{b}{2i} \int_{\gamma(j)} e^{2i\xi z} \frac{1}{z} P_j(z)\, dz/z, \ j = 1, 2.$$

THEOREM 4.   Let $P_1(z)$, $P_2(z)$ be the phase amplitudes of $H_\nu^{(1)}(z)$, $H_\nu^{(2)}(z)$ with Fourier-like transforms $F_1(\xi)$, $F_2(\xi)$ defined by (13.2). Then

(i) the only finite singular point of both analytic functions $F_1(\xi)$, $-F_2(-\xi)$ is $\xi = 1$

(ii) the limiting values of $F_1(\xi)$, $-F_2(-\xi)$ at $\xi = 1$ exist and are equal to connection coefficients

$$(13.3) \qquad \lim_{\xi \to 1} F_1(\xi) = T_1, \ \lim_{\xi \to 1}\left(-F_2(-\xi)\right) = T_2$$

(iii)

$$(13.4) \qquad T_j = (-1)^j \frac{b}{2i} \int_{\gamma^*(j)} e^{(-1)^{j-1} 2iz} \frac{1}{z} P_j(z)\, dz/z, \ j = 1, 2$$

where $\gamma^*(j)$ are obtained by rotating $\gamma(j)$ into positions where functions $e^{(-1)^{j-1} 2iz}$ are decreasing for $z \in \gamma^*(j)$, $j = 1, 2$ respectively.

(iv) Moreover, let $f_{j,k}$ be coefficients of power series in (10.2) for $H(z)$ and $C(z)$ given by (13.1). Then

$$(13.5) \qquad T_1 = \lim_{\xi \to 1 - 0}\left(\sum_{k=0}^{\infty} f_{1,k} \xi^k\right)$$

$$(13.6) \qquad T_2 = \lim_{\xi \to 1-0} \left(\sum_{k=0}^{\infty} (-1)^{k+1} f_{2,k} \xi^k \right).$$

It follows from Theorems 1 and 2 that for $|\xi| < 1$

$$(13.7) \qquad \begin{aligned} F_1(\xi) &= -2\pi bi \sum_{m=0}^{\infty} \frac{1}{(m+1)!} \frac{\left(\frac{1}{2} - \nu\right)_m \left(\frac{1}{2} + \nu\right)_m}{m!} \xi^m \\ F_2(\xi) &= 2\pi bi \sum_{m=0}^{\infty} \frac{(-1)^k}{(m+1)!} \frac{\left(\frac{1}{2} - \nu\right)_m \left(\frac{1}{2} + \nu\right)_m}{m!} \xi^m \end{aligned}$$

hence

$$(13.8) \qquad F_1(\xi) = -F_2(-\xi) = -2\pi ibF\left(\frac{1}{2} - \nu, \frac{1}{2} + \nu, 2, \xi\right).$$

Substituting $\xi = 1$ yields

$$(13.9) \qquad T_j = -2\pi biF\left(\frac{1}{2} - \nu, \frac{1}{2} + \nu, 2, 1\right), \, j = 1, 2$$

which, using Gauss' formula, reduces to

$$(13.10) \qquad T_j = \frac{-2\pi bi}{\Gamma\left(1 + \frac{1}{2} + \nu\right)\Gamma\left(1 + \frac{1}{2} - \nu\right)}, \, j = 1, 2,$$

and finally

$$(13.11) \qquad T_j = 2i\cos\pi\nu, \, j = 1, 2.$$

It is worth noting that generally it is impossible to express T_j in terms of known fuctions.

Their integral representation should be used to evaluate them asymptotically for extremal values of parameters of the differential equation.

Their Taylor series representation should be used for their numerical evaluation.

14. Conclusions

We have shown that the Stokes Structure $\mathfrak{S}$ is of fundamental importance. Starting with Bessel's equation (1.1) we derived $\mathfrak{S}$ and introduced and studied Fourier-like transforms adjusted to $\mathfrak{S}$. These yielded formal power series that are in fact formal solutions of (1.7). Furthermore, as shown by (12.9) *the phase amplitudes and their respective formal series are generated in the same manner by different branches of the same hypergeometric function.* These provide the basis for a systematic chain of steps to answer questions (1), (2), (3), (4) above, and an approach which can be extended to matrix equations with many applications.

References

[1] Braaksma B.L.J., G. Immink and M. van der Put, (eds) *The Stokes Phenomenon and Hilbert's 16th Problem*, Singapore: World Scientific, 1996, 326 pp.

[2] Coddington E. A. and N. Levinson. *Theory of ordinary differential equations*. New York: McGraw Hill, 1955, 441p.

[3] Gurarii V. and V. Katsnelson. The Stokes Structure for the Bessel Equation and the Monodromy of the Hypergeometric Equation. Preprint 3/2000, NTZ, Universität Leipzig, Preprint is available from the WEB site http:// www.uni-leipzig.de/˜ntz/prentz.htm

[4] Gurarii V. and V. Matsaev. The generalized Borel transform and Stokes multipliers. In: *Theoretical and Mathematical Physics*, Vol.100, No 2, 173-182 pp., Moscow, 1994.

[5] Gurarii V. and V. Matsaev. The Generalized Borel Transform in Asymptotic Analysis. In: *The Role of Mathematics in Modern Engineering*. (eds. A.K. Easton and J.M. Steiner) 585-597 pp. Sweden, Lund:Chartwell-Bratt, 1996.

[6] Reed M. and B. Simon. *Methods of Modern Mathematical Physics, IV, Analysis of Operators* Academic Press, New York San Francisco London, 1978, 428p.

[7] Watson W. *A Theory of Asymptotic Series*. In: Phil. Trans. Roy. Soc., Ser. A 211, 279-313. London, 1911.

Index

B-functions, 184
G_δ set, 58
G_δ set, 141
M sets, 67
M_α sets, 67
X-dominated, 243
X-spectral hull, 249
ε-compression, 13
a-pair, 8
a-pair, strong, 9
n-resolvent, 242
n-visible, 237
p-oscillation, 133, 134
z-transform, 381
hard summation, 103

A/D, 298
Abiguity Problem, 312, 314
abiguity problem, 312–314
absorbtion condition, 259
admissibility condition, 334–338, 363
Admissible, 365
admissible, 337, 338, 340
Affine Dual, 364
affine dual, 341
Affine Group, 335
affine group, 174, 333, 335, 338
Affine System, 366
affine system, 339–341
Airy's differential equation, 392
Aleksandrov, A.B., 14
almost periodicity, 274, 275
$\{\alpha, \beta\}$function, 224
Amplifiers, 296
amplifiers, 296
amplitude modulation, 298
analytic bootstrapping, 213
Annihilating, 8
annihilating, 8
antenna, 296, 299, 300, 316
antilocality, 24–27
approximation theory, 4, 203
arithmetic, 7
atomic positions, 164

atomicity, 163, 164
attentuation, 299, 325
Auscher, 360
Auslander, 303, 305, 328

Baire, 57–59, 61, 65, 70, 71
Baire's space, 141
Balian-Low Theorem, 179, 180
Balian-Low theorem, 76, 93, 101
band limited, 5, 17, 20, 23
bandwidth, 296, 299, 300, 309, 310, 325, 327
Bari, 15, 28
Barker codes, 320, 322
Barker on Barker, 321, 323
Barker sequences, 226, 320, 321
Bastiaans, 74, 75, 94, 99
beams, 296
Beck, 284, 292
Belov, 15
Benedicks, 13, 14, 23
Bernstein norm estimate, 14
Bernstein's inequality, 202
Bernstein, S.N., 202
Besov, 362
Besov spaces, 127, 128, 130–132, 135, 144
Bessel family, 340
Bessel functions, 145, 146, 148, 149, 388, 398
Bessel's equation, 387, 392, 396, 398, 400
Bessel's inequality, 285
best polynomial approximations, 247
Beurling, 18, 21, 23, 28, 272, 273, 292
Beurling algebras, 247
Beurling, A., 40
Beurling-Malliavin majorant, 20
Beurling-Sobolev algebras, 235, 240
Bezout equations, 235
bi-unimodular, 220
bi-unimodular polynomial, 228
binary function, 223
bistatic, 295
Björck, 219
Björk, J.-E., 241
block matrices, 47
block shift, 38–40, 42, 45

BM-majorant, 20
Bochner's theorem, 31, 37
Bohr compactification, 257
Borden's, 298
Borel, 62, 63
Borel measure, 14
Borel measures, 253
Borel transforms, 394, 397
Borgain, 66, 67, 70
Borichev Theorem, 22
Borwein, P., 217
bounded action, 239
bounded character, 254
bounded support, 5
Bourbaki, 266
Bourgain, 27, 28, 288, 292
Bownik, M., 342
Brennan, J., 22
Brownian motion, 61, 62, 68, 69
Bueckner, 312
butterfly, 151, 152

Calderón, 334
Calderón reproducing formula, 335
Calderon, A., 241
canonical dual system, 76, 78, 80, 82, 99
Cantor, 58, 68
Cantor measure, 15
Cantor subsets, 15
Carleman formula, 18
carrier frequency, 298, 300
carrier signal, 296
Cartwright, 21
Cartwright class, 25
category theorem, 57, 58, 71
Cauchy potential, 17, 22
causal, 6
causality principle, 6
CC-condition, 87
Chebishev, 203
Chen, 283, 292
Cherednik, 145
Chinese remainder theorem, 152
Chirp, 319, 321
chirp, 319, 320, 323

Clairaut, 220
Classical PONS, 326
clutter, 296, 298
Cohen, 196, 199
Cohen, P., 241
Cohen, Paul, 231
Coherent States, 364, 365
coherent states, 338
coin tossing, 107
combinatorial designs, 201
combined spaces, 173, 197
commutant lifting theorem, 31, 33, 52
complementary pairs, 325
completely δ-visible, 238
completely antilocal, 26
completely local, 24, 25
completely visible, 237
Complex Analysis, 14–17
complex analysis, 3–6, 21
complexity, 381
condensation of singularities, 109, 116
Connection Coefficients, 399
connection coefficients, 393, 399
connectivity, 357, 360
constrained global optimization, 163
converters, 298
convolution multiplier norm, 246
Cooley, James, 152
Cooley-Tukey, 151, 152
corona problem, 237
correlation, 301, 317, 320–323
Costas arrays, 322, 323, 327
crest factor, 218
cross ambiguity function, 306
crystal structure, 163–165, 384
Curtis, 190, 198
cyclic difference sets, 201
cyclic groups, 260
cyclotomy, 201

Dai, Xingde, 357
Daubechies, 175, 198
dB plot, 319
De Branges, 23, 25, 26, 28
De Branges Theorem, 25

De Buda, 313
de Jeu, 145, 148, 149
De Leeuw-Katznelson Theorem, 15
decibels, 319
deep zero, 4, 15, 18, 19
demodulated, 300
Denjoy, 57, 58, 70
DFT, 220
dichotomy problem, 66
difference sets, 222
diffraction experiment, 385
diffraction experiments, 384
diffraction maxima, 163
diffraction pattern, 164
dilation, 44
dilation matrix, 339, 362
dilation operator, 329, 346
dilation subgroup, 335
dilation theory, 33, 43, 47, 52–54
dimension function, 356
Direct Methods, 171, 384
direct methods, 164
disc algebra, 381
discrete Fourier transform, 220
discrete logarithmic integral, 5
distance controlled resolvent growth, 242
Distribution, 28
distribution, 3, 4, 6, 16, 19, 22–24
Divergence, 125
divergence, 119, 125
Diximier, 377
Doppler, 298
doppler, 295, 296, 298, 300–302, 305, 309, 310, 315–320, 325, 327
dual frame, 73–75, 78, 83, 84, 179
dual group, 260
Dunkl, 145, 149
duplexer, 296
Dyn'kin, 22
Dyn'kin, E., 241
Dyson, Freeman, 289

EFET, 182–184, 195
efficiency hull, 235
efficient resolvent complement, 249

Ehrling, G., 241
El-Fallah, O., 241
electronic counter-measures, 296
Enflo, Per, 220
entropy, 24, 26
Erdös-Newman conjectures, 231
Erdős, 272
Erdelyi, T., 217
essential spectrum, 8
essential support, 8, 10
Exponential Sum, 276, 292
exponential sum, 271, 276, 278, 281
extremal, 311, 312
extremal problems, 201

F. and M. Riesz Theorem, 14
Fabry Gap Theorem, 282
fast bilateral decay, 4
Fast Fourier transform, 151
Fatou, 57, 71
Fejér, 204
Fejer, 247
Fekete, 204
Fekete polynomials, 230
Filters, 296
filters, 297, 301
finite energy signals, 305
first category, 59
flatness, 218
Foias-Sz.-Nagy result, 52
Fourier, 220
Fourier transform, 4, 9, 29
Fourier-like transform, 387, 393, 394, 396, 399, 400
Fourier-Stieltjes transform, 173, 183, 190
FPGA, 151
fractal dimensions, 127, 128, 132
fractional brownian motions, 136
frame, 78
frame bounds, 178
frame lower bound, 78, 81, 85, 89, 92, 93, 96
frame operator, 75, 76, 78–87, 89–92, 96, 97, 99
frame upper bound, 77, 78, 80, 82–85, 87–94, 96, 97
Frappier, C., 218

Frisch-Parisi formula, 136
full spectral hull, 249
function spaces, 127–129
functional calculus, 33, 38, 46
Fundamental maximal property, 170

Gabor, 74–76, 84, 98–101, 173–177, 179–181, 196–
    199
Gabor frame, 73, 75, 76, 79, 84–86, 88, 89, 92–95,
    97–99
Gabor frame operator, 73, 76, 100
Gabor frame operators, 73
Gabor frames, 173
Gabor system, 73–77, 79, 83, 85, 87–93, 98, 99
Gabor systems, 363
Gabor theory, 73, 75–77, 83
Gallagher, 286, 290, 292
Galois field, 323
Garrigós, G., 360
Gauss, 151, 155, 220
Gaussian process, 299
Gaussian sequence, 220
Gaussian series, 62
Gelfand, 241
Gelfand transform, 237
Gelfand-Naimark theorem, 37
generalized functions, 32
geometric mean, 5
geometric means, 7
GHzs, 298, 301, 319
Golay, 325, 326
Golay sequences, 226
Goldston, 290, 292
Golomb, 323, 327
Gram-Schmidt orthonormalization, 129
Gu, Q., 359

Hölder exponent, 136–138
Haagerup, 222
Haar basis, 129, 130
Haar measure, 260, 334, 335, 338, 340
Haar system, 105, 106, 109, 120, 122, 124
Haar wavelet, 329, 360
Haar, A., 221
Hadamard Circulant Conjecture, 219
Hadamard difference sets, 227

Hadamard factorisation theorem, 313
Hadamard lacunary condition, 65
Hadamard-Menon difference sets, 227
Hadamard-Paley cyclic difference set, 221
Hadamard-Paley difference sets, 227
Halász, 283
Halmos, 39–41, 46, 55
Halton, 283, 292
Hankel expansions, 393, 398
Hankel functions, 388, 389, 392, 396, 397
hard summation, 121
Hardy, 310, 316
Hardy class, 6, 19, 28
Hardy spaces, 32, 41
Hardy-Littlewood maximal operator, 124
Hardy-Littlewood unimodular polynomials, 230
Hardy-Weierstrass function, 60
Harmonic sliding analysis, 375
Hausdorff dimension, 60, 68
Hausdorff dimensions, 127, 128
Hayes, 189, 191, 193–195, 198, 199
Heckman, 145, 149
Heisenberg, 174
Heisenberg group, 173, 303, 305–308, 311, 312
Heisenberg inequality, 3
Heisenberg related, 313, 314
Helson sets, 67, 68
Helson, H., 241
Herglotz-F. Riesz-Toeplitz theorem, 37
Hermite functions, 308–310
HF, 295
Hilbert basis, 304
Hilbert space, 31–33, 35, 36, 38–40, 43–46, 50, 52,
    53
Hilbert transform, 14, 26, 184–186, 192, 301
Hilbert's Inequality, 286–288
histogram, 127, 128, 133, 134, 138–140, 142
Holder singularities, 127
homogeneous Galilei group, 338
horodiscs, 250
Hruščev theorem, 24, 26
HSA, 375
Huffman sequences, 219
hull-kernel topology, 377
hypergeometric, 387, 389–391, 397, 398, 400

hypergeometric series, 145

I and Q channels, 298, 300, 301
IF, 297
Image enhancement, 168
impulse response, 381
induced representations, 307
inhomogeneous Galilei group, 338
inner function, 41
intensities, 163, 164
interpolation formula, 217
intertwining, 306–308
intrinseque, 134, 136, 140
invariant subspace theorem, 33
invariant subspaces, 32, 40, 43, 53–55
invisible spectrum, 235
irreducible representations, 303, 305, 312
isometric lifting, 45
Ivashev-Musatov Theorem, 15

jamming, 296
Jensen inequality, 6, 7
Journée, J-L, 360

Körner, T., 68
Körner, Thomas, 70
Kahane and Katznelson, theorem of, 119
Kahane, J-P, 241
Kahane, J.-P., 207
Kargaev, 21, 23, 28
Kargayev, 23, 25, 27
Katzenelson, 198
Katznelson, 58, 66, 71
Katznelson, Y., 240
Kaufman, Robert, 70
Kazhdan, 377, 378
Khinchin-Ostrowski theorem, 24
Kislyakov, 23
Klemes, Ivo, 231
Kolmogorov, 14, 104, 121, 122
Koornwinder, 145
Koosis, 21, 23, 28
Kronecker sets, 67
Kronecker's Theorem, 274
Kuratowski-Ulam theorem, 59

Lévy flight, 68
Lévy processes, 68
lacunary, 12, 13
lacunary Taylor series, 62
Lagrange interpolation, 229
Laguerre function, 309
Landau, 228
Landau, E., 204
Laplace transform, 18, 19
Laplace transforms, 389, 394
Large Sieve, 285
large sieve, 272, 285–287, 292, 293
LCAO method, 386
Lebesgue, 57, 58, 62, 65, 69, 72
Lebesgue measure, 4, 8, 9, 11, 26
Leblanc, N., 241
left invariant, 303
Legendre symbol, 221
Lemarié, 360
Levin, 18, 23, 28
Levinson, 21
Liang, Rufeng, 357
lift, 43
lifting, 33, 43, 44, 46, 52, 53, 55
lifting problem, 129
Lim, 189–191, 193, 195, 198, 199
line bundle, 307, 308
linear chirp, 319
Linnik, 285, 292, 293
Logan, 184, 186, 187, 190, 193–195, 198
logarithmic integral, 5, 8, 20, 22
Logvinenko, 12, 23
Lorentz, G.G., 205
Lorenz norm, 382
Lovász, 221
low pass filter multiplier, 357
low pass filters, 296
lower box dimension, 134

M. Riesz potential, 26
Möbius, 7
Macdonald, 145
Mackey, 307
Mackey theory, 307
Makarov, N., 24

Malliavin, 66, 69, 71
Mandelbrojt theorem, 15
Mandelbrot, 68, 71, 72
Marcus, 62, 65
Markov's inequality, 202
Markov, A.A, 202
matched filter, 298, 302, 317, 325
matrix-valued functions, 31
Maximal Function, 170
maximal function, 122, 123, 170
maximum likelihood, 301
meager sets, 59
mean ergodic theorem, 33
Menchoff universality property, 65
Mendeleiev, 202
Menon difference sets, 227
Meyer, 360, 365, 366
microlocal critical constant, 249
microlocal upper bound, 249
Minimal Function, 171
minimal function, 166, 168
Minimal Principle, 166
minimal principle, 166
minimal spectral hull, 239
Minimally Supported Frequency, 333
model operator, 252
Modified Bessel Function, 165, 166, 169
modulation, 298
moment problem, 18
Moment Theorem, 54
moment theorem, 31–33, 53
momentum space, 385
monostatic, 295
Morgan, 18
Moyal's identity, 306
MRA, 340, 350, 353–362, 366
MRA wavelet multiplier, 357
MSF wavelets, 333, 355, 360
multifractal analysis, 127, 136
multifractal formalism, 128, 135, 137–141, 143
multiresolution analysis, 350
multistatic, 295

narrow band, 300, 301, 303, 315, 316
Nazarov, 13, 14, 18, 21, 23, 29

Nestoridis universality property, 64
Newman, D.J., 230, 240
Newton convolution, 27
Newton-Kantorovich method, 23
Neymann-Pearson Lemma, 301
nilpotent semigroups, 260
noise process, 299
non-unimodular group, 316
norm-controlled, 249
normalized structure factors, 163–165, 168, 386
Nyquist's theorem, 74

Olevskiĭ, 103, 108
Opdam, 145, 148, 149
Oppenheim, 189–191, 193, 195, 198, 199
opposite phase, 298
optimal detector, 301
oscillation spaces, 132–135
oscillator, 296
Ostrowski, 283, 292
Oversampling, 364, 365
oversampling, 339

Paley, 61–63, 67, 223
Paley-Wiener theorem, 182, 313
Paneah Theorem, 11
parameter-shift method, 167
peak factor, 218
Peetre theorem, 24
Perelomov, 74, 75, 100
perfectly flat, 218
periodic electron density, 384
periodization, 13
permutation matrix, 322, 323
Perrin, Jean, 61
Petrov, A., 241
phase amplitudes, 388–390, 392, 396, 399
phase coded pulses, 298
phase modulation, 298
Phase Problem, 163
phase problem, 163, 164, 166
phase relations, 384
phased array, 296
Phragmen-Lindelöf theorem, 18
Pichorides S. K., 212
Pichorides, S.K., 205

Pick-Nevanlinna theorem, 33
Piecewise Zak Transform, 176
Pigno, 14
Pisier, 62, 65, 66, 72
Pisier algebra, 65
Pitt, R., 236
Plancherel theorem, 330, 331
plus-charge, 14–16
plus-function, 6–8, 16, 17, 20, 27
PN problem, 48
Poggio, 184, 198
Poincaré asymptotic expansions, 395–397
Poisson measure, 5
Poisson summation formula, 152, 273
Poisson-Jensen formula, 19, 20
Pollard, 23
Pollington, 283
polynomial, 203
PONS, 326, 327
PONS sequences, 226
Porat, 196, 198, 199
positive definite, 311, 312
positive definiteness, 32
potential function, 385
potential theory, 6, 23, 24
Poussin C. de La Vallée, 217
Poussin, De la Vallée, 203
power spectral density, 299
Power Sums, 281
power sums, 281, 285
prediction theory, 42, 53, 55
Prime k-tuple Conjecture, 290
Prime Number Theorem, 275, 281
Principle of Functional Closure, 397
probability measure, 6
property (T), 377
pseudo-random codes, 317
pseudofunctions, 65
pseudomeasures, 65
pulse compression waveforms, 320

quadratic non-residues, 221
quadratic residues, 223
Quantum Mechanics, 385
quasi all, 59–61, 64, 65

quasi everywhere, 59
quasi sure, 59, 61, 64
quasi-affine dual, 341
quasi-affine system, 340, 341
quasianalyticity, 4, 6, 18
Quincunx wavelet, 361

Rényi, 285
radar ambiguity function, 302, 327, 328
Radar System, 295, 328
radar system, 295–297, 299, 301
Rademacher Taylor series, 62, 63
radially symmetric, 309
Rahman Q. I., 217
Rajchman, 15
Rajchman measures, 66
Random Series, 71
random series, 62
range estimator, 301
rearrangement, 103, 104, 121
receiver, 295–297, 299, 301, 302, 316
receiver noise, 299
reciprocal lattice vector, 165, 169
reflection principle, 107
reflectivity kernel, 302
Reiter's condition, 377
relatively dense, 11
Requicha, 185, 198, 199
resolution ellipse, 309
resolution factor, 309, 310
resolvable, 309
resolvent complement, 249
restricted ambiguity partners, 314
return signal, 299–302
RF, 296
RF mixers, 296
Riemann, 57, 58, 72
Riemann Hypothesis, 275, 289, 290
Riemann Lebesgue lemma, 123
Riemann-Lebesgue lemma, 256
Riemann-Mellin inversion formula, 19
Riesz basis, 105, 110–113, 115, 117–119
Riesz constant, 105, 114
Riesz, F., 205
Riesz, M., 205

Riesz, Marcel, 215
Riesz-Dunford calculus, 250
Riesz-Dunford functional calculi, 241
right invariant, 303
RISC, 151–153
rolling hump, 109, 116
Ron-Shen theory, 73
Rotem, 185, 186, 198
Roth, 283–286, 292, 293
Rudin, W., 241
Ryser, 219
Rzeszotnik, Z., 342

S-band, 300
Salem, 15, 28
Salem sets, 67
Salem, Raphaël, 230
sample values, 381
Sapogov, 23
Sarason, 31, 48, 50–53, 55, 56
SAS tangent equations, 169, 170
scaling function, 350, 353, 355, 357–360, 362, 364
scaling function multiplier, 357
Schaake-van der Corput inequality, 208
Schauder basis, 129
Schempp, 303, 309, 328
Schmeisser, G., 217
Schmidt, 283, 284, 293
Schmidt, B., 230
Schrödinger representation, 305
Schrodinger equation, 385
Schur, 206
Schur's algorithm, 50
Schur's Lemma, 304
Schur, I., 48
second category, 59
Selberg, 272, 273, 287, 293
self-inversive, 212
sembounded support, 5
semicharacter, 254
semigroup, 254
separable, 332, 355, 364
separation property, 254
separation rays, 391
Sereda, 12

series, divergence of, 103
series, divergence of, 125
series, rearrangement of, 103–125
set of uniqueness, 68
Shake-and-Bake algorithm, 163
Shamai, 194, 195
Shannon wavelet, 329, 333, 359, 360, 362
Shapiro, 325
Shapiro sequences, 226
Shapiro, H.S., 230, 240
Shapiro, J., 14
shift operator, 38, 40, 53
shift-invariant systems, 73, 76, 77, 79, 80, 83, 93
Shimorin, S., 39, 43
Shitz, 190, 191, 199
short-time Fourier transform, 74
sidelobes, 296, 319, 327
Sidon set, 66, 67, 70
signal-to-noise ratio, 296, 301
simple spectrum, 35, 36
Single Wavelength Anomalous Scattering, 168
single wavelength anomalous scattering, 163
Slepian-Pollack inequality, 11
Smith, 14
Sobolev space, 25
Sobolev spaces, 129, 131, 135, 263
Soundararajan, 290, 292
spacious, 23, 24
SPARSE, 12
sparse representations, 131
sparse support, 5, 23, 24
spectra of singularities, 127
spectral gap, 23
spectral hulls, 235
spectral magnitude, 369
spectral sets, 45
spectral synthesis, 69
spectral theorem, 32, 34, 35, 37
spectrally closed, 253
spectrum, 4, 6, 10, 14, 15, 17–20, 296, 300, 315
spectrum of singulgularities, 128, 135–137, 141
Speegle, D., 360
square integrable representation, 306
Sreider measures, 240
Sreider, Yu., 236

Stafney, J., 241
Steinhaus, 58, 62, 72
stepped frequency radars, 322
Stokes Phenomenon, 387, 393
Stokes Structure, 387, 390, 391, 393–396, 400
Stone's theorem, 31
Stone-von Neumann theorem, 305
Strömberg, 130
strong a-pair, 9
strong asymptotic expansions, 395, 397
Strong Pair Correlation Conjecture, 289, 290
strongly admissible, 141
Structure Invariants, 165
structure invariants, 163, 165, 167, 170
sub-representations, 304
subharmonic, 17
subnormal operators, 32
symmetrised form, 308
synthetic aperture radar, 320
Szegö's inequality, 208

tangent formula, 163
Tao, 103
Target, 327
target, 295, 296, 298–302, 305, 309, 315–317, 319, 325
Taylor series, 61–65, 67, 394, 396, 400
Tchebyshev, 203
tempered distribution, 7
tensor orthonormal basis, 311
theorem of Kahane and Katznelson, 119
thin sets, 57, 65, 68, 70, 72
thumbtack, 316, 317, 325
tight frame, 178, 332, 339, 340, 343–345, 349, 364
time dilation, 300
time duration, 309
time-frequency localization, 375, 376
time-invariant linear system, 381
Toeplitz matrix, 34
Toeplitz operator, 50, 51, 53, 54
Tolimieri, 303, 305, 328
Tolimieri-Orr-Janssen representation, 76
totalization, 58
transfer functions, 382
translation operator, 329, 346

translation subgroup, 335
transmitter, 295, 296, 299
Triebel-Lizorkim spaces, 362
triplets, 165, 166
Turán, 272
Turán's First Main Theorem, 281
Turan, 14
Turyn, R., 230
TWIN algorithm, 385

uncertainty inequality, 74
Uncertainty Principle, 3, 386
uniform spectral radius, 241
uniformly distributed, 272, 274, 276, 277
unilateral decay, 21–24
unit cell, 164–166, 168
unital Banach algebra, 243
unitarily equivalent, 35, 36, 38–40, 45, 46
Unitary Dilation, 56
unitary dilation, 31, 33, 44, 45
unitary operator, 304
Unitary Representation, 304
unitary representation, 303, 304, 306, 307
Urieli, 196, 199

Vaaler, 274, 287, 292, 293
van der Corput, 276–279
van der Corput set, 277
Van Hove, 189, 199
Van-der-Corput lemmas, 15
Vaughan, 287, 291, 292
VHF, 295
Vinogradov, S., 241
Voelecker, 185, 190, 194, 199
Volberg Theorem, 22
Volterra operator, 32
von Mangoldt's lambda function, 275
von Neumann, 31, 33, 37, 45, 74, 76, 99
von Neumann's inequality, 38, 44

Wagner, 284, 293
Walnut's representation, 76
wandering space, 45
Watson's Theorem, 395
wave functions, 384
Waveform, 298, 316, 323

waveform, 296, 298–300, 302, 309, 310, 312, 314, 316–323, 325, 326
wavelet coefficients, 127, 128, 130–137, 139, 140, 142–144
Wavelet expansions, 127
wavelet expansions, 127, 144
wavelet multiplier, 357–359
wavelet sets, 333
Wavelet Transform, 365
wavelet transform, 334, 335
wavelets, 103–125
Weak Pair Correlation Conjecture, 289
Weil-Brezin formula, 308
Wermer, J., 262
Wexler-Raz biorthogonality condition, 76
Weyl operational calculus, 174
Weyl sum, 276, 278
Weyl's Criterion, 272, 274, 277
Weyl, H., 271
Weyl-Heisenberg frames, 175
Weyl-Heisenberg group, 74
Weyl-Heisenberg transform, 363
wide band cross ambiguity function, 315
wide-band ambiguity function, 316
wide-sense stationary random sequences, 41
Wiener, 58, 61, 62, 65
Wiener algebra, 58, 239
Wiener theorem, 236
Wiener, N., 236
Wiener-Lévy theorem, 241
Wiener-Pitt phenomenon, 236
Wiener-Pitt-Sreider phenomenon, 235
Wigner distribution, 175
Wigner-Ville distribution, 74
Wilcox, 303, 308–311, 327
Wilson systems, 76
windowed Fourier transforms, 173
Wold decomposition, 38, 42, 45
Wolff, 27–29
Woodward, 315
worst-case identification, 381
Wutam Consortium, 357

Zak Transform, 173
Zak transform, 73–76, 90–93, 98, 100, 101, 308

Zeevi, 175, 177–180, 185, 186, 190, 191, 194–196, 198, 199
Zhang, L., 362
Zibulski, 175, 177–180, 199
Zibulski-Zeevi description, 76
Zygmund, 61, 62, 67, 72, 215
Zygmund sets, 66, 67